Electromagnetic Interactions of Hadrons
Volume 1

NUCLEAR PHYSICS MONOGRAPHS

Series Editors: **Erich W. Vogt**, *University of British Columbia*
John W. Negele, *Massachusetts Institute of Technology*

GROUP SYMMETRIES IN NUCLEAR STRUCTURE
Jitendra C. Parikh

ELECTROMAGNETIC INTERACTIONS OF HADRONS, VOLUME 1
A. Donnachie and G. Shaw

ELECTROMAGNETIC INTERACTIONS OF HADRONS, VOLUME 2
A. Donnachie and G. Shaw

A Continuation Order Plan is available for this series. A continuation order will bring delivery of each new volume immediately upon publication. Volumes are billed only upon actual shipment. For further information please contact the publisher.

Electromagnetic Interactions of Hadrons
Volume 1

Edited by

A. Donnachie

and

G. Shaw
University of Manchester, England

Plenum Press · New York and London

Library of Congress Cataloging in Publication Data

Main entry under title:

Electromagnetic interactions of hadrons.

(Nuclear physics monographs)
Bibliography: p.
Includes index.
1. Hadrons. 2. Electromagnetic interactions. I. Donnachie, A. II. Shaw, G. III. Series.

QC793.5.H328E44   539.7'5   77-17811
ISBN 0-306-31052-X

© 1978 Plenum Press, New York
A Division of Plenum Publishing Corporation
227 West 17th Street, New York, N.Y. 10011

Printed in the United States of America

Contributors

A. Donnachie, Department of Theoretical Physics, University of Manchester, Manchester, England

A. J. G. Hey, Department of Physics, University of Southampton, Southampton, England

D. W. G. S. Leith, Stanford Linear Accelerator Center, Stanford University, Stanford, California

D. H. Lyth, Department of Physics, University of Lancaster, Lancaster, England

R. G. Moorhouse, Department of Natural Philosophy, University of Glasgow, Glasgow, Scotland

G. Shaw, Department of Theoretical Physics, University of Manchester, Manchester, England

J. K. Storrow, Department of Theoretical Physics, University of Manchester, Manchester, England

Preface

While electromagnetic interactions were first used to probe the structure of elementary particles more than 20 years ago, their importance has only become fully evident in the last 10 years. In the resonance region, photo-production experiments have provided clear evidence for simple quark model ideas, and confirmed the Melosh-transformed $SU(6)_W$ as a relevant symmetry classification. At higher energies, their most striking feature is their similarity to hadron-induced reactions, and they have provided fresh insight into the ideas developed to explain strong-interaction physics. New dimensions are added by taking the photon off mass shell, both in the spacelike region, where the development of high-energy electron and muon beams has led to the discovery and study of scaling, and the introduction of "partons," and even more dramatically in the timelike region, where the development of high-energy electron–positron storage rings has led to the exciting discoveries of the last four years.

In view of the immense interest stimulated by these developments, an extensive review of our present state of knowledge is both timely and useful. Because of the very wide range of the subject, a cooperative venture presents itself as the most suitable format and is the one we have adopted here. The emphasis throughout is primarily, but not entirely, on phenomenology, concentrating on describing the main features of the experimental data and on the theoretical ideas used directly in their interpretation. As such we hope that it will be of interest and of use to all practicing physicists in the field of elementary particles, including graduate students.

The work is in two volumes. This volume deals with photoproduction and electroproduction in the resonance region and at medium energies, treating mainly two-body and quasi-two-body final states. The companion volume first considers multiparticle production and inclusive reactions, and then goes on to tackle deep inelastic scattering and electron–positron annihilation.

We are deeply indebted to the many authors who have contributed to this work. Their adherence to the proposed guidelines greatly eased the problems of editing, and contributed significantly towards achieving a balanced presentation.

We would like to thank Mrs. S. A. Lowndes of Daresbury Laboratory, for her invaluable assistance in the technical editing of the articles in both this and the companion volume.

Manchester, 1978

A. Donnachie
G. Shaw

Contents

Chapter 2

Electromagnetic Excitation and Decay of Hadron Resonances

R. G. Moorhouse

Chapter 3

Low-Energy Pion Photoproduction

A. Donnachie and G. Shaw

Chapter 4

Exclusive Electroproduction Processes

D. H. Lyth

Chapter 5
Form Factors and Electroproduction
A. Donnachie, G. Shaw, and D. H. Lyth

Chapter 6
High-Energy Photoproduction: Nondiffractive Processes
J. K. Storrow

Chapter 7

High-Energy Photoproduction: Diffractive Processes

D. W. G. S. Leith

Contents of Volume 2

Quarks and Symmetries

A. J. G. Hey

1. Introduction

Quarks were first invoked by Gell-Mann and Zweig over ten years ago (Gell-Mann, 1964; Zweig, 1964). They provided some sort of "explanation" for the success of the Eightfold Way based on $SU(3)$ symmetry. One of the novel features about $SU(3)$ symmetry was that the smallest representation, the basic triplet, did not seem to be realized by nature. Gell-Mann and Zweig proposed that in fact it was, but perhaps only in the sense of a building block, q, with the baryons composed of three quarks $(3q)$ and mesons of a quark and antiquark $(q\bar{q})$. However, there are several peculiar features about such a scheme. To overcome the difficulty of the earlier Sakata model (Sakata, 1956) the quarks were assigned nonintegral charge and baryon number. Such curious properties have stimulated a painstaking search for quarks as physical particles in their own right, but so far, at least, with no success. One possibility is, of course, that free quarks do not exist and the nonintegral quantum numbers are a flag to tell us that quarks only exist in bound states. Theorists have been trying to embody this "unobservability" in a field-theoretic framework, and one hopeful avenue stems from another curious feature of the quarks. To build up hadrons of integral and half-integral spin, the most economical scheme is to endow the quarks with spin $\frac{1}{2}$. Thus one expects them to be fermions— but if so, they appear to obey "funny" statistics. For this and for other theoretical reasons, one popular hypothesis is that the three quarks come

A. J. G. Hey • Department of Physics, University of Southampton, Highfield, Southampton SO9 5NH, England

in three colors and are triplets of an $SU(3)$ color group. Nonobservability of quarks then finds a simple expression in the statement that all observed hadrons must be color singlets. Non-Abelian gauge theories based on this $SU(3)$ color local gauge group may provide quark confinement.

For the present article it will not be necessary to commit oneself to one or other particular scheme. Instead, the purpose will be to demonstrate that the quark hypothesis has definite phenomenological validity in that there is "something more" than $SU(3)$ present in nature. However, there are many approaches to quarks and quark models, and the one we shall adopt in this article attempts to avoid detailed dynamical assumptions and concentrates on symmetries and algebraic structures. We begin by describing a simple model for the spectrum of excited baryon and meson states based on an $SU(6) \otimes O(3)$ symmetry. Since this is a classification group, these quarks are dubbed "constituent" or "classification" quarks. After introducing the mechanics of $SU(6)$, we discuss the extension of such symmetries to decay processes. The problems here lead us rapidly to the algebraic quark predictions of current algebra, whose postulates may be phrased in terms of "current" quark fields. A discussion of current algebra sum rules and infinite momentum leads naturally to an $SU(6)$ algebra of the so-called good null-plane charges. Instead of discussing some of the other possible approaches to quarks and quark predictions [for comprehensive reviews see Lipkin (1973) and Rosner (1974)], we describe recent attempts by Melosh and others (Melosh 1973; Bucella *et al.* 1970) to knit together the "constituent" and "current" quark approaches to $SU(6)$. Although there are many dynamical questions yet to be answered, the use of the Melosh transformation leads to a unifying and extremely useful framework for algebraic $SU(6)$ phenomenology. We review and contrast the success and scope of this scheme with the predictions of some less formal, more intuitive quark models.

Finally, in the last section we arrive at full circle. In 1964, $SU(3)$ and quarks were introduced to bring order and simplicity to the excited hadron spectrum. The spectrum of new particles discovered at Brookhaven and SLAC in 1974 may upset our preconceived notions about quarks and perhaps generate some real progress in quark spectroscopy!

2. Constituent Quarks

2.1. Introduction to SU(3) and SU(6)

The search for regularities in the hadron resonance spectrum is the starting point of many symmetry schemes. The baryons and mesons are observed to fall into very approximately degenerate $SU(3)$ multiplets. The

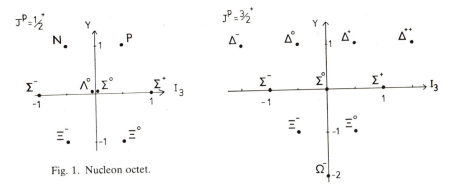

Fig. 1. Nucleon octet.

Fig. 2. Delta decuplet.

observed representations may be summarized by the rule

Baryons: **10, 8, 1**

Mesons: **8, 1**

Resonances not falling into these representations are called "exotic," and so far no exotic multiplets have definitely been observed. Figures 1–4 show the lowest-lying baryon and meson $SU(3)$ multiplets—namely, the nucleon octet, delta decuplet, and the pion and rho meson "nonets"— octets and singlets. The multiplets are plotted on an $SU(3)$ weight diagram with I_3 versus Y ($Y = B + S$, the hypercharge). Clearly $SU(3)$ is a much more approximate symmetry than isospin, since the mass splitting within an $SU(3)$ multiplet is much greater.

How can one explain that only **1**'s, **8**'s and **10**'s of $SU(3)$ have been observed? The constituent quark model provides a mnemonic that accounts for this in a very simple way.

Before we discuss combining $SU(3)$ quarks, it is helpful to obtain some insight from the more familiar $SU(2)$ case of angular momentum or

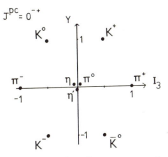

Fig. 3. Pseudoscalar meson nonet.

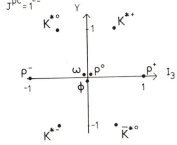

Fig. 4. Vector meson nonet.

isospin. The fundamental representation of $SU(2)$ is the spin-$\frac{1}{2}$ doublet: all other spins may be made up by combining two or more of these fundamental objects.

Consider combining two spin-$\frac{1}{2}$ representations. From the usual angular momentum theory we have the result

$$\tfrac{1}{2} \otimes \tfrac{1}{2} = 1 \oplus 0 \qquad (2.1)$$

i.e., the product is the sum of two irreducible representations of spin 1 and spin 0. In terms of the dimensionality of the $SU(2)$ multiplets, Eq. (2.1) may be rewritten

$$2 \otimes 2 = 3 \oplus 1 \qquad (2.2)$$

since there are three spin-1 states (spin projections ± 1 and 0) and one spin 0. Furthermore, we know something about the symmetry of the wave functions. Labeling the two spin-$\frac{1}{2}$ representations 1 and 2, the wave functions are explicitly

$$S = 1: \quad \begin{cases} \uparrow_1\uparrow_2 \\ 2^{-1/2}(\uparrow_1\downarrow_2 + \downarrow_1\uparrow_2) \\ \downarrow_1\downarrow_2 \end{cases} \qquad (2.3)$$

$$S = 0: \qquad 2^{-1/2}(\uparrow_1\downarrow_2 - \downarrow_1\uparrow_2)$$

i.e., the spin-1 wave functions are symmetric under the exchange $1 \leftrightarrow 2$ and the spin-0 wave function is antisymmetric.

All these properties may be elegantly summarized using a technique invented by an English clergyman named Young (Young, 1901). The fundamental spin-$\frac{1}{2}$ representation corresponds to the "Young diagram" (box) shown below:

$$\tfrac{1}{2}: \quad \square \quad \text{dimensionality 2}$$

To obtain the product of two spin-$\frac{1}{2}$ representations, just add a second box to make all the possible allowed Young tableaux. The rules are given explicitly in the Appendix, but it is plausible that we can make just two diagrams:

$$\square \otimes \square = \square\square \oplus \begin{array}{c}\square\\\square\end{array} \qquad (2.4)$$

$$\quad\; 2 \qquad 2 \qquad\quad 3 \qquad\quad 1$$

The dimensionalities of the tableaux are given beneath them and may be calculated easily using the "Hook" rule (see Appendix at the end of the chapter). The tableaux, however, have more significance than just giving the dimensions of the irreducible representations. Adding the second box on the same row corresponds to symmetrizing 1 and 2, while adding in a second column corresponds to antisymmetrizing. Thus the diagrams correspond exactly to the explicit wave functions we wrote down. For $SU(2)$, a column of two boxes is the maximum length of column allowed and corresponds to the singlet or spin-0 representation. Now add a third spin-$\frac{1}{2}$ particle. We know that the answer is

$$1 \otimes \tfrac{1}{2} = \tfrac{3}{2} \oplus \tfrac{1}{2}$$

and

$$0 \otimes \tfrac{1}{2} = \tfrac{1}{2}$$

(2.5)

Thus

$$\tfrac{1}{2} \otimes \tfrac{1}{2} \otimes \tfrac{1}{2} = \tfrac{3}{2} \oplus \tfrac{1}{2} \oplus \tfrac{1}{2}$$

or

$$2 \otimes 2 \otimes 2 = 4 \oplus 2 \oplus 2$$

(2.6)

What about the symmetries of these wave functions? Clearly the spin-$\frac{3}{2}$ wave function is "symmetric" between spin 1 and spin $\frac{1}{2}$ (the "stretched" state is symmetric under $j_1 \leftrightarrow j_2$). Since the spin-1 state is symmetric between 1 and 2, the spin-$\frac{3}{2}$ state is in fact totally symmetric between 1, 2, and 3. For example:

$$|\tfrac{3}{2}\ \tfrac{3}{2}\rangle = \uparrow_1\uparrow_2\uparrow_3$$
$$|\tfrac{3}{2}\ \tfrac{1}{2}\rangle = 3^{-1/2}(|11\rangle\downarrow_3 + 2^{1/2}|10\rangle\uparrow_3)$$
$$= 3^{-1/2}(\uparrow_1\uparrow_2\downarrow_3 + \uparrow_1\downarrow_2\uparrow_3 + \downarrow_1\uparrow_2\uparrow_3)$$

(2.7)

The spin-$\frac{1}{2}$ state obtained from the spin 1 and $\frac{1}{2}$ is "antisymmetric" between spin 1 and spin $\frac{1}{2}$ ($j_1 \leftrightarrow j_2$) and we have

$$|\tfrac{1}{2}\ \tfrac{1}{2}\rangle = 3^{-1/2}(2^{1/2}|11\rangle\downarrow_3 - |10\rangle\uparrow_3)$$
$$= 3^{-1/2}(2\uparrow_1\uparrow_2\downarrow_3 - \uparrow_1\downarrow_2\uparrow_3 - \downarrow_1\uparrow_2\uparrow_3)$$

(2.8)

i.e., symmetric between 1 and 2 but not totally symmetric between 1, 2, and 3. This is known as a state of mixed symmetry. From spin 0 with spin $\frac{1}{2}$ we obtain

$$|\tfrac{1}{2}\ \tfrac{1}{2}\rangle = |00\rangle\uparrow_3$$
$$= 2^{-1/2}(\uparrow_1\downarrow_2\uparrow_3 - \downarrow_1\uparrow_2\uparrow_3)$$

(2.9)

i.e., a spin-$\frac{1}{2}$ state of different mixed symmetry—antisymmetric between 1 and 2. All this is again evident from the corresponding Young tableaux

$$1 \otimes \tfrac{1}{2} = \tfrac{3}{2} \oplus \tfrac{1}{2}$$

(2.10a)

and

$$0 \otimes \tfrac{1}{2} = \tfrac{1}{2}$$

$$\square \!\!\!\!\begin{array}{c}\square\\\square\end{array} \otimes \square = \begin{array}{c}\square\square\end{array} \tag{2.10b}$$

1   2   2

[For $SU(n)$, n boxes in a column corresponds to the singlet and the column can be effectively removed. No more than n boxes can appear in a column.]
Thus for our three spin-$\tfrac{1}{2}$ decomposition we have in $SU(2)$

$$\square \otimes \square \otimes \square = \square\square\square \oplus \begin{array}{c}\square\square\\\square\end{array} \oplus \begin{array}{c}\square\square\\\square\end{array} \tag{2.11}$$

2   2   2   4   2   2

where the two spin-$\tfrac{1}{2}$ combinations clearly have different symmetry properties with respect to the three basic objects. They are mixed symmetry states and the spin-$\tfrac{3}{2}$ state is totally symmetric.

Now let us imitate this with $SU(3)$ using the basic triplet representation (which we may as well call a quark) as a building block (see Fig. 5). Using the standard graphical or tensor methods, one can show that the two-quark product decomposes into the sum of two irreducible representations [this may most easily be proved by a generalization of the usual raising and lowering technique that one uses in $SU(2)$]:

$$\mathbf{3} \otimes \mathbf{3} = \mathbf{6} \oplus \bar{\mathbf{3}} \tag{2.12}$$

The members of the $\bar{\mathbf{3}}$ representation are illustrated in Fig. 5. Again the explicit wave functions show that the $\mathbf{6}$ is symmetric between 1 and 2 and the $\bar{\mathbf{3}}$ is antisymmetric. For example,

$$\psi\{\mathbf{6}: I = 1 \;\; I_3 = 1\} = u_1 u_2$$

$$\psi\{\mathbf{6}: I = 1 \;\; I_3 = 0\} = 2^{-1/2}(u_1 d_2 + d_1 u_2) \tag{2.13}$$

$$\psi\{\mathbf{6}: I = \tfrac{1}{2} \;\; I_3 = \tfrac{1}{2}\} = 2^{-1/2}(u_1 s_2 + s_1 u_2)$$

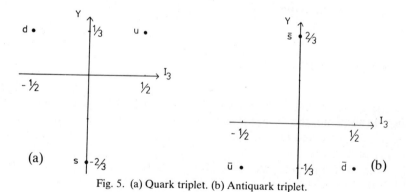

Fig. 5. (a) Quark triplet. (b) Antiquark triplet.

and so on for the **6**, while the $\overline{\mathbf{3}}$ states are explicitly

$$\psi\{\overline{\mathbf{3}}: I = 0\} = 2^{-1/2}(u_1 d_2 - d_1 u_2)$$

$$\psi\{\overline{\mathbf{3}}: I = \tfrac{1}{2} \; I_3 = \tfrac{1}{2}\} = 2^{-1/2}(u_1 s_2 - s_1 u_2) \tag{2.14}$$

$$\psi\{\overline{\mathbf{3}}: I = \tfrac{1}{2} \; I_3 = -\tfrac{1}{2}\} = 2^{-1/2}(d_1 s_2 - s_1 d_2)$$

Again this may be represented by Young diagrams. For $SU(3)$ a column of three boxes is the singlet representation and thus it is plausible that a column of two boxes (which can combine with the **3** to produce a singlet) is the $\overline{\mathbf{3}}$ representation

$$\square \otimes \square = \square\square \oplus \begin{smallmatrix}\square\\\square\end{smallmatrix} \tag{2.15}$$
$$\quad\; 3 \qquad 3 \qquad\; 6 \qquad\; \overline{3}$$

Again the tableaux show also the symmetry of the $SU(3)$ wave functions **6** symmetric—same row—and $\overline{\mathbf{3}}$ antisymmetric—same column.

Now add a third quark. From the Young diagrams we obtain

$$\square\square \otimes \square = \square\square\square \oplus \begin{smallmatrix}\square\square\\\square\end{smallmatrix}$$
$$\quad\; 6 \qquad\quad 3 \qquad\quad 10 \qquad\qquad 8$$
$$\tag{2.16}$$
$$\begin{smallmatrix}\square\\\square\end{smallmatrix} \otimes \square = \begin{smallmatrix}\square\square\\\square\end{smallmatrix} \oplus \begin{smallmatrix}\square\\\square\\\square\end{smallmatrix}$$
$$\quad\; \overline{3} \qquad\; 3 \qquad\quad 8 \qquad\quad 1$$

These decompositions may be explicitly verified and the wave functions constructed by the usual ladder methods. As in $SU(2)$ we obtain a totally symmetric representation, the **10**, plus two mixed symmetry **8**'s. In $SU(3)$, also, a column of three boxes is allowed and corresponds to the singlet: This was not present for $SU(2)$, where the singlet is a column of two boxes.

From all this group theory emerges a simple mnemonic to reproduce the allowed baryon and meson $SU(3)$ representations:

Baryons—three quarks:

$$\mathbf{3} \otimes \mathbf{3} \otimes \mathbf{3} = \mathbf{10} \oplus \mathbf{8} \oplus \mathbf{8} \oplus \mathbf{1} \tag{2.17}$$

Mesons—quark–antiquark:

$$\mathbf{3} \otimes \overline{\mathbf{3}} = \mathbf{8} \oplus \mathbf{1} \tag{2.18}$$

With this scheme, the "Gell-Mann–Zweig quarks" must have the unusual fractional charge and baryon number shown in Table 1. At present, these quarks may be regarded merely as the basis of a prescription for the observed $SU(3)$ representations, although they could equally well be real physically observable particles, but with rather peculiar properties.

Table 1. *Quantum Numbers of Gell-Mann–Zweig Quarks*

Quark	Quantum numbers					
	B	I	I_3	Y	S	Q
"up" u	1/3	1/2	1/2	1/3	0	2/3
"down" d	1/3	1/2	−1/2	1/3	0	−1/3
"strange" s	1/3	0	0	−2/3	−1	−1/3

What other regularities are observed in the spectrum? In the spectrum of baryon resonances, to a first approximation, one may group the resonances in mass bands with alternating parity (Fig. 6). This is strongly reminiscent of some rotational excitation and suggests some sort of orbital angular momentum. However, the prediction of spins and parities of resonances clearly needs *more* than $SU(3)$. The most natural step is to endow the quarks with spin $\frac{1}{2}$. The basic quark wave function now has six components:

$$q\{SU(6)\} \sim \{SU(3): u, d, s\} \cdot \{SU(2): \uparrow, \downarrow\}$$

which leads us to consider $SU(6)$ symmetry. Combining three quarks for baryons leads to quark spin-$\frac{3}{2}$ and -$\frac{1}{2}$ states, which could correspond neatly with the Δ and nucleon. Similarly for quark–antiquark, one expects spin 1 and spin 0—again intriguing, since the lowest meson multiplets have these spins. If we define an orbital angular momentum **L** according to the rule

$$\mathbf{J} = \mathbf{L} + \mathbf{S} \tag{2.19}$$

one naively expects for a three-fermion system

$$P = (-1)^L \tag{2.20}$$

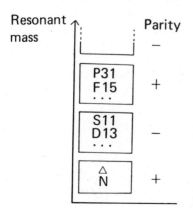

Fig. 6. Idealized view of the baryon spectrum.

and for a fermion–antifermion system

$$P = (-1)^{L+1} \tag{2.21}$$

Thus $L = 0$ produces positive-parity baryons and negative-parity mesons—which corresponds exactly to the physical situation. It is therefore worthwhile to consider what $SU(6)$ tells us are the allowed representations.

What happens when we combine two of these $SU(6)$ quarks? The decomposition may be worked out laboriously by the usual methods or very quickly via Young diagrams:

$$\mathbf{6} \otimes \mathbf{6} = \mathbf{21} \oplus \mathbf{15}$$

$$\square \otimes \square = \square\square \oplus \begin{array}{c}\square\\\square\end{array} \tag{2.22}$$

The result may be made more transparent by looking at the $SU(2) \otimes SU(3)$ decomposition of these $SU(6)$ representations and their symmetry properties. One must remember the rules for combining two states of different symmetry:

$$|a\rangle_S |b\rangle_S = |\,\rangle_S$$
$$|a\rangle_A |b\rangle_A = |\,\rangle_S$$
$$|a\rangle_S |b\rangle_A = |\,\rangle_A$$
$$|a\rangle_A |b\rangle_S = |\,\rangle_A \tag{2.23}$$

where S and A stand for symmetric and antisymmetric, respectively. In this case, $|a\rangle$ represents the spin wave function of the two objects and $|b\rangle$ the $SU(3)$ unitary spin wave function. We have

$$SU(2) \qquad \mathbf{2} \otimes \mathbf{2} = \mathbf{3} \oplus \mathbf{1}$$
$$SU(3) \qquad \mathbf{3} \otimes \mathbf{3} = \mathbf{6} \oplus \bar{\mathbf{3}} \tag{2.24}$$

Since the $\mathbf{21}$ ($\square\square$) is symmetric and $\mathbf{15}$ ($\begin{array}{c}\square\\\square\end{array}$) antisymmetric we arrive at the decomposition

$$\left.\begin{array}{l}\mathbf{21} = (\mathbf{3}, \mathbf{6}) + (\mathbf{1}, \bar{\mathbf{3}})\\[4pt] \mathbf{15} = (\mathbf{3}, \bar{\mathbf{3}}) + (\mathbf{1}, \mathbf{6})\end{array}\right\} \quad (SU(2), SU(3)) \tag{2.25}$$

Now add a third quark. The result is

$$\mathbf{6} \otimes \mathbf{6} \otimes \mathbf{6} = \mathbf{56} \oplus \mathbf{70} \oplus \mathbf{70} \oplus \mathbf{20}$$

$$\square \otimes \square \otimes \square = \square\square\square \oplus \begin{array}{c}\square\square\\\square\end{array} \oplus \begin{array}{c}\square\square\\\square\end{array} \oplus \begin{array}{c}\square\\\square\\\square\end{array} \tag{2.26}$$

[Note that exactly the same Young diagrams occur in $SU(3)$.] This comes from

$$\mathbf{21} \otimes \mathbf{6} \qquad \mathbf{56} \ \oplus\ \mathbf{70}$$

$$\square\square \otimes \square = \square\square\square \oplus \begin{matrix}\square\square\\\square\end{matrix} \tag{2.27}$$

and

$$\mathbf{15} \otimes \mathbf{6} = \mathbf{70} \oplus \mathbf{20}$$

$$\begin{matrix}\square\\\square\end{matrix} \otimes \square \qquad \begin{matrix}\square\square\\\square\end{matrix} \oplus \begin{matrix}\square\\\square\\\square\end{matrix} \tag{2.28}$$

Just as in $SU(3)$, there are two different mixed-symmetry states—**70** in this case—which differ in their permutation symmetry. It is not difficult to break these $SU(6)$ representations down into their spin and unitary-spin decompositions—although one must know how to combine two mixed symmetry states to make states of three particles that are totally symmetric, mixed symmetric, or totally antisymmetric. In the Appendix at the end of the chapter is a handbook on how to do this, adapted from Feynman, Kislinger, and Ravndal's quark model paper (Feynman *et al.*, 1971).

The important result of all this is

$$\mathbf{56} = (4, \mathbf{10}) + (2, \mathbf{8})$$

$$\mathbf{70} = (4, \mathbf{8}) + (2, \mathbf{8}) + (2, \mathbf{10}) + (2, \mathbf{1}) \tag{2.29}$$

$$\mathbf{20} = (2, \mathbf{8}) + (4, \mathbf{1})$$

We can already see that the $\frac{1}{2}^+$ nucleon octet and the $\frac{3}{2}^+\Delta$ decuplet fit magically into the **56**. Before looking at the baryons in more detail, let us consider what quark–antiquark gives us for the mesons.

The antiquark, $\bar{\mathbf{6}}$, is represented by a column of five boxes—since a column of **6** is now the singlet representation. By the now standard methods

$$\bar{\mathbf{6}} \otimes \mathbf{6} = \mathbf{35} \oplus \mathbf{1}$$

$$\begin{matrix}\square\\\square\\\square\\\square\\\square\end{matrix} \otimes \square = \begin{matrix}\square\square\\\square\\\square\\\square\\\square\end{matrix} \oplus \begin{matrix}\square\\\square\\\square\\\square\\\square\\\square\end{matrix} \tag{2.30}$$

The spin and $SU(3)$ decomposition of the **35** is

$$\mathbf{35} = (1, \mathbf{8}) + (3, \mathbf{8}) + (3, \mathbf{1})$$

$$\mathbf{1} = (1, \mathbf{1}) \tag{2.31}$$

Again this "**36**" obviously accommodates the $0^-\pi$ and $1^-\rho$ nonets, but before we look closely at the hadron spectrum and how it compares with our $SU(6)$ model, a final word about Young diagrams. They are a very useful tool to calculate more complicated products of representations. In $SU(3)$ for example, combining

□□   □□
□   and   □

to make all allowed Young tableaux, and calculating the dimensions with the "Hook" rule gives very quickly the well-known result

$$8 \otimes 8 = 27 \oplus 10 \oplus \overline{10} \oplus 8 \oplus 8 \oplus 1 \qquad (2.32)$$

However, the result of $\mathbf{70} \otimes \mathbf{70}$ in $SU(6)$ is less well known (and much less useful), but it may be calculated just as easily. A brief summary of the rules for using Young diagrams is given in the Appendix.

Now let us forget group theory and get back to physics. The comparison of the allowed $SU(6)$ representations with the spectrum is astonishingly successful!

2.2. Baryon Spectroscopy and $SU(6) \otimes O(3)$

If we define **L** according to the rule

$$\mathbf{J} = \mathbf{L} + \mathbf{S} \qquad (2.19)$$

and define* the parity of the representation to be

$$P = (-1)^L \qquad (2.20)$$

presumably then, the ground state will have $L = 0$ and positive parity. The $\{\mathbf{56}, L^P = 0^+\}$ representation accommodates the nucleon and Δ beautifully!

$$\{\mathbf{56}, L^P = 0^+\}$$

$\Delta:$ $^4\mathbf{10} = 40$ $\Big\}$ 56 spin and unitary spin states

$N:$ $^2\mathbf{8} = 16$

The rather curious number 56 appears naturally in this scheme, and it is this agreement which encourages us to search for further $SU(6)$ representations.

The negative parity resonances may all be assigned to a $\{\mathbf{70}: L^P = 1^-\}$ representation, and there are no unwanted negative parity resonances in the mass range below 1800 MeV or so. Figure 7 summarizes the status of the required $S = 0$ and $S = -1$ resonances: Only a few Y^*'s remain to be found.

---

*For a system of three physical particles, this rule is not so obvious. One must consider in detail the symmetries of the spatial wave functions. At present, however, this rule may be regarded as a fairly ad hoc prescription.

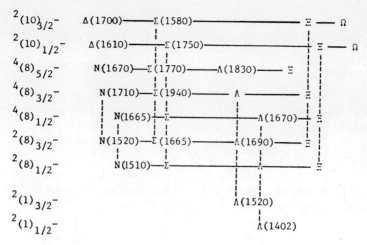

Fig. 7. Status of $S = 0$ and $S = -1$ members of the $\{\mathbf{70}, 1^-\}$.

Relatively higher in mass than the S and D wave negative parity resonances, the P and F resonances can be assigned to a $\{\mathbf{56}; L^P = 2^+\}$ multiplet, with again few states missing among the $S = 0$ and -1 resonances (see Fig. 8).

In summary then, the $SU(6)$ rules provide a good description of the observed low mass resonance spectrum (Litchfield, 1974). As always, however, a few comments are in order at this point.

(i) Quark Statistics. The assignment of $\{\mathbf{56}, 0^+\}$ as the ground state of a three-quark system seems to be at variance with Fermi statistics for the quarks. This is because the $\mathbf{56}$ representation is symmetric under permutation of its spin and unitary spin labels

$$\mathbf{56} \sim \square\square\square$$

and one expects an $\mathbf{L} = 0$ ground-state *spatial* wave function also to be

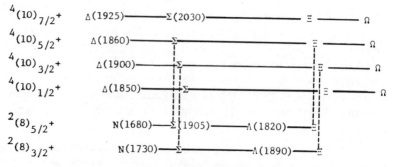

Fig. 8. Status of $S = 0$ and $S = -1$ members of the $\{\mathbf{56}, 2^+\}$.

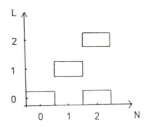

Fig. 9. Three-dimensional harmonic oscillator spectrum.

symmetric. This leads then to the statistics paradox for the quarks, and to the idea that they either have "funny" statistics rules or some unobserved (as yet) degree of freedom (like color). The alternatives are clearly summarized in a recent article by Dalitz (1974). Here we can remain agnostic on this point and avoid specifying how the problem is to be circumvented since we are at present treating the quark rules as fairly ad hoc prescriptions.

(*ii*) *The Roper Resonance.* The Roper resonance p_{11} (1470) is a positive-parity nucleon resonance in amongst our "band" of negative-parity resonances. It is often "explained" as a "radial excitation" or "breathing mode" by analogy with a harmonic oscillator spectrum (Fig. 9). At the $n = 2$ level one has two L values, $L = 2$ and $L = 0$. Thus the Roper is usually assigned to another $\{\mathbf{56}, 0^+\}$ multiplet—a radial excitation of the ground state. Recently a Δ partner of the Roper has been plausibly identified (Herndon *et al.*, 1975).

(*iii*) *Are There* **20** *Representations?* So far there is no evidence for extra resonances that cannot be classified in a **56** or **70** representation. However, what are the distinguishing characteristics of a **20** resonance? [One of Feynman's problems for the student! (Feynman, 1973)]. If one can treat $SU(6)$ like $SU(3)$ (or if one believes in explicit quark models with single-quark interactions) one expects they may be difficult to produce. The available beams (π's and K's) belong to a **35** of $SU(6)$ and the available targets (p, n) to a **56**. Their product is easily shown *not* to contain a **20**:

$$\mathbf{35} \otimes \mathbf{56} = \mathbf{700} \oplus \mathbf{1134} \oplus \mathbf{56} \oplus \mathbf{70} \qquad (2.33)$$

(*iv*) *Harmonic Oscillator Quark Shell Model Spectrum.* Realistic quark models involving three spin-$\frac{1}{2}$ particles moving in a harmonic oscillator potential (Greenberg, 1964; Dalitz 1965) predict a fairly rich spectrum, and a more complicated parity rule:

$$n = 0 \quad \{\mathbf{56}, 0^+\}$$

$$n = 1 \quad \{\mathbf{70}, 1^-\} \qquad\qquad\qquad (2.34)$$

$$n = 2 \quad \{\mathbf{56}, 2^+\}, \{\mathbf{70}, 2^+\}, \{\mathbf{56}, 0^+\}, \{\mathbf{70}, 0^+\}, \{\mathbf{20}, 1^+\}$$

At present there is little hint of the two **70**, even parity multiplets predicted at the $n = 2$ level.

2.3. Meson Spectroscopy and $SU(6) \otimes O(3)$

Treating mesons as bound states of a fermion–antifermion pair leads to the rules

$$\mathbf{J} = \mathbf{L} + \mathbf{S}$$
$$P = (-1)^{L+1} \tag{2.35}$$
$$C = (-1)^{L+S}$$

In Table 2 we show the expected J^{PC} structure and the tentative assignments. The pseudoscalar and vector mesons fall into a $\{35 + 1: L^P = 0^-\}$ representation, and some of the positive-parity $L = 1$ states have been identified. However, the status of the $A_1(1^{++}; I = 1)$ meson and the two strange axial vector mesons (the Q bump in $K\pi\pi$) has remained confused for many years. The trouble is that most, if not all, of the features of these diffractive three-meson bumps can be explained by a nonresonant mechanism, usually called the Deck effect. This is illustrated for the "A_1" bump in Fig. 10. This mechanism produces $\rho\pi$, predominantly in an S state with $J^P = 1^+$, and hence the confusion. It is noteworthy that the Deck argument for producing mainly S-wave ($\rho\pi$) depends crucially on the "Pomeron" intercept being unity. Thus in charge or strangeness exchange reactions the Deck model should have less effect on the 1^+ system. At present the nondiffractive reactions seem not very encouraging for the A_1, but the definitive experiment remains to be done. A new analysis (Brandenburg et al., 1975) with enormous statistics for diffractive Q production

Table 2. Present Status of $N = 0$ and $N = 1$ Multiplets^a

Oscillator quantum numbers				$I = 1$	$I = 1/2$	$I = 0$		
N	L	S	J^{PC}	**8**	**8**	**8**	**1**	Mixing
0	0	0	0^{-+}	π	K	η	η'	U
		1	1^{--}	ρ	K^*	ω	ϕ	M
1	1	0	1^{+-}	B	Q_B?	$[B_\eta]$	$[B'_\eta]$	$[U]$
		1	2^{++}	A_2	K^{**}	f	f'	M
			1^{++}	A_1?	Q_A?	D?	$[D']$	$[M]$
			0^{++}	δ?	κ?	ε?	S^*?	M?

^{a}Notation: States that are in some doubt are marked with a question mark; states for which there is little or no evidence are enclosed in square brackets. The mixing is indicated by U (unmixed) or M ("magic" or ideal mixing), whichever is nearer to the physical situation.

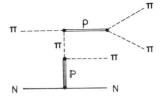

Fig. 10. Deck diagram for "A_1" production.

does, however, appear to show resonant effects. The status of the 0^{++} nonet is also unclear and $\pi\pi$ experts wrangle over the meaning of the phase shift passing through 90°, the effect of thresholds, and so on. We postpone further discussion to Section 5. Nevertheless, although meson spectroscopy is clearly much less advanced than that for baryons, there does seem to be the same alternating parity structure as observed in the baryon spectrum.

2.4. *Conclusions from Hadron Spectroscopy*

From the empirical spin and parity regularities in the resonance spectrum we conclude that it is useful to classify resonances with respect to a higher symmetry than $SU(3)$. In particular, there seems rather strong evidence for $SU(6)$ supermultiplets in the baryon and meson spectra. We shall return to discuss in more detail some aspects of $SU(6)$ spectroscopy in Section 5. So far we have only attempted to enlarge the classification group $SU(3)$ by incorporating quark spin. Another possibility is to increase the number of quarks—in current jargon, to increase the number of quark flavors. The earliest and most conservative extension is to enlarge $SU(3)$ to $SU(4)$ by adding a fourth, "charmed" quark, to the basic triplet (Bjorken and Glashow, 1964). Just as the strange quark is presumably heavier than the nonstrange u and d quarks, so the charmed quark c is plausibly heavier than all of these. This idea of a fourth quark was combined with the Weinberg–Salam unified gauge field model of the weak and electromagnetic interactions (Weinberg, 1967; Salam, 1968) via the so-called GIM mechanism, (Glashow *et al.*, 1970) to produce an attractive model for hadronic weak interactions. This $SU(4)$ quark model has received much popularity as the most likely "solution" to the riddle of the new particles $J/\psi, \psi', \psi'' \ldots$ found in e^+e^- reactions. In Section 6 we shall make some comments on the spectroscopy of these new particles and their relevance to quark classification schemes. Here we just remark that the $SU(4)$ multiplet structure and the $SU(3)$ decomposition may be obtained by a straightforward application of the group theory rules developed in this section. General references on group theory that may be helpful include Carruthers (1966), Gasiorowicz (1967), and Lipkin (1966). Lipkin's book also contains a remarkably far-sighted discussion of $SU(4)$ symmetry, and also a clear, pedestrian introduction to Young tableaux.

2.5. Vertex Symmetries and SU(6)$_{W,constituents}$

Resonance couplings are observed to obey an approximate $SU(3)$ invariance, i.e., the couplings of resonances within the same $SU(3)$ multiplets are related by a calculable $SU(3)$ Clebsch–Gordan coefficient. For example

$$g(Y_{10}^* \Lambda_8 \pi) \sim g(\Delta_{10} N_8 \pi)$$

Experimentally, $SU(3)$ is not an exact symmetry, ($M_\eta \neq M_\pi$ for example), which leads to some ambiguity as to how best to take account of the mass differences in the phase space and reduced matrix elements. Nevertheless, despite such worries, $SU(3)$ is a useful vertex symmetry correlating couplings. The next question that obviously arises concerns $SU(6)$. If $SU(6)$ is a larger symmetry in the spectrum, is there a larger symmetry for vertices?

Treating $SU(6)$ like $SU(3)$ quickly leads to nonsense. Quark spin conservation immediately forbids all the well-known decays. For example

$$\rho \not\to \pi + \pi$$
$$\scriptstyle s=1 \qquad s=0 \qquad s=0$$

$$\Delta \not\to N + \pi$$
$$\scriptstyle s=\frac{3}{2} \qquad s=\frac{1}{2} \qquad s=0$$

However, this is not surprising since these processes involve particles in motion for which spin is not a good quantum number. Thus to demand that the "spin part" of the angular momentum be conserved is not a Lorentz-invariant requirement, and consequently not a sensible restriction.

Nevertheless, is it possible to define some sort of relativistic spin symmetry that could be conserved in decay processes? Lipkin and Meshkov (1965) were led to suggest $SU(6)_W$ by considering the rotation and reflection invariance properties of a particle in motion. The algebra of $SU(6)_W$ had been suggested somewhat earlier in connection with nucleon form factors (Barnes *et al.*, 1965).

Consider a spin-$\frac{1}{2}$ particle moving in the z direction (Fig. 11). Clearly its momentum **p** is not left invariant by rotations about the x or y axes.

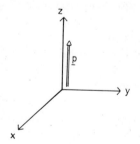

Fig. 11. Reflection invariance of a moving particle.

However, under the reflections $x \to -x$ and $y \to -y$, **p** is unchanged. Since a reflection may be written as the product of the parity operator P and a π rotation about the appropriate axis, the momentum is left invariant by

$$R_x = Pe^{i\pi J_x}, \qquad R_y = Pe^{i\pi J_y} \tag{2.36}$$

Rotations about the z-axis also leave the momentum unchanged

$$R_z = e^{i\phi J_z}$$

In the nonrelativistic limit, the action of these three operators on the internal degrees of freedom of the particle is

$$(R_x)_{\text{int}} = P_{\text{int}} e^{i\pi S_x}$$
$$(R_y)_{\text{int}} = P_{\text{int}} e^{i\pi S_y} \tag{2.37}$$
$$(R_z)_{\text{int}} = e^{i\phi S_z}$$

where P_{int} is the intrinsic parity of the particle. If we consider spin-$\frac{1}{2}$ particles, using the properties of the σ matrices, and $(P_{\text{int}})^2 = +1$ in the expansion of the exponential, these operators may be rewritten as

$$(R_x)_{\text{int}} = e^{i\pi P_{\text{int}}\sigma_x/2}$$
$$(R_y)_{\text{int}} = e^{i\pi P_{\text{int}}\sigma_y/2} \tag{2.38}$$
$$(R_z)_{\text{int}} = e^{i\pi\sigma_z/2}$$

If we demand invariance under an $SU(2)$ group of spin transformations that includes this rotation–reflection symmetry as an element, the natural candidate is W-spin—the $SU(2)$ group generated by

$$W_x = P_{\text{int}}\sigma_x/2, \qquad W_y = P_{\text{int}}\sigma_y/2, \qquad W_z = \sigma_z/2 \tag{2.39}$$

Furthermore, the W-spin of a spin-$\frac{1}{2}$ particle is, in fact, perfectly respectable as a relativistic spin symmetry. In Dirac algebra language the operators

$$\beta\sigma_x/2, \qquad \beta\sigma_y/2, \qquad \sigma_z/2 \tag{2.40}$$

commute with α_3, the generator of boosts in the z direction, and define an $SU(2)$ for a moving spin-$\frac{1}{2}$ particle. The operator β is just the intrinsic parity operator in the rest frame, according to the usual conventions,

$$\beta = \begin{pmatrix} 1 & 0 \\ 0 & -1 \end{pmatrix}$$

If we now add $SU(3)$ and describe hadrons as composites of spin-$\frac{1}{2}$ quarks, we can classify not only particles at rest, but also moving particles under our $SU(6)_W$ symmetry. This is in contrast to our old $SU(6)$ scheme, which was only appropriate for particles at rest. Note that this does, however, involve the assumption that for hadrons at rest the quark motion may be neglected—as indicated by the success of the simple nonrelativistic $SU(6)$ classification.

Baryons: 3q and SU(6)$_W$ Classification. As for ordinary fermions, assume

$$P_{int}(q) = +1$$

Then for a three-quark system the W spin and spin classifications are identical

$$\mathbf{W} = \mathbf{S} \quad \text{for baryons} \tag{2.41}$$

Thus a proton that was classified as

$$\{\mathbf{56}: S = \tfrac{1}{2}\} \quad \text{under } SU(6)$$

is classified as

$$\{\mathbf{56}: W = \tfrac{1}{2}\} \quad \text{under } SU(6)_W$$

where this latter classification is supposed valid for a moving proton.

Mesons: q$\bar{q}$ and SU(6)$_W$ Classification. As in the case of the $N\bar{N}$ system we assume

$$P_{int}(\bar{q}) = -1$$

i.e., that an antiquark has the opposite intrinsic parity to a quark. This leads to a subtlety in classifying mesons under W spin. Let us construct explicitly the W spin states of the $q\bar{q}$ system. Since $W_z = S_z$ we have

$$|S = 1 \ S_z = 1\rangle = |q\uparrow; \bar{q}\uparrow\rangle = |W = 1 \ W_z = 1\rangle$$

Now apply the W-spin lowering operator to this state to generate the $|W = 1 \ W_z = 0\rangle$ state:

$$W_- = W_x - iW_y = W_-(q) + W_-(\bar{q}) \tag{2.42}$$

For the quark

$$W_{x,y}(q) = S_{x,y}(q)$$

but for the antiquark

$$W_{x,y}(\bar{q}) = -S_{x,y}(\bar{q})$$

because of the intrinsic parity difference. Thus

$$W_- = S_-(q) - S_-(\bar{q}) \tag{2.43}$$

Applying this operator to $|W = 1 \ W_z = 1\rangle$ gives

$$|W = 1 \ W_z = 0\rangle = 2^{-1/2}\{|q\downarrow: \bar{q}\uparrow\rangle - |q\uparrow: \bar{q}\downarrow\rangle\}$$
$$= -|S = 0 \ S_z = 0\rangle$$

i.e., the $S_z = W_z = 0$ states of the $q\bar{q}$ system undergo the so-called *W–S* flip. In full, we have the correspondence

$$|W = 1 \ W_z = 1\rangle \leftrightarrow |S = 1 \ S_z = 1\rangle$$

$$|W = 1 \ W_z = 0\rangle \leftrightarrow -|S = 0 \ S_z = 0\rangle \qquad (2.44)$$

$$|W = 1 \ W_z = -1\rangle \leftrightarrow -|S = 1 \ S_z = -1\rangle$$

$$|W = 0 \ W_z = 0\rangle \leftrightarrow -|S = 1 \ S_z = 0\rangle$$

How does all this help us? It helps since collinear decay processes *could* be $SU(6)_W$ invariant. Consider $\rho \to \pi\pi$ again, now demanding W-spin conservation instead of quark spin:

$$\pi \quad S = 0 \ S_z = 0 \text{ is } W = 1 \ W_z = 0$$

$$\rho(\lambda = 0) \quad S = 1 \ S_z = 0 \text{ is } W = 0 \ W_z = 0$$

so

$$\underset{W=0}{\rho} \to \underset{W=1}{\pi} + \underset{W=1}{\pi}$$

and the process *is* allowed by W-spin conservation. So too is $\Delta \to N\pi$. Thus the W-spin classification allows us to consider $SU(6)_W$ as a possible vertex symmetry. We must remember, however, that this postulated $SU(6)_W$ invariance for collinear decay processes is not really well motivated. This is because $SU(6)_W$ is *not* a symmetry in the sense of $SU(3)$. For $SU(3)$ we may consider an idealization where the generators of $SU(3)$, Q_α, commute with the world Hamiltonian and $SU(3)$ is exact:

$$[Q_\alpha, H] = 0$$

Thus, the S matrix commutes with $SU(3)$ and vertices and scattering amplitudes will have $SU(3)$ symmetry. An analogous "symmetry limit" for $SU(6)_W$ is almost inconceivable. One is liable to run into some of the famous $SU(6)$ "no go" theorems and obtain results such as the S matrix being unity (i.e., no interactions!), troubles with unitarity, and so on. A readable review of some of these "no go" theorems is contained in the lectures of Bell (1966). The most powerful "no go" theorem is that of Coleman and Mandula, which concerns possible symmetries of the S matrix (Coleman and Mandula, 1967; see also Coleman, 1968). Given that $SU(6)_W$ is not a symmetry like $SU(3)$, two-particle states, which necessarily involve interactions via the S matrix, need not be simply classified by $SU(6)_W$, even though the single-particle states are! Thus in a decay process

$$A \to B + C$$

although this is a collinear process and the single-particle states of A, B, and C are classified in $SU(6)_W$ representations, the decay vertex need not have $SU(6)_W$ invariance. Having said all this, a postulate of $SU(6)_W$ invariance for three-point functions is nevertheless the simplest sensible guess for a higher symmetry, and its predictions can be compared with experiment. This we call the "naive" $SU(6)_W$ model for vertices. For decays involving only particles with $L = 0$ everything is in reasonable agreement. It is at the $L = 1$ decays that the model fails disastrously—both for the baryon decays and for the mesons. The most celebrated example is the decay

$$B \to \omega + \pi$$

"Naive" $SU(6)_W$ predicts that the transverse ($\lambda = \pm 1$) decay is forbidden:

$$B(\lambda = \pm 1) \;\nrightarrow\; \omega(\lambda = \pm 1) \;+\; \pi$$

$$
\begin{array}{lll}
S = 0 \; S_z = 0 & S = 1 \; S_z = \pm 1 & S = 0 \; S_z = 0 \\
W = 1 \; W_z = 0 & W = 1 \; W_z = \pm 1 & W = 1 \; W_z = 0 \\
L_z = \pm 1 & L_z = 0 & L_z = 0
\end{array}
\tag{2.45}
$$

W_z can only be conserved in the $\lambda = 0$ decay mode. Unfortunately experiment is unanimous that $\lambda = \pm 1$ is the dominant mode! (Particle Data Group, 1974).

These troubles with the $L = 1$ decays led to many phenomenological "broken" $SU(6)_W$ schemes (for a review, see Rosner, 1974). What we shall attempt to do in the next sections is to describe a more theoretically motivated approach to $SU(6)_W$ vertex models via current algebra.

2.6. Summary

By considering $SU(6)_W$ rather than $SU(6)$ we are able to classify single-particle states in motion in $SU(6)_W$ supermultiplets. If we suppose the $SU(6)_W$ to be defined by some generators, $\hat{W}_\alpha$, which close on the $SU(6)_W$ algebra, then these generators, to a good approximation, take one resonant state to another resonant state within the same $SU(6)_W$ multiplet, i.e.,

$$\hat{W}_\alpha |a\rangle \approx |a'\rangle$$

where $|a\rangle$ and $|a'\rangle$ belong to the same $SU(6)_W$ multiplet. [This is analogous to the familiar $SU(2)$ case, where the generators, $J_\pm, J_3$, do not change the j values of the state $|jm\rangle$.]

One outstanding question concerns the generators $\hat{W}_\alpha$. Are they related to other physically measurable charges or currents? Another question is whether or not one can obtain some sort of $SU(6)_W$ model for vertices, given the failure of our "naive" $SU(6)_W$ model. In order to try and provide a coherent theoretical framework for answering these questions, we now make a long detour and consider quarks as they appear in current algebra, infinite momentum sum rules, and so on. The quarks we have considered so far are the building blocks for hadrons and provide a framework for the allowed hadron resonances. Thus the $SU(6)_W$ algebra of the spectrum is often called "$SU(6)_{W,\text{constituents}}$", since we are concerned with "constituent quarks." From current algebra we shall see that another $SU(6)_W$ arises, which we shall call "$SU(6)_{W,\text{currents}}$"—since it depends on the use of quarks as a model for the transformation properties of currents. These are the so-called "current quarks," and we will see that the distinction is both important and necessary—if one wishes to make a connection between $SU(6)_W$ multiplet structure and current algebra.

3. Introduction to Current Algebra and Infinite Momentum

3.1. Introductory Remarks

The essential idea of current algebra is that matrix elements of various "current" operators are measurable in physical experiments. For example, photoexcitation of a nucleon resonance in principle allows the measurement of the matrix element of the electromagnetic current, taken between the nucleon and the resonant state, i.e., of the amplitude

$$M = \varepsilon^\mu \langle N^* | J_\mu | N \rangle \tag{3.1}$$

representing the process

$$\gamma + N \to N^*$$

The $SU(3)$ properties of the electromagnetic current are summarized by writing an unintegrated form of the Gell-Mann–Nishijima relation

$$Q = I_3 + Y/2 \tag{3.2}$$

namely,

$$J_\mu^{\text{e.m.}} = F_\mu^3 + 3^{-1/2} F_\mu^8 \tag{3.3}$$

where the F^i ($i = 1, 2 \ldots, 8$) are $SU(3)$ currents, i.e., which transform as an octet under $SU(3)$ transformations. The charge operators are given by the space integral of the time component, e.g.,

$$Q = \int d^3 x J_0^{\text{e.m.}}(x) \tag{3.4}$$

which is just the familiar electromagnetic charge operator. The charges Q^i

$$Q^i = \int d^3 x F_0^i(x) \tag{3.5}$$

are the generators of the $SU(3)$ algebra

$$[Q^i, Q^j] = if^{ijk}Q^k \tag{3.6}$$

where the f^{ijk} are the usual $SU(3)$ structure constants.

Other current matrix elements are measured in weak decays and in neutrino reactions. Our present understanding of the $SU(3)$ nature of the charged weak current is embodied in the Cabibbo theory, in which the hadronic weak current is written as follows: First the weak current is decomposed into its vector and axial pieces

$$J_\mu^{wk} = V_\mu + A_\mu \tag{3.7}$$

where these are to be understood to refer to the charge-changing weak current. The $SU(3)$ properties of these are again displayed in terms of octet currents—vector currents—whose charges are the $SU(3)$ generators (which, of course, include isospin)—and axial vector currents:

$$V_\mu = F_\mu^{1+i2} \cos \theta + F_\mu^{4+i5} \sin \theta \tag{3.8}$$

$$A_\mu = F_\mu^{(5)1+i2} \cos \theta + F_\mu^{(5)4+i5} \sin \theta \tag{3.9}$$

The angle θ is the Cabibbo angle whose smallness accommodates the suppression of strangeness-changing processes relative to strangeness-conserving reactions. The vector currents are approximately conserved by Nature, whereas the axial currents are not. By definition, the vector charges Q^i satisfy an $SU(3)$ algebra. What can one say about the axial charges?

$$Q_5^i = \int F_0^{(5)i}(x) \, d^3x \tag{3.10}$$

It is at this point that the current algebra postulate enters. Gell-Mann suggested that the physical currents ($J_\mu^{e.m.}$, J_μ^{wk}, etc.) have the same *algebraic* properties under commutation as the vector currents of a simple class of field theories. In particular, field theories based on spin-$\frac{1}{2}$ quark fields: This will be the origin of the $SU(6)_W$ of currents and the notion of current quarks. For simplicity we shall usually restrict ourselves to free-quark field theory, but it is important to note that most current algebra results may also be plausibly derived from interacting quark theories. In the following sections, we will sketch the derivation of current algebra and infinite momentum sum rules. No attempt is made to be rigorous, and this outline is intended merely to serve as an heuristic introduction to more detailed and mathematically careful treatises on current algebra (for example, Adler and Dashen, 1968; Bernstein, 1968; de Alfaro *et al.*, 1973; these contain very complete references to the literature).

3.2. Free-Quark Field Theory

In this context, quark denotes an $SU(3)$ triplet of spin-$\frac{1}{2}$ fields $q(x)$, with twelve components—four for Dirac spinor indices times three for $SU(3)$. The Lagrangian is

$$\mathscr{L} = \bar{q}(x)(i\slashed{\partial} - m)q(x) \tag{3.11}$$

and we impose the canonical quantization rules on the anticommutators (ACR's)

$$\{q(x), q^{\dagger}(x')\}_{x^0=x^{0\prime}} = i\delta^3(\mathbf{x} - \mathbf{x}') \tag{3.12}$$

with all other anticommutators zero. In this model, currents have very simple forms:

$$J_\alpha(x) = \bar{q}(x)M_\alpha q(x) \tag{3.13}$$

where M_α is any product of $SU(3)$ λ matrix and Dirac Γ matrix. With equal quark masses there is an exact $SU(3)$ invariance, and the Noether conserved currents are just the $SU(3)$ vector currents

$$F_\mu^i(x) = \bar{q}(x)\gamma_\mu \frac{\lambda^i}{2} q(x) \tag{3.14}$$

Clearly one may also form an octet of axial vector currents

$$F_\mu^{(5)i}(x) = \bar{q}(x)\gamma_\mu\gamma_5 \frac{\lambda^i}{2} q(x) \tag{3.15}$$

which, however, are not conserved. Gell-Mann's postulate was that, conservation or nonconservation notwithstanding, the physical charges

$$Q^\alpha = \int J_0^\alpha(x)\, d^3x \tag{3.16}$$

have the *same* algebraic structure under commutation at equal times as do the corresponding charges in this quark field theory. Using the explicit forms for the charges and the canonical ACR's for the quark fields, one finds that the algebra formed by commuting vector and axial-vector charges *closes*:

$$[Q_i, Q_j] = if_{ijk}Q_k \tag{3.17}$$

$$[Q_i, Q_j^5] = if_{ijk}Q_k^5 \tag{3.18}$$

$$[Q_i^5, Q_j^5] = if_{ijk}Q_k \tag{3.19}$$

The first two commutators are obvious—the first (3.17) defines the $SU(3)$ algebra (3.6) and the second (3.18) contains the information that the axial vector charges transform as an $SU(3)$ octet. The content of current algebra is in the last relation (3.19). This has set the *scale* of the weak charges, since matrix elements of the vector charges are known constants. The structure of the algebra is more clearly identified by defining "chiral" charges

$$Q_i^{\pm} = \tfrac{1}{2}[Q_i \pm Q_i^5] \qquad (3.20)$$

In terms of these, the commutation relations are

$$[Q_i^+, Q_j^+] = if_{ijk}Q_k^+$$

$$[Q_i^-, Q_j^-] = if_{ijk}Q_k^- \qquad (3.21)$$

$$[Q_i^+, Q_j^-] = 0$$

which are just those of two commuting $SU(3)$ algebras. The algebra is referred to as (chiral) $SU(3) \otimes SU(3)$. Now although these commutation relations were derived by performing simple exercises with free-quark fields, Gell-Mann suggested that the resulting structure has in fact more general validity. He proposed that physically measurable charges obey the same equal-time commutation relations—despite the obvious fact that physical currents cannot be written in terms of free-quark fields! Gell-Mann used an analogy from French cuisine to illustrate this "principle of abstraction": "A piece of pheasant meat is cooked between two slices of veal, which are then discarded." Here the algebraic structure is retained while free-quark fields and the specific quark model are discarded.

3.3. Current Algebra Sum Rules and Infinite Momentum

Although current algebra did indeed set the scale for the weak interactions it was some years before Adler (1965) and Weisberger (1965) used it successfully to calculate the value of $|g_A/g_V|$. In this section we shall outline the derivation of such current algebra sum rules and introduce the $P_z \to \infty$ limit. We shall not do justice to the many other successful applications of current algebra, for which the reader is referred to the standard references. We shall also not concern ourselves with the subtle aspects of the PCAC hypothesis. In the applications we discuss, this amounts to approximating the matrix element of the pion axial charge with the corresponding pion matrix element—symbolically

$$\langle A|Q_\pi^5|B\rangle \sim \langle \pi A|B\rangle \qquad (3.22)$$

The essential idea of a sum rule is very simple. Consider the matrix element of the commutator of two charges which has the general form

$$\langle A|\left[\int d^3x J^\alpha(x), \int d^3x J^\beta(x)\right]|B\rangle \sim \langle A|\int d^3x J^\gamma(x)|B\rangle \qquad (3.23)$$

Roughly speaking this represents a connection between a current–particle

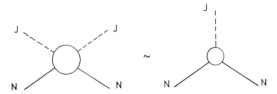

Fig. 12. Schematic relation of current algebra sum rule.

scattering process and a vertex function or form factor: see Fig. 12. A sum rule is derived by inserting a complete set of hadronic intermediate states $|\alpha\rangle$ into the commutator

$$1 = \sum_{\alpha} |\alpha\rangle\langle\alpha| \qquad (3.24)$$

As we shall see, it requires some ingenuity and manipulation to convert the resulting expression into a relation involving physical quantities.

Consider the derivation of the Adler–Weisberger relation. For this we need the proton matrix element of the $SU(2) \otimes SU(2)$ relation

$$[Q_+^5, Q_-^5] = 2Q_3 \qquad (3.25)$$

where $+$ and $-$ refer to the usual $1 + i2$ and $1 - i2$ spherical components and Q_3 is the third component of isospin. Since the proton is an eigenstate of isospin with $I_3 = +\frac{1}{2}$ the right-hand side may be immediately evaluated:

$$\langle P|[Q_+^5, Q_-^5]|P\rangle = 1 \qquad (3.26)$$

Now insert the complete set of states $|\alpha\rangle$ (3.24), and consider the matrix element

$$\langle \alpha; \mathbf{p}_\alpha|Q_\pm^5|P; \mathbf{p}\rangle \qquad (3.27)$$

Since the charge is the space-integral of a scalar density, translational invariance demands that

$$\mathbf{p}_\alpha = \mathbf{p} \qquad (3.28)$$

In terms of the invariant four-momentum transfer q,

$$q = p_\alpha - p$$

this yields

$$q^2 = (p_\alpha - p)^2 = q_0^2$$

i.e., the invariant four-momentum transfer squared is timelike:

$$q^2 = [(M^2 + \mathbf{p}^2)^{1/2} - (M_\alpha^2 + \mathbf{p}^2)^{1/2}]^2 \gtrsim 0 \qquad (3.29)$$

where M and M_α are the masses of the proton and intermediate state α, respectively. Thus the sum rule sums over matrix elements of currents with timelike q^2 which—in the absence of $\bar{\nu}e$ colliding beams for example—are not physically accessible. Furthermore, the q^2 clearly varies with the intermediate state mass M_α. In order to avoid these problems Fubini and Furlan (1965) pointed out that since the equal-time commutation relations are not covariant, one in fact generates a whole family of sum rules corresponding to different choices of external momenta. They stressed the virtue of sum rules in the $p_z \to \infty$ frame, since in this frame, the four-momentum transfer squared is zero and independent of the mass of the intermediate state: From (3.29)

$$\lim_{p_z \to \infty} q^2 = 0 \tag{3.30}$$

In this frame therefore one has the possibility of a fixed q^2 sum rule involving a physically accessible value of q^2. It is clear, however, that this limit is rather delicate and one should not be too surprised that on occasions, depending on the specific commutator, it leads to invalid results. These troubles we postpone to the next section. For the commutator (3.25) the final step to physics was taken by Adler and Weisberger, combining current algebra (3.25), the infinite momentum limit (3.30), together with PCAC (3.22) to relate the sum over states to $\pi^\pm p$ cross sections. The Born term, with a neutron intermediate state, is separated from this sum and related to the axial-vector coupling constant g_A/g_V of neutron β-decay. Their result is

$$1 = |g_A/g_V|^2 + |f_\pi|^2 \int ds \, \frac{[\sigma^{\pi^- p} - \sigma^{\pi^+ p}]}{(s - M^2)^2} \cdot \frac{2P_\pi M}{\pi} \tag{3.31}$$

where f_π is the pion decay constant, s the total invariant mass squared, and $\sigma^{\pi p}$ the pion nucleon cross sections. Numerical evaluation of the sum rule yielded a value $|g_A/g_V| \sim 1.15$, in remarkable agreement with the experimental value $|g_A/g_V| = 1.25 \pm 0.09$.

3.4. Current Algebra Sum Rules and Light-Cone Current Algebra

It is instructive to consider current algebra sum rules with reference to inclusive deep inelastic electron scattering:

$$e + p \to e' + \text{unobserved hadrons}$$

Such experiments measure the four-dimensional Fourier transform of a current commutator. In a fairly standard notation (for details see Landshoff and Osborn, in the companion volume) one measures the tensor $W_{\mu\nu}^{ab}$

$$W_{\mu\nu}^{ab}(p, q) = \frac{1}{2\pi} \int d^4x \, e^{iq \cdot x} \langle P|[J_\mu^a(x), J_\nu^b(0)]|P\rangle \tag{3.32}$$

where p and q are the four-momenta of the proton and virtual photon, respectively, and the indices a and b refer to the $SU(3)$ current components. $W^{ab}_{\mu\nu}$ is decomposed in standard fashion into the available tensors

$$W^{ab}_{\mu\nu} = [-g_{\mu\nu} + q_\mu q_\nu/q^2] W^{ab}_1(p \cdot q, q^2)$$
$$+ \left[p_\mu - \frac{p \cdot q}{q^2 q_\mu} \right] \left[p_\nu - \frac{p \cdot q}{q^2 q_\nu} \right] \frac{1}{M^2} W^{ab}_2(p \cdot q, q^2) \qquad (3.33)$$

The structure functions W^{ab}_1 and W^{ab}_2 are functions of the two independent invariants $p \cdot q$ and q^2. To derive sum rules we shall assume for this example that not only may the charge algebra (3.21) be abstracted from the quark model but also the following current density commutation relations:

$$[J_0(x), J^b_\nu(x')]_{x^0 = x^{0'}} = if^{abc} \delta^3(\mathbf{x} - \mathbf{x}') J^c_\nu(x) \qquad (3.34)$$

The derivation of sum rules is now deceptively simple. Choose a frame in which

$$\mathbf{p} \cdot \mathbf{q} = 0$$

so that

$$p \cdot q = p^0 q^0 = \nu \qquad (3.35)$$

Consider the integral

$$\int_{-\infty}^{\infty} dq^0 W^{ab}_{\mu\nu}(p, q) = \frac{1}{2\pi} \int dq^0 \int d^4 x e^{iq \cdot x} \langle P|[J^a_\mu(x), J^b_\nu(0)]|P \rangle$$
$$= \int d^4 x \delta(x^0) e^{-i\mathbf{q} \cdot \mathbf{x}} \langle P|[J^a_\mu(x), J^b_\nu(0)]|P \rangle \qquad (3.36)$$

where we have assumed that the order of integration may be interchanged. This integral therefore samples only the equal time commutator. Defining the form factor normalization of a single current matrix element by

$$\langle P|J^c_\nu(0)|P \rangle \equiv p_\nu F_c \qquad (3.37)$$

and specializing to the 0ν components, one obtains the result

$$\int dq^0 W^{ab}_{0\nu}(p, q) = if^{abc} F_c p_\nu \qquad (3.38)$$

using the commutation relation (3.34). In the integral, the mass squared of the virtual photon

$$q^2 = q_0^2 - \mathbf{q}^2 \qquad (3.39)$$

is varying and the sum rules are not useful. However, consider the limit $p_z \to \infty$ again, or equivalently $p_0 \to \infty$. Changing variables to ν (3.35) and specializing to the 00 component converts (3.38) to

$$\int d\nu \, \frac{W_{00}^{ab}}{p_0^2} = i f^{abc} F_c \tag{3.40}$$

and in terms of the structure functions the integrand is

$$\frac{W_{00}^{ab}}{p_0^2} = \frac{1}{p_0^2} \left(-1 + \frac{\nu^2}{p_0^2 q^2} \right) W_1^{ab}(\nu, q^2) + \frac{1}{M^2} \left[1 - \frac{\nu^2}{p_0^2 q^2} \right]^2 W_2^{ab}(\nu, q^2) \tag{3.41}$$

The limit $p_0 \to \infty$ at fixed ν leads to the results

$$q^2 \to -\mathbf{q}^2$$

and

$$\frac{W_{00}^{ab}}{p_0^2} \to \frac{1}{M^2} \, W_2^{ab}(\nu, -\mathbf{q}^2) \tag{3.42}$$

Thus, if the interchange of this limit with the integral over ν is permissible, the result is a fixed q^2 sum rule—the Adler–Dashen–Fubini–Gell-Mann sum rule

$$\int d\nu \, \frac{W_2^{ab}(\nu, -\mathbf{q}^2)}{M^2} = i f^{abc} F_c \tag{3.43}$$

This may be rewritten in terms of a virtual-current–hadron cross section.

From this example, it is evident that the $p_z \to \infty$ limit is crucial in order to obtain fixed q^2 sum rules but also that its use is fraught with dangers. By considering all the possible current components one can derive in this way a whole set of sum rules. For some of them, inserting conventional Regge asymptotics for the relevant amplitude leads to divergent integrals—which are apparently equated with some finite quantity, sometimes zero. For these sum rules, the derivation must be invalid! In this way we can make a purely empirical classification of the various current components: Those leading to convergent sum rules are "good" and the others, "bad." Before looking at good and bad currents in more detail, it is an interesting digression to rederive these sum rules without the infinite-momentum limit but using the more powerful quark light-cone current-algebra assumption.

Define null-plane variables

$$a^{\pm} = 2^{-1/2}(a^0 \pm a^3), \qquad \mathbf{a}_{\perp} = (a^1, a^2) \tag{3.44}$$

with the scalar product accordingly

$$a \cdot b = a^+ b^- + a^- b^+ - \mathbf{a}_{\perp} \cdot \mathbf{b}_{\perp} \tag{3.45}$$

Again the derivation of fixed q^2 sum rules is deceptively simple. In terms of null plane variables,

$$q^2 = 2q^+q^- - \mathbf{q}_\perp^2 \tag{3.46}$$

$$q \cdot x = q^+x^- + q^-x^+ - \mathbf{q}_\perp \cdot \mathbf{x}_\perp \tag{3.47}$$

Select a frame in which q^- vanishes so that $q^2 = -\mathbf{q}_\perp^2$. Now integrate $W_{\mu\nu}^{ab}$ with respect to q^+ and interchange the integrals as before to obtain

$$\int dq^+ W_{\mu\nu}^{ab}(p, q)\Big|_{\substack{q^-=0 \\ q^2=-\mathbf{q}_\perp^2}} = \int d^4x\, e^{-i\mathbf{q}_\perp \cdot \mathbf{x}_\perp}\delta(x^-)\langle P|[J_\mu^a(x), J_\nu^b(0)]|P\rangle \tag{3.48}$$

By causality the commutator must vanish for $x^2 < 0$ and thus with the δ-function restriction on x^-, since

$$x^2 = 2x^+x^- - \mathbf{x}_\perp^2 \tag{3.49}$$

the only contributions are from the light cone $x^2 = 0$—where there is in general a singularity. If, as Fritzsch and Gell-Mann (1971) suggested, the leading light-cone singularity structure is abstracted from the quark model, one rederives the current-algebra fixed-q^2 sum rules. Dicus *et al.* (1971) discussed the sum rules in this manner and showed that while the "good" sum rules were unchanged, in general "bad" sum rules were modified—so that at least a divergent integral was now equated to another presumably divergent integral! However, even this is not the whole story since integrals of singular functions must be treated with great care. These problems are discussed by Dicus *et al.* (1971) and also by Jersak and Stern (1969), who first pointed out the delicate nature of the fixed-q^2 limit.

3.5. Good and Bad Currents

The nature of the $P_z \to \infty$ limit is illuminated by a study of field-theory examples. In the simplest case, that of free-quark field theory, one can explicitly examine the problem of interchanging limit and integral. In this noninteracting theory only two types of intermediate state can contribute in the sum: the "direct" graph (Fig. 13) and the pair or Z diagram (Fig. 14). For the Z diagrams, the intermediate state "mass" tends to ∞ as $P_z \to \infty$ (see Adler and Dashen, 1968, for details), and the interchange of the limit with the integral is valid only if the contribution from these pair

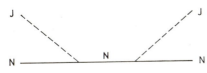
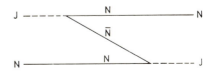

Fig. 13. Direct contribution.   Fig. 14. Pair or Z-diagram contribution.

states is "damped out" as $P_z \to \infty$. Explicit calculation shows that there are two classes:

Good components. The 0 and 3 components of V_μ and A_μ pick up $1/P_0$ factors, which suppress the Z diagrams as $P_z \to \infty$.

Bad components. The 1 and 2 components do not have such factors and the Z diagrams are not suppressed and in fact are essential to satisfy the sum rule at $P_z = \infty$.

In less trivial theories, this classification remains true: The details with the concomitant subtleties of "class II" states are discussed by Adler and Dashen (1968).

One way of understanding the origin of the P_0 factors is as follows. Under z-boosts the current components transform as

$$
\begin{aligned}
J_0' &= \cosh \xi J_0 + \sinh \xi J_3 \\
J_3' &= \sinh \xi J_0 + \cosh \xi J_3 \\
J_1' &= J_1 \\
J_2' &= J_2
\end{aligned}
\tag{3.50}
$$

where the boost is from $P^0 = M$ to $P^0 = M \cosh \xi$. The limit $P_z \to \infty$ demands that both $\cosh \xi$ and $\sinh \xi$ tend to ∞ like P_0. No such factors arise for the 1 and 2 components. (Using noncovariant normalization of states, the classification of good components is sometimes defined as those currents whose matrix elements are nonvanishing at infinite momentum: This is equivalent to the above.)

In terms of light plane variables one may distinguish also "terrible" components:

$$
\begin{aligned}
J_+' &= e^\xi J_+ &\quad \text{good} \\
\mathbf{J}_\perp' &= \mathbf{J}_\perp &\quad \text{bad} \\
J_-' &= e^{-\xi} J_- &\quad \text{terrible}
\end{aligned}
\tag{3.51}
$$

These have simple forms in terms of free-quark fields quantized on the null plane $x^+ = 0$ (see Section 4).

3.6. The "Maximal" Good Quark Algebra

The algebra formed by commuting all 144 current components—vector, tensor, axial vector, and so on—of the quark model closes on $U(12)$. However, the largest "good" infinite-momentum algebra is smaller than $U(12)$ but larger than $SU(3) \otimes SU(3)$: It is in fact an $SU(6)_W$ algebra

$$
J_M \sim q^+ M q
$$

where

$$M = (1, \lambda') \otimes [\beta\sigma_x, \beta\sigma_y, \sigma_z]$$

This is the algebra $SU(6)_{W,\text{currents}}$ at infinite momentum. Its significance lies in the fact that for this algebra, in model field theories at least, it is a sensible approximation to ignore the multiparticle, Z type diagrams in the sum over intermediate states at infinite momentum. One can hope therefore to give a definition to an approximate $SU(6)_W$ symmetry that leads to no theoretical conflicts. If the sum rule may be approximately saturated by only a finite number of single-particle (or resonant) intermediate states, then this implies that these states at infinite momentum may be classified by $SU(6)_W$ representations.

$SU(6)_W$ is the largest single-particle symmetry that one can abstract from the quark model—since only for good currents at infinite momentum can one avoid the problem of multiparticle states. The generators of this $SU(6)_W$ algebra are the integrals of local, measurable currents. It is therefore a natural question to ask whether these $SU(6)_{W,\text{currents}}$ generators are related to the generators of the spectrum symmetry $SU(6)_{W,\text{constituents}}$. The Melosh transformation is an attempt to formulate a connection. Before discussing his attempts, it is helpful to reformulate current algebra at infinite momentum in terms of null plane charges.

4. The Melosh Transformation

4.1. Approximate Symmetries

Consider again the free-quark model, defined by the Hamiltonian

$$H_0 = \int d^4x \delta(t) q^\dagger(x)(\boldsymbol{\alpha} \cdot \mathbf{p} + \beta M)q(x) \tag{4.1}$$

(where $\mathbf{p}$ is shorthand for $-i\partial$). Equal-time charges $F(\Lambda)$ are defined by

$$F(\Lambda) = \int d^4x \delta(t) q^\dagger(x)\Lambda q(x) \tag{4.2}$$

where $\Lambda \equiv \lambda^i \otimes \Gamma$—a product of $SU(3)$ and Dirac matrices. Using the canonical equal-time ACR's for the quark fields [Eq. (3.12)] we find in general

$$[F(\Lambda), H_0] \neq 0 \tag{4.3}$$

with the exception of the $SU(3)$ generators Q_i corresponding to $\Lambda = \lambda^i \otimes 1$. These Q_i are the only conserved charges: Even this symmetry is easily broken by the introduction of unequal quark masses via the quark mass matrix M. However, in the case of degenerate quark masses there is an octet of conserved $SU(3)$ vector currents

$$\partial^\mu F_\mu^i = 0 \tag{4.4}$$

but the corresponding $SU(3)$ axial vector currents are not conserved,

$$\partial^\mu F_\mu^{(5)i} \neq 0 \tag{4.5}$$

and the divergence is proportional to the mass matrix M. Despite the lack of any exact symmetry beyond $SU(3)$, Gell-Mann attempted to use the charges $F(\Lambda)$ to formulate a precise definition of approximate symmetries. The first such attempt has been called "little leakage" (see Bell, 1966).

4.1.1. Little Leakage and Coleman's Theorems

We consider the case where the generators F do not commute with the Hamiltonian

$$[F, H_0] \neq 0 \tag{4.3}$$

and therefore do not generate an exact symmetry group. Nevertheless, suppose that they generate an approximate symmetry in the following sense. If their action on a one-particle state transforms it, to a good approximation, to another one-particle (or resonant) state

$$F|1\rangle = |1'\rangle \tag{4.6}$$

then there is the possibility that a limited number of resonant states form an approximate representation of the symmetry group generated by the F algebra. In particular, the hope was that $SU(6)_W$ could be defined in this way. Two theorems of Coleman eliminated this method of defining approximate symmetries. The physics content of these theorems is sketched below: For mathematical details the reader is referred to the original papers (Coleman, 1965, 1966).

Theorem 1. "An invariance of one-particle states is an invariance of the vacuum," i.e.,

$$F|1\rangle = |1'\rangle$$

implies

$$F|0\rangle = 0$$

Bell (1966, 1974) gives the following intuitive explanation of the theorem: A localized one-particle state consists mainly of vacuum:

$$|\text{one-particle state}\rangle \sim \left|\begin{matrix}\text{one particle at} \\ \text{point } x\end{matrix} + \begin{matrix}\text{vacuum} \\ \text{elsewhere}\end{matrix}\right\rangle$$

Thus, in order for a one-particle state to remain a one-particle when acted on by an operator F, the operator F must leave the vacuum as vacuum.

In free-field theory, the nonconserved equal-time charges (with $\Gamma \neq 1$) do not annihilate the vacuum:

$$F_\Lambda |0\rangle \neq 0$$

There are cross terms in fermion and antifermion creation operators (e.g., $\sim a^+ b^+$) which create pairs

$$F_\Lambda |0\rangle \rightarrow (q\bar{q}) \text{ pairs}$$

Since the vacuum has infinite extent these unwanted pair states have infinite norm, and the leakage is always infinite! In free-field theory at least, a "little leakage" approximation makes no sense. Conceivably, however, all could be well in an interacting theory with

$$H = H_0 + \text{interactions}$$

and F_Λ's that do not excite the vacuum. Here Coleman's second theorem blocks the way:

Theorem 2. "An invariance of the vacuum is an invariance of the world," i.e.,

$$F|0\rangle = 0$$

implies

$$[F, H] = 0 \tag{4.7}$$

The essence of the proof is as follows. Since we require $F|0\rangle = 0$ and the vacuum is an eigenstate of H, we have

$$[F, H]|0\rangle = 0$$

Thus for any state $|n\rangle$

$$\langle n|[F, H]|0\rangle = 0$$

However, in relativistic field theories all matrix elements of $[F, H]$ are related by crossing to those exciting the vacuum, e.g.,

$$0 = \langle 4|[F, H]|0\rangle \rightarrow \langle 2|[F, H]|2\rangle$$

By demonstrating that all matrix elements are zero, one deduces

$$[F, H] = 0 \tag{4.8}$$

i.e., the equal-time charges must commute with the Hamiltonian in the same way as for an exact symmetry. Thus, demanding a symmetry of one-particle states and that the equal-time generators F do not commute with H leads, via Theorems 1 and 2, to a contradiction.

One must conclude that nonconserved equal-time generators and little leakage are not compatible: There is always infinite leakage unless $[F, H] = 0$. Nevertheless, it is evident that the cause of the problems is the infinite extent of the vacuum and uninteresting pair states—not the initial one-particle state. This implies that a sensible definition of an approximate symmetry must avoid the complications due to the vacuum. One way to avoid this problem of vacuum excitation is to ensure that the vacuum break-up terms must cancel. Gell-Mann therefore considered the saturation of commutators by a finite number of states—"swift saturation"—as a possible definition of approximate symmetries.

4.1.2. Swift Saturation

Consider the one-particle matrix element of the commutator

$$[F(\Lambda_1), F(\Lambda_2)] = F([\Lambda_1, \Lambda_2]) \tag{4.9}$$

Taken between single-particle states $|a\rangle$ and $|b\rangle$ and inserting a complete set of states $|n\rangle$ one obtains

$$\sum_n \langle a|F(\Lambda_1)|n\rangle\langle n|F(\Lambda_2)|b\rangle - (1 \leftrightarrow 2) = \langle a|F([\Lambda_1, \Lambda_2])|b\rangle \tag{4.10}$$

The specification of the one-particle matrix element of $F([\Lambda_1, \Lambda_2])$ ensures that vacuum break-up terms, involving intermediate states $|n\rangle$ with additional particle–antiparticle pairs, cancel in the difference between the two terms. Gell-Mann conjectured that the nonconserved charges could define an approximate symmetry if the states $|n\rangle$ could be restricted to a small number of one-particle (or resonance) states. This leads to the classification of one-particle states into multiplets of the approximate symmetry group. However, the equal-time charges

$$F(\Lambda) = \int d^4x \delta(t) q^\dagger(x) \Lambda q(x) \tag{4.11}$$

are clearly not Lorentz invariant, and there is a family of different sum rules corresponding to every external momentum **P**. The validity of the "swift saturation" approximation may therefore depend on the external momentum. In particular, the free field theory example of Section 3 showed that the essentially *multiparticle Z* diagrams are, in general, *necessary* to saturate the sum rules. However, for the integrals of good currents these multiparticle contributions are suppressed at $P_z = \infty$. Thus resonance saturation without multiparticle states is most plausible for $P_z = \infty$ external momenta. Note that for the bad and terrible current components, multiparticle contributions are necessary even at $P_z = \infty$ in order to satisfy the sum rules. The swift saturation approach therefore leads to the conclusion that the maximal approximate symmetry algebra that may sensibly be abstracted from the quark model is the $SU(6)_{W,currents}$ algebra of good currents at infinite momentum.

4.2. Null-Plane Charges

Consider the matrix element of the equal-time charges $F(\Lambda)$ taken between infinite-momentum states

$$\left\langle b, P_z = \infty \left| \int d^4x \delta(t) q^\dagger(x) \Lambda q(x) \right| a, P_z = \infty \right\rangle \qquad (4.12)$$

The infinite-momentum states are obtained as the limit of finite boosted states

$$|a, P_z = \infty\rangle = \lim_{\omega \to \infty} e^{-i\omega K_3}|a\rangle \qquad (4.13)$$

where K_3 is the generator of z boosts. In (4.12), let these boost operators act on the quark fields instead of the states: The quark field transforms as

$$e^{i\omega K_3} q(x) e^{-i\omega K_3} = e^{+\omega \alpha_3/2} q(x') \qquad (4.14)$$

with

$$x = x', \qquad z = z' \cosh \omega + t' \sinh \omega$$
$$y = y', \qquad t = t' \cosh \omega + z' \sinh \omega$$

One is therefore concerned with matrix elements, between finite-momentum states, of modified charges

$$e^{i\omega K_3} F(\Lambda) e^{-i\omega K_3} = \int d^4x \delta(t) q^\dagger(x') e^{\omega \alpha_3/2} \Lambda e^{\omega \alpha_3/2} q(x') \qquad (4.15)$$

in the limit $\omega \to \infty$. Write

$$e^{\omega \alpha_3/2} = \cosh \omega/2 [1 + \alpha_3 \tanh (\omega/2)]$$

and then change variables $x_\mu \to x'_\mu$ and, formally at least, take the limit $\omega \to \infty$. The result is

$$\lim_{\omega \to \infty} e^{i\omega K_3} F(\Lambda) e^{-i\omega K_3} = 2 \int d^4x \delta(t + z) q^\dagger(x) \Lambda_g q(x) \qquad (4.16)$$

where

$$\Lambda_g = \tfrac{1}{2}(1 + \alpha_3) \Lambda \tfrac{1}{2}(1 + \alpha_3) \qquad (4.17)$$

It will be convenient to introduce the "good" components of quark fields by

$$q_+(x) = \tfrac{1}{2}(1+\alpha_3)q(x) \tag{4.18}$$

and rewrite

$$q^\dagger \Lambda_g q = q_+^{\ \dagger}\Lambda q_+ \tag{4.19}$$

Thus, the equal-time charges $F(\Lambda)$ when subjected to an infinite boost lead to the null-plane charges $\hat{F}(\Lambda)$

$$\hat{F}(\Lambda) \equiv 2 \int d^4x \delta(t+z)q_+^{\ \dagger}(x)\Lambda q_+(x) \tag{4.20}$$

The charges $F(\Lambda)$ are obtained by integrating a local density over the spacelike surface $t = 0$, whereas these objects $\hat{F}(\Lambda)$ are obtained by integrating over the null plane $t + z = 0$ (Fig. 15). As emphasized by Bell, there is nothing unnatural about this integration along a null plane—it is after all, what we see!

One great advantage of the null-plane formulation is that manipulations "in the $P_z = \infty$ frame" are much easier to perform. For example, the good $SU(6)_W$ algebra arises very simply in this formalism. Since

$$(1+\alpha_3)(1-\alpha_3) = 0 \tag{4.21}$$

if

$$\{\Lambda, \alpha_3\} = 0 \tag{4.22}$$

then

$$\Lambda_g = 0$$

In other words, only Λ's that commute with α_3 may be restricted to the null plane, i.e.,

$$\Lambda_g \equiv (1, \lambda^i) \otimes (\beta\sigma_x, \beta\sigma_y, \sigma_z)$$

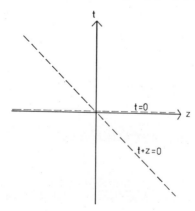

Fig. 15. Equal-time and null-plane quantization surfaces.

Thus the null plane charges commute to form an $SU(6)_W$ algebra,

$$[\hat{F}(\Lambda_\alpha), \hat{F}(\Lambda_\beta)] = c_{\alpha\beta\gamma}\hat{F}(\Lambda_\gamma) \tag{4.23}$$

4.3. Free-Quark Fields on the Null Plane

It is helpful to consider free-quark theory quantized not on the usual equal-time surface but rather on a null plane (Fig. 15). It is convenient to define null-plane coordinates [as in (3.43) and (3.44)]

$$a^\pm = 2^{-1/2}(a^0 \pm a^3), \qquad \mathbf{a}_\perp = (a^1, a^2) \tag{4.24}$$

The scalar product $a \cdot b$ may be written

$$a \cdot b = a^+ b^- + a^- b^+ - \mathbf{a} \cdot \mathbf{b}_\perp$$
$$= g^L_{\mu\nu} a^\mu b^\nu \tag{4.25}$$

where now the Greek indices μ and ν refer to components $(+, 1, 2, -)$ and not the usual $(0, 1, 2, 3)$. Thus the metric tensor in null-plane coordinates is off-diagonal:

$$g_L^{\mu\nu} = \begin{pmatrix} 0 & 0 & 0 & 1 \\ 0 & -1 & 0 & 0 \\ 0 & 0 & -1 & 0 \\ 1 & 0 & 0 & 0 \end{pmatrix} \tag{4.26}$$

In the null-plane basis, Dirac γ matrices satisfy

$$\{\gamma^\mu, \gamma^\nu\} = 2g_L^{\mu\nu} \tag{4.27}$$

leading to results such as

$$(\gamma^+)^2 = 0 \tag{4.28}$$

Moreover, care must be taken with upper and lower indices since

$$a_- = a^+, \qquad a_+ = a^-$$

and

$$\partial^- = \frac{\partial}{\partial x^+}; \qquad \partial^+ = \frac{\partial}{\partial x^-}$$

The Dirac equation takes the form

$$(\gamma^+\partial^- + \gamma^-\partial^+ - \boldsymbol{\gamma}_\perp \cdot \boldsymbol{\partial}_\perp)q(x) = iMq(x) \tag{4.29}$$

For theories quantized on equal-time surfaces, all dynamics is contained in the Hamiltonian H, which generates the evolution in time t. Correspondingly, on the null plane the role of the Hamiltonian is played by the dynamical operator P^- which generates displacements in x^+, away from the initial value $x^+ = 0$. Multiplying (4.29) by γ^+ and using (4.28) yields the relation

$$\gamma^+\gamma^-\partial^+q(x) = (iM - \boldsymbol{\gamma}_\perp \cdot \boldsymbol{\partial}_\perp)\gamma^+q(x) \tag{4.30}$$

Since neither ∂^+ nor $\boldsymbol{\partial}_\perp$ displace $q(x)$ out of the null plane $x^+ = 0$, this is therefore a constraint equation on the field $q(x)$. Defining the so-called "good" and "bad" components of the quark field, q_+ and q_-, respectively, by

$$q_\pm \equiv \tfrac{1}{2}(1 \pm \alpha_3)q \equiv \mathbb{P}_\pm q \tag{4.31}$$

the constraint equation reads

$$\partial^+q_-(x) = (iM - \boldsymbol{\gamma}_\perp \cdot \boldsymbol{\partial}_\perp)\gamma^+q_+(x) \tag{4.32}$$

This may formally be solved by

$$q_-(x) = \int dy\varepsilon(x^- - y)(iM - \boldsymbol{\gamma}_\perp \cdot \boldsymbol{\partial}_\perp)\gamma^+q_+(y, \boldsymbol{x}_\perp, x^+)$$

to show that q_- is not independent of q_+. Correspondence with the equal-time theory is obtained with the use of null-plane ACR's for good components alone:

$$2^{1/2}\delta(x^+)\{q_+(0), q_+^\dagger(x)\} = \delta^4(x)$$
$$2^{1/2}\delta(x^+)\{q_+(0), q_+(x)\} = 0 \tag{4.33}$$

From their definition (4.31), good and bad fields behave simply under z boosts:

$$q_+ \to e^{+\omega/2}q_+$$
$$q_- \to e^{-\omega/2}q_- \tag{4.34}$$

Before looking again at the question of good and bad currents it is helpful to choose a particular representation of the γ matrices. From (4.31) it is clearly appropriate to choose α_3 diagonal

$$\alpha_3 = \begin{pmatrix} 1 & 0 \\ 0 & -1 \end{pmatrix} \tag{4.35}$$

whence

$$\mathbb{P}_+ = \begin{pmatrix} 1 & 0 \\ 0 & 0 \end{pmatrix}; \qquad \mathbb{P}_- = \begin{pmatrix} 0 & 0 \\ 0 & 1 \end{pmatrix} \tag{4.36}$$

The remaining γ matrices are conveniently chosen to be (Bell, 1974)

$$\boldsymbol{\alpha}_\perp = \begin{pmatrix} 0 & -\boldsymbol{\sigma}_\perp \\ -\boldsymbol{\sigma}_\perp & 0 \end{pmatrix}, \qquad \beta = \begin{pmatrix} 0 & \sigma_3 \\ \sigma_3 & 0 \end{pmatrix} \tag{4.37}$$

By virtue of (4.36) the matrices Λ_g (4.17) connect only good components q_+

$$\Lambda_g = \left(\begin{array}{c|c} /\,/\,/ & 0 \\ \hline 0 & 0 \end{array}\right)$$

Good currents therefore only involve the good quark fields q_+ obeying the canonical ACR's (4.33), which one hopes remain unchanged by inter-actions. The constraint equation for q_- fields would, however, involve the interaction explicitly, and thus abstractions involving bad fields are presumably model dependent. Bad currents contain one q_- field, and terrible currents involve two: Thus in accordance with our earlier dis-cussion of their Lorentz boost properties [(3.50) and (4.34)] we have the following:

good: $\qquad J_g \sim q_+{}^\dagger \Lambda q_+$

bad: $\qquad J_b \sim q_-{}^\dagger \Lambda q_+ + q_+{}^\dagger \Lambda q_-$

terrible: $\qquad J_t \sim q_-{}^\dagger \Lambda q_-$

Only the good currents have simple commutation properties, and the charges $\hat{F}$ (4.20), together with the ACR's (4.33), lead to the $SU(6)_W$ algebra. In phenomenological applications, it is therefore safer to restrict oneself to predictions involving only good currents and charges.

4.4. Properties of the Null-Plane Charges $\hat{F}(\Lambda)$

(1) In the *free*-quark model (with no mass splitting) the Hamiltonian may be written (de Alwis, 1973)

$$H_0 = \int d^2\mathbf{x}_\perp \, dx^- \int dy \, \varepsilon(x^- - y)q_+{}^\dagger(\partial_\perp^2 + m^2)q_+ \qquad (4.38)$$

Thus we have immediately

$$[\hat{F}, H_0] = 0 \qquad (4.39)$$

(2) The null-plane charges annihilate the vacuum

$$\hat{F}|0\rangle = 0 \qquad (4.40)$$

Because the region of integration for a null-plane charge is restricted by $\delta(t+z)$ instead of the usual $\delta(t)$, the statement of translational in-variance for spacelike charges

$$[P^i, F] = 0$$

is replaced by the statements

$$[P_\perp, \hat{F}] = 0 \qquad (4.41)$$

and

$$[P^+, \hat{F}] = 0 \qquad (4.42)$$

The combination $P^+ \sim P^0 + P^3$ of displacement generators has eigenvalues positive and greater than zero for physical states (excluding the possibility of zero-mass particles and infinite-momentum states). The relation (4.42) therefore shows that $\hat{F}$ does not connect the vacuum to any physical states, and (4.40) follows. Since this is a trivial kinematic property of null-plane charges, it is true even if, as in the case of quark mass splitting, $[F, H_0] \neq 0$. Thus, by construction, these null-plane charges avoid the vacuum troubles encountered with nonconserved equal-time charges. It is for this reason that they are serious candidates for the generators of an approximate symmetry.

(3) Again by virtue of the integration region the null-plane charges can be shown to commute with the generator K_3 of z boosts

$$[K_3, \hat{F}] = 0 \qquad (4.43)$$

(4) Under rotations about the z direction the components involving 1 and σ_3 are invariant, while the components $(\beta\sigma_1, \beta\sigma_2)$ transform like a two-vector.

(5) Galilean invariance: The charges $\hat{F}(\Lambda)$ are not invariant under rotations about the x and y directions

$$[J_\perp, \hat{F}(\Lambda)] \neq 0$$

Since the constraint $t + z = 0$ is not invariant under such rotations, these operators generate translations out of the null plane and are therefore *dynamical* rather than *kinematical*. However, combinations of transverse rotations and transverse boosts do leave the null plane invariant, transforming

$$z \to z' = z + \delta$$

$$t \to t' = t - \delta$$

By explicit construction the Galilean boosts

$$E_1 = K_1 + J_2$$
$$E_2 = K_2 - J_1 \qquad (4.44)$$

are found to commute with the null-plane charges

$$[\mathbf{E}_\perp, \hat{F}] = 0 \qquad (4.45)$$

To summarize: The free-quark model contains an $SU(6)_W$ algebra of null-plane charges

$$\hat{F}(\Lambda) = 2^{1/2} \int d^4x \, \delta(x^+) q_+^\dagger \Lambda q_+$$
$$[\hat{F}(\Lambda_\alpha), \hat{F}(\Lambda_\beta)] = c_{\alpha\beta\gamma} \hat{F}(\Lambda_\gamma) \qquad (4.46)$$

where

$$\Lambda \equiv (1, \lambda) \otimes (\beta\sigma_1, \beta\sigma_2, \sigma_3)$$

This is the so-called $SU(6)_{W,\text{currents}}$: It contains a W-spin $SU(2)$ subalgebra generated by

$$\mathbf{W} \equiv \{\hat{F}(\beta\sigma_x), \hat{F}(\beta\sigma_y), \hat{F}(\sigma_z)\} \tag{4.47}$$

Current algebra postulates that physical good charges and currents have the same algebraic properties as the corresponding quark model charges and currents.

4.5. The Necessity of a Melosh Transformation

The null-plane charges $\hat{F}_\alpha$ are candidates for generators of an approximate $SU(6)_W$ symmetry of states since they do not excite the vacuum. They already generate the $SU(6)_{W,\text{currents}}$ symmetry of physical currents, and the Melosh transformation addresses itself to the question of whether they are also generators of $SU(6)_{W,\text{constituents}}$. If we denote the spectrum (constituent) symmetry generators by $\hat{W}_\alpha$, can one identify $\hat{F}_\alpha$ and $\hat{W}_\alpha$? Gell-Mann (1972) suggested that the two sets of generators are related by a unitary transformation V:

$$\hat{W}_\alpha = V\hat{F}_\alpha V^{-1} \tag{4.48}$$

(where V is now usually referred to as the Melosh transformation). In this framework then, can V be unity?

This question may best be answered by assuming the truth of the proposition and then deriving an absurdity. Assume therefore that particles of momentum $P \equiv (P^+, \mathbf{P}_\perp)$ fall into $SU(6)_W$ multiplets classified by the $\hat{F}$'s. Since $\hat{F}(\frac{1}{2}\sigma_z) \equiv W_z = S_z$, the z component of total quark spin, commutes with J_z, an orbital angular momentum component L_z may be defined by the relation

$$J_z = L_z + \hat{F}(\tfrac{1}{2}\sigma_3) \tag{4.49}$$

In the following examples we shall only consider states for which $L_z = 0$ so that the z component of W spin may be identified with the total angular momentum. In defining spin states for moving particles one must take care to use only boost operators that commute with the $\hat{F}$'s in order to preserve the $SU(6)_W$ classification of the state. Thus it is convenient to define "null-plane helicity" states by (Kogut and Soper, 1970)

$$|\{n\}; P^+, \mathbf{P}_\perp; m\rangle = e^{-i\boldsymbol{\omega}_\perp \cdot \mathbf{E}_\perp} e^{-i\omega_3 K_3} |\{n\}; P_0; m\rangle \tag{4.50}$$

where m is the eigenvalue of W_z and $\{n\}$ represents other conserved quantum numbers such as B, I, Y, etc. The final-state momentum is given by $P^+ = (m/2^{1/2})e^{\omega_3}$ and $\mathbf{P}_\perp = \omega_\perp P^+$. In this way, one can construct states with any momentum having the same classification with respect to the $SU(6)_W$ algebra of the $\hat{F}$ charges.

Consider two of the simplest applications of such a scheme in which $V = 1$.

4.5.1. Magnetic Moments of the $\{56; L = 0\}$

Consider the $SU(6)_W$ transformation properties of the good component of the electromagnetic current

$$J^+ = 2^{-1/2}(J^0 + J^3) \qquad (4.51)$$

In quark field theory this is simply

$$J^+ = 2^{-1/2}q^+(1 + \alpha_3) \cdot Qq = 2^{-1/2}q_+^+ \cdot 1Qq_+ \qquad (4.52)$$

which evidently commutes with the W-spin generators (4.47), demonstrating that J^+ has W-spin zero.

Nucleons are assigned to a W-spin-$\frac{1}{2}$ doublet belonging to a **56** representation with $L_z = 0$. The proton matrix element of J^+

$$\langle \mathbf{P}_L', m'|J^+|\mathbf{P}_L, m\rangle$$

where $\mathbf{P}_L \equiv (P^+, \mathbf{P}_\perp)$, and m and m' are the eigenvalues of both J_z and W_z, yields immediately the $SU(6)_W$ prediction

$$\langle \mathbf{P}_L', m'|J^+|\mathbf{P}_L, m\rangle \alpha \delta_{m'm} \qquad (4.53)$$

To obtain a prediction in terms of the usual form factors F_1 and F_2

$$\langle \mathbf{P}_L', m'|J_\mu|\mathbf{P}_L, m\rangle = \bar{u}(\mathbf{P}_L', m')[(F_1 + F_2)\gamma_\mu - (F_2/2m)(P + P')_\mu]u(\mathbf{P}_L, m) \qquad (4.54)$$

one must evaluate the J^+ matrix element using null-plane helicity spinors

$$u(\mathbf{P}_L, m) = \frac{1}{(P^+)^{1/2}}\begin{pmatrix} P^+\phi(m) \\ (m\sigma_3 - \boldsymbol{\sigma}_\perp \cdot \mathbf{P}_\perp)\phi(m) \end{pmatrix} \qquad (4.55)$$

By direct evaluation one finds a term proportional to $\delta_{mm'}$ as required—but also a spin-flip term proportional to the form factor F_2. The $SU(6)_W$ symmetry prediction for $V = 1$ is therefore that

$$F_2 = 0 \qquad (4.56)$$

which is the famous Dashen–Gell-Mann (1966) result. However, it is possible to argue that this is not a theoretical contradiction—merely not a very good phenomenological approximation to our world, in which

$$F_2 \sim \mu_{\text{Dirac}}$$

It is not difficult, however, to remove this hope and destroy the possibility that $V = 1$ for any world except that of free quarks.

4.5.2. Form Factors of the {35; L = 0}

In a $q\bar{q}$ W-spin classification scheme for the ground-state mesons, the helicity states $\lambda = 0, \pm 1$ of the ρ meson correspond to $J_z = W_z = 0, \pm 1$. Consider the $SU(6)_W$ prediction for the matrix element of J^+ between charged ρ^+ mesons: As before one has

$$\langle \rho^+; \mathbf{P}_L', \lambda' | J^+ | \rho^+; \mathbf{P}_L, \lambda \rangle \propto \delta_{\lambda'\lambda} \tag{4.57}$$

One must now translate this into relations for form factors.

From Lorentz and time-reversal invariance, plus current conservation, the spin-one electromagnetic vertex function is characterized by three form factors:

$$\langle P', \lambda' | J^\mu | P, \lambda \rangle = I_1^\mu F_1(t) + I_2^\mu F_1(t) + I_3^\mu F_3(t) \tag{4.58}$$

where $t = (P' - P)^2 = q^2$ and the kinematic functions may be chosen as

$$I_1^\mu = -(P + P')^\mu \varepsilon'^* \cdot \varepsilon$$
$$I_2^\mu = (\varepsilon'^* \cdot q)\varepsilon^\mu - (\varepsilon \cdot q)\varepsilon'^{*\mu} \tag{4.59}$$
$$I_3^\mu = (P + P')^\mu (\varepsilon'^* \cdot q)(\varepsilon \cdot q)$$

where ε' and ε are the polarization vectors corresponding to λ' and λ. Using the appropriate null-plane polarization vectors

$$\varepsilon^\mu (P, \lambda = \pm 1) = \mp 2^{-1/2}\left(0, 1, \pm i, \frac{P^1 \pm iP^2}{P^+}\right)$$

$$\varepsilon^\mu (P, \lambda = 0) = \frac{1}{M}\left(P^+, P^1, P^2, \frac{\mathbf{P}_\perp^2 - M^2}{2P^+}\right) \tag{4.60}$$

the $SU(6)_W$ prediction (4.57) leads to the conditions

$$F_3(t) = 0 \tag{4.61}$$

$$F_1(t)(1 + P^+/P^{+'}) + F_2(t) = 0 \tag{4.62}$$

for all $t \neq 0$. However, since the ratio $P^+/P^{+'}$ is *not* determined by t

$$t = M^2\left(2 - \frac{P^+}{P^{+'}} - \frac{P^{+'}}{P^+}\right) - \left(\frac{P^+\mathbf{P}_\perp'^2}{P^{+'}} - 2\mathbf{P}_\perp' \cdot \mathbf{P}_\perp + \frac{P^{+'}\mathbf{P}_\perp^2}{P^+}\right)$$

then (4.62) together with (4.61) require that for $t \neq 0$

$$F_1(t) = F_2(t) = F_3(t) = 0 \tag{4.63}$$

A quite minimal analyticity assumption leads to the result

$$F_1(0) = 0 \qquad (4.64)$$

in contradiction with the charge normalization

$$F_1(0) = 1 \qquad (4.65)$$

Our hypothesis that $V = 1$ is therefore too strong to accommodate any sensible dynamics, and must therefore be rejected. Some nontrivial "Melosh transformation" is required to relate $SU(6)_{W,\text{currents}}$ to $SU(6)_{W,\text{constituents}}$ (Bell and Hey, 1974).

It is interesting to remark that the results on ρ-meson form-factors are exactly those expected for a "ρ-meson" made from free quarks—any nonzero momentum transfer disintegrates the "meson" since there are no forces to bind the quarks.

4.6. Melosh's Transformations

In his thesis, Melosh (1973) was able to derive an explicit form for V in the free-quark model. This was essentially the result of combining the Foldy–Wouthuysen transformation with the demand of z-boost invariance, and specializing to the null plane. In the free-quark model, it is well known that it is possible to use the Foldy–Wouthuysen transformation to define a set of equal-time charges $W(\Lambda)$ that commute with the free Hamiltonian

$$[W(\Lambda), H_0] = 0 \qquad (4.66)$$

The W charges may be written as

$$W(\Lambda) = \int d^4x \, \delta(t) \phi^\dagger(x) \Lambda \phi(x) \qquad (4.67)$$

where the field ϕ is related to the field q by a unitary transformation

$$\phi = e^{iS} q \qquad (4.68)$$

with

$$S = \frac{-i\boldsymbol{\gamma} \cdot \hat{\mathbf{P}}}{2} \arctan\left\{\frac{|\mathbf{P}|}{m}\right\} \qquad (4.69)$$

As emphasized by Bell (1974), the largest possible symmetry one may obtain in this way is a $U(6) \otimes U(6) \otimes O(3)$ algebra. Melosh considered the $SU(6)_W$ subalgebra and the transformation

$$\phi_m = e^{iS_m} q \qquad (4.70)$$

where

$$S_m = \frac{-i}{2} \boldsymbol{\gamma} \cdot \hat{\mathbf{P}}_\perp \arctan\left\{\frac{|\mathbf{P}_\perp|}{m}\right\} \qquad (4.71)$$

In second-quantized form the transformation may be written

$$\phi_m = VqV^{-1} \tag{4.72}$$

where

$$V = \exp i Y_m \tag{4.73}$$

$$Y_m = \int d^4x \delta(t) q^\dagger(x) S_m q(x) \tag{4.74}$$

In terms of charges, the relation is

$$W(\Lambda) = VF(\Lambda)V^{-1} \tag{4.75}$$

where the W charges are expressed in terms of ϕ_m fields (4.72).

Historically, however, there were serious objections to this type of $SU(6)_W$ symmetry (Jordan, 1965; Bell, 1974), and Melosh evaded these troubles by taking the infinite-momentum limit of these charges. Since $\gamma_\perp$ is a good operator, one obtains

$$\hat{W}(\Lambda) = \hat{V}\hat{F}(\Lambda)\hat{V}^{-1} \tag{4.76}$$

where

$$\hat{V} = \exp(i\hat{Y}) \tag{4.77}$$

$$\hat{Y} = 2^{1/2} \int d^4x \delta(x^+) q_+^\dagger(x) S_m q_+(x) \tag{4.78}$$

This proposal has been criticized by de Alwis and Stern (1974), Eichten *et al.* (1973), and Osborn (1974). The essential point is that the null-plane charges $\hat{F}$ already commute with the free Hamiltonian (de Alwis, 1973), and thus, on the null-plane, the exact form for any transformation is poorly motivated. These authors propose general criteria that lead to a very similar form for V with identical algebraic properties. Instead of the specific function of Melosh

$$\theta(|\mathbf{P}_\perp|) = \arctan\left\{\frac{|\mathbf{P}_\perp|}{m}\right\} \tag{4.79}$$

a more general function of $|\mathbf{P}_\perp|$ is allowed.

In his published paper, Melosh (1974) attempted to motivate the need for a transformation by very different arguments, based on the rotational properties of free-quark wave packets. On the null plane, rotations are complicated dynamical operators, and the problem of solving the constraints of rotational invariance is directly analogous to Gell-Mann and Dashen's angular condition for current algebra. Since rotations mix good and bad field components, the constraint equation for q_- must be used before the effect of rotations on good components q_+ alone may be obtained. The essentials of Melosh's arguments are discussed by Bell (1974), who demonstrates that this second transformation is equivalent in the case of free quarks to the old Foldy–Wouthuysen transformation specialized to good components.

Before we survey the various attempts to improve on these free-quark transformations, it is worthwhile demonstrating how such ideas have been applied to develop phenomenological $SU(6)_W$ models for pion and photon transitions.

4.7. Algebraic Properties of the Melosh Transformation

In discussing the algebraic properties of the Melosh transformation it is sufficient to consider Melosh's first proposal (4.78). Since this does not commute with $\mathbf{E}_\perp$ boosts, the problems of Section 4.5 are avoided, since the predictive power is reduced. What results can one derive? Consider the single-particle matrix element of a current operator O

$$\langle A|O|B\rangle \tag{4.80}$$

The states $|A\rangle$ and $|B\rangle$ are classified into $SU(6)_W$ multiplets by the $\hat{W}$ generators, but the operator, by the $\hat{F}$ charges. The Melosh transformation V may be regarded as describing the change of basis from the constituent representation to the current representation. For $SU(6)_W$ symmetry predictions, both the operator *and* the states must be referred to the *same* basis

$$_{\hat{W}}\langle A|V^{-1}O_{\hat{F}}V|B\rangle_{\hat{W}} \tag{4.81}$$

In general, the states, $V|A\rangle_{\hat{W}} \equiv |A\rangle_{\hat{F}}$, in the current basis will be complicated mixtures of many representations. Melosh observed that it was simpler to rewrite these mixing schemes in terms of the mixed operators:

$$(O)_{\hat{W}} = V^{-1}O_{\hat{F}}V \tag{4.82}$$

Let us demonstrate this approach with two examples of physical interest.

4.7.1. The Axial Vector Charges

The null-plane axial charge $\hat{Q}_\alpha^5$ is

$$\hat{Q}_\alpha^5 \equiv 2\int d^4x\delta(t+z)q^\dagger(1+\alpha_3)\gamma_5 q$$

$$= 2\int d^4x\delta(t+z)q_+^\dagger\sigma_3 q_+ \equiv \hat{F}_\alpha(\sigma_3) \tag{4.83}$$

using the null-plane representation of γ matrices (4.35) and (4.37). Thus, $\hat{Q}_\alpha^5$ is a generator of the $\hat{F}$ algebra and its transformation properties may be specified as

$$(\hat{Q}_\alpha^5)_{\hat{F}} \sim \{\mathbf{35};\ W = 1\ \ W_z = 0;\ L_z = 0\} \tag{4.84}$$

For phenomenological applications we require

$$(\hat{Q}^5)_{\hat{W}} = V^{-1}Q_\alpha^5 V \qquad (4.85)$$

This may be calculated explicitly using Melosh's form for V (4.78), and the result of its action on the good-quark field

$$V^{-1}q_+ V = (\cos\tfrac{1}{2}\theta + \boldsymbol{\gamma}_\perp \cdot \hat{\mathbf{P}}_\perp \sin\tfrac{1}{2}\theta)q_+ \qquad (4.86)$$

where, in Melosh's first example, θ is given by (4.79). In obtaining the action on the conjugate field $q_+{}^\dagger$, care must be taken with derivatives, since $\mathbf{P}_\perp = -i\partial_\perp$ is Hermitian only when acting under a space integral. One arrives at the result

$$(\hat{Q}_\alpha^5)_{\hat{W}} = 2 \int d^4x\delta(t+z)q_+{}^\dagger[(\cos 2\theta)\sigma_3 + (\sin 2\theta)\,\boldsymbol{\gamma}_\perp \cdot \hat{\mathbf{P}}_\perp\sigma_3]q_+ \qquad (4.87)$$

which can be simplified since $\boldsymbol{\gamma}_\perp\sigma_3 = \boldsymbol{\sigma}_\perp$ between good operators.

The $SU(6)_W$ representation of $(\hat{Q}_\alpha^5)_{\hat{W}}$ is now clear: It contains two terms with different W-spin properties and neither term a generator of the algebra. This result may be written

$$(\hat{O}_\alpha^5)_{\hat{W}} = \alpha\{35; W=1 \quad W_z = 0; L_z = 0\}$$
$$+ \beta\{35; W=1 \quad W_z = \pm 1; L_z = \mp 1\} \qquad (4.88)$$

where the L_z properties arise since $J_z = 0$ for this operator. It is important to note that the transformed operator is still a $q\bar{q}$ **35**: This is a result of the very simple form for the transformation in the free-quark model. Nevertheless, since only good operators are involved, it is tempting to abstract this property from the model and apply it to the real world. Hopefully, this will result in an $SU(6)_W$ phenomenology that is a good approximation to nature.

To obtain a model for pion transitions one must relate the matrix elements of $\hat{Q}_\alpha^5$ to physical pion amplitudes. This may be done via PCAC, either by an heuristic dispersion relation argument, or more carefully, as done by Carlitz *et al.* (1975). The result is

$$\langle B\pi_\alpha|A\rangle \sim \frac{M_A^2 - M_B^2}{f_\pi}\langle B|\hat{Q}_\alpha^5|A\rangle \qquad (4.89)$$

In Section 5, we describe the phenomenology resulting from such a model.

4.7.2. The Dipole Operator

For applications to electromagnetic transitions it is useful to consider the dipole operator

$$\mathbf{D}_\perp = \int d^4x\delta(x^+)\mathbf{x}_\perp J^+(x) \qquad (4.90)$$

Under the $\hat{F}$'s this transforms as

$$(\mathbf{D}_\perp)_{\hat{F}} \sim \{\mathbf{35}; \ W = 0 \ \ W_z = 0; L_z = \pm 1\} \tag{4.91}$$

Applying the explicit Melosh transformation, one obtains the result that $(\mathbf{D}_\perp)_{\hat{W}}$ transforms as the sum of four $SU(6)_W$ representations

$$(\mathbf{D})_{\hat{W}} \sim A\{\mathbf{35}; \ W = 0 \ \ W_z = 0; L_z = \pm 1\} + B\{\mathbf{35}; \ W = 1 \ \ W_z = \pm 1; L_z = 0\}$$

$$+ C\{\mathbf{35}; \ W = 1 \ \ W_z = 0; L_z = \pm 1\}$$

$$+ D\{\mathbf{35}; \ W = 1 \ \ W_z = \mp 1; L_z = \pm 2\} \tag{4.92}$$

Again notice that $(\mathbf{D}_\perp)_{\hat{W}}$ contains only $q\bar{q}$ **35** operators. For phenomenology, one must relate matrix elements of $\mathbf{D}_\perp$ to the general photoproduction matrix element

$$F_\perp \equiv \langle B; P', \lambda' | J_\perp | A; P, \lambda \rangle \tag{4.93}$$

This involves some manipulation, starting from the following matrix element taken between unequal mass states:

$$\langle B; P^+, \mathbf{P}_\perp; \lambda' | \left. \left| \int d^4 x \delta(x^+) J^+(x) e^{i\mathbf{P}_\perp \cdot \mathbf{x}_\perp} \right| A; P^+, \mathbf{0}_\perp; \lambda \rangle \right|_{q^+=0} \tag{4.94}$$

where $q^+ = (P' - P)^+ = 0$. From this one may derive the relation (valid for $M_A \neq M_B$)

$$\langle B; P', \lambda' | \mathbf{D}_\perp | A; P, \lambda \rangle \Big|_{\substack{q^+=0 \\ \mathbf{q}_\perp = 0}}$$

$$= \left\{ \frac{\partial}{\partial \mathbf{P}_\perp} \langle B; P', \lambda' | \left| \int d^4 x \delta(x^+) J^+(x) e^{i\mathbf{P}_\perp \cdot \mathbf{x}_\perp} \right| A; P, \lambda \rangle \right\}_{\substack{q^+=0 \\ \mathbf{q}_\perp = 0}}$$

$$= (2\pi)^3 \delta_L^3(0) \frac{1}{q} \langle B; P', \lambda' | \mathbf{J}_\perp(0) | A; P, \lambda \rangle \Big|_{\substack{q^+=0 \\ \mathbf{q}_\perp = 0}} \tag{4.95}$$

where $\delta_L^3(0)$ is a normalization factor and current conservation

$$q^+ J^- + q^- J^+ = \mathbf{q}_\perp \cdot \mathbf{J}_\perp \tag{4.96}$$

has been used. This is the required relation between $F_\perp$ and the matrix elements of the dipole operator.

4.8. Further Theoretical Work

It is impossible here to attempt a detailed review of subsequent theoretical work on the Melosh transformation—even the appellation "Melosh transformation" is somewhat unfair since a similar mixing operator V was considered earlier by Bucella *et al.* (1970). It is clear, however, that there are many problems remaining: For a satisfactory transformation one must go beyond the free-quark model. Perhaps one will be able to make connection with theories of quark confinement and asymptotic freedom, but so far, this remains just a hope. So far too, the connection of spin-$\frac{1}{2}$ quark partons and the Melosh approach remains rather obscure, and one cannot help feeling uneasy about the marriage of chiral symmetry and $SU(6)_W$, in which π's and ρ's play very different roles. To see how far such questions have been answered (if at all) the reader is referred to the Bibliography, which contains (hopefully) a comprehensive guide to recent work on $SU(6)_W$ and current and constituent quarks.

5. Hadronic Decays and SU(6)ᵥᵥ Phenomenology

5.1. Algebraic SU(6)ᵥᵥ Models

5.1.1. Preamble

In Section 2.5 we saw how the first attempt to define an $SU(6)_W$ vertex symmetry, treating $SU(6)_W$ in the same way as $SU(3)$, failed on phenomenological grounds for the decays of L-excited states. Moreover, we have seen that there are likely to be theoretical difficulties in treating $SU(6)_W$ as an "ordinary" symmetry. Clearly, direct application of $SU(6)_W$ symmetry to vertices is at best poorly motivated. How then, can one construct models exhibiting some kind of $SU(6)_W$ structure in a sensible manner? There are two basic approaches.

Firstly, an approximate $SU(6)_W$ structure for vertices could arise from an explicit dynamical model based on spin-$\frac{1}{2}$ quarks. Many people have followed this approach, and, in most cases, the crucial assumption for $SU(6)_W$ structure is that current interactions are represented by one-quark transition operators.

Alternatively, we have seen in Section 4 that, via null-plane current algebra and the existence of a Melosh transformation, the $SU(6)_W$ properties of single-particle matrix elements of current operators may be precisely defined. The use of PCAC or vector meson dominance then leads to an $SU(6)_W$ algebraic structure for decays involving the emission of a pion or rho meson. This is the approach we shall describe in the next sections, although the predictions of some dynamical $SU(6)_W$ models will also be discussed. The Melosh approach is thus an attempt to avoid detailed dynamical assumptions by making instead well-defined, minimal assumptions about the nature of the Melosh transformation.

5.1.2. Pionic Decays

As discussed in Section 4.7, Melosh's transformation suggested that the pionic axial charge, $\hat{Q}_\pi^5$, transforms as the sum of two **35** representations under $SU(6)_{W,\text{constituents}}$:

$$(\hat{Q}_\pi)_{\hat{W}} \sim \alpha\{35;\ W = 1\quad W_z = 0;\ L_z = 0\}$$
$$+ \beta\{35;\ W = 1\quad W_z + \pm 1;\ L_z = \mp 1\} \qquad (5.1)$$

or in an obvious shorthand

$$(\hat{Q}\pi)_{\hat{W}} \sim \alpha\{35;\ `\pi'\} + \beta\{35;\ `A_1'\} \qquad (5.2)$$

In principle there could be many other terms present—exotic **405** representations for example—but this is the most general form within **35** representations. This assumption reflects our hope that exotic pieces are relatively unimportant—but this may not be exact (see the papers of Osborn, 1974, in this connection). However, it is ultimately up to experiment to tell us whether such an assumption is valid: If it is not, then $SU(6)_W$ models will not be useful to classify transitions. To apply these ideas about $SU(6)_W$ symmetry to pionic decays, $A \rightarrow B + \pi$, requires the use of the PCAC approximation in the form (4.89)

$$\langle B\pi | A \rangle \underset{\text{PCAC}}{\sim} \langle B | \hat{Q}_\pi | A \rangle \qquad (5.3)$$

The $SU(6)_W$ representations of resonances A and B are presumed known from the spectrum, and the resulting $SU(6)_W$ algebraic structure may be calculated using standard Clebsch–Gordan techniques. [The $SU(6)$ Clebsch–Gordan factors are tabulated by Cook and Murtaza, 1965 and Carter *et al.*, 1965.] The predictions are phrased in terms of a small number of unknown reduced matrix elements, which are the parameters of the model.

5.1.3. Photon Transitions

Photon transition matrix elements may be subjected to a similar algebraic $SU(6)_W$ analysis, via the dipole operator as in Section 4.7. The photoproduction amplitudes are described by helicity amplitudes $F_\pm$:

$$F_\pm \sim \langle B | J_\pm | A \rangle \qquad (5.4)$$

where $J_\pm$ represents the electromagnetic current operator corresponding to $\lambda = \pm 1$ photons. The Melosh transformation suggests the following

$SU(6)_W$ structure for $(J_{\pm})_{\hat{w}}$ (4.92):

$$(J_{\pm})_{\hat{w}} \sim A\left\{35; \begin{array}{l} W=0 \\ W_z=0 \end{array}; L_z=\pm 1\right\} + B\left\{35; \begin{array}{l} W=1 \\ W_z=\pm 1 \end{array}; L_z=0\right\}$$

$$+ C\left\{35; \begin{array}{l} W=1 \\ W_z=0 \end{array}; L_z=\pm 1\right\} + D\left\{35; \begin{array}{l} W=1 \\ W_z=\mp 1 \end{array}; L_z=\pm 2\right\} \quad (5.5)$$

The resulting $SU(6)_W$ predictions are just straightforward Clebsch–Gordan algebra as before.

5.1.4. Rho Decays

Amplitudes for rho meson emission may be treated within this $SU(6)_W$ framework by relating them, via vector meson dominance, to matrix elements of the electromagnetic current. The algebraic structure of $\lambda = \pm 1$ isovector photon transitions must therefore be extrapolated from $q^2 = 0$ to $q^2 = M_\rho^2$. In this approach to $SU(6)$ the π and ρ decays of baryons involve assumptions about different current operators, and their decay parameters are therefore unrelated, even though, in the spectrum, both π and ρ are in the same $SU(6)_W$ multiplet. For meson decays into a $\pi\rho$ final state, this asymmetry is highlighted: The parameters of the two approaches must be related in an ad hoc fashion to avoid inconsistencies.

Longitudinal rho amplitudes F_0, may be related to matrix elements of the good component of the electromagnetic current $J^+ \sim J^0 + J^3$:

$$F_0 \sim \langle B|J^+|A\rangle \quad (5.6)$$

Taking account of the parity constraint, $(J^+)_{\hat{w}}$ is assumed to transform as (Close *et al.*, 1974; Cashmore *et al.*, 1975)

$$(J^+)_{\hat{w}} \sim a^0\left\{35; \begin{array}{l} W=0 \\ W_z=0 \end{array}; L_z=0\right\} + a^1\left\{35 \begin{array}{l} W=1 \\ W_z=\pm 1 \end{array}; L_z=\mp 1\right\} \quad (5.7)$$

5.1.5. Explicit Quark Models

It is interesting to contrast this algebraic approach with the example of an explicit quark model. In many such models the quarks are considered to interact via an harmonic oscillator potential. Decay transitions are then characterized by one-quark operators—which automatically restricts the operators to **35** representations. For example, for pion decays the pion-emission operator H_π for $L = 1 \rightarrow L = 0$ transitions has the simple form

$$H_\pi = A\sigma_z L_z + B\sigma_{\pm} L_{\mp}$$

where the $L_{z,\pm}$ operators are orbital excitation operators and $\sigma_{z,\pm}$, quark spin operators. The similarity with $(\hat{Q}_\pi)_{\hat{W}}$ (5.1) is obvious, and both models in fact yield the same algebraic structure (Lipkin, 1974). Explicit quark models, however, make further, more specific predictions. Firstly, A and B are known functions of the resonance masses involved in the decay, and consequently incorporate $SU(6)$ [and $SU(3)$] mass-splitting effects in a much more detailed way than the Melosh approach. Secondly, harmonic oscillator models also predict the relative magnitudes of the reduced matrix elements A and B (corresponding to α and β of the Melosh model), for each $SU(6)$ multiplet transition. Two levels of predictions may therefore be distinguished; first the $SU(6)_W$ algebraic structure and then, if this is successful, the more detailed intermultiplet relations of more explicit models. In these models similar single-quark transition operators are constructed for real and virtual photon transitions. However, for such transitions most of these explicit models do not assume the most general **35** structure. In what follows we shall confront the more general algebraic $SU(6)_W$ structures with the data before restricting to the more specific parameterizations of various models.

5.2. Baryon Decays

5.2.1. Pionic Decays

The $SU(6)_W$ model based on the Melosh transformation and PCAC is still a very predictive symmetry. For decays of the {**70**, 1^-} states to the {**56**, 0^+} ground state via pseudoscalar meson emission, the 21 independent $SU(3)$ couplings are related to only two unknown parameters via calculable $SU(6)$ coupling coefficients. These two parameters may be chosen as an S wave and a D wave amplitude. Similarly, the {**56**, 2^+} and {**56**, 0^+}$_R$ decays to the ground state are described in terms of P and F amplitudes, and a P' amplitude, respectively.

Unfortunately, this simple picture is somewhat marred by symmetry-breaking effects. Before the model may be extended to include Y^* decays, some decision must be made as to how obvious $SU(3)$ breaking effects—such as the unequal masses of the π, K, and η mesons—are to be incorporated. Furthermore, since the $SU(6)$ multiplets are by no means mass degenerate, the question of barrier factors is important for quantitative numerical agreement. For example, Gilman *et al.* (1974) retain in the amplitude a factor $(M_A^2 - M_B^2)$ arising from PCAC (4.89), and use the same barrier factor for decays with different partial waves. A recent $SU(6)_W$ analysis (Hey *et al.*, 1975) found that best agreement is obtained with a barrier factor that varies with the partial wave angular momentum l. A form such as p^l, where p is the three-momentum of the decay, which is favored in $SU(3)$ analyses seems to be preferred. This was first noticed by Faiman and Plane (1972) in their earlier $SU(6)_W$ fit.

A final complication is the possibility of mixing between the pure $SU(6)$ states—since there are N^* or Y^* states with the same spin and parity in both the $\{70, 1^-\}$ and $\{56, 2^+\}$ multiplets. In the $SU(6)_W$ decay analysis it is found that a good fit to the $\{56, 2^+\}$ could be obtained without mixing but that the fit for the $\{70, 1^-\}$ was improved when mixing was allowed. One must be careful in interpreting these results, since the $SU(6)_W$ model with PCAC is not expected to be exact. A good measure of its validity may be gauged from the $\frac{5}{2}^- \Lambda^* d_{05}(1830)$ predictions. This state cannot mix within the $\{70, 1^-\}$ and is predicted to decouple from the $\bar{K}N$ channel. However, this resonance is clearly seen in $\bar{K}N \to \Sigma \pi$ and accounts for a χ^2 contribution of 36 to the fit! Nevertheless, this large χ^2 contribution is somewhat misleading since the elasticity of this resonance is actually small and less than 10%—to be compared with the zero prediction of $SU(6)_W$. It seems most reasonable to view this as the level of accuracy of the model—until there is a viable theory of $SU(3)$ symmetry-breaking effects. Alternatively, one may hope that the $SU(6)_W$ predictions should be better than 10% and make an attempt to "explain" the discrepancy by intermultiplet mixing with, say, a Λ^* from a possible $\{70, 3^-\}$ multiplet. However, since $SU(6)$ representation mixing is also a symmetry-breaking effect this seems a somewhat circular procedure. In the decay fit of Hey *et al.* (1975), it was found that, apart from $SU(3)$ singlet–octet mixing among the Λ^*'s, to a reasonable approximation the states could be regarded as unmixed—corresponding to one dominant $SU(6)$ quark spin assignment. Even in this approximation of zero mixing, the $SU(6)_W$ fits were not much worse than $SU(3)$ fits (see below). However, other workers (Horgan, 1974; Jones *et al.*, 1974) have determined mixing matrices from an $SU(6)$ fit to resonance masses, and the mixing matrices, for the Y^*'s in particular, from such fits do not show very good agreement with those from the decay analysis. Moreover, Jones *et al.* (1974) have attempted to fit the masses and decays simultaneously and were unable to find any consistent set of mixing matrices! In terms of a Melosh approach, the $SU(6)_W$ properties of the mass operator are highly dependent on dynamics, and, furthermore, it is unclear whether an $SU(6)_W$ symmetry limit of the mass operator can exist. Thus in the algebraic approach the exact status of such mass fits is not clear—it is an interesting problem!

After all these caveats, one is nevertheless impressed by the quality of the $SU(6)_W$ fits. For full details the reader is referred to the original papers. Here we shall consider some qualitative indications of the success of the model (taken from Hey *et al.*, 1975, and Cashmore *et al.*, 1975).

(i) *Amplitude Signs.* For two resonances, N_1^* and N_2^*, in the same

$SU(6)$ multiplet, the model predicts the relative signs of inelastic amplitudes, i.e., A_1/A_2 where, e.g.,

$$A_1 = \text{amplitude for } \pi N \to N_1^* \to \pi\Delta$$

$$A_2 = \text{amplitude for } \pi N \to N_2^* \to \pi\Delta$$

A check of the number of nontrivial sign predictions for the $\{70, 1^-\}$ and $\{56, 2^+\}$ decays yields a total of 16 in agreement with the $SU(6)_W$ model and at present no obvious discrepancies.

(ii) *Magnitudes.* The detailed fits to experimental amplitudes $(xx')^{1/2}$ are generally quite good, and this can be demonstrated in two ways: (a) In order to assess the significance of the $SU(6)_W$ Clebsch–Gordan coefficients, one can attempt fits with random numbers (normalized between ± 1) in their place. The $SU(3)$ relations are of course retained. The results of a number of such fits for the $\{56, 2^+\}$ are shown in Fig. 16: The $SU(6)$ coefficients gave a χ^2 of 34 to be compared with the best "random" fit with a χ^2 of 94. (b) Consider the relative success of the $SU(6)_W$ analysis compared to standard $SU(3)$ analyses. The ratio

$$R = g_{\text{exp}}^{SU(3)} / g_{\text{model}}^{SU(3)}$$

gives some idea of this, and this is plotted in Fig. 17a for members of both the $\{70, 1^-\}$ and $\{56, 2^+\}$ multiplets. The coupling constants $g_{\text{expt}}^{SU(3)}$ are taken from independent $SU(3)$ analyses (Plane *et al.*, 1970; Litchfield *et al.*, 1971; and Barbaro–Galtieri, 1972), of N^*, Σ^*, and Λ^* resonances in various J^P multiplets. The $g_{\text{model}}^{SU(3)}$ are calculated from the $SU(6)_W$ model using the fitted best values of the parameters S, D, P, and F together with the predicted $SU(6)$ f/d ratios (ignoring mixing). From the agreement in Fig. 17a, where the results for the various ratios cluster around unity, it is clear that the $SU(6)$ fit (of two parameters to 55 data points for the 70 and

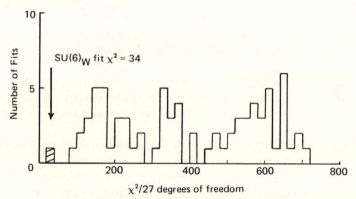

Fig. 16. Random number fits to $\{56, 2^+\}$ π decays.

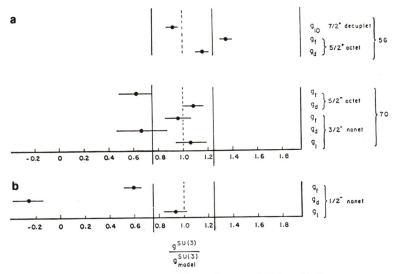

Fig. 17. Comparison of $SU(3)$ couplings vs. $SU(6)$ predictions.

two parameters to 27 data points for the **56**) fares quite well compared with the much less ambitious $SU(3)$ fits. The errors shown are those quoted by the $SU(3)$ analysis and are possibly a little optimistic. However, in Fig. 17b, the $\frac{1}{2}^-$ states do not show good agreement. The $SU(3)$ predictions here were calculated as if the three $\frac{1}{2}^-$ resonances used in the $SU(3)$ analyses were actually the $^2\mathbf{8}$ and $^2\mathbf{1}$ $SU(6)$ states. In fact the full $SU(6)_w$ fit classifies these states, the $N^*(1520)$, $\Lambda^*(1670)$, and $\Sigma^*(1740)$, as predominantly $^2\mathbf{8}$, $^4\mathbf{8}$, and $^2\mathbf{10}$ respectively! Thus the $SU(3)$ analysis of these states is inappropriate.

In summary we must conclude that there is strong evidence for $SU(6)_w$ structure in these pion decays. The amplitude signs may be categorized as follows:

For the {**70**, 1^-}: "Anti-$SU(6)_w$"
(signs as when $\alpha_{70}(L_z = 0)$ term is absent)

For the {**56**, 2^+}: "$SU(6)_w$-like"
(signs as when $\beta_{56}(L_z = \pm 1)$ term is absent)

The result for the {**56**, 2^+} is important in the context of various explicit models, so it is worth commenting on here. The "$SU(6)_w$-like" sign is determined only by the $N^*f_{15}(1680)$ sign in the Berkeley-SLAC $\Delta\pi$ isobar analysis, where, however, it is considered a large and well-determined amplitude. Furthermore, it is supported by the independent CHS $\Sigma^*\pi$ isobar analysis, where the sign of $\Lambda f_{05}(1820)$ is also "$SU(6)_w$-like". In the framework of an isobar analysis, therefore, the d_{15} sign is well-determined and not easily dismissed. Anyone wishing to cast doubt on this sign, because

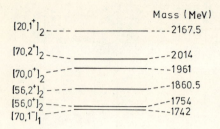

Fig. 18. Mass spectrum of harmonic oscillator quark model.

of unitarity corrections, limitations of the isobar model, and so on, must in all conscience disregard the successful agreement of the other $\pi\Delta$ signs—at least until these corrections have been shown to affect only the f_{15} amplitude significantly! This is unfortunate for some of the more explicit quark models, which predict that both multiplets should have the "anti-$SU(6)_W$" sign. Two examples are the $^3P_0$ quark-pair creation model of Le Yaouanc *et al.* (1974, 1975), which incorporates explicit quark orbital wave functions for the states, and also the Feynman *et al.* (1971) version of the harmonic oscillator quark model.

Finally, some comments on possible radial excitations—$\{56, 0^+\}_R$ multiplets. These multiplets contain $^2\mathbf{8}$ and $^4\mathbf{10}$ states and there are two candidates for $^2\mathbf{8}$ p_{11} nucleon states—the $Np_{11}(1430)$ "Roper" resonance and an $Np_{11}(1750)$. It is interesting that $SU(6)_W$ fits to other multiplets can shed some light on their assignment. In the $SU(6)_W$ analysis of the $\{70, 1^-\}$ multiplet, after allowing for the freedom of an overall sign in inelastic amplitudes—one for $\pi\Delta$ amplitudes from Berkeley–SLAC, one for $\pi\Sigma^*$ amplitudes, etc.—and for the relative signs of S and D, ten well-determined correct sign predictions were obtained. However, for the $\{56, 2^+\}$, $\{56, 4^+\}$, and $\{56, 0^+\}_R$ fits one is more tightly constrained since there is no longer a sign ambiguity remaining in the inelastic amplitudes. Thus one is able from the observed $\pi\Delta$ signs to determine that neither the $p_{11}(1430)$ nor the $p_{11}(1750)$ may be classified in possible $\mathbf{70}$ multiplets: Assignments in a $\{70, 0^+\}$ or a $\{70, 2^+\}$ both predict the opposite $\pi\Delta$ sign to experiment. Thus if these p_{11} resonances are to be classified as bona-fide quark resonances in $SU(6)$ multiplets, then they must both belong to $\mathbf{56}$'s. In the absence of positive-parity $L = 1$ multiplets, the only possible assignments are in $\{56, 0^+\}$ multiplets. In fact such assignments for the $p_{11}(1430)$ and $p_{11}(1750)$ are weakly supported by an independent $SU(6)_W$ analysis of photoproduction data (see below). Nevertheless, a quantitative fit to the Roper multiplet including the $p_{33}(1690)$ is not very satisfactory. The **56** assignment for the $p_{11}(1750)$, which in the past has been assigned (Heusch and Ravndal, 1970) to a harmonic oscillator $\{70, 0^+\}$ multiplet, must, if confirmed, cause problems for explicit energy level calculations of harmonic oscillator quark models (see Fig. 18, adapted from the fit of Horgan, 1974).

5.2.2. Photon Transitions

Resonance photoproduction may be subjected to a similar algebraic $SU(6)_W$ analysis. The $SU(6)_W$ model for $\gamma N \rightarrow N^*$ is described in Section 5.1. Combining these results with the predictions for $N^* \rightarrow N\pi$ decays (Section 5.2.1) leads to amplitudes for $\gamma N \rightarrow N^* \rightarrow \pi N$. The results of such fits to the photoproduction analysis of Metcalf and Walker (1974) are shown in Table 3. The conclusions may be summarized as follows:

For the $\{70, 1^-\}$: $A_{70} \neq C_{70}; C_{70} \neq 0$
for the $\{56, 2^+\}$: $A_{56} \sim C_{56}; C_{56} \neq 0; D_{56} \neq 0$

These conclusions agree well with those of Babcock and Rosner (1975), who have recently performed a critical appraisal of three recent photoproduction analyses. All lead to the conclusion that for reasonable numerical agreement the "C terms" must be present and important. This result is sufficient to show that the simplest versions of harmonic oscillator quark models, which have $C = D = 0$ for all multiplets, cannot give good numerical agreement. The $^3P_0$ model (Rosner and Petersen, 1973), which predicts $A = C$ and $D = 0$, fares somewhat better, but in the case of the

Table 3. Photoproduction Fits to $\{70, 1^-\}$ and $\{56, 2^+\}$

(a) $\{70, 1^-\}$	χ^2	A_{70}	B_{70}	C_{70}	
1. Melosh $SU(6)_W$ fit					
A_{70}, B_{70}, C_{70} free	25.4	8.3	2.2	4.2	
2. Quark model fit					
$C_{70}=0$	73.4	8.9	3.9	0	
3. $^3P_0$ model fit					
$A_{70} = C_{70}$	47.3	6.6	0.6	6.6	
(b) $\{56, 2^+\}$	χ^2	A_{56}	B_{56}	C_{56}	D_{56}
1. Melosh $SU(6)_W$ fit					
$A_{56}, B_{56}, C_{56}, D_{56}$ free	10.7	−6.7	−1.1	−6.8	4.6
2. Quark model fit					
$C_{56} = D_{56} = 0$	65.2	−4.8	−1.8	0	0
3. $^3P_0$ model fit					
$A_{56} = C_{56}; D_{56} = 0$	32.8	−5.9	−1.1	−5.9	0

$\{56, 2^+\}$ it seems that D_{56} is probably nonzero. The parameter B_{56} is rather ill-determined at present, and the exact status of these results may need some reappraisal in the light of new data.

Carlitz and Weyers (1975) exploit the idea of an expansion of a nonlocal Melosh-type transformation in powers of a fundamental length $a = 1/M$. This leads to predictions for the relative importance of the various $SU(6)_{W,\text{constituent}}$ representations in the transformed current operators. In particular, it predicts that the pionic transitions of the $\{70, 1^-\}$ should be "anti-$SU(6)_W$", whereas those of the $\{56, 2^+\}$ should be "$SU(6)_W$-like"—in agreement with experiment. In photon transitions, while this approach predicts that A_{70} and C_{70} should dominate in the $\{70, 1^-\}$ transitions, it also predicts that B_{56} and D_{56} should dominate for the $\{56, 2^+\}$. At present, therefore, the status of this proposal is unclear.

Photoproduction can also give information on possible $\{56, 0^+\}_R$ multiplets. The $SU(6)_W$ structure implies the following predictions for p_{33} and p_{11} amplitudes (in the notation of Metcalf and Walker, 1974):

$$p_{33}: \qquad A_{1+}/B_{1+} = -\tfrac{1}{2}$$

$$p_{11}: \qquad A_{1-}^p/A_{1-}^n = +\tfrac{3}{2}$$

Metcalf and Walker quote the value

$$p_{33}(1238): \qquad A_{1+}/B_{1+} = -0.48 \pm 0.02$$

confirming its well-known assignment in the ground state $\{56, 0^+\}$. Likewise, for the two p_{11} resonances, they give the values

$$p_{11}(1430): \qquad A_{1-}^p/A_{1-}^n = +1.62 \pm 1.4$$

and

$$p_{11}(1650): \qquad A_{1-}^p/A_{1-}^n = +1.42 \pm 1.3$$

Within the very large errors, this seems to give weak support to their assignment in **56**'s that was independently deduced from their pion decays: The ratio for $\{70, 0^+\}$ or $\{70, 2^+\}$ assignments would be 3 or zero, respectively. Some explicit quark models apparently have a problem with the sign of the $p_{11}(1430)$ amplitude.

The algebraic approach can also be extended to spacelike virtual photon transitions with the addition of a corresponding smoothness assumption for the reduced matrix elements.

A longitudinal amplitude would also be allowed. Such algebraic models for electroproduction while they could accommodate some rapid q^2 variation in the helicity structure, are not able to predict such behavior— unlike the harmonic oscillator quark model (Close and Gilman, 1972). Up to now the data have not allowed a very meaningful test of the $SU(6)_W$ structure (see the article by Donnachie, Lyth, and Shaw).

5.2.3. Rho Decays

In the Melosh algebraic approach there are many free parameters to fit the very modest number of $N\rho$ amplitudes determined by the Berkeley–SLAC isobar analysis (Longacre *et al.*, 1975). A direct vector dominance extrapolation from photoproduction values of the reduced matrix elements leads to an acceptable fit, although the inclusion of longitudinal rho amplitudes improves the fit. At present, numerical values are not of much significance since there are not enough well-determined amplitudes, but there appear to be no obvious inconsistencies. This is in contrast to some explicit quark models, which predict some wrong signs. The so-called l-broken $SU(6)_W$ approach to π and ρ vertex symmetries yields an identical algebraic structure for π decays (up to barrier factor ambiguities and so on), as the Melosh model. However, for ρ decays this model has only two independent parameters (since π and ρ are treated on the same footing) and the model is correspondingly much more predictive than Melosh treatments. There are indications of possible problems for this model for consistency of the π and ρ decays of the $\{\mathbf{56}, 2^+\}$ (Faiman, 1975).

5.3. Meson Decays

5.3.1. Preamble

For the baryons the richness of data on $N\pi$, $\Delta\pi$, $N\gamma$, and $N\rho$ decays meant that $SU(6)_W$ models are quite well constrained. For the mesons, the resonances themselves are not yet known, let alone their decay systematics! (Tables 2 and 4 give an idea of the parlous state of meson spectroscopy.) In this situation, the algebraic approach via the Melosh transformation is not very predictive since it requires a modicum of reasonable data to determine the reduced matrix elements of the model.

Table 4. Present Status of N = 2 Multiplets[a]

| | | | | $I=1$ | $I=1/2$ | $I=0$ | | |
| | | | | 8 | 8 | 8 | 1 | |
N	L	S	J^{PC}	8	8	8	1	Mixing
2	0	0	0^{-+}	$[\pi^*]$	$K_{\pi^*}]$	$[\eta^*]$	$[\eta^{*\prime}]$	$[U]$
		1	1^{--}	ρ^*?	$[K_{\rho^*}]$	$[\omega^*]$	$[\phi^*]$	$[M]$
2	2	0	2^{-+}	A_3?	L?	$[\eta_{A_3}]$	$[\eta_{A_3}]$	$[U]$
		1	3^{--}	g	K_g	ω_g	$[\phi_g]$	M?
			2^{--}	$[X]$	$[K_x]$	$[\omega_x]$	$[\phi_x]$	$[M]$
			1^{--}	$[\rho^{**}]$	$[K_{\rho^{**}}]$	$[\omega^{**}]$	$[\phi^{**}]$	$[M]$

[a] Notation as in Table 2.

Even for the $N = 1$ excited states there are barely enough data to do this, and even then numerical predictions for missing resonances will depend on unknown masses and mixing. Explicit quark models are more predictive, but again one has no reliable values for missing resonance masses. In the following sections we briefly compare and contrast the predictions of these two approaches for the "well-known" $N = 0$ and $N = 1$ multiplets before concluding with a few remarks about "unknown" $N = 2$ resonances. The radiative transitions of the ground-state $N = 0$ multiplet are discussed by Moorhouse in this volume: Here the discussion is restricted to the hadronic decays.

5.3.2. $N = 0$ and $N = 1$ Meson Multiplets

The $SU(6)_W$ model for pseudoscalar emission was outlined in Section 5.1.2. As usual, the predictions are subject to barrier factor uncertainties and singlet–octet mixing. In explicit quark models such as that of Feynman *et al.* (FKR), the decay $\phi \to \rho\pi$ is forbidden by Zweig's rule, in the approximation that ϕ is a pure $s\bar{s}$ quark–antiquark state. This is clear in the model since H_π as a single-quark operator cannot connect ϕ to ρ. In the algebraic Melosh approach, this rule—no disconnected quark diagrams—

Table 5a. *Decay Rates for Pseudoscalar Emission of Mesons in the $N = 0$ and $N = 1$ Supermultiplet Transitions*[a]

	Particle decay			Partial width in MeV		Helicity ratios			Total width	
J^{PC}	Mass	Mode	Exp.	Quark model FKR	Melosh fit to data	0	1	2	Exp.	FKR
1^{--}	ρ 770	$\to \pi\pi$	*150 ± 10	150	149				150 ± 10	158
	K^* 892	$\to K\pi$	*50 ± 1	65	48				50 ± 1	65
	ϕ 1019	$\to K\bar{K}$	*3 ± 0	11	5				4 ± 0	11
1^{+-}	B 1235	$\to \omega\pi$	*120 ± 20	84	117	0.02	0.98		*120 ± 20	100
						(*mainly $\lambda = 1$)				
		$\delta\pi$		12	16					
		also $A_1\pi$								
	Q_B 1350	$\to \omega\pi$		76		0.19	0.81			157
		$K^*\pi$		55		0.01	0.99			
		ωK		23		0.21	0.79			
		also $\kappa\pi$								
	η_B 1250	$\to \rho\pi$		94		0.01	0.99			94
	η'_B 1350	$\to \rho\pi$		256		0.00	1.0			268
		$\omega\eta$		12		0.30	0.70			

Table 5a (continued)

J^{PC}	Particle decay Mass	Mode	Exp.	Quark model FKR	Melosh fit to data	0	1	2	Total width Exp.	FKR
2^{++}	A_2 1310	$\to KK$	*5 ± 1	19	11				*100 ± 10	111
		$\to \eta\pi$	*15 ± 1	23	17					
		$\to \rho\pi$	*72 ± 2	67	53					
		also $\eta'\pi$								
	K^{**} 1420	$\to K\pi$	*55 ± 3	88	57				*100 ± 10	128
		$\to \eta K$	*2 ± 2	4	2					
		$\to \rho K$	*9 ± 2	8	5					
		$\to K^*\pi$	*30 ± 3	24	17					
		$\to \omega K$	*4 ± 2	2	1					
		also $Q\pi$								
	f 1270	$\to \pi\pi$	*150 ± 20	241	146				*170 ± 20	264
		$\to K\bar{K}$	*8 ± 6	14	8					
		also $A_1\pi, \eta\eta$								
	f' 1514	$\to K\bar{K}$	40 ± 10	105	52				40 ± 10	152
		$\to \eta\eta$		33						
		$\to K^*\bar{K}$		14						
		also $\eta'\eta$								
1^{++}	A_1 1070	$\to \rho\pi$		151		0.46	0.54			151
	Q_A 1240	$\to K^*\pi$		59		0.52	0.48			59
	D 1285	$\to A_1\pi$		20						
		$\to \delta\pi$		79					30 ± 20	99
	D' 1416	$\to K^*\bar{K}$		211	147	0.36	0.64			211
0^{++}	δ 970	$\to \eta\pi$		99	174				50 ± 2	99
	κ 1100	$\to K\pi$		333	419					362
		$\to \eta K$		29	19					
	ε 1200	$\to \pi\pi$		1230					large	1549
		$\to K\bar{K}$		292						
		$\to \eta\eta$		27						
	S^* 1000	$\to K\bar{K}$		178					40 ± 8	178

[a] The data preceded by * are the reliable data used in the plot of Fig. 19.

*Table 5b. Standard Resonance Mass and Mixing Assumptions Used in the
Calculations*

J^{PC}	$I=1$		$I=\frac{1}{2}$		$I=0$		$I=0$		Mixing
0^{-+}	π	140	K	494	η	549	η'	958	U
1^{--}	ρ	770	K^*	892	ω	783	ϕ	1019	M
1^{+-}	B	1235	Q_B	1350	η_B	1250	η'_B	1350	U
2^{++}	A_2	1310	K^{**}	1420	f	1270	f'	1514	M
1^{++}	A_1	1070	Q_A	1240	D	1285	D'	1416	M
0^{++}	δ	970	κ	1100	ε	1200	S^*	1000	M

must be inserted as a requirement on the ratio of singlet to octet decay amplitudes.

The predictions of a Melosh fit and a quark model (FKR) calculation for the $N = 1$ to $N = 0$ transitions are given in Table 5a. For ill-established states these predictions depend on the masses and mixing assumed. The "conventional" choice used in Table 5a is listed above in Table 5b. A fuller discussion, including the effect of other choices, may be found in the paper of Burkhardt and Hey (1975). It should be noted that the $N = 0$ decays do not test the $SU(6)$ nature of the model. For $N = 1$ to $N = 0$ transitions all decays of the four J^{PC} multiplets are related to two Melosh parameters—and the helicity structure of the decay $B \to \omega\pi$ has shown that the presence of the "anti-$SU(6)_W$" term is important. The FKR model, however, makes absolute predictions for all these decays. Note the following points:

(i) For the baryon decays, the FKR model was seen to fail in some details and thus its predictions should be regarded as at best a *qualitative* guide to the decay systematics of the mesons.

(ii) In this qualitative spirit, the enormous width predicted for the $0^{++}\varepsilon$ resonance by the FKR model ($\Gamma \sim 1500\,\text{MeV}$) should be regarded as a success: The ε is clearly *not* a narrow resonance—as assumed in the calculations. Moreover, the model also predicts "normal" widths ($\Gamma \sim 100\,\text{MeV}$) for other resonances in the same multiplet—the difference seems to be more than just a barrier factor effect.

(iii) Figures 19 and 20 show a "Feynman" plot of Melosh and FKR predictions compared with the available reliable data. As remarked by Feynman *et al.* (1971), their quark model predicts widths a factor ~ 2 too large whenever kaon PCAC has to be used. In contrast, the simple kinematic $SU(3)$ barrier factor of the Melosh approach seems to cope better with these $SU(3)$-breaking effects than does the explicit mass-splitting factors in the FKR matrix elements.

In conclusion, we should remark that the $SU(6)$ character of these models has not been very stringently tested: Removing $SU(3)$-related results leaves only three or four $SU(6)$ agreements. It is nevertheless remarkable that such a simple explicit quark model can do so well, since it

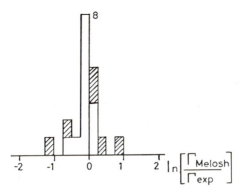

Fig. 19. "Feynman" plot of Melosh model structure against reliable data (marked with * in Table 5). The shaded regions correspond to use of K-PCAC; the g multiplet is excluded.

makes absolute predictions for both $N = 1$ *and* $N = 2$ transitions (the *g* multiplet), which fare really very well. There are some hopeful signs that there is some experimental improvement for the situation of the strange axial-vector mesons. The recent analysis of Brandenburg *et al.* (1975) confirms the existence of *two* resonances, but unfortunately the extraction of reliable branching ratios is still obscured by Deck diagram effects and possible mixing of the two Q's. Any information on the $I = 0 \, 1^+$ resonances will be very useful both for $SU(6)$ decay systematics and singlet–octet mixing systematics. For the 0^{++} resonances, and in particular the $\pi\pi$ channel, little is clear except for the fact that the situation is obviously much more complicated than assumed in these simple narrow-resonance, magic mixing quark models, e.g., how do $SU(3)$ breaking and the existence of the $K\bar{K}$ threshold affect these naive predictions?

5.3.3. $N = 2$ Meson Multiplets

A glance at Table 4 shows that, with the exception of the $g \, 3^{--}$ nonet, very little is known about the $N = 2$ $SU(6)$ supermultiplet. For example,

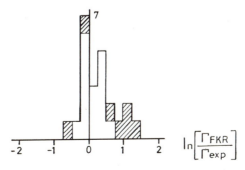

Fig. 20. "Feynman" plot of FKR model predictions against reliable data (including *g* multiplet). The shaded regions correspond to use of K-PCAC.

the existence of meson radial excitations is often taken for granted by theorists, yet an objective look at the data shows little reason for such faith. The $\rho'(1600)$ is certainly not the most reliable resonance on which to base one's theory, and there have been for several years persistent rumors of a possible ρ' resonance at 1250 MeV. Where also is its partner the π'—the radial excitation of the pion? Why has this not been seen? In fact there is a recent report of a possible 0^- $K\pi\pi$ resonance around 1450 MeV, decaying primarily via the $K\varepsilon$ channel (Brandenburg *et al.*, 1975).

The existence of all these J^{PC} multiplets is almost mandatory for a quark theory of mesons, and their continued nonobservation would be seriously disturbing. Recently, the character of these missing multiplets has been explored within the context of the FKR quark model (Burkhardt and Hey, 1975). For theorists at least there appears to be some grounds for cautious optimism. The quark model presents very plausible explanations for the nonobservance of the missing multiplets: Most such resonances are predicted to be very broad and they decay only into multibody final states. Such states, e.g., the $I = 1$ member of the 2^{--} with a predicted width ~ 400 MeV, decaying primarily into 4π's, would not have been found by present analyses. However, the situation seems not without hope experimentally. For example, the calculations suggest that "cascade" decays of well-known $N = 2$ states may be useful in finding elusive $N = 1$ mesons such as the A_1. For theorists, however, it is clear that for the present an algebraic $SU(6)_W$ approach to the $N = 2$ decays is of little use.

6. Conclusions

For the baryons, $SU(6)_W$ symmetry has proved very successful both in the spectrum classification and for analyzing π, γ, and ρ transitions. Nevertheless, there are many problems still to be resolved, in particular the radial excitation pattern and the existence or nonexistence of even parity **70** multiplets at relatively low masses. The answers to these experimental questions will impose severe constraints on any theoretical model. However, in "conventional" $SU(6)$ spectroscopy it is probably fair to say that the most pressing experimental questions concern the mesons. Before the resonance spectrum can be disentangled, however, it will be necessary to resolve the "A_1 crisis" (and related Q, A_3, and L crises)—both theoretically and experimentally. One hopeful avenue is that of A_1 production in nondiffractive reactions, and in particular A_1 photoproduction. Such experiments are under way and hopefully we shall soon have some answers. The systematics of meson radial excitations are also of great interest and will be an important hurdle for theories of the new particles. Only when the spectrum is established will it be possible to examine the systematics of the various decay modes for information on $SU(6)_W$ vertex structure. Similarly, the mixing patterns of the $I = 0$ mesons for the different J^P multiplets will provide clues to the presumed underlying quark dynamics. Decays of an excited ϕ meson, such as $\phi^* \to \phi\pi\pi$, are relevant to understanding Zweig's rule and especially its variation with the mass of the initial state. It is clear that more knowledge of these "conventional" mesons must provide important information on quark dynamics—be they Han–Nambu or Gell-Mann–Zweig.

All of these above problems have been around a long time but they are nonetheless important. Most of the excitement, it must be admitted, rests with the symmetries and spectroscopy of the new particles—the J/ψ, ψ', $\psi''\cdots$ sequence found in e^+e^- annihilation. Some new degree of freedom is being excited, but its precise character is still debated. In almost any theoretical scheme, new, weakly decaying baryons and mesons appear, and their spectroscopy and decay systematics are of intense interest. Many symmetries are now under serious discussion, ranging from four-quark models based on $SU(4)$—with its two uneasy bed-fellows $SU(8)$, a spin symmetry, and $SU(4) \otimes SU(4)$, a chiral symmetry—to less familiar groups such as $O(9)$, $E(6)$, $SU(3, 1)$, and so on. These symmetries, however, are badly broken: A realistic theory of symmetry breaking is needed for them to be really useful. The theoretical challenge is to incorporate the established $SU(6)_W$ successes together with a charmlike (or colorlike) degree of freedom in a realistic dynamical model, presumably based on some sort of quarks. This is the same problem as that of constructing a realistic Melosh transformation—nontrivial dynamics must be incorporated into theories exhibiting current algebras. Two popular constraints on the dynamics are that the theory should be asymptotically free at short distances, yet the quarks be confined at large distances. The stage seems set for significant theoretical progress—were it not for the perfidious suspicion that quarks might not be so fundamental after all!

ACKNOWLEDGMENTS

I would like to thank Shanta de Alwis, John Bell, and Jacques Weyers for many enjoyable and educational discussions on the subject of $SU(6)$, null-planes, and the Melosh transformation. Thanks too to Roger Cashmore and Peter Litchfield for their guidance in the real world of experimental data; to my wife Jessie Hey for preparing the bibliography, and to Jan Jones for her patience and care in typing this manuscript.

Appendix

A.1. Young Diagrams

To obtain the decomposition into irreducible representations of the product of two representations of $SU(n)$, one starts with the corresponding Young tableaux. Since we are only concerned with the simplest products it suffices to know that

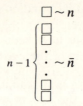

and a column of n boxes is the singlet. [More complete rules may be found elsewhere (Hamermesh, 1962)]. For the product of two tableaux, one must add to one of them, taken as fixed, all the boxes of the other in all possible ways compatible with the following restrictions:

(i) The boxes of the multiplying diagram must be attached, to the fixed one, row by row starting from the first (upper) one.

(ii) The resulting tableaux must be again a Young diagram. This means that the number of boxes in any row must not be increasing from the upper to the lower row.

(iii) Two boxes from the same row of the multiplying diagram can never be placed in the same column.

(iv) The maximum allowed number of rows for the resulting diagrams is n, with the convention that every column with n boxes must be dropped out of the final diagrams (just corresponding to a singlet).

(v) In the multiplying diagram, put an index a into every box in the first row, b into every box of the second row, etc. For every diagram constructed according to the previous rules, one can read off the sequence of a's, b's, c's etc., starting from the *right* of the top row, the *right* of the second row, and so on. The only allowed diagrams are the ones in which this sequence is a *lattice* permutation of the a's, b's, etc.

For example, a lattice permutation of $a^n b^m c^l$ is a sequence of n a's, m b's, l c's like

$$aaabbac \cdots$$

such that to the left of any point in the sequence there are not less a's than b's, and not less b's than c's and so on.

For $a^2 bc$, the lattice permutations are

$$aabc, \quad abac, \quad abca$$

Consider the example in $SU(3)$ of $\mathbf{8}\otimes\mathbf{8}$—in terms of Young diagrams we have

How does one calculate the dimension of the representation corresponding to a given Young diagram? The answer is either a long formula or a simple rule—the "Hook" or Littlewood–Richardson (1934) rule. This is best illustrated by an example. For a diagram in $SU(n)$, assign numbers to each box as shown in Figure 21—n's down the leading diagonal; $n + 1, n + 2$, etc., for diagonals to the right; $n - 1, n - 2$, etc., for diagonals to the left.

The product of these numbers forms the numerator. The denominator is given by the "product of the hooks." The "hook" of each box is how many squares one must pass through, entering along the appropriate row from the right and leaving the diagram from the bottom of the appropriate column. For example, in $SU(3)$ consider the $\mathbf{8}$ representation. The rule says that the dimension of

is given in $SU(3)$ by

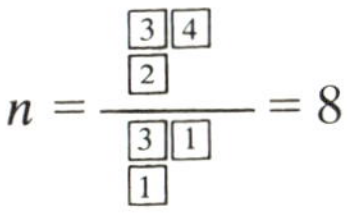

$$n = \frac{\boxed{3}\,\boxed{4}}{\boxed{3}\,\boxed{1}} = 8$$

and in $SU(6)$ by

$$n = \frac{\boxed{6}\,\boxed{7}}{\boxed{3}\,\boxed{1}} = 70$$

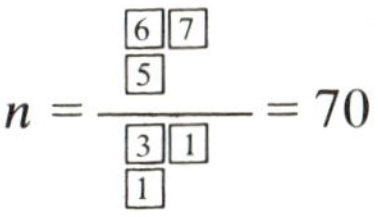

In this way we can easily find in our $SU(3)$ example the product $\mathbf{8}\otimes\mathbf{8}$:

$$\mathbf{8}\otimes\mathbf{8} = \mathbf{27}\oplus\mathbf{10}\oplus\overline{\mathbf{10}}\oplus\mathbf{8}\oplus\mathbf{8}\oplus\mathbf{1}$$

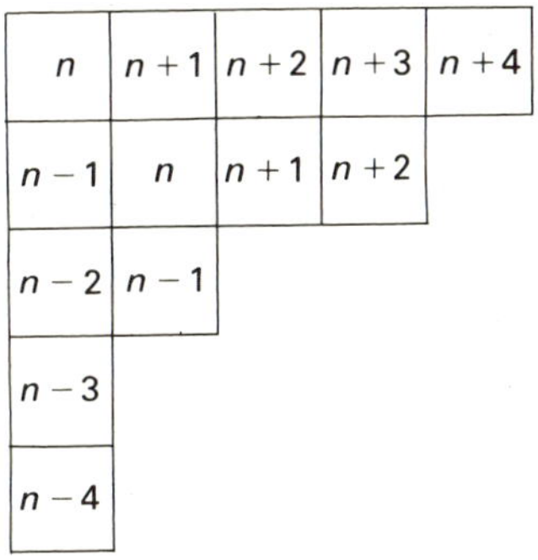

n	$n+1$	$n+2$	$n+3$	$n+4$
$n-1$	n	$n+1$	$n+2$	
$n-2$	$n-1$			
$n-3$				
$n-4$				

Fig. 21. Assignment of numbers to diagonals in a Young diagram for $SU(n)$.

(One must decide on whether or not it is the conjugate representation from the diagram.)

These rules may easily be extended for calculations of the irreducible representations of $O(n)$ and $Sp(n)$.

A.2. Wave Functions for Three-Quark States

(This section of the appendix is adapted from the paper of Feynman *et al.*, 1971.)

A.2.1. The Wave-Function Symmetries of Three Objects

If an object can be in one of a number of conditions $x, y, z, \ldots$ we can, when we have three such objects, form states of four kinds of symmetry, which we call S, α, β, A, symmetric (S), mixed-symmetric (α, β), and antisymmetric (A),

$$|S\rangle = |xyz\rangle_S = 6^{-1/2}(|xyz\rangle + |xzy\rangle + |yxz\rangle + |yzx\rangle + |zxy\rangle + |zyx\rangle)$$

$$|\alpha\rangle = |xyz\rangle_\alpha = \tfrac{1}{2}3^{-1/2}(|xyz\rangle + |xzy\rangle + |yxz\rangle + |yzx\rangle - 2|zxy\rangle - 2|zyx\rangle)$$

$$|A\rangle = |xyz\rangle_A = 6^{-1/2}(-|xyz\rangle + |xzy\rangle - |yzx\rangle + |yxz\rangle - |zxy\rangle + |zyx\rangle) \tag{A.1}$$

$$|\beta\rangle = |xyz\rangle_\beta = \tfrac{1}{2}(|xyz\rangle - |xzy\rangle + |yxz\rangle - |yzx\rangle)$$

where $|zxy\rangle$ means that the first object is in state z, the second in x, and the third in y. If, say, x, and y are the same state $y = x$, we must replace $|xyz\rangle + |yxz\rangle$ by $2^{1/2}|xxz\rangle$. If x, y, z are all the same, only the S state survives as $|xxx\rangle_s = |xxx\rangle$. The state α has been chosen to be symmetric in the last two quarks, the state β is antisymmetric. If we combine two states of these kinds, say $|1\rangle$ and $|2\rangle$, states of varying symmetry may be formed according to the rules

$$|1\rangle_S|2\rangle_S = |\rangle_S, \qquad |1\rangle_S|2\rangle_\alpha = |\rangle_\alpha$$

$$|1\rangle_S|2\rangle_\beta = |\rangle_\beta, \qquad |1\rangle_S|2\rangle_A = |\rangle_A$$

$$|1\rangle_A|2\rangle_S = |\rangle_A, \qquad |1\rangle_A|2\rangle_\alpha = |\rangle_\beta$$

$$-|1\rangle_A|2\rangle_\beta = |\rangle_\alpha, \qquad |1\rangle_A|2\rangle_A = |\rangle_S$$

$$2^{-1/2}(+|1\rangle_\alpha|2\rangle_\alpha + |1\rangle_\beta|2\rangle_\beta) = |\rangle_S \tag{A.2}$$

$$2^{-1/2}(-|1\rangle_\alpha|2\rangle_\alpha + |1\rangle_\beta|2\rangle_\beta) = |\rangle_\alpha$$

$$2^{-1/2}(+|1\rangle_\alpha|2\rangle_\beta + |1\rangle_\beta|2\rangle_\alpha) = |\rangle_\beta$$

$$2^{-1/2}(-|1\rangle_\alpha|2\rangle_\beta + |1\rangle_\beta|2\rangle_\alpha) = |\rangle_A$$

A.2.2. Dependence on Spin

If we combine spins so that x, y, and z must either be $+\frac{1}{2}$ or $-\frac{1}{2}$ (written simply $+$, $-$), we find that $|\ \rangle_S$ is spin $\frac{3}{2}$, $|\ \rangle_{\alpha,\beta}$ are spin $\frac{1}{2}$ and $|\ \rangle_A = 0$. We have four states of spin $\frac{3}{2}$:

$$|\tfrac{3}{2}, +\tfrac{3}{2}\rangle_S = |+++\rangle_S$$

$$|\tfrac{3}{2}, +\tfrac{1}{2}\rangle_S = |++-\rangle_S$$

$$|\tfrac{3}{2}, -\tfrac{1}{2}\rangle_S = |+--\rangle_S$$

$$|\tfrac{3}{2}, -\tfrac{3}{2}\rangle_S = |---\rangle_S$$

and two α states of spin $\frac{1}{2}$,

$$|\tfrac{1}{2}, +\tfrac{1}{2}\rangle_\alpha = +|++-\rangle_\alpha$$

$$|\tfrac{1}{2}, -\tfrac{1}{2}\rangle_\alpha = -|--+\rangle_\alpha$$

(A.3)

and the corresponding β states,

$$|\tfrac{1}{2}, +\tfrac{1}{2}\rangle_\beta = +|++-\rangle_\beta$$

$$|\tfrac{1}{2}, -\tfrac{1}{2}\rangle_\beta = -|--+\rangle_\beta$$

A.2.3. Dependence on Unitary Spin

If the objects can have three values as for unitary spin, we have several possibilities. The totally symmetric state $|\ \rangle_S$ is a decuplet, $|\mathbf{10}\rangle_S$; the α and β states are octets, $|\mathbf{8}\rangle_\alpha$ and $|\mathbf{8}\rangle_\beta$; and the antisymmetric state is a singlet, $|\mathbf{1}\rangle_A$. The wave functions for these are evident. For example, the quantum numbers for a Σ^0 require it to be made of an s, u, d quark. Therefore Σ^0 in a state $|\mathbf{8}\rangle_\alpha$ is $|s, u, d\rangle_\alpha$, in a decuplet it is $|s, u, d\rangle_S$, where we mean to substitute into (A.1) s, u, d for x, y, z, respectively. A neutron is d, d, u, so if it is in an octet it is $|d, d, u\rangle_{\alpha,\beta}$, if it is in a decuplet it is $|d, d, u\rangle_S$ and is called a Δ^0, etc.

Combining these two(spin and unitary spin) by the multiplication table (A.2), we can make the following **56** symmetrical states $|\mathbf{56}\rangle_S$ [we use "quartet" and "doublet" symbols $^4(\ )$ and $^2(\ )$ to represent total spin-$\frac{3}{2}$ and -$\frac{1}{2}$ states; there is no $|\text{spin}\rangle_A$]:

$$|\mathbf{56}\rangle_S: \quad {}^4(\mathbf{10}) = |\tfrac{3}{2}\rangle_S|\mathbf{10}\rangle_S$$

$$\quad {}^2(\mathbf{8}) = 2^{-1/2}(|\tfrac{1}{2}\rangle_\alpha|\mathbf{8}\rangle_\alpha + |\tfrac{1}{2}\rangle_\beta|\mathbf{8}\rangle_\beta)$$

(A.4)

the following **70** states of type α:

$$|70\rangle_\alpha: \qquad {}^4(\mathbf{8})_\alpha = |\tfrac{3}{2}\rangle_S |\mathbf{8}\rangle_\alpha$$

$$\qquad {}^2(\mathbf{10})_\alpha = |\tfrac{1}{2}\rangle_\alpha |\mathbf{10}\rangle_S$$

$$\qquad {}^2(\mathbf{8})_\alpha = 2^{-1/2}(-|\tfrac{1}{2}\rangle_\alpha |\mathbf{8}\rangle_\alpha + |\tfrac{1}{2}\rangle_\beta |\mathbf{8}\rangle_\beta) \qquad (A.5)$$

$$\qquad {}^2(\mathbf{1})_\alpha = -|\tfrac{1}{2}\rangle_\beta |\mathbf{1}\rangle_A$$

and **70** of type β:

$$|70\rangle_\beta: \qquad {}^4(\mathbf{8})_\beta = |\tfrac{3}{2}\rangle_S |\mathbf{8}\rangle_\beta$$

$$\qquad {}^2(\mathbf{10})_\beta = |\tfrac{1}{2}\rangle_\beta |\mathbf{10}\rangle_S$$

$$\qquad {}^2(\mathbf{8})_\beta = 2^{-1/2}(|\tfrac{1}{2}\rangle_\alpha |\mathbf{8}\rangle_\beta + |\tfrac{1}{2}\rangle_\beta |\mathbf{8}\rangle_\alpha) \qquad (A.6)$$

$$\qquad {}^2(\mathbf{1})_\beta = |\tfrac{1}{2}\rangle_\alpha |\mathbf{1}\rangle_A$$

and the following **20** antisymmetric states

$$|20\rangle_A: \qquad {}^4(\mathbf{1})_A = |\tfrac{3}{2}\rangle_S |\mathbf{1}\rangle_A$$

$$\qquad {}^2(\mathbf{8})_A = 2^{-1/2}(-|\tfrac{1}{2}\rangle_\alpha |\mathbf{8}\rangle_\beta + |\tfrac{1}{2}\rangle_\beta |\mathbf{8}\rangle_\alpha) \qquad (A.7)$$

A.2.4. *Dependence on Space: Harmonic Oscillator Wave Functions*

For harmonic oscillator quark model calculations we must work out the orbital states and combine them with these $SU(6)$ states to produce purely overall symmetrical states, for we assume the baryon resonances are pure $|\rangle_S$ states only. This is the so-called "symmetric quark model" assumption, which is incorporated in the "hidden color" approach.

N = 0. The ground state is symmetric. We call it $|0\rangle$. Therefore, it combines only with $|56\rangle_S$ to make a totally symmetrical state,

$$|56, 0\rangle = |56\rangle_S |0\rangle \qquad (A.8)$$

yielding ${}^4(\mathbf{10})_{3/2}$ for the spin-$\tfrac{3}{2}$ decuplet and ${}^2(\mathbf{8})_{1/2}$ for the spin-$\tfrac{1}{2}$ octet.

N = 1. The states are $a^*|0\rangle$ or $b^*|0\rangle$. The first is an α state, the second a β state.

For each case, choosing a^* spacelike, we have three states forming the components of an $L = 1$ state. If we select those components having a z component, $+1, 0, -1$, we have for these orbital states

$$|1, +1\rangle_\alpha^1 = a_+^*|0\rangle$$

$$|1, 0\rangle_\alpha^1 = a_z^*|0\rangle \qquad (A.9)$$

$$|1, -1\rangle_\alpha^1 = a_-^*|0\rangle$$

where $a_\pm^* = \mp 2^{-1/2}(a_x^* \pm i a_y^*)$, and the superscript 1 in the state system is to clarify that it is an orbit state of $N = 1$. Corresponding $|1\rangle_\beta$ states are generated by b^*. Symmetric states are now formed by combining

$$|\mathbf{70}, 1\rangle = 2^{-1/2}(|\mathbf{70}\rangle_\alpha|1\rangle_\alpha^1 + |\mathbf{70}\rangle_\beta|1\rangle_\beta^1) \tag{A.10}$$

Each state, say the $^4(\mathbf{8})$, being obtained by substituting the expression for that state in (A.5) for $|\mathbf{70}\rangle_\alpha$, and in (A.6) for $|\mathbf{70}\rangle_\beta$.

Which components of angular momentum, of spin, and of orbit are to be combined in (A.10)? That depends on which total orbital angular momentum you wish to work out. Thus, for the $N(1700)$ $^4(\mathbf{8})_{1/2}$ we must combine the $S = \frac{3}{2}$ in the expressions (A.5) or (A.6) for $^4(\mathbf{8})$ to the $L = 1$ of the orbit states in (A.10), to compound a suitable (say, $+\frac{1}{2}$) component of total spin $J = \frac{1}{2}$, using a linear combination with the correct Clebsch–Gordan coefficients for these values of angular momentum. In this way the wave functions for all of the $|\mathbf{70}, 1\rangle$ states of various J are worked out (they are all of negative parity, because of the a^*).

$N = 2$. Here we can have two excitations of orbital motion and thus we can combine them, using the rules (A.2) to make up space states of various symmetry. The parity is now positive:

$$N = 2: \qquad |2\rangle_S^2, |0\rangle_S^2 = 2^{-1/2}(|1\rangle_\alpha^1|1\rangle_\alpha^1 + |1\rangle_\beta^1|1\rangle_\beta^1)$$

$$|2\rangle_\alpha^2, |0\rangle_\alpha^2 = 2^{-1/2}(-|1\rangle_\alpha^1|1\rangle_\alpha^1 + |1\rangle_\beta^1|1\rangle_\beta^1)$$

$$|2\rangle_\beta^2, |0\rangle_\beta^2 = 2^{-1/2}(|1\rangle_\alpha^1|1\rangle_\beta^1 + |1\rangle_\beta^1|1\rangle_\alpha^1) \tag{A.11}$$

$$|1\rangle_A = 2^{-1/2}(-|1\rangle_\alpha^1|1\rangle_\beta^1 + |1\rangle_\beta^1|1\rangle_\alpha^1)$$

Here again, we shall have to use Clebsch–Gordan coefficients to make an $L = 2$ or 0 or 1 out of our two $L = 1$ pieces. For example, since

$$|2, 0\rangle = 6^{-1/2}(|+1\rangle|-1\rangle + |-1\rangle|+1\rangle + 2|0\rangle|0\rangle)$$

we have

$$|2, 0\rangle_\beta^2 = 6^{-1/2}(|+1\rangle_\alpha^1|-1\rangle_\beta^1 + |-1\rangle_\alpha^1|+1\rangle_\beta^1 + 2|0\rangle_\alpha^1|0\rangle_\beta^1)$$

$$= 6^{-1/2}(a_+^*b_-^* + a_-^*b_+^* + 2a_z^*b_z^*)|0\rangle$$

$$|2, 0\rangle_S^2 = 6^{-1/2}(|+1\rangle_\alpha^1|-1\rangle_\alpha^1 + 2^{1/2}|0\rangle_\alpha^1|0\rangle_\alpha^1 + |+1\rangle_\beta^1|-1\rangle_\beta^1 + 2^{1/2}|0\rangle_\beta^1|0\rangle_\beta^1)$$

$$= 6^{-1/2}(a_+^*a_-^* + a_z^*a_z^* + b_+^*b_-^* + b_z^*b_z^*)|0\rangle$$

We take $|0\rangle_\alpha^1|0\rangle_\alpha^1$ to mean the double excitation of a_z^* or $a_z^*a_z^*|0\rangle$, but it is normalized, so it is

$$|0\rangle_\alpha^1|0\rangle_\alpha^1 = 2^{-1/2}a_z^*a_z^*|0\rangle$$

whereas

$$|0\rangle_\alpha^1|0\rangle_\beta^1 = a_z^*b_z^*|0\rangle$$

These new states must now be combined with the unitary states (A.4) to (A.7) to form, finally, symmetrized states, and the correct Clebsch–Gordan coefficients used to combine substates of various z components of angular momentum together to form states of definite J and J_z.

We need not discuss the higher states, as the principles are always the same. The meson states are self-evident. These wave functions are the basis of all harmonic oscillator quark model calculations for transition matrix elements.

References

Adler, S. L. (1965), *Phys. Rev. Lett.* **14**, 1051; *Phys. Rev.* **140B**, 736.

Adler, S. L., and Dashen, R. (1968), *Current Algebras* (New York, W. A. Benjamin).

Babcock, J., and Rosner, J. L. (1975), Caltech preprint CALT-68-485.

Barbaro-Galtieri, A. (1972), *in Proceedings of the XVI International Conference on High Energy Physics.* Ed. J. D. Jackson and A. Roberts. (NAL), Vol. 1, p. 159.

Barnes, K. J., Carruthers, P. A. and von Hippel, F. (1965), *Phys. Rev. Lett.* **14**, 82.

Bell, J. S. (1966), *in Recent Developments in Particle Symmetries*: 1965 Int. School of Physics 'Ettore Majorana.' Ed. A. Zichichi (New York, Academic Press), p. 138.

Bell, J. S. (1974), *Acta Phys. Austriaca Suppl.* **13**, 395.

Bell, J. S., and Hey, A. J. G. (1974), *Phys. Lett.* **51B**, 365.

Bernstein, J. (1968), *Elementary Particles and Their Currents* (San Francisco, W. H. Freeman).

Bjorken, J. D., and Glashow, S. L. (1964), *Phys. Lett.* **11**, 255.

Brandenburg, G. W., Carnegie, R. K., Cashmore, R. J., Davier, M., Dunwoodie, W. M., Lasinski, T. A., Leith, D. W. G. S., Matthews, J. A. J., and Walden, P. (1974), Stanford Linear Accelerator Center preprints SLAC-PUB-1697, -1698.

Bucella, F., Kleinert, H., Savoy, C. A., Celeghini, E., and Sorace, E. (1970), *Nuovo Cim.* **69A**, 133.

Burkhardt, G. H., and Hey, A. J. G. (1975), University of Birmingham preprint BUMP 7601.

Carlitz, R., and Weyers, J., (1975), *Phys. Lett.* **56B**, 154.

Carlitz, R., Heckathorn, D., Kaur, J., and Tung, W. K. (1975), *Phys. Rev.* **D11**, 1234.

Carruthers, P. A. (1966), *Introduction to Unitary Symmetry* (N.Y., London, Wiley-Interscience).

Carter J., Coyne, J., and Meshkov, S. (1965), *Phys. Rev. Lett.* **14**, 525.

Cashmore, R. J., Hey, A. J. G., and Litchfield, P. J. (1975), *Nucl. Phys.* **B98**, 237.

Close, F. E., and Gilman, F. J. (1972), *Phys. Lett.* **38B**, 541.

Close, F. E., Osborn, H., and Thomson, A. M. (1974), *Nucl. Phys.* **B77**, 281.

Coleman, S. (1965), *Phys. Lett.* **19**, 144.

Coleman, S. (1966), *J. Math. Phys.* **7**, 787.

Coleman, S. (1968), *in Hadrons and Their Interactions*, 1967 Int. School of Physics 'Ettore Majorana.' Ed. A. Zichichi (New York, Academic Press), p. 9.

Coleman, S., and Mandula, J. E. (1967), *Phys. Rev.* **159**, 1251.

Cook, C. L.., and Murtaza, G. (1965), *Nuovo Cim.* **39**, 352.

Dalitz, R. H. (1965), *in High Energy Physics*: Lectures delivered at Les Houches during 1965 session of the Summer School on Theoretical Physics. Eds. E. Witt and M. Jacob (New York, Gordon and Breach), p. 253.

Dalitz, R. H. (1974), *in Hadron Interactions at Low Energies, Physics and Applications*, Proc. Triangle Meeting. Eds. D. Krupa and J. Pisut (Bratislava, VEDA), Vol. 1, p. 145.

Dashen, R., and Gell-Mann, M. (1966), *in Symmetry Principles at High Energy*, Proc. 3rd Coral Gables Conference Jan. 19–21, 1966. Eds. A.Perlmutter *et al.* (San Francisco, W. H. Freeman), p. 168.

de Alfaro, V., Fubini, S., Furlan, G., and Rossetti, C. (1973), *Currents in Hadron Physics* (Amsterdam, North-Holland).

de Alwis, S. P. (1973), *Nucl. Phys.* **B55**, 427.

de Alwis, S. P., and Stern, J. (1974), *Nucl. Phys.* **B77**, 509.

Dicus, D., Jackiw, R., and Teplitz, V. L. (1971), *Phys. Rev.* **D4**, 1733.

Eichten, E., Feinberg, F., and Willemsen, J. F. (1973), *Phys. Rev.* **D8**, 1204.

Faiman, D. (1975), Weizmann preprint WIS-75/53-Ph.

Faiman, D., and Plane, D. E. (1972), *Nucl. Phys.* **B50**, 379.

Feynman, R. P. (1973), *Photon-Hadron Interactions* (New York, Benjamin).

Feynman, R. P., Kislinger, M., and Ravndal, F. (1971), *Phys. Rev.* **D3**, 2706.

Fritzsch, H., and Gell-Mann, M. (1971), *Contribution to the International Conference on Duality and Symmetry in Hadron Physics*. Ed. E. Gotsman (Jerusalem, Weizmann Science Press).

Fubini, S., and Furlan, G. (1965), *Physics* **1**, 229.

Gasiorowicz, S. (1967), *Elementary Particle Physics* (New York, Wiley).

Gell-Mann, M. (1964), *Phys. Lett.* **8**, 214.

Gell-Mann, M. (1972), *Acta Phys. Austriaca Suppl.* **9**, 733.

Gilman, F. J., Kugler, M., and Meshkov, S. (1974), *Phys. Rev.* **D9**, 715.

Glashow, S. L., Iliopoulous, J., and Maiani, L. (1970), *Phys. Rev.* **D2**, 1874.

Greenberg, O. W. (1964), *Phys. Rev. Lett.* **13**, 598.

Hamermesh, M. (1962), *Group Theory and its Applications to Physical Problems* (London, Addison-Wesley).

Herndon, D. J., Longacre, R., Miller, L. R., Rosenfeld, A. H., Smadja, G., Soding, P., Cashmore, R. J., and Leith, D. W. G. S. (1975), *Phys. Rev.* **D11**, 3183.

Heusch, C., and Ravndal, F. (1970), *Phys. Rev. Lett.* **25**, 253.

Hey, A. J. G., Litchfield, P. J., and Cashmore, R. J. (1975), *Nucl. Phys.* **B95**, 547.

Horgan, R. (1974), *Nucl. Phys.* **B71**, 514.

Jersak, J., and Stern, J. (1969), *Nuovo Cim.* **59**, 315.

Jones, M., Levi-Setti, R., and Lasinski, T. (1974), *Nuovo Cim.* **19A**, 365.

Jordan, T. F. (1965), *Phys. Rev.* **B139**, 148; **B140**, 766.

Kogut, J. B., and Soper, D. E. (1970), *Phys. Rev.* **D1**, 2901.

Le Yaouanc, A., Olivier, L., Pene, O., and Raynal, J. C. (1974), *Phys. Rev.* **D9**, 1415.

Le Yaouanc, A., Olivier, L., Pene, O., and Raynal, J. C. (1975), *Phys. Rev.* **D11**, 1272.

Lipkin, H. J. (1966), *Lie Groups for Pedestrians* 2nd ed. (Amsterdam, North-Holland).

Lipkin, H. J. (1973), *Phys. Rep.* **8C**, 174.

Lipkin, H. J. (1974), *Phys. Rev.* **D9**, 1579.

Lipkin, H. J., and Meshkov, S. (1965), *Phys. Rev. Lett.* **14**, 670.

Litchfield, P. J. (1974), Proc. XVII International Conference on High Energy Physics, London, July 1974. Ed. J. R. Smith (Chilton, SRC Rutherford Laboratory), II-65.

Litchfield, P. J., Bacon, T. C., Butterworth, I., Smith, J. R., Lesquoy, E., Strub, R., Berthon, A., Vrana, J., Meyer, J., Pauli, E., Tallini, B., and Zatz, J. (1971), *Nucl. Phys.* **B30**, 125.

Littlewood, D. E., and Richardson, A. R. (1934), *Phil. Trans. R. Soc. London* **A233**, 99.

Longacre, R. S., Rosenfeld, A. H., Lasinski, T., Smadja, G., Cashmore, R. J., Leith, D. W. G. S. (1975), *Phys. Lett.* **55B**, 415.

Melosh, H. J. (1973), Caltech. Ph.D. thesis.

Melosh, H. J. (1974), *Phys. Rev.* **D9**, 1095.

Metcalf, W. J., and Walker, R. L. (1974), *Nucl. Phys.* **B76**, 253.

Osborn, H. (1974), *Nucl. Phys.* **B80**, 90, 113.

Particle Data Group (1974), *Phys. Lett.* **50B**, 1.
Plane, D. E., Baillon, P., Bricman, C., Ferro-Luzzi, M., Meyer, J., Pagiola, E., Schmitz, N., Burkhardt, E., Filthuth, H., Kluge, E., Oberlack, H., Barloutaud, R., Granet, P., Porte, J. P., and Prevost, J. (1970), *Nucl. Phys.* **B22**, 93.
Rosner, J. L. (1974), *Phys. Rep.* **11C**, 189.
Rosner, J. L., and Petersen, W. P. (1973), *Phys. Rev.* **D7**, 747.
Sakata, S. (1956), *Prog. Theor. Phys.* **16**, 686.
Salam, A. (1968), in *Elementary Particle Theory*, Ed. N. Svartholm (Stockholm, Almquist and Forlag).
Weinberg, S. (1967), *Phys. Rev. Lett.* **19**, 1264.
Weisberger, W. I. (1965), *Phys. Rev. Lett.* **14**, 1047.
Young, A. (1901), *Proc. London Math. Soc.* **33**, 97.
Zweig, G. (1964), CERN preprint TH-401.

Bibliography for $SU(6)_W$ and the Melosh Transformation between Current and Constituent Quarks

Compiled by Jessie M. N. Hey

The entries have been arranged under the following headings:
  Reviews of early work on relativistic and nonrelativistic $SU(6)$
  Periodical articles (arranged by year)
  Preprints, reports, and theses (arranged by year)
  Published papers from conferences, summer institutes, and schools (arranged by year of publication)

The references in each section (or subsection) are arranged in alphabetical order by author. While conferences, summer institutes, and schools are arranged by year of publication of the proceedings, the place and date of the meetings are also indicated.

Special thanks are due to the Library Staff at SLAC for their valuable assistance with the Stanford Public Information Retrieval System (SPIRES) data base. This provided a basis for this compilation of papers arising from Melosh's work. The reviews of earlier work have been included for perspective.

The following abbreviations are used:

Caltech:	California Institute of Technology
CERN:	European Organization for Nuclear Research
DESY:	Deutsches Elektronen-Synchrotron
M.I.T.:	Massachusetts Institute of Technology
SLAC:	Stanford Linear Accelerator Center

Reviews of Early Work on Relativistic* and Nonrelativistic SU(6)

Difficulties of Relativistic $U(6)$. J. S. Bell (CERN, Geneva). In: *Recent Developments in Particle Symmetries*, International School of Physics "Ettore Majorana," held in Erice, 1965, pp. 138–175. New York, Academic Press (1966). A popular account of early difficulties with relativistic $SU(6)$.
$SU(6)$ in Particle Physics. B. W. Lee (University of Pennsylvania, Philadelphia). In: *Particle Symmetries*, Brandeis University Summer Institute in Theoretical Physics, held in Bran-

---

*$SU(6)_W$ is sometimes referred to as relativistic $SU(6)$.

deis, 1965, pp. 327–451. New York, Gordon and Breach (1966). A series of lectures giving a critical review. 155 references.

Dynamical Symmetry in Particle Physics. A. Pais (Rockefeller Institute, New York). *Rev. Mod. Phys.* **38**, 215–253 (1966). Reviews those applications of symmetry arguments in particle physics that deal with developments starting with $SU(6)$.

The $SU(6)$ Model and Its Relativistic Generalizations. H. Ruegg (International Centre for Theoretical Physics, Trieste, and CERN, Geneva), W. Rühl (CERN, Geneva), and T. S. Santhanam (International Centre for Theoretical Physics, Trieste). *Helv. Phys. Acta* **40**, 9–134 (1967). A more technical review. 403 references.

Periodical Articles

1972

An $SU(6)_W$ Fit to the Negative-Parity Baryon Decay Rates. D. Faiman (CERN, Geneva) and D. E. Plane (Max Planck Institute, München). *Nucl. Phys.* **B50**, 379–406 (1972).

Symmetries and Nonsymmetries of the Relativistic Quark Model. J. L. Rosner (Minnesota University, Minneapolis). *Phys. Rev.* **D6**, 1781–1796 (1972).

1973

$SU(6)_W$ Breaking Pattern Implied by the Transformation from Constituent to Current Quarks. F. Buccella (Rome University), F. Nicolo, and A. Pugliese (Bari University). *Lett. Nuovo Cimento* **8**, 244–248 (1973).

The Transformation from Constituent to Current Quarks in the Symmetric Quark Model. F. Buccella (Rome University) and C. A. Savoy (Geneva University). *Lett. Nuovo Cimento* **8**, 569–574 (1973).

$SU(6)_W$ on the Light Cone. S. P. De Alwis (CERN, Geneva). *Nucl. Phys.* **B55**, 427–435 (1973).

Current and Constituent Quarks in the Light Cone Quantization. E. Eichten, J. F. Willemsen (SLAC, Stanford), and F. Feinberg (MIT, Cambridge). *Phys. Rev.* **D8**, 1204–1219 (1973).

Comment on $U(6) \times O(2)$ Symmetry Breaking. P. G. O. Freund (Chicago University). *Phys. Rev.* **D8**, 4631–4634 (1973).

Pionic Transitions as Tests of the Connection between Current and Constituent Quarks. F. J. Gilman (Caltech, Pasadena, and SLAC, Stanford), M. Kugler (SLAC, Stanford), and S. Meshkov (Caltech, Pasadena, and National Bureau of Standards, Washington). *Phys. Lett.* **45B**, 481–486 (1973).

Algebra of Strengths Generated by Bilocal Currents Acting on Constituent States. L. Gomberoff, L. P. Horwitz (Tel Aviv University), and Y. Ne'eman (Tel Aviv University and University of Texas, Austin). *Phys. Lett.* **45B**, 131–135 (1973).

Constituent Quarks, Current Quarks, and Pionic Decays of Resonances. A. J. G. Hey and J. Weyers (CERN, Geneva). *Phys. Lett.* **44B**, 263–265 (1973).

Current Quarks, Constituent Quarks, and Symmetries of Resonance Decays. A. J. G. Hey, J. L. Rosner, and J. Weyers (CERN, Geneva). *Nucl. Phys.* **B61**, 205–229 (1973).

$SU(3) \times SU(3)$ of Currents, $SU(6)_W$ of Constituents, and Electromagnetic Decays. A. Love and D. V. Nanopoulos (University of Sussex, Brighton), *Phys. Lett.* **45B**, 507–509 (1973).

Melosh Transformation for Interacting Quarks. S. Machida (Kyoto University). *Prog. Theor. Phys.* **50**, 343–344 (1973).

$SU(6)_W$ Model of $\Delta S = 0$ and $\Delta S = 1$ Nonleptonic Parity-Violating Weak Interactions. B. H. J. McKellar and P. Pick (Sydney University). *Phys. Rev.* **D7**, 260–266 (1973).

The Current Quark Content of Hadrons. C. A. Savoy (Geneva University). *Lett. Nuovo Cimento* **7**, 841–845 (1973).

1974

$SU(6)_W$ Analysis of the Electroproduction of Pseudoscalar Mesons in the Second Resonance Region. C. Avilez (Instituto Politecnio Nacional, Mexico), and G. Cocho (Mexico University). *Phys. Rev.* **D10**, 3638–3644 (1974).

A Theoretical Argument for Something Like the Second Melosh Transformation. J. S. Bell and A. J. G. Hey (CERN, Geneva). *Phys. Lett.* **51B**, 365–366 (1974).

Current Quarks, Constituent Quarks, and the Poincaré Group. F. Buccella (Rome University), C. A. Savoy (Geneva University), and P. Sorba (Marseille University). *Lett. Nuovo Cimento* **10**, 455–460 (1974).

Current and Constituent Quarks—Their Implications for Resonance Excitations and Polarized and Unpolarized Inelastic Structure Functions. F. E. Close (CERN, Geneva), H. Osborn (Caltech, Pasadena), and A. M. Thomson (Durham University). *Nucl. Phys.* **B77**, 281–308 (1974).

On the Transformation between Current and Constituent Quarks and Consequences for Polarized Electroproduction Structure Functions. F. E. Close (CERN, Geneva). *Nucl. Phys.* **B80**, 269–298 (1974).

Why and How to Make Constituent and Current Quarks Different. S. P. De Alwis (CERN, Geneva) and J. Stern (Institut de Physique Nucléaire, Orsay). *Nucl. Phys.* **B77**, 509–544 (1974).

How Large is the ρ-Meson Longitudinal Coupling? D. Faiman (Weizmann Institute, Rehovoth). *Phys. Lett.* **49B**, 365–368 (1974).

Phases of Resonant Amplitudes $\pi N \to \rho N$. D. Faiman (Weizmann Institute, Rehovoth). *Nucl. Phys.* **B77**, 443–452 (1974).

Current Quarks and Constituent Quarks—Symmetry Breaking and Interaction. D. H. Fuchs (Purdue University, Lafayette). *Phys. Rev.* **D10**, 1280–1283 (1974).

Chiral Breaking and Low-Energy Pion Electroproduction. G. Furlan, N. Paver, C. Verzegnassi (Trieste University, International Centre for Theoretical Physics, Trieste). *Nuovo Cimento* **20A**, 295–306 (1974).

Photon Amplitudes Predicted by the Transformation between Current and Constituent Quarks. F. J. Gilman and I. Karliner (SLAC, Stanford). *Phys. Rev.* **D10**, 2194–2211 (1974).

Transformation between Current and Constituent Quarks and Transitions between Hadrons. F. J. Gilman (SLAC, Stanford), M. Kugler (Caltech, Pasadena), and S. Meshkov (National Bureau of Standards, Washington). *Phys. Rev.* **D3**, 715–735 (1974).

A Unified Picture of the Algebra of Strengths, $SU(6)_W$ (Currents) and $SU(6)_W$ (Strong). L. Gomberoff, L. P. Horwitz (Tel Aviv University), and Y. Ne'eman (Tel Aviv University and University of Texas, Austin). *Phys. Rev.* **D9**, 3545–3561 (1974).

Coplanar Breaking of $SU(6)_W$ Applied to Pionic Decays of the Baryons ($L = 1$). H. Haut, C. Leroy, and J. Van Parijs (Louvain University, Belgium). *Phys. Rev.* **D9**, 2385–2390 (1974).

Quarks and the Helicity Structure of Photoproduction Amplitudes. A. J. G. Hey and J. Weyers (CERN, Geneva). *Phys. Lett.* **48B**, 69–72 (1974).

Algebraic Realization of $SU(6)_W$ Currents, Charge—Algebra Sum Rules. B. Horwitz (Pennsylvania University, Philadelphia). *Phys. Rev.* **D9**, 996–1003 (1974).

Helicity Amplitudes Constructed by $SU(6) \times O(3)$ Spectrum with Broken $SU(6)_W$. K. Inoue and M. Uehara (Kyushu University, Fukuoka). *Prog. Theor. Phys.* **52**, 731–733 (1974).

Momentum-Dependent Transformation between Bare and Constituent Quarks. T. Kobayashi (University of Education, Tokyo). *Lett. Nuovo Cimento* **11**, 345–351 (1974).

Naive Quark-Pair-Creation Model and Baryon Decays. A. Le Yaouanc, L. Oliver, O. Pene, and J. C. Raynal (Laboratoire de Physique Théorique et Hautes Energies, Orsay). *Phys. Rev.* **D9**, 1415–1419 (1974).

General Formulation of Pionic Decays in the Quark Model. H. J. Lipkin (National Accelerator Laboratory, Batavia; Argonne National Laboratory). *Phys. Rev.* **D9**, 1579–1583 (1974).

Quarks—Currents and Constituents. H. J. Melosh (Chicago University). *Phys. Rev.* **D9**, 1095–1112 (1974). Postulates the transformation between the current and constituent quarks later frequently referred to as the second Melosh transformation. The first Melosh transformation is postulated in his thesis and is still discussed and cited.

Angular Constraints and the Melosh Transformation. H. Osborn (Caltech, Pasadena). *Nucl. Phys.* **B80**, 90–112 (1974).

On the Presence of Nonadditive Contributions of Constituent Quark Calculations of Resonance Photoproduction. H. Osborn (Caltech, Pasadena). *Nucl. Phys.* **B80**, 113–126 (1974).

Canonical Transforms between Current and Constituent Quarks. W. F. Palmer and V. Rabl (Ohio State University, Columbus). *Phys. Rev.* **D10**, 2554–2566 (1974).

The Classification and Decays of Resonant Particles. J. L. Rosner (SLAC, Stanford; Minnesota University; Tel Aviv University). *Phys. Rep.* **11C**, 189–326 (1974). A review. Sections 4–6 (pp. 224–263) are particularly relevant to $SU(6)_W$. A fairly comprehensive bibliography on other types of $SU(6)$ models is also included.

The Representation Mixing and the KSFR Relation. S. Wada (Tokyo University). *Lett. Nuovo Cimento* **9**, 381–384 (1974).

1975

Properties of Hadrons as Quark Bound-States on the Null-Plane. M. Abud, R. Lacaze, and C. A. Savoy (Geneva University). *Nucl. Phys.* **B98**, 215–236 (1975).

High-Energy Behaviour, Duality, and the $SU(6)_W$ Structure of the Electromagnetic and Mesonic Transition Operators. C. Avilez, G. Cocho, M. Dubovoy (Instituto Politecnico Nacional, Mexico). *Phys. Rev.* **D11**, 555–565 (1975).

Hydrogen Atom on Null-Plane and Melosh Transformation. J. S. Bell (CERN, Geneva) and H. Ruegg (Geneva University). *Nucl. Phys.* **B93**, 12–22 (1975).

Chiral Symmetry and the Quark Model. R. Carlitz, D. Heckathorn, J. Kaur, and W. K. Tung (Chicago University). *Phys. Rev.* **D11**, 1234–1250 (1975).

Constituent Quark Analysis of Hadronic Transitions. R. Carlitz (Chicago University) and J. Weyers (CERN, Geneva). *Phys. Lett.* **56B**, 154–158 (1975).

Baryon Resonance Couplings in the Reactions $\pi N \to \pi \Delta$ and $\pi N \to \rho N$: Comparison with Theory and Related Reactions. R. J. Cashmore, D. W. G. S. Leith (SLAC, Stanford), R. S. Longacre, and A. H. Rosenfeld (Lawrence Berkeley Laboratory, Berkeley). *Nucl. Phys.* **B92**, 37–50 (1975).

Further Applications of $SU(6)_W$ Decay Schemes for Baryons. R. J. Cashmore (Oxford University), A. J. G. Hey (Southampton University), and P. J. Litchfield (Rutherford Laboratory, Chilton). *Nucl. Phys.* **B98**, 237–258 (1975).

Current Algebra—Real Quarks and Poincaré Invariance. E. Celeghini, L. Lusanna, and E. Sorace (Florence University). *Nuovo Cimento* **25A**, 331–352 (1975).

The Components of νW_2 and $SU(6)_W$ Breaking in Deep Inelastic Scattering. B. Flume-Gorczyca and S. Kitakado (DESY, Hamburg). *Nuovo Cimento* **28A**, 321–336 (1975).

How to Construct a Relativistic $SU(6)$ Classification Symmetry Group. N. H. Fuchs (Purdue University, Lafayette). *Phys. Rev.* **D11**, 1569–1579 (1975).

$SU(6)_W$ and Decays of Baryon Resonances. A. J. G. Hey, P. J. Litchfield (CERN, Geneva), and R. J. Cashmore (Oxford University). *Nucl. Phys.* **B95**, 516–546 (1975).

Quark Pairs inside Hadrons. H. Kleinert (Freie Universität, Berlin). *Phys. Lett.* **59B**, 163–167 (1975).

$SU(6)$ Strong Breaking: Structure Functions and Static Properties of the Nucleon. A. Le Yaouanc, L. Oliver, O. Pene, and J. C. Raynal (Laboratoire de Physique Théorique et Hautes Energies, Orsay). *Phys. Rev.* **D12**, 2137–2156 (1975).

N^* Resonance Parameters and K-Matrix Fits to the Reactions $\pi N \rightarrow \Delta \pi$, ρN, εN. R. S. Longacre, A. H. Rosenfeld, T. Lasinski, G. Smadja, R. J. Cashmore, D. W. G. S. Leith (SLAC, Stanford). *Phys. Lett.* **55B**, 415–419 (1975). Contains mainly the results of an experimental analysis that are then compared to $SU(6)_W$ theoretical predictions.

Determination of the Chiral Mixing Angle through Resonance Decays. Y. Okamoto and T. Kunimasa (Osaka University). *Prog. Theor. Phys.* **53**, 1197–1199 (1975).

Duality, Current and Constituent Quarks in Meson–Baryon Reactions. H. Ruegg and C. A. Savoy (Geneva University). *Nuovo Cimento* **26A**, 243–258 (1975).

Traces of Chiral Symmetry on Light Planes. H. Sazdjian and J. Stern (Institut de Physique Nucléaire, Orsay). *Nucl. Phys.* **B94**, 163–188 (1975).

Some Formal Aspects of the Melosh and Generalized Unitary Transformations of a Class of Massive-Particle Wave Equations. D. L. Weaver (Tufts University, Medford, Massachusetts). *Phys. Rev.* **D12**, 2325–2329 (1975).

Preprints, Reports, and Theses

1973

$SU(6)_W \times O(2)_{L_z}$ Symmetry in Baryon Decays. C. T. Chen-Tsai and T. P. Pai (Taiwan National University). No number. 12 pp.

$SU(6)_W$ for Baryon Decay Widths. An $SU(6)$ 20-Plet? A Quark-Eye View of Exotics. D. Faiman (CERN, Geneva). CERN-TH-1678. Nonconsecutive pagination.

Non-Conserved $SU(3)$ Currents in the Presence of Exact $SU(6)_W$ Symmetry. N. H. Fuchs (CERN, Geneva). CERN-TH-1648. 7 pp.

Hadronic Decay Processes and the Transformation from Current to Constituent Quarks. F. J. Gilman (SLAC, Stanford). SLAC-PUB-1256. 12 pp. Lectures.

A Comparison of Models for Resonance Decays. M. Kugler (Stanford University). (Weizmann Institute, Rehovoth). WIS-73/42-PH. 9 pp.

Constituent Quarks and Meson Decays. T. Kuroiwa (Kyushu University, Fukuoka). KYUSHU-HE-73-11. 10 pp.

Quarks: Current and Constituents. H. J. Melosh (Caltech, Pasadena). Caltech Ph.D. Thesis, submitted February 26, 1973, 68 pp. Postulates the transformation between current and constituent quarks later frequently referred to as the first Melosh transformation.

1974

Light Cone Description of $O(3)$ Excited $SU(6)_W$ Multiplets. S. P. De Alwis and J. Stern (Laboratoire de Physique Théorique et Hautes Energies, Orsay). CERN-TH-1783. 29 pp.

Symmetries of Currents as Seen by Hadrons. F. J. Gilman (SLAC, Stanford). SLAC-PUB-1436. 18 pp. Talk.

Polarization Effects in Electron Proton Scattering—Form Factors, Scaling, Quarks and $SU(6)_W$. A. J. G. Hey (Nuclear Physics Laboratory, Daresbury). Daresbury Laboratory Report, DL/R33 (1974). 103 pp.

Lightlike Chiral Algebra in the Quark Model. M. Ida (Kyoto University), RIFP-210. 28 pp.

A Simple Interpretation of the Melosh Transformation. N. Marinescu and M. Kugler (Weizmann Institute, Rehovoth). WIS-74/42-PH. 10 pp.

Supermultiplet Interactions among Hadrons and Resonances. A. N. Mitra (Delhi University). No number. Nonconsecutive pagination. Lectures. Bibliography.

$3P_0$ Model and Helicity Structure of Photoproduction Couplings. W. Petersen (Nuclear Physics Laboratory, Daresbury). Daresbury Laboratory Preprint, DL/P-201. 23 pp.

Quarks from the Viewpoint of a Phase Transition Model for Hadronic Processes. S. Sen (Trinity College, Dublin). TCD-1974-6. 15 pp.

1975

Multipole Analysis of Resonance Photoproduction. J. Babcock (Minnesota University, Minneapolis) and J. L. Rosner (Caltech, Pasadena). CALT-68-485. 66 pp.

A New Approach to Relativistic Composite Model of Hadrons. M. Bando, S. Tanaka, and M. Toya (Kyoto University). KUNS 319. 47 pp.

Current Quarks, Constituent Quarks and the Reactions $K^-P \to \pi^0 \Lambda^*$. F. Buccella (Rome University), C. A. Savoy (Geneva University), and P. Sorba (Centre de Physique Théorique, Marseille). CPT-75/P-710. 23 pp.

Three Particle Poincaré States and $SU(6) \times SU(3)$ as a Classification Group for Baryons. F. Buccella, A. Sciarrino, and P. Sorba (Rome University and Centre de Physique Théorique, Marseille). CPT-75/P-729. 39 pp.

Construction of a Quark Spin Operator. R. Carlitz and Wu-Ki Tung (Enrico Fermi Institute and University of Chicago). No number. 47 pp.

$SU(6)$ Classification for "In" and "Out" States. N. H. Fuchs (Purdue University, Lafayette). No number. 16 pp.

Nonleptonic Weak Decays and the Melosh Transformation. M. Machacek and Y. Tomozawa (Michigan University, Ann Arbor). UM-HE-75-7. 56 pp. *Bull. Am. Phys. Soc.* **20** (1975); p. 551 contains the abstract of a talk presented at the 1975 Spring meeting of the APS in Washington.

$SU(6)$ Isoscalar Factors for the product $\mathbf{405} \times \mathbf{56} \to \mathbf{56, 70}$. M. Machacek and Y. Tomozawa (Michigan University, Ann Arbor). UM-HE-75-5. 22 pp.

1976

On the Melosh Transformations. V. Aldaya (Salamanca University) and J. de Azcarraga (Oxford University). Oxford preprint 4/76. 19 pp.

Books

1973

Currents in Hadron Physics. V. De Alfaro, S. Fubini, G. Furlan, and C. Rossetti, Amsterdam, North-Holland Publishing (1973). Chapter 8 (pp. 467–570) gives a very good description of the current quark approach to $SU(6)_W$. Most of the remainder of the book is relevant background. Contains material available before the end of 1970 (with a few exceptions).

1974

Phenomenological Aspects of Localizability. L. P. Horwitz. In: *Physical Reality and Mathematical Description* (C. P. Enz and J. Mehra, eds.), pp. 331–356. Boston, Reidel (1974). A mathematical treatment of the Melosh transformation. Bibliography.

Published Papers from Conferences, Summer Institutes, and Schools

1972

Quarks. M. Gell-Mann (CERN, Geneva). In: *Proceedings*, XI Internationale Universitätswochen für Kernphysik, held February 21–March 4, 1972 in Schladming. Vienna, Springer-Verlag (1972). [*Acta Phys. Austriaca Suppl.* **9**, 733–761 (1972)].

1973

Classifying Baryons. D. Faiman (CERN, Geneva). In: *Proceedings*, 2nd International Conference on Elementary Particles, held September 6–12, 1973 in Aix-en-Provence. Paris, La Société Française (1973). [*J. Phys. Paris Suppl.* **34**, C1, 167–170 (1973).] Talk.

Hadron Symmetry and Quarks. S. Matsuda (Kyoto University). In: *Proceedings*, KEK Summer School, held July 17–21, 1973 in Ibaraki. Ibaraki, National Laboratory for High Energy Physics (1973), KEK-73-10, pp. 239–271. Lectures.

Higher Symmetries and Baryon Resonances. J. L. Rosner (Minnesota University, Minneapolis). In: *Proceedings*, 16th International Conference on High Energy Physics, held September 6–13, 1972 in Chicago, pp. 149–163. Batavia, National Accelerator Laboratory (1973), Vol. 3. Appendix.

Symmetries Beyond *SU*(3) (Numbers 365, 650, 790). J. L. Rosner (Minnesota University, Minneapolis). In: *Proceedings*, 16th International Conference on High Energy Physics, held September 6-13, 1972 in Chicago, pp. 189–190. Batavia, National Accelerator Laboratory (1973), Vol. 1. Talk.

1974

The Melosh Transformation and the Pryce–Tani–Foldy–Wouthuysen Transformation. J. S. Bell (CERN, Geneva). In: *Proceedings*, XIII Internationale Universitätswochen für Kernphysik, held February 4-15, 1974 in Schladming. Vienna, Springer-Verlag (1974). [*Acta Phys. Austriaca Suppl.* **13**, 395–445 (1974).]

Chiral Symmetry and the Hadron Spectrum. R. Carlitz (Chicago University). In: *Particle Interactions at Very High Energies*, symposium held in Louvain, 1973 (E. Halzen, D. Spieser, and J. Weyers, eds.), pp. 97–114. NATO Advanced Study Institute Series, Vol. 4B. New York and London, Plenum Press (1974).

The Current Status of Constituent Quarks in Resonance Photo- and Electroproduction. F. E. Close (CERN, Geneva). In: *Proceedings*, XVIIth International Conference on High Energy Physics, held July 1–10, 1974 in London, pp. II-157–II-160. Chilton, S.R.C. (1974). Talk.

Algebraic Properties of Electromagnetic and Weak Currents and Their Representation of Hadron States. F. J. Gilman (SLAC, Stanford). In: *Phenomenology of Particles at High Energies* (R. L. Crawford and R. Jennings, eds.), 14th Scottish Universities Summer

School in Physics, held in Edinburgh, 1973, pp. 553–610. New York and London, Academic Press (1974). Lectures. Bibliography.

Resonances: A Quark View of Hadron Spectroscopy and Transitions. F. J. Gilman (SLAC, Stanford). Summer Institute on Particle Physics, held July 29–Aug. 10, 1974 in Stanford. SLAC-179, Vol. 1, *Strong Interactions*, pp. 307–326.

$SU(6)_W$ and Decays of Baryon Resonances. A. J. G. Hey (CERN, Geneva). In: *Proceedings*, XVIIth International Conference on High Energy Physics, held July 1–10, 1974 in London, pp. II-120–II-133. Chilton, S.R.C. (1974). Talk.

Baryon Resonances. P. J. Litchfield (CERN, Geneva). In: *Proceedings*, XVIIth International Conference on High Energy Physics, held July 1–10, 1974 in London, pp. II-65–II-100. Chilton, S.R.C. (1974). Plenary report. Review of the experimental situation with regard to $SU(6)_W$ baryon supermultiplets.

Current Quarks and Constituent Quarks. S. Meshkov (National Burea of Standards, Washington). In: *Proceedings*, XVIIth International Conference on High Energy Physics, held July 1–10, 1974 in London, pp. II-101–II-112. Chilton, S.R.C. (1974). Talk.

Are Non-**35** Terms Needed in Resonance Couplings? H. Osborn (Cambridge University). In: *Proceedings*, XVIIth International Conference on High Energy Physics, held July 1–10, 1974 in London, pp. II-162–II-163. Chilton, S.R.C. (1974).

Resonance Spectroscopy Theory. J. L. Rosner (Minnesota University, Minneapolis). In: *Proceedings* XVIIth International Conference on High Energy Physics, held July 1–10, 1974 in London, pp. II-171–II-199. Chilton, S.R.C. (1974). Plenary report.

Constituent Quarks and Current Quarks. J. Weyers (CERN, Geneva). In: *Particle Interactions at Very High Energies* (F. Halzen, D. Spieser, and J. Weyers, eds.), pp. 43–95. NATO Advanced Study Institute Series, Vol. 4B. New York and London, Plenum Press (1974).

1975

Baryonic Spectroscopy and Its Immediate Future. R. H. Dalitz (Oxford University). In: *New Directions in Hadron Spectroscopy*, proceedings of a summer symposium held July 7–10, 1975 in Argonne (S. L. Kramer and E. L. Berger, eds.), pp. 383–406. Argonne National Laboratory (1975), ANL-HEP-CP-75-58.

Comparison of Theory and Experiment in Photoproduction. A. Donnachie (Manchester University). International Symposium on Lepton and Photon Interactions at High Energies, held August 21–27, 1975 in Stanford. Stanford, California, Stanford Linear Accelerator Center, Stanford University (1975).

Lost States of the Quark Model and How to Find Them. A. J. G. Hey (Southampton University). In: *New Directions in Hadron Spectroscopy*, proceedings of a summer symposium held July 7–10, 1975 in Argonne (S. L. Kramer and E. L. Berger, eds.), pp. 1–18. Argonne National Laboratory (1975), ANL-HEP-CP-75-58.

The Melosh Transformation; Theory and Experiment. A. J. G. Hey (Southampton University). In: *Proceedings*, XVth Cracow School of Theoretical Physics, held June 5–18, 1975 in Zakopane, Poland. *Acta Phy. Polonica* **B6**, 831–849 (1975).

Spectroscopy after the New Particles. H. J. Lipkin (Weizmann Institute). In: *New Directions in Hadron Spectroscopy*, proceedings of a summer symposium held July 7-10, 1975 in Argonne (S. L. Kramer and E. L. Berger, eds.), pp. 96–114. Argonne National Laboratory (1975), ANL-HEP-CP-75-58.

Current Status of Baryon Spectroscopy. K. C. Wali (Syracuse University). In: *New Directions in Hadron Spectroscopy*, proceedings of a summer symposium held July 7-10, 1975 in Argonne (S. L. Kramer and E. L. Berger, eds.), pp. 296–308. Argonne National Laboratory (1975), ANL-HEP-CP-75-58.

2

Electromagnetic Excitation and Decay of Hadron Resonances

R. G. Moorhouse

1. Introduction

In investigating atomic and molecular, and to a lesser extent, nuclear structure, the emission and absorption of electromagnetic radiation has been a key tool, and the same seems likely to hold true for hadron structure. There are two roles for electromagnetic spectroscopy; the first is in the establishment of the states, their spin parity, and their energy; the second is to give further information on the wave functions of the states from the electromagnetic transition rates or, better, if experimentally attainable, the electromagnetic matrix elements. Generally, in hadrons, contrary to the atomic case, electromagnetic radiation does not play the first role, this energy-level information being usually given by the strong formation or the production and decay of hadron resonances; notable exceptions to this general rule being the observation, first in photopion production off nucleons, of the prominent isospin-$\frac{1}{2}$ nucleon resonance at around 1500 MeV (Wilson, 1958; Peierls, 1958; Sakurai, 1958) and second, the observation of narrow-width mesons at ~3.4 and ~3.5 GeV energy from the radiative decay of the ψ' (3.7) (Braunschweig *et al.*, 1975; Feldman *et al.*, 1975; Tanenbaum *et al.*, 1975).

R. G. Moorhouse • Department of Natural Philosophy, University of Glasgow, Glasgow, G12 8QQ, Scotland

It is in the second role, that of establishing the hadron wave functions, that electromagnetic transitions are most important. In mesons we have only information on transition rates (reviewed in Section 4), but in nucleons we have information on transition matrix elements—that is, not only their magnitude, which is equivalent to transition rates, but also their sign. This is because nearly all information on electromagnetic transitions of nucleon resonances comes from pion photoproduction (with a real photon), $\gamma + N \to \pi + N$, or pion electroproduction $e^- + N \to e^- + \pi + N$, this latter being equivalent to photoproduction with a virtual photon, $\gamma_v + N \to \pi + N$. In either case a dominant part of the process (except in the extreme forward direction) proceeds by resonance formation $\gamma + N \to N^* \to \pi + N$, where N^* is a nucleon resonance of definite isospin, I, spin angular momentum, J, and parity, P. A partial wave analysis of π^+, π^-, and π^0 photoproduction over a range of energies gives the magnitudes and relative signs of the matrix elements $\langle \gamma N | \pi N^*(I, J, P) \rangle$ of the various definite values of I, J, P; from the energy dependence it is possible to separate out the (dominant) part, due to resonance formation: $\langle \gamma N | N^*(I, J, P) \rangle \langle N^*(I, J, P) | \pi N \rangle$. We know the magnitude of $\langle N^* | \pi N \rangle$ from partial wave analysis of $\pi + N \to \pi + N$, so photoproduction experiment and analysis determine the magnitude of the amplitude $\langle \gamma N | N^* \rangle$ and the sign of the product $\langle \gamma N | N^* \rangle \cdot \text{sign}[\langle N^* | \pi N \rangle]$. The extraction of this knowledge from the data by means of partial wave analysis will be reviewed in Section 2.

Such knowledge obviously provides a considerable basis for comparison with any theory of hadron structure—the prime candidate for a theory, and the one which will be considered in this article, being the constituent quark model (Morpurgo, 1965; Dalitz, 1965; Greenberg, 1964). The correspondence of this theory with the observed baryon and meson spectrum (apart from one or two residual problems, of which the question of the existence of the A_1 meson is the best known) was well established many years ago (Dalitz, 1965, 1968) and remains valid, and some approximate agreement with electromagnetic decay rates such as $\omega \to \pi + \gamma$ (Becchi and Morpurgo, 1965b) and $\Delta(1236) \to N + \gamma$ [including the magnetic dipole selection rule (Becchi and Morpurgo, 1965a)] were also established. A constituent quark model also predicts the sign and magnitude of the products $\langle \gamma N | N^* \rangle \cdot \text{sign}[\langle N^* | \pi N \rangle]$, and in its simplest version these predictions are essentially without adjustable parameters and the predicted signs agree with all the well-determined partial wave analysis results for the resonances of the **56**, $L = 0^+$, **56**, $L = 2^+$, and **70**, $L = 1^-$ multiplets (Moorhouse and Oberlack, 1973). The significance of this agreement is that it holds without adjustable parameters, both for inter-multiplet and intramultiplet comparisons, and the process $\gamma + N \to \pi + N$ being "$SU(3)$ inelastic" [γ and π belonging to different $SU(3)$ multiplets], the agreement is not due to overall isospin symmetry of structureless hadrons, but depends on the quark charges, the nucleon quark wave functions and the assumed simple structure of the current interaction with quarks. These results could well be regarded as a validation of the constituent quark model for baryons. There remain quantitative disagreements in the magnitude of matrix elements (though, happily, theoretically large matrix elements are also large experimentally), which may be connected with faults in constituent quark models as at present formulated and with details of the effective current. Such questions and the detailed comparisons with the partial wave analysis results are discussed in Section 3. These discussions form a comparative basis for our assessment of the experimental and theoretical radiative meson transition results in Section 4.

2. Resonance Photocouplings through Partial Wave Analysis of Pion Photoproduction

Pion photoproduction, $\gamma + N \to \pi + N$, which in the resonance region is dominated (except in the extreme forward direction) by resonance formation, $\gamma + N \to N^* \to \pi + N$, gives a vital part of resonance systematics. The photon has two helicities, and isospin 0 or 1, so that it has two independent helicity couplings to isospin-$\frac{3}{2}$ resonances ($N^*_{3/2}$) and four independent couplings to isospin-$\frac{1}{2}$ resonances ($N^*_{1/2}$). Moreover, for a given resonance, N^*, any one of these two or four couplings $\langle \gamma N | N^* \rangle$ when multiplied by the unique coupling $\langle N^* | \pi N \rangle$ to give the product $\langle \gamma N | N^* \rangle \langle N^* | \pi N \rangle$ forms a pion photoproduction amplitude that can be determined (by partial wave analysis) relative to the Born approximation in *sign* as well as in *magnitude*:

This contrasts with the determination of πN couplings, $\langle N^* | \pi N \rangle$, from πN elastic scattering, where, for each of these resonances, we can only determine the magnitude of *one* number. However, the fact is that we do know the magnitude of $\langle \gamma N | N^* \rangle$ alone, as well as the sign of the product $\langle \gamma N | N^* \rangle \langle N^* | \pi N \rangle$.

Thus, from photoproduction we obtain knowledge of numbers associated with each resonance which can be a test of, or a guide to, theories of particle structure. None of the numbers are predicted by $SU(3)$ alone, since the photon belongs to a different $SU(3)$ multiplet from the pion, and the process $\gamma + N \rightarrow \pi + N$ is "$SU(3)$ inelastic." More particularly, $SU(3)$, even together with the F/D ratio and vector dominance evaluation of photon couplings (with quark model signs), only predicts the ratio of the two isospin couplings of any one isospin-$\frac{1}{2}$ resonance in any one helicity transition. Much more powerfully, a quark model will predict every number in magnitude and sign (the signs in the quark model also being defined with respect to the Born terms). These predictions may be compared with the numbers obtained from a partial wave analysis of the data.

The quark model theory can be contrasted with prescriptions for the matrix elements based on the algebra of $SU(6)_W$ [but not having $SU(6)_W$ symmetry, which gives completely incorrect results], such as those phenomenologies (Gilman *et al.*, 1973; Hey *et al.*, 1973; Gilman and Karliner, 1973, 1974; Cashmore *et al.*, 1975) inspired by the Melosh transformation (Melosh, 1974) and described by Hey in this volume. These latter phenomenologies have a number of arbitrary parameters, different for each supermultiplet such as **56**, $L = 0^+$, **70**, $L = 1^-$, etc., and so do not predict complete intermultiplet signs, nor are the predictions related to the Born approximation.

In this section we shall describe the various methods of partial wave analysis, including the degree of reliability of the results currently available. For reasons that will be discussed, the reliability of the numbers obtained can be very different, even for resonances of about the same energy.

But first, in order to make this article a self-contained reference work, we write down again the formalism for pion photoproduction. This is done in the Section 2.1. A full discussion of the general formalism for pion electroproduction (of which pion photoproduction is a special case) is given in Chapter 4 by Lyth.

2.1. The Formalism of Pion Photoproduction

2.1.1. The Expression of Invariant Amplitudes

The matrix element for any single-pion photoproduction process such as $\gamma p \rightarrow \pi^+ n$ may be written $\bar{u}(p_2) T u(p_1)$, $u(p_1)$ and $u(p_2)$ being, respectively, the initial and final free-nucleon Dirac spinors corresponding to nucleon four-momenta p_1 and p_2. The T matrix may be expressed (in the Pauli metric) as

$$T = [\tfrac{1}{2}i\gamma_5\gamma_\mu\gamma_\nu A_1(s, t) + 2i\gamma_5 P_\mu(q - \tfrac{1}{2}k)_\nu A_2(s, t) + \gamma_5\gamma_\mu q_\nu A_3(s, t)$$
$$+ \gamma_5\gamma_\mu(2P_\nu - iM\gamma_\nu)A_4(s, t)](\varepsilon_\mu k_\nu - \varepsilon_\nu k_\mu) \tag{2.1}$$

where q_μ, k_μ are the four-momenta of the pion and photon, respectively, $P_\mu = \tfrac{1}{2}(p_1 + p_2)_\mu$, ε_μ is the photon polarization vector, and the $A_i(s, t)$ ($i = 1, 2, 3, 4$) are invariant functions of the usual Mandelstam invariants s and t. The nucleon mass is M.

In Eq. (2.1) the form of the dependence on the photon polarization, $(\varepsilon_\mu k_\nu - \varepsilon_\nu k_\mu)$, follows from gauge invariance, which requires that the matrix element vanish when $\varepsilon_\mu = k_\nu$; the form of the expression in square brackets is uniquely determined by the requirements of Lorentz invariance and parity conservation of the transition matrix (Chew *et al.*, 1957). Of course, it is open to choose as independent invariants any linear combinations of A_1, A_2, A_3, A_4, which would then have to be multiplied by appropriately complicated combinations of γ matrices and four-vectors to form the correct T matrix; the particular form (2.1) was chosen so that the A_i ($i = 1, 2, 3, 4$) would have simple properties under crossing, given in Section 2.3 below, and a reasonably simple correspondence with the center-of-mass Pauli matrix form.

For the calculation of experimental quantities, it is usual and easier to work in the center-of-mass system and reduce the T matrix to form $\mathcal{F}$, where

$$\bar{u}(p_2)Tu(p_1) = \frac{4\pi E}{M}\chi_f^\dagger \mathcal{F}\chi_i \tag{2.2}$$

where χ_f, χ_i are the Pauli two-component spinors of the initial and final states quantized along the z axis. The factor $4\pi E/M$ is a conventional normalization with E being the center-of-mass energy. Using projection operators to express Dirac spinors in terms of Pauli spinors, one finds the standard expression

$$\mathcal{F} = i\boldsymbol{\sigma}\cdot\boldsymbol{\varepsilon}\mathcal{F}_1 + (\boldsymbol{\sigma}\cdot\mathbf{q})\boldsymbol{\sigma}\cdot(\mathbf{k}\times\boldsymbol{\varepsilon})\mathcal{F}_2$$
$$+ i(\boldsymbol{\sigma}\cdot\mathbf{k})(\mathbf{q}\cdot\boldsymbol{\varepsilon})\mathcal{F}_3 + i(\boldsymbol{\sigma}\cdot\mathbf{q})(\mathbf{q}\cdot\boldsymbol{\varepsilon})\mathcal{F}_4 \tag{2.3}$$

where $\mathbf{k}$, $\mathbf{q}$ are the center-of-mass three-momenta, and $\boldsymbol{\varepsilon}$ is the polarization of the photon in the radiation gauge. The normalization of the T matrix is such that the differential cross section from an initial nucleon spinor χ_i to a final nucleon spinor χ_f is

$$\frac{d\sigma}{d\Omega} = \frac{q}{k}|\langle\chi_f|\mathcal{F}|\chi_i\rangle|^2 \tag{2.4}$$

The relationship between the A_i and $\mathcal{F}_i$ is

$$\mathcal{F}_1 = \frac{E-M}{8\pi E}(D_1 D_2)^{1/2}\left[A_1 + (E-M)A_4 - \frac{k_0 q_0 - \mathbf{k}\cdot\mathbf{q}}{E-M}(A_3 - A_4)\right]$$

$$\mathcal{F}_2 = \frac{E-M}{8\pi E}\left(\frac{D_2}{D_1}\right)^{1/2}q\left[-A_1 + (E+M)A_4 + \frac{k_0 q_0 - \mathbf{k}\cdot\mathbf{q}}{E+M}(A_3 - A_4)\right]$$

$$\mathcal{F}_3 = \frac{E-M}{8\pi E}(D_1 D_2)^{1/2}q[(E-M)A_2 + A_3 - A_4]$$

$$\mathcal{F}_4 = \frac{E-M}{8\pi M}\left(\frac{D_2}{D_1}\right)^{1/2}q^2[-(E+M)A_2 + A_3 - A_4]$$

with

$$D_1 = (M^2 + \mathbf{k}^2)^{1/2} + M, \qquad D_2 = (M^2 + \mathbf{q}^2)^{1/2} + M \qquad (2.5)$$

From Eqs. (2.3) and (2.4) one can readily deduce the expressions for the experimental quantities in terms of the $\mathcal{F}_i$ as given, for example, by Chew *et al.* (1957). However, we prefer rather to work in terms of helicity amplitudes, which are given by writing in the center-of-mass system

$$A_{\mu\lambda}(\theta, \phi) = \frac{M}{4\pi E}\bar{u}(p_2, \lambda_2)T(\lambda_\gamma)u(p_1, \lambda_1) \qquad (2.6a)$$

$$= \chi_2^\dagger(\lambda_2)\mathcal{F}(\lambda_\gamma)\chi_1(\lambda_1) \qquad (2.6b)$$

with

$$\lambda = \lambda_\gamma - \lambda_1, \qquad \mu = -\lambda_2 \qquad (2.7)$$

where $u(p_i, \lambda_i)$ is a spinor representing a nucleon of four-momentum p_i, helicity λ_i; $T(\lambda_\gamma)$ is such that in Eq. (2.1) the photon has helicity λ_γ; and $\chi_i(\lambda_i)$, $\mathcal{F}(\lambda_\gamma)$ are the corresponding quantities in the two-component spinor expression. Equation (2.7) defines the initial-state helicity λ and the final-state helicity μ. In Eqs. (2.6) we have conventionally omitted functional dependence on energy of the helicity amplitudes.

There are four independent helicity amplitudes, the dependence of the other four being expressed through the relationship

$$A_{-\mu,-\lambda}(\theta, \phi) = -e^{i(\lambda-\mu)(\pi-2\phi)}A_{\mu\lambda}(\theta, \phi) \qquad (2.8)$$

We choose the four independent amplitudes to be those with $\lambda_\gamma = +1$, and by separating the phase factor $e^{i(\lambda-\mu)\phi}$ we define amplitudes $H_N(\theta)$, $H_{SP}(\theta)$, $H_{SA}(\theta)$, $H_D(\theta)$ (where, if "flip" means the change in units of total s-channel helicity between the initial and final states, H_N is no-flip, H_{SA}

and H_{SP} are single flip, and H_D is a double flip) through

$$H_N(\theta) \equiv A_{1/2,1/2}(\theta, \phi) = 2^{1/2} \cos \tfrac{1}{2}\theta [(\mathscr{F}_2 - \mathscr{F}_1) + \tfrac{1}{2}(1 - \cos \theta)(\mathscr{F}_3 - \mathscr{F}_4)]$$

$$(2.9a)$$

$$H_{SP}(\theta) \equiv e^{-i\phi} A_{1/2,3/2}(\theta, \phi) = -2^{-1/2} \sin \tfrac{1}{2}\theta \, [(1 + \cos \theta)(\mathscr{F}_3 + \mathscr{F}_4)] \quad (2.9b)$$

$$H_{SA}(\theta) \equiv e^{-i\phi} A_{-1/2,1/2}(\theta, \phi)$$

$$= 2^{1/2} \sin \tfrac{1}{2}\theta [(\mathscr{F}_1 + \mathscr{F}_2) + \tfrac{1}{2}(1 + \cos \theta)(\mathscr{F}_3 + \mathscr{F}_4)] \quad\quad\quad (2.9c)$$

$$H_D(\theta) \equiv e^{-2i\phi} A_{-1/2,3/2}(\theta, \phi)$$

$$= 2^{-1/2} \cos \tfrac{1}{2}\theta [(1 - \cos \theta)(\mathscr{F}_3 - \mathscr{F}_4)] \quad\quad\quad\quad (2.9d)$$

Experimental quantities can be expressed in terms of the H_N, H_{SA}, H_{SP}, H_D as described and given in the Appendix. For example, the differential cross section is

$$\frac{d\sigma}{d\Omega} = \frac{1}{2}\frac{q}{k}(|H_N|^2 + |H_D|^2 + |H_{SP}|^2 + |H_{SA}|^2) \quad\quad (2.10)$$

2.1.2. The Expression of Helicity Amplitudes in Terms of Partial Wave Amplitudes

The helicity amplitudes have the expansion

$$A_{\mu\lambda}(\theta, \phi) = \sum_j A^j_{\mu\lambda}(2j + 1)d^j_{\lambda\mu}(\theta)e^{i(\lambda - \mu)\phi} \quad\quad (2.11a)$$

where $A^j_{\mu\lambda}$ comprises four independent amplitudes of total angular momentum j, which can be combined into four independent partial wave amplitudes (proportional to $A^j_{\mu\lambda} \pm A^j_{-\mu\lambda}$) of good parity and total angular momentum j. These are defined by

$$C^{l+}_\lambda(W) = 2^{-1/2}\{A^j_{\lambda/2}(W) + A^j_{-\lambda/2}(W)\}$$

$$\quad\quad (2.12)$$

$$C^{(l+1)-}_\lambda(W) = 2^{-1/2}\{A^j_{\lambda/2}(W) - A^j_{-\lambda/2}(W)\}$$

where $\lambda = \tfrac{1}{2}, \tfrac{3}{2}$; the superscripts $l \pm$ refer to the two states with pion orbital angular momentum l and total angular momentum $j = l \pm \tfrac{1}{2}$. Equation (2.11a) can be rewritten in terms of the amplitudes H_N, H_{SP}, H_{SA}, H_D defined by Eq. (2.9) on the left-hand side, and the amplitudes $C_{1/2}$, $C_{3/2}$

instead of $A_{\mu\lambda}^i$ on the right-hand side:

$$H_N(\theta) = 2^{1/2} \cos \tfrac{1}{2}\theta \sum_{l=0}^{\infty} (C_{1/2}^{l+} + C_{1/2}^{(l+1)-})(P_{l+1}' - P_l')$$

$$H_{SA}(\theta) = 2^{1/2} \sin \tfrac{1}{2}\theta \sum_{l=0}^{\infty} (C_{1/2}^{(l+1)-} - C_{1/2}^{l+})(P_{l+1}' + P_l')$$

$$\text{(2.11b)}$$

$$H_{SP}(\theta) = \left[\frac{2}{l(l+1)}\right]^{1/2} \sin \tfrac{1}{2}\theta(1 + \cos\theta) \sum_{l=1}^{\infty} (C_{3/2}^{(l+1)-} + C_{3/2}^{l+})(P_l'' - P_{l+1}'')$$

$$H_P(\theta) = \left[\frac{2}{l(l+1)}\right]^{1/2} \cos \tfrac{1}{2}\theta(1 - \cos\theta) \sum_{l=1}^{\infty} (C_{3/2}^{l+} - C_{3/2}^{(l+1)-})(P_l'' + P_{l+1}'')$$

2.1.3. Charge and Isospin Amplitudes

The formalism of Sections 2.1.1 and 2.1.2 applies to amplitudes for any of the processes $\gamma p \to \pi^+ n$, $\gamma n \to \pi^- p$, $\gamma p \to \pi^0 p$, and $\gamma n \to \pi^0 n$. For none of these amplitudes is isospin a good quantum number, but the πN interaction conserves isospin and πN resonances are in states of definite isospin. Consequently, it is necessary to distinguish the isospin properties of the amplitudes. If we consider the amplitude A_i as referring to the emission of a pion of isospin index a, we obtain the well-known formula

$$A_i = A_i^{(+)}\delta_{a3} + A_i^{(-)}\tfrac{1}{2}[\tau_a, \tau_3] + A_i^{(0)}\tau_a \tag{2.13}$$

Equation (2.13) only assumes that the photon interaction with hadrons occurs through isoscalar and isovector parts, so that $A^{(0)}$ is an isoscalar amplitude and $A^{(+)}$, $A^{(-)}$ are isovector amplitudes. The evidence for this assumption is discussed by Donnachie and Shaw in Chapter 3 of this volume. By taking pion–nucleon final states of definite isospin, we find that the combinations $A_i^{(+)} + 2A_i^{(-)}$, $A_i^{(+)} - A_i^{(-)}$ lead to isospins in the final state of value $\tfrac{1}{2}, \tfrac{3}{2}$, respectively, and we define

$$A^S = -3^{1/2}A^{(0)} \qquad \text{(isoscalar amplitude)}$$

$$A^{V1} = (\tfrac{1}{3})^{1/2}(A_i^{(+)} + 2A_i^{(-)}) \qquad \begin{array}{l}\text{(isovector amplitude}\\ \text{leading to isospin } \tfrac{1}{2} \text{ in } \pi N)\end{array}$$

$$A^{V3} = (\tfrac{2}{3})^{1/2}(A_i^{(+)} - A_i^{(-)}) \qquad \begin{array}{l}\text{(isovector amplitude leading}\\ \text{to isospin } \tfrac{3}{2} \text{ in } \pi N)\end{array}$$

The expression for the transitions $\gamma p \to \pi^+ n$, $\gamma n \to \pi^- p$, $\gamma p \to \pi^0 p$ and $\gamma n \to \pi^0 n$ in terms of the isoscalar and isovector amplitudes, Eq. (2.14) are, from Eq. (2.13),

$$A_{i+} \equiv \langle \pi^+ n | A_i | \gamma p \rangle = -(\tfrac{1}{3})^{1/2} A^{V3} + (\tfrac{2}{3})^{1/2} (A^{V1} - A^S)$$
$$A_{i0} \equiv \langle \pi^0 p | A_i | \gamma p \rangle = (\tfrac{2}{3})^{1/2} A^{V3} + (\tfrac{1}{3})^{1/2} (A^{V1} - A^S)$$
$$A_{i-} \equiv \langle \pi^- p | A_i | \gamma n \rangle = (\tfrac{1}{3})^{1/2} A^{V3} - (\tfrac{2}{3})^{1/2} (A^{V1} + A^S) \qquad (2.15a)$$
$$A_{in0} \equiv \langle \pi^0 n | A_i | \gamma n \rangle = (\tfrac{2}{3})^{1/2} A^{V3} + (\tfrac{1}{3})^{1/2} (A^{V1} + A^S)$$

Exactly corresponding linear relations to Eq. (2.15) hold between helicity amplitudes and partial wave amplitudes with corresponding isospin properties, and in what follows we will use the indices of the left- or right-hand sides of Eq. (2.15) for any amplitudes without further explanation. [We have no further use for the amplitudes $A_i^{(+)}$, $A_i^{(-)}$, and $A_i^{(0)}$ of Eq. (2.13), so no confusion will arise.]

Since the combination $(A^{V1} \mp A^S)$ gives the coupling of photons to positive and neutral isospin-$\tfrac{1}{2}$ states, respectively, we define explicitly

$$A^p \equiv -(A^{V1} - A^S)$$
$$A^n \equiv +(A^{V1} + A^S) \qquad (2.15b)$$
$$A^\Delta = A^{V3}$$

2.2. Isobar Model Analysis

2.2.1. Resonance Dominance of Amplitudes

Having set out the basic formalism we can now begin the main business of Section 2, which is to recount the methods of partial wave analysis used in pion photoproduction. Since even without benefit of partial wave analysis, but just in looking at the cross sections, the principal resonances of pion–nucleon scattering are evident, it is natural to proceed using an *isobar model*. By an *isobar model* we mean that the principal component of the amplitude is assumed to be a sum over Breit–Wigner resonance amplitudes:

The resonance and their total widths are known beforehand from pion–nucleon scattering analysis, so that the advantage of this method is that the nature of the energy dependence of the resonance part of the amplitude is known. Of course, the more important the resonance part of the amplitude the more effective the method is, and it fortunately turns out that, apart from the known Born terms (as discussed below), the resonances are much the greater part of the photoproduction amplitude. It is for this reason that the *isobar model* is so powerful in partial wave analysis of pion photoproduction; the fixed-t dispersion relation method, to be described in the next major subsection, also makes full use of this *resonance dominance* property of low-energy pion photoproduction, and can be regarded in practice as a sophisticated version of the isobar model.

In the isobar model analysis (Gourdin and Salin, 1963; Salin, 1963; Chau *et al.*, 1967; Walker, 1969) the amplitudes are parameterized as functions of energy, the parameterized cross sections and polarizations, say $[d\sigma(\theta, W)/d\Omega]_{par}$, are deduced therefrom as functions of energy and angle, and the χ^2 sum is formed:

$$\chi^2 = \sum_{\theta, W, \sigma} \frac{|[d\sigma(\theta, W)/d\Omega]_{par} - [d\sigma(\theta, W)/d\Omega]_{exp}|^2}{\Delta^2_{exp}(\theta, W)} \tag{2.16}$$

where Δ_{exp} is the experimental error and the sum runs over all experimental quantities σ and all energies W and angles θ where experimental quantities are available. The parameters are then varied so as to minimize χ^2, and the values of the parameters at that minimum are taken as the best values for the model and, when substituted in the expression for the amplitudes, give the model amplitudes that best fit the experiments. These amplitudes, and consequent cross sections and polarizations, also give the predictions of the fit at places where experiments are not available. The process involves fitting experiments over a large range of energies *simultaneously*, and this is the way in which the assumed Breit–Wigner energy dependence is used.

Thus, in the isobar model one writes the partial wave amplitudes of Eqs. (2.12) and 2.11b) as

$$C_\lambda^{l\pm}(W) = \sum_{N^*} \varepsilon \left[\frac{\Gamma^\lambda(N^* \to \gamma N)\Gamma(N^* \to \pi N)}{kq} \right]^{1/2} \frac{W}{W^2 - M^{*2} - iW\Gamma}$$

$$+ \text{background} \tag{2.17}$$

where $\varepsilon = \pm 1$ is the sign of the amplitude, M^* the resonance energy, k the photon, and q the pion center-of-mass momenta. The resonances N^*, of course, have spin $j = l \pm \frac{1}{2}$.

2.2.2. Resonance Couplings

In the following discussion of resonance couplings we use the tilde notation $\tilde{X}(M^*)$ for the imaginary part of the *resonance* contribution to the amplitude $X(W)$ evaluated at resonance; thus, from Eq. (2.17),

$$\tilde{C}_\lambda^{l\pm}(M^*) = \varepsilon \left[\frac{\Gamma_\gamma^\lambda \Gamma_\pi}{k(M^*)q(M^*)\Gamma^2} \right]^{1/2} \tag{2.18}$$

For shortness we drop the M^*, only inserting it when there is danger of ambiguity. We introduce helicity amplitudes A_λ^{jP} (Moorhouse and Oberlack, 1973) for the decay $N^*(jP) \rightarrow (\gamma N)_\lambda$ (where j, P labels the spin and parity of the N^*) in terms of which the radiative width Γ_γ^λ is given by (Copley *et al.*, 1969a, 1969b)

$$\Gamma_\gamma^\lambda = \frac{k^2}{\pi} \frac{M}{M^*} \frac{1}{2j+1} (\tilde{A}_\lambda^{jP})^2 \tag{2.19}$$

Putting this expression into Eq. (2.18) we find

$$\tilde{C}_\lambda^{l\pm} = \left[\frac{1}{(2j+1)\pi} \frac{k}{q} \frac{M}{M^*} \frac{\Gamma_\pi}{\Gamma^2} \right]^{1/2} \tilde{A}_\lambda^{jP} \tag{2.20}$$

The total radiative width of the resonance, Γ_γ, is given by

$$\Gamma_\gamma = \sum_{\lambda=-3/2}^{3/2} \Gamma_\gamma^\lambda = \frac{k^2}{\pi} \frac{M}{M^*} \frac{2}{2j+1} [(\tilde{A}_{1/2}^{jP})^2 + (\tilde{A}_{3/2}^{jP})^2] \tag{2.21}$$

We will often quote results of partial wave analyses as the signed number A_λ^{jP} in units of $GeV^{-1/2}$ (or, in order to have an integer number, $GeV^{-1/2} \times 10^{-3}$); this is a standard measure of the $N^* \rightarrow N\gamma$ coupling, but note that by Eq. (2.20) it also incorporates the sign (only) of the $N^* \rightarrow \pi N$ coupling.

We should make quite clear what the above amplitudes mean as to charge and isospin. In Eq. (2.17) the amplitude C is the amplitude for the phototransition of a nucleon in a given charge state (proton or neutron) into a pion–nucleon state (πN) of corresponding charge and good total isospin, $I = \frac{1}{2}$ or $I = \frac{3}{2}$. So $\Gamma(N^* \rightarrow \pi N) \equiv \Gamma_\pi$ is the usually quoted partial width of N^* for decay into a (πN) state of good total isospin, such as $(\frac{1}{3})^{1/2}|\pi^0 p\rangle - (\frac{2}{3})^{1/2}|\pi^+ n\rangle$. If we like, we can add to $C_\lambda^{l\pm}$ (or to $\tilde{A}_\lambda^{jP}$) a further superscript to denote which charge-isospin amplitude it is, thus: $C_\lambda^{l\pm,p}$, $C_\lambda^{l\pm,n}$, $C_\lambda^{l\pm,\Delta}$. So far as charge and isospin are concerned, these latter correspond to the amplitudes of Eq. (2.15b) and the $C_\lambda^{l\pm}$ amplitudes corresponding to the actual physically observed charge states are obtained by substitution into Eq. (2.15a); thus, for example,

$$C_\lambda^{l\pm}(\gamma p \rightarrow \pi^+ n) = -(\tfrac{1}{3})^{1/2} C_\lambda^{l\pm,\Delta} - (\tfrac{2}{3})^{1/2} C_\lambda^{l\pm,p}$$

2.2.3. Born Terms in the Amplitudes

In pion photoproduction, terms of first order in the electromagnetic coupling constant, e, and the pion–nucleon coupling constant, g, are known as Born terms (since they are the only terms that would occur in the Born approximation to pion photoproduction). The terms can be pictured by their Feynman diagrams (Fig. 1). One of them (Fig. 1a), the one-pion exchange for charged pion photoproduction, is dominant for high partial waves—that is, those waves which at a given energy W have large enough angular momentum l [$l > L(W)$, say] to be small and resonant free; the one-pion exchange dominates what is usually called the peripheral interaction. This term is not gauge invariant by itself, but it becomes so on the addition of the nucleon exchange terms (Figs. 1b and 1c), where the e at the electromagnetic vertex signifies that in these diagrams the nucleon is taken to interact electromagnetically as a pure Dirac particle without anomalous magnetic moments. (The anomalous magnetic moment Born terms are discussed below.) These Feynman diagrams picture the minimal set of Born terms which both include the one-pion exchange term and are gauge invariant; we will follow a standard nomenclature and call them the *electric Born terms*. They give rise to the forward peak, at angles such that $|t| < m_\pi^2$, in charged pion photoproduction for center-of-mass energy $W >$ 1400 MeV, and thus it is necessary to add either the whole or part of these terms to the amplitude. It is simplest to add the whole of the electric Born terms to the amplitudes, in addition to the resonances, as in the work of Walker (1969, 1970) and Metcalf and Walker (1974). However, this raises a question concerning the lower partial waves which contain resonances—namely, has this procedure distorted the resonance parameters which we wish to find by already including some of their real part through the Born

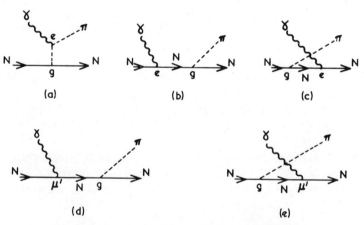

Fig. 1. The Born terms for pion photoproduction.

approximation? [The same question would be raised in a different form, if we only included the higher partial wave projection of the electric Born terms (Chau *et al.*, 1967)]. We shall see that all such questions are neatly resolved by the fixed-*t* dispersion relation method described in Section 2.3.

In addition to the *electric Born terms* (Figs. 1a–1c), there are also *magnetic Born terms*, in which the photon interacts with the nucleon by the anomalous nucleon magnetic moment, μ' (Figs. 1d and 1e). These are usually omitted from isobar model analysis, since they do not contribute to high partial waves, and are assumed to be subsumed in the resonance and background terms.

The importance of the Born terms for the partial wave analysis in the isobar model is that they are the only significant contributors to the high partial waves in charged pion photoproduction and as such form a known background, interfering with the resonances in the expression for the cross section and thus helping to fix the resonance parameters. The Born terms are real and the signs of the resonance couplings are determined by the interference. We shall see that the Born terms are even more important in the fixed-*t* dispersion relation analysis and, in Section 2.3, we give the full expression for the Born term contribution to the invariant amplitudes. Here we write down just the contribution of the electric Born terms (Figs. 1a–1c) to the "Pauli" amplitudes $\mathscr{F}_1$, $\mathscr{F}_2$, $\mathscr{F}_3$, $\mathscr{F}_4$ of Eqs. (2.3) and (2.5):

$$\mathscr{F}_1 = -\frac{eG_\pi}{4\pi}\frac{D_1 D_2}{W+M}\left(\frac{\varepsilon_i}{2W}+\varepsilon_f\frac{k}{u-M^2}\right)$$

$$\mathscr{F}_2 = \frac{eG_\pi}{4\pi}\frac{D_1}{D_2}\frac{q}{W+M}\left(\frac{\varepsilon_i}{2W}+\varepsilon_f\frac{k}{u-M^2}\right)$$

$$\mathscr{F}_3 = \frac{eG_\pi}{4\pi}\frac{D_2}{D_1}\frac{qk}{W}\left(\frac{\varepsilon_\pi}{t-m^2}-\varepsilon_f\frac{1}{u-M^2}\right)$$

$$\mathscr{F}_4 = -\frac{eG_\pi}{4\pi}\frac{D_1}{D_2}\frac{q^2}{W}\left(\frac{\varepsilon_\pi}{t-m^2}-\varepsilon_f\frac{1}{u-m^2}\right)$$

(2.22)

The amplitudes here are for definite charged states of the pion and nucleon (such as $\gamma p \to \pi^+ n$) and the G_π, ε_π, ε_i, ε_f are according to those charges: ε_π, ε_i, ε_f are the charges in units of e of the pion, initial nucleon, and final nucleon, respectively (for example, $\varepsilon_\pi = -1$ for π^-) and $G_\pi = 2^{1/2}g$, $-2^{1/2}g$, $-g$, g ($g^2/4\pi \sim 14$) according as the πNN vertex is ($\pi^+ pn$), ($\pi^- pn$), ($\pi^0 pn$), or ($\pi^0 nn$), respectively. The other quantities are as defined in Section 2.1.1; u is the third Mandelstam invariant quantity $u = -(k - p_f)_\mu (k - p_f)^\mu$, $p_{f\mu}$ being the final nucleon four-momentum and k_μ (here) the photon four-momentum.

2.2.4. Background in the Amplitudes

As explained in Section 2.2.1, the isobar model assumes that resonances have a major role in the amplitudes and the model is even better if this role turns out to be dominant, as it does in photoproduction. But in addition to the Born term there is always some background in the amplitudes, for example, that coming from the tail of u-channel resonances, just as in the Born terms (2.22) there are contributions from the u-channel nucleon mass pole, $1/(u - M^2)$. This background may be small, but an accurate analysis demands its inclusion. It may be parameterized as a continuous function of energy, or added pinch by pinch at each experimental energy to improve the fit (Walker, 1969), some regard being paid to energy continuity.

It turns out that a good fit can be obtained using purely real background; this points towards the resonances nearly saturating the *imaginary* part of the amplitudes. This remark leads directly to the more sophisticated method of analysis described in the next section.

2.3. Fixed-t Dispersion Relation Analysis

2.3.1. Resonance Dominance of the Imaginary Part of the Amplitudes

The fixed-t dispersion relations for the invariant amplitudes of Eq. (2.15a) A_{i+}, A_{i-}, and A_{i0}, corresponding, respectively, to $\gamma p \to \pi^+ n$, $\gamma n \to \pi^- p$, and $\gamma p \to \pi^0 p$, can be written

$$\operatorname{Re} A_{i\pm,0}(s, t) = B_{i\pm,0}(s, t) + \int_{(M+m)^2}^{\infty} ds' \left[\frac{\operatorname{Im} A_{i\pm,0}(s', t)}{s' - s} + \xi_i \frac{\operatorname{Im} A_{i\pm,0}(s', t)}{s' - u} \right]$$

where the Born terms are given by

$$(2.23)$$

$$B_{1+}(s, t) = 2^{1/2} GS, \qquad B_{2+}(s, t) = -2^{1/2} GST$$

$$B_{3+}(s, t) = -\frac{2^{1/2} G}{2M} (\mu_p' S - \mu_n' U)$$

$$B_{4+}(s, t) = -\frac{2^{1/2} G}{2M} (\mu_p' S + \mu_n' U)$$

$$B_{i-}(s, t) = \xi_i B_{i+}(u, t) \qquad (i = 1, 2, 3, 4) \qquad (2.24)$$

$$B_{10}(s, t) = \tfrac{1}{2} G(S + U), \qquad B_{20}(s, t) = -\tfrac{1}{2} G(S + U)T$$

$$B_{30}(s, t) = -\tfrac{1}{2} G \frac{\mu_p' - \mu_n'}{2M} (S - U)$$

$$B_{40}(s, t) = -\tfrac{1}{2} G \frac{\mu_p' + \mu_n'}{2M} (S + U)$$

Here $G = ge/4\pi$, $S = 1/(s - M^2)$, $U = 1/(u - M^2)$, $T = 1/(t - m^2)$, and μ'_p, μ'_n are the proton and neutron anomalous magnetic moments, respectively. We might take $g^2/4\pi = 14.7$, $e^2/4\pi = 1/137$, $\mu'_p = 1.793$, $\mu'_n = -1.913$. In Eqs. (2.23) and (2.24) $\xi_i = +1$ if $i = 1, 2, 4$ and $\xi_i = -1$ if $i = 3$.

The crossing relations

$$A_{i,\pm}(s, t) = \xi_i A_i(u, t), \qquad A_{i,0}(s, t) = A_{i,0}(u, t) \qquad (2.25)$$

give rise to the form of the last term in brackets in the integrand in Eq. (2.23). These relations follow (Chew *et al.*, 1957) from the definition of the invariant amplitudes in Eq. (2.1).

In the analysis of pion photoproduction we see from Eq. (2.23) that if we have an ansatz, in terms of certain parameters to be determined by the fitting process, for the *imaginary part* of the amplitudes only, then this is sufficient to give the real part of the amplitudes in terms of the same parameters and to use the resulting whole amplitude to fit to experiment and to determine the parameters. Thus, the fixed-t dispersion relation transfers the whole problem of parameterization to the imaginary part only. One of the consequences in analysis of the resonance region is that one also requires, in principle, a form of the imaginary part for energies above the resonance region; this is not necessarily a disadvantage, as will be discussed below. We attend for the moment to the resonance region only, where we have seen that the Born terms and the real tail of u-channel resonances (which, of course, is more extensive than the imaginary tail) necessarily provides a considerable background to the real part of the amplitudes. There is no such necessary contribution to the imaginary parts and these *can* be resonance saturated or nearly so. The evidence from isobar model analysis that this be the case was one of the principle motivations for the introduction of the use of fixed-t dispersion relations into photoproduction analysis (Moorhouse and Oberlack, 1973; Moorhouse *et al.*, 1974; Devenish *et al.*, 1972, 1973; Devenish *et al.*, 1971). We should emphasize that it is by no means necessary to the success of the method that the imaginary part be nearly saturated by resonances; resonance dominance, which means that a large part of the energy variation is specified, is quite sufficient for success. *In fact, it turns out that the fixed-t dispersion relations analysis agrees with the isobar model analysis in finding that the imaginary parts of the amplitudes in the resonance region are nearly saturated by resonances* (Moorhouse *et al.*, 1974).

We may summarize the power of the fixed-t dispersion analysis as follows:

(i) The imaginary part of the amplitude is resonance dominated in the resonance region.

(ii) The masses and widths of the majority of the most important resonances in the resonance region are already known and fixed from

$\pi N \to \pi N$ partial wave analysis and so largely predetermine the form of the imaginary part with the $N^*N\gamma$ coupling constants as the only free parameters.

(iii) It connects the amplitudes for $\gamma p \to \pi^+ n$ and $\gamma n \to \pi^- p$ relations.

2.3.2. Born Term Problem Solved

In the isobar model we saw a problem of distinguishing between the real part of resonances (and background) and the Born term. As the Born term is purely real, no such problem arises with fixed-t dispersion relation analysis; Re $A_{i\pm,0}(s, t)$ is given by Eq. (2.23) once the imaginary part of the amplitude has been specified. The question in the isobar model is how much of the Born term is built up from the real parts of Breit–Wigners (+ background) *or* vice versa; in the fixed-t dispersion relation the real part due to the Born term and that due to resonances are separately and uniquely specified by analyticity.

2.3.3 Construction of the Imaginary Parts in the Resonance Region

The amplitudes Im $A_{i\pm,0}(s', t)$ for s' in the resonance region are to be constructed as the sum of a partial wave series, cut off at some upper limit of angular momentum. As an illustration we give the simplest possible ansatz:

Take the expression (2.17) for the partial wave amplitudes and assume that the background contains no imaginary part; then

$$\text{Im } C_\lambda^{l\pm}(W) = \sum_{N^*} \varepsilon \left[\frac{\Gamma_\gamma^\lambda \Gamma_\pi}{kq}\right]^{1/2} \frac{W^2\Gamma}{(W^2 - M^{*2})^2 + W^2\Gamma^2}$$

$$= \sum_{N^*} \tilde{C}_\lambda^{l\pm}(M^*) \left[\frac{k(M^*)q(M^*)}{kq}\right]^{1/2} \frac{W^2\Gamma^2}{(W^2 - M^{*2})^2 + W^2\Gamma^2} \quad (2.26)$$

This expresses the imaginary part of the partial wave amplitudes for s' in the resonance region in terms of known resonance masses and widths with $\tilde{C}_\lambda^{l\pm}(M^*)$ as the only unknown parameters. To find the invariant amplitudes Im $A_{i\pm,0}(s', t)$, insert (2.26) in (2.11b) and invert Eqs. (2.9) and (2.5).

This simple ansatz can be used with success as in the work of Devenish *et al.* (1973). Moorhouse and Oberlack (1973), Moorhouse *et al.* (1974) and Knies *et al.* (1974) used a more complicated K-matrix fit to the πN amplitudes, giving some background as well as resonance in the imaginary part. Probably a small proportion, generally perhaps up to 10%, imaginary background to the resonance is necessary for the best possible fit.

2.3.4. Watson Theorem

One region of energy where the imaginary background, though small, represents nearly all the imaginary part, is in s-waves and p-waves from threshold up to the lower tail of the first resonance in those waves. The real parts of these waves are bigger, unitarity constraining them to satisfy the Watson theorem up to the two-pion threshold, so that in terms of the partial wave amplitudes of Eq. (2.12)

$$\text{Im } C_\lambda^{l\pm}(W) = \tan \delta^{l\pm} \cdot \text{Re } C_\lambda^{l\pm}(W) \qquad (2.27)$$

where $\delta^{l\pm}$ is the phase of the corresponding elastic πN scattering amplitude. These are small, $<20^0$ in the energy regions under consideration. This relation should be put in as a constraint, and Moorhouse and Oberlack (1973), Moorhouse *et al.* (1974), and Knies *et al.* (1974) do this, though probably from faults in the parameterization they are not able to satisfy the constraints very well. Devenish *et al.* (1973) apparently have a zero imaginary part in this region (or one just consisting of the extreme resonance tail). Since the imaginary parts are small, the results on the fit to experiment are not disastrous.

2.3.5. A Problem of Principle Not Solved

We have formed Im $A_{i\pm,0}(s', t)$ for s' in the resonance region, in terms of parameters, as the sum of a partial wave series, cut off at some upper limit of angular momentum. However, the convergence of the partial wave series is not proved, except for certain processes within certain regions (ellipses of convergence in the $\cos \theta$ plane, Lehmann, 1954). Nevertheless, convergence can hold outside this region, and it is possible that a cutoff series provides a good approximation in a considerably extended region. Devenish *et al.* (1972) have argued on the basis of the Mandelstam double-spectral representation that the cutoff series for Im $A_i(s', t)$ is good for $-t \le 1.0$ $(\text{GeV}/c)^2$ in $\pi^\pm$ photoproduction and $-t \le 1.5$ $(\text{GeV}/c)^2$ in π^0 photoproduction. Some authors (Devenish *et al.*, 1973, 1974; Crawford, 1975) have been quite cautious by limiting themselves to values of $-t < 1.0$ $(\text{GeV}/c)^2$, while others (Moorhouse and Oberlack, 1973; Moorhouse *et al.*, 1974; Knies *et al.*, 1974) have been more adventurous, proceeding in some cases to values of $-t$ as large as 2.5 $(\text{GeV}/c)^2$, with the superficial precaution of additional parameters to hopefully take care of any nonconvergence (Moorhouse *et al.*, 1974). It is resonances of high mass in which such differences of approach are likely to show themselves in differences of the coupling constants found, because it is for high values of $s' = M^{*2}$ that $-t$ is large in the backward direction, and we will comment later (Section 2.4) on the case of the f_{37} (1950).

This is obviously a critical point of fixed-t dispersion relation partial wave analysis and deserves further study through techniques of analytic continuation.

2.3.6. The Imaginary Parts at Higher Energies

We can divide the integral in the dispersion relation into two parts:

$$\text{Re } A_{i\pm,0}(s, t) = B_{i\pm,0}(s, t) + \int_{(M+m)^2}^{\Lambda^2} ds' \left[\frac{\text{Im } A_{i\pm,0}(s', t)}{s' - s} + \xi_i \frac{\text{Im } A_{i\pm,0}(s', t)}{s' - u} \right]$$

$$+ \int_{\Lambda^2}^{\infty} ds' \left[\frac{\text{Im } A_{i\pm,0}(s', t)}{s' - s} + \xi_i \frac{\text{Im } A_{i\pm,0}(s', t)}{s' - u} \right] \qquad (2.28)$$

where the interval $(M + m)^2 \leqslant s' < \Lambda^2$ is the resonance region with which we have dealt in Section 2.3.3.

The parameterization of $\text{Im } A_{i\pm,0}(s', t)$ for $s' > \Lambda^2$ depends on the data range fitted, and one can distinguish two cases:

(i) *Data Only Fitted in the Resonance Region.* In this case the form of $\text{Im } A(s', t)$ for $s' > \Lambda^2$ should be guided by its expected form, but by no means need be totally realistic; the only requirement it need fulfill is that the last integral $(s' > \Lambda^2)$ of Eq. (2.27) effectively simulates the contribution of the integral $(s' > \Lambda^2)$ over the true amplitude to $\text{Re } A(s, t)$ for s in the resonance region, $s < \Lambda^2$. This is the approach adopted by Moorhouse and Oberlack (1973), Moorhouse *et al.* (1974), Knies *et al.* (1974) and Devenish *et al.* (1973), who simulate the contribution of $\text{Im } A(s', t)$ for $s' > \Lambda^2$, by resonance or pseudoresonance poles in $\text{Im } A(s', t)$, $s' > \Lambda^2$.

(ii) *Data Fitted Above the Resonance Region.* More recently, analyses have been made that fit data simultaneously in both the resonance region and the higher-energy region. This demands a realistic representation of the imaginary part of the amplitudes in the higher-energy region, with the energy dependence being guided by Regge theory. [Fits to the higher-energy data *only* using fixed-t dispersion relations and a Regge parameterization of the high-energy imaginary part and known resonance couplings in the low-energy imaginary part previously existed (Barbour and Moorhouse, 1974; Argyres *et al.*, 1973; Hontebeyrie *et al.*, 1973).]

So far as fitting the resonance region is concerned, the idea of the more complete analyses is to get a better representation of $\text{Im } A(s', t)$ for $s' > \Lambda^2$ (by including the high-energy data) and thus eliminating any distortion of $\text{Im } A(s', t)$ for $s' < \Lambda^2$ which might be induced in method (i) by $\text{Im } A(s', t)$ for $s' < \Lambda^2$ having itself to simulate the role of the higher-energy imaginary parts. [Of course, a corresponding danger in method (ii) is that the resonance region amplitudes will distort themselves in an effort to help fit the high-energy data; so it is important to use a correct high-energy ansatz.]

Devenish *et al.* (1974) used the following high-energy parameterization

$$\text{Im } A(s, t) = \frac{Z^7}{1 + Z^7} [(a + bt)e^{ct} + (a' + b't)e^{c't}]Z^{\alpha(t)-1} \tag{2.29}$$

where $Z = (s - s_0)/(s_1 - s_0)$, s_0 being the pion threshold and s_1 being $(2 \text{ GeV})^2$ with

$$\alpha(t) = 0.5 + 0.8t \qquad \text{for } A_1, A_2, \text{ and } A_4 \tag{2.30a}$$

$$\alpha(t) = -0.5 + 0.8t \qquad \text{for } A_3 \tag{2.30b}$$

the amplitudes A_1, A_2, A_4 corresponding to t-channel natural parity exchange (ρ, ω, A_2 trajectories) and A_3 to unnatural parity exchange (π, B trajectories). The expression (2.29) is *added* to the resonance region contribution $\text{Im } A^{\text{res}}(s, t)$ of Section 2.3.3 to make the total imaginary amplitude

$$\text{Im } A(s, t) = \text{Im } A^{\text{res}}(s, t) + \text{Im } A^{\text{Regge}}(s, t) \tag{2.31}$$

$\text{Im } A^{\text{Regge}}(s, t)$ of Eq. (2.28) is of such a form that it falls rapidly to zero below 2 GeV c.m. energy, and $\text{Im } A^{\text{res}}(s, t)$ contains no contribution for $s > (2 \text{ GeV})^2$. In Eq. (2.28) the variable parameters are a, b, c, a', b', c' (c, c' being constrained to be the same for all A_i).

A similar analysis with different high-energy parameterization has been carried out by Barbour and Crawford (1976).

Some of the results of the work described in this whole section (Section 2.3) will be given in Section 2.4 below and discussed in relation to theory in Section 3.

2.4. Discrete Energy Analysis

2.4.1. Difficulties with Energywise Smoothness

In discrete energy analysis the experimental data at each discrete energy (where there is a considerable amount available) is fitted independently by varying the partial wave amplitudes expected to be important at that particular energy. Thus, it may seem that the analyst finds at each energy a set of partial wave amplitudes, each set of which has been obtained independently of the other. This seems at first sight a very objective and assumption-free way of proceeding. However, the analyst is faced with difficulties. The quality of the data is generally such that ambiguities exist and more than one set of partial waves is compatible with the data. The choice between these sets can only be made using analyticity or smoothness, and the application of this becomes a matter of art, in which sometimes subjective and unstated criteria are used.

However, in the case of photoproduction, especially, the situation can be alleviated. We know that the (gauge invariant) pion pole dominates the individually relatively small higher partial waves, and that the *sum* of those partial waves is important. One can (and must) specify those higher waves theoretically, for example, by the electric Born terms discussed in Section 2.2. Within the assumption this immediately eliminates *ambiguities of principle* (Bowcock and Burkhardt, 1975) that would otherwise occur (because these sort of ambiguous solutions are generated by transferring parts of lower partial waves into higher partial waves and vice versa) and can lessen, as discussed below, the ambiguities due to inadequate data.

This compromise with theory is necessary for the success of a discrete energy analysis even in a situation of very good data. The much stronger, but (as we have discussed) largely reasonable assumptions of the isobar or fixed-t dispersion relation methods enable them to operate in an energy region where data is sparser than that required for discrete energy analysis, and insofar as their energy dependence assumptions are correct these two former methods (especially the fixed-t method) can ride smoothly over patches of bad data.

However, the discrete energy analysis has some advantages and will hopefully be pursued further as data accumulate. At present, analysis has reached to ~ 1600 MeV center-of-mass energy for photoproduction from protons only.

2.4.2. Advantages of the Method

Among the advantages are the following:

(i) Resonances are "discovered" in the photoproduction analysis rather than assumed *a priori*. The main significance of this is that a separate determination of the resonance masses and widths is made in the photoproduction process (though it should be borne in mind that some of the fixed-t dispersion relation analyses vary the masses and widths of the resonances). The isobar and dispersion analyses *might* possibly be biased to finding too large photocouplings of weakly photocoupled resonances.

(ii) The fixed-t dispersion relation difficulty with the summation of the partial wave series for large t is avoided.

(iii) It is easy to impose the unitarity constraint (Watson's theorem) at low energies.

2.4.3. Multipole Amplitudes

Before describing in somewhat more detail the presently most advanced discrete energy analysis (Berends and Donnachie, 1975*b*) we write down here the electric and magnetic multipole amplitudes for $\gamma N \to \pi N$ in terms of the helicity amplitudes defined in Section 2.1, Eq. (2.12). These amplitudes are eigenamplitudes of photon angular momentum (and also, of course, like the helicity amplitudes, eigenamplitudes of πN orbital angular momentum). They are much used in the low-energy region, which is featured prominently in discrete energy analysis. We give the Chew–Low–Goldberger–Nambu (Chew *et al.*, 1957) version of the electric and magnetic multipole amplitudes:

$$E_{l+}(W) = \left[-C_{1/2}^{l+}(W) + \left(\frac{l}{l+2}\right)^{1/2} C_{3/2}^{l+}(W) \right] \Big/ (l+1)$$

$$M_{l+}(W) = -\left[C_{1/2}^{l+}(W) + \left(\frac{l+2}{l}\right)^{1/2} C_{3/2}^{l+}(W) \right] \Big/ (l+1)$$

$$E_{(l+1)-}(W) = -\left[C_{1/2}^{(l+1)-}(W) + \left(\frac{l+2}{l}\right)^{1/2} C_{3/2}^{(l+1)-}(W) \right] \Big/ (l+1)$$

$$M_{(l+1)-}(W) = \left[C_{\frac{1}{2}}^{(l+1)-}(W) - \left(\frac{l}{l+2}\right)^{1/2} C_{3/2}^{(l+1)-}(W) \right] \Big/ (l+1)$$

(2.32)

The angular momentum of the photon in E_{l+}, M_{l+}, $E_{(l+1)-}$, $M_{(l+1)-}$ is $l+1$, l, l, $l+1$, respectively, and that of the pion is (as previously noted for the helicity amplitudes) l, l, $l+1$, $l+1$, respectively.

2.4.4. Large-Energy-Range Analysis

Till recently discrete energy analyses (Berends and Weaver, 1971; Noelle *et al.*, 1971; Pfeil and Schwela, 1972; Berends and Donnachie, 1975*a*) have limited themselves to the low-energy region and the first resonance $p_{33}(1230)$, so that much of the above discussion of possible advantages and disadvantages of discrete analysis would have been rather academic. Recently, however, Berends and Donnachie (1975*b*) have performed a discrete energy analysis of $\gamma p \to \pi^+ n$ and $\gamma p \to \pi^0 p$ up to center-of-mass energy $W = 1540\,\text{MeV}$, thus covering the second resonance region of the $p_{11}(1450)$, $d_{13}(1510)$ and the $s_{11}(1530)$ resonances. Since they do not analyze $\gamma n \to \pi^- p$ because of lack of data, the information on the "neutron" charge state of these $I = \frac{1}{2}$ resonances is lacking.

For those partial waves that were not varied in their fit (that is, the higher waves) they used as amplitudes the full Born terms of Eqs. (2.23) and (2.24), that is, both electric and magnetic Born terms; they also included in these higher partial waves a contribution from the second term in the integrand of (2.23) whose dominant contribution was the p_{33} resonance, whose parameters are already well known. We do see here an obeisance in the direction of the fixed-t dispersion relation method, and a rather strong constraint which would help continuity of the varied amplitudes.

They varied mainly the multipoles leading to states of $j = \frac{3}{2}$ and below, that is, E_{0+}, M_{1-}, M_{1+}, E_{1+}, E_{2-}, M_{2-} [taking full account of unitarity and Watson's theorem, Eq. (2.27)], but at the higher end of the range some variation of the E_{2+}, M_{2+}, E_{3-}, M_{3-} multipoles was found to be necessary, reflecting the influence of the $d_{15}(1670)$ and $f_{15}(1690)$ resonances.

They rediscovered the $p_{11}(1450)$, $d_{13}(1510)$, and $s_{11}(1530)$ resonances. Since the authors have as yet no resonance parameters from an analysis of their partial wave amplitudes we will not further discuss their work in the results, Section 2.5 below. However, we show here (Fig. 2)

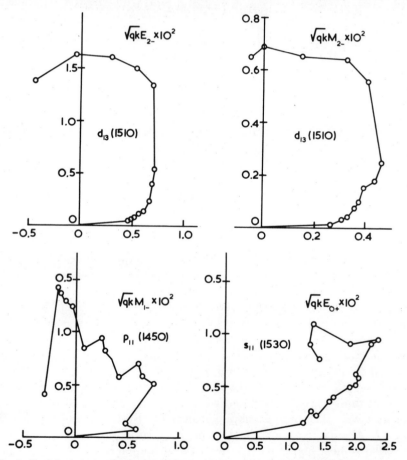

Fig. 2. Partial wave amplitudes associated with the resonances $p_{11}(1450)$, $d_{13}(1510)$, and $s_{11}(1530)$ from the energy-independent analysis of Berends and Donnachie (1975*b*).

Argand diagrams of the partial wave amplitudes pertaining to the resonances, which substantiate the authors' claim to have "rediscovered" these resonances, but at the same time show some residual problems with smoothness.

2.5. Results from Partial Wave Analysis

2.5.1. Critical and Historical Survey

In Table 1 we display the resonance parameter results from certain selected representative partial wave analyses covering the years 1969–1976. The resonance couplings are those defined in Sections 2.2 and 2.3 [see, for example, Eqs. (2.12), (2.18), (2.20), (2.26)] and are given in units of $(GeV)^{-1/2} \times 10^{-3}$. As stated in Sections 2.2 and 2.3, the suffix $\frac{1}{2}$ or $\frac{3}{2}$ on the resonance coupling refers to s-channel helicity, and the superfix p or n in the case of isospin-$\frac{1}{2}$ resonances refers to coupling with the proton state or the neutron state, respectively. In the name of the resonance, for example, $d_{13}(1520)$, the mass (1520) is at or near that assigned in the Data Card Listings of the Particle Data Group; the actual masses used or fitted in any of the analyses may vary from this mass and from each other; for details, the reader is referred to the original papers.

The year 1969 was chosen as a starting point, because it was the date of publication of the first extensive analysis (Walker, 1969) of pion photoproduction analyzing both photoproduction from protons ($\gamma p \to \pi^+ n, \pi^0 p$) and from neutrons ($\gamma n \to \pi^- p$), thus establishing some knowledge for the first time of the neutron couplings of the $d_{13}(1520)$, $s_{11}(1535)$, and $f_{15}(1690)$. This isobar model analysis of Walker was preceded by similar isobar analyses by Gourdin and Salin (1963), Salin (1963), and Chau *et al.* (1967), which had analyzed photoproduction from protons only, establishing knowledge of the proton couplings of $d_{13}(1520)$, $s_{11}(1535)$, and $p_{11}(1470)$.

Column 2 of the table represents the first extensive analysis (Moorhouse and Oberlack, 1973), using fixed-t dispersion relations, the fresh results being the signs and some knowledge of the magnitudes of the couplings of the $s_{31}(1650)$, $d_{33}(1670)$, $s_{11}(1700)$, $d_{15}(1670)$, $f_{37}(1950)^*$ as well as the smaller amplitudes of the $f_{15}(1690)$. The following columns illustrate subsequent development, including the development of analysis of low and high energies simultaneously. Column 3 gives the results of a fixed-t dispersion relation fit by Devenish *et al.* (1973); it did not include as much data as the preceding comparable analysis in column 2. Column 4 presents the results of a fixed-t dispersion relation fit by Knies *et al.* (1974)

*Some knowledge of the $f_{37}(1950)$ was previously available (Devenish *et al.*, 1972).

Table 1. The Results for Resonance Couplings of Some Selected and Representative Partial Wave Analyses during the Period 1969–1976 [a]

Multiplet	Resonance J^P	Coupling	1 1969	2 1972	3 1973	4 1974	5 1974	6 1974	7 1976
$\{56, 0^+\}_0$	$p_{33}(1230)$ $\frac{3}{2}^+$	$\tilde{A}^{\Delta}_{1/2}$	−140	−142±6	−144±14	−138±4	−140±6		−129
		$\tilde{A}^{\Delta}_{3/2}$	−240	−259±16	−262±15	−253±2	−254±7		−251
	$s_{11}(1535)$ $\frac{1}{2}^-$	$\tilde{A}^{p}_{1/2}$	80	53±20	42±23	56±20	63±13	78±20	63
		$\tilde{A}^{n}_{1/2}$	−100	−48±21	−26±29	−52±5	−51±21	−37±23	−109
	$d_{13}(1520)$ $\frac{3}{2}^-$	$\tilde{A}^{p}_{1/2}$	−30	−26±15	10±22	−19±8	−6±6	−8±15	−12
		$\tilde{A}^{p}_{3/2}$	150	194±31	180±17	169±12	165±11	171±12	158
		$\tilde{A}^{n}_{1/2}$		−85±13	−75±37	−77±5	−66±10	−89±19	−56
		$\tilde{A}^{n}_{3/2}$	−130	−124±13	−126±28	−120±10	−118±13	−155±19	−136
	$s_{31}(1650)$ $\frac{1}{2}^-$	$\tilde{A}^{\Delta}_{1/2}$		90±76	4±33	33±15	105±38	−10±17	55
$\{70, 1^-\}_1$	$d_{33}(1670)$ $\frac{3}{2}^-$	$\tilde{A}^{\Delta}_{1/2}$		68±42	36±52	78±9	0±48	54±29	120
		$\tilde{A}^{\Delta}_{3/2}$		22±52	110±39	70±9	0±41	72±14	−117
	$s_{11}(1700)$ $\frac{1}{2}^-$	$\tilde{A}^{p}_{1/2}$		66±42	24±33	58±18	12±15	29±18	44
		$\tilde{A}^{n}_{1/2}$		−72±66	10±43	−15±35	−19±22	−6±31	−22
	$d_{13}(1700)$ $\frac{3}{2}^-$	$\tilde{A}^{p}_{1/2}$			−103±130	−15±40	0±34	−48±50	−5
		$\tilde{A}^{p}_{3/2}$			55±65	30±40	0±29	−6±14	−9
		$\tilde{A}^{n}_{1/2}$			13±220	−36±40	0±34	−21±98	17
		$\tilde{A}^{n}_{3/2}$			−88±87	24±24	0±44	−26±67	22
	$d_{15}(1670)$ $\frac{5}{2}^-$	$\tilde{A}^{p}_{1/2}$	0	11±12	27±30	13±14	10±13	19±21	8
		$\tilde{A}^{p}_{3/2}$		21±20	36±30	14±8	42±24	14±4	21
		$\tilde{A}^{n}_{1/2}$	40?	10±40	−60±62	−43±6	4±15	−29±23	−58
		$\tilde{A}^{n}_{3/2}$		−35±14	−72±22	−71±30	−9±29	−68±20	−80

$f_{15}(1690)\frac{5}{2}^+$	$\tilde{A}^p_{1/2}$	0	-8 ± 4	15 ± 23	-16 ± 14	-8 ± 11	27 ± 19	-4
	$\tilde{A}^p_{3/2}$	140	100 ± 12	146 ± 31	97 ± 7	129 ± 16	163 ± 11	132
	$\tilde{A}^n_{1/2}$	0	17 ± 14	35 ± 49	23 ± 5	8 ± 18	31 ± 28	34
	$\tilde{A}^n_{1/2}$		-5 ± 18	-18 ± 39	1 ± 18	0 ± 30	-21 ± 28	-28
$p_{13}(1800)\frac{3}{2}^+$	$\tilde{A}^p_{1/2}$			-22 ± 94	-4 ± 32	0 ± 25		86
	$\tilde{A}^p_{3/2}$			-1 ± 106	-6 ± 30	0 ± 22		-60
	$\tilde{A}^n_{1/2}$			132 ± 173	14 ± 14	0 ± 50		-20
	$\tilde{A}^n_{3/2}$			80 ± 133	-8 ± 25	0 ± 44		46
$\{56,2^+\}_2$ $p_{31}(1900)\frac{1}{2}^+$	$\tilde{A}^n_{1/2}$				10 ± 12	-32 ± 65	0 ± 25	-31
$p_{33}(1900)\frac{3}{2}^+$								a
$f_{33}(1890)\frac{5}{2}^+$	$\tilde{A}^\Delta_{1/2}$		$-60?$		42 ± 16	47 ± 67	19 ± 27	35
	$\tilde{A}^\Delta_{3/2}$		$-100?$		-22 ± 20	-28 ± 66	78 ± 20	-13
$f_{37}(1950)\frac{7}{2}^+$	$\tilde{A}^\Delta_{1/2}$		-133 ± 46		-70 ± 12	-59 ± 29	-88 ± 25	-76
	$\tilde{A}^\Delta_{3/2}$		-100 ± 41		-78 ± 10	-93 ± 24	-80 ± 21	-65
$\dagger\{56,0^+\}_2$ $p_{11}(1470)\frac{1}{2}^+$	$\tilde{A}^p_{1/2}$		-55 ± 28	-96 ± 22	-66 ± 13	-70 ± 23	-79 ± 12	-53
	$\tilde{A}^n_{1/2}$		2 ± 25	89 ± 56	0 ± 13	43 ± 35	41 ± 25	58
$p_{11}(1780)\frac{1}{2}^+$	$\tilde{A}^p_{1/2}$		26 ± 28	22 ± 57	22 ± 15	-68 ± 24	-14 ± 21	
	$\tilde{A}^n_{1/2}$		27 ± 22	-28 ± 67	27 ± 15	48 ± 45	-60 ± 61	

a The energies of p_{33} resonances higher than the $p_{33}(1230)$ are so uncertain, and their inclusion and treatment in the various photoproduction analyses is so variable, that we do not list any higher p_{33} photon couplings.

extending the analysis of column 2 through the fourth resonance region (~1950 MeV), and column 5 the results of an isobar model analysis by Metcalf and Walker (1974), using the methods of Walker (1969) as in column 1. Finally, column 6 contains the results of a fixed-t dispersion relation fit by Devenish *et al.* (1974); the first analysis to fit high- and low-energy photoproduction data simultaneously. In column 7 the results

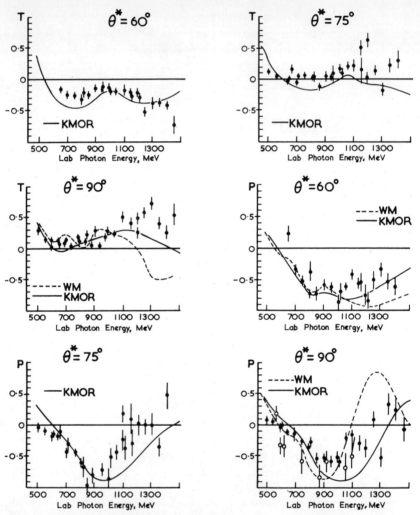

Fig. 3. Photoproduction polarization data (target asymmetry T and recoil neutron polarization P) from the Glasgow–Liverpool–Sheffield collaboration and from Bonn (see Lyth, 1974). Also shown are the *predictions* of previous analyses: WM = Metcalf and Walker (1974), column 5 of Table 1; KMOR = Knies, Moorhouse, Oberlack, and Rosenfeld (see Lyth, 1974; Rutherglen, 1974), similar to column 4 of Table 1.

of an analysis by Barbour and Crawford (1976) are given for comparison with the sixth, since it is also a fixed-fit dispersion relation fit to both high and low energies.

As one reads along the rows one can follow the historical development of that particular photon coupling.

Table 1 establishes a large number of important facts, some of which are pointed out in Section 2.5.2 below, and all of which are discussed in Section 3. However, so far as the historical development shown in the table is concerned, the author's reaction is one of disappointment at the development subsequent to column 2 (1972); the consensus of the later analyses (5–7) is not markedly firmer than that of the earlier (1–3).

The errors shown in the different analyses reflect the flexibility possible in fitting the data, which often have quite large experimental error and may display inconsistencies from one set of experimental data to another. In Figures 3 and 4 we display some data which appeared in 1974 (Rutherglen, 1974), in some cases along with earlier data; these show the above points of inconsistency and fairly large errors (even in recent data). Also displayed in the figures are the predictions of some of the partial wave analyses of the table—predictions because the analyses were performed before the data were available. Of course, much of the experimental data

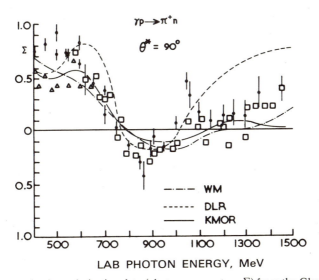

Fig. 4. Photoproduction polarization data (photon asymmetry—Σ) from the Glasgow–Liverpool–Sheffield collaboration (see Lyth, 1974) compared with earlier data. Also shown are the *predictions* of some previous analyses: WM = Metcalf and Walker (1974), column 5 of Table 1; DLR = Devenish *et al.* (1973), column 3 of Table 1; KMOR = Knies, Moorhouse, Oberlack, and Rosenfeld (see Lyth, 1974; Rutherglen, 1974), similar to column 4 of Table 1.

on unpolarized differential cross sections, particularly that on $\gamma p \to \pi^+ n$, is quite consistent and has small experimental errors.

The *differences* between the results of different analyses, sometimes outside the errors shown [$A_{3/2}^p$ of $f_{15}(1690)$ is a particularly striking example] are probably largely due to the different ansätze used in fitting the data. These include not only the general nature of the ansatz (isobar, dispersion relations with low energy only, dispersion relations with low and high energy), but also important details. For example, the use of imaginary background (Section 2.3.3) *or* the position and total width of the resonances, which in some analyses are fixed beforehand and in others allowed to vary, *or* whether a K-matrix or a T-matrix form of resonance parameterization be used. Largely because the hadronic widths enter into the definition of the resonance couplings through Eq. (2.20) the (hadronic) resonance parameters Γ_π, Γ, and W_R (mass) are particularly important. These are all *systematic* differences between the various analyses.

A comparison of the photoproduction amplitudes, rather than the extracted resonance couplings, would eliminate *some* of the *systematic* differences. Unfortunately, it is not always possible to reconstruct the amplitudes from the information in the paper, and even if possible it is laborious. Publication of amplitudes would be valuable. At the moment these have only been given by Walker (1969) (column 1, Table 1) and Moorhouse *et al.* (1974).

Different results, outside the given errors, can also be due to the use of different data sets, a particularly important consideration when some data may be wrong. The errors shown, which might be called *statistical errors*, can also be underestimated by the authors.

However, the outlook for the future is probably hopeful for a more precise determination of the coupling constants. Many new data have been acquired from 1974 onward, most of which have not yet been adequately incorporated into the analyses, and more data, including double polarization data [Eqs. (A9)–(A12)] for photoproduction from protons, will be produced. This should both help to show which method is most adequate for representing the data, *and* help to reduce the *statistical* error within a given method.

2.5.2. Resonance Coupling Certitudes

Despite the uncertainties and disagreements that we have just noted, there are many interesting, and as we shall discuss in Section 3, also significant, certitudes in the couplings. Some of these are the following:

(i) Dominance of the magnetic multipole in $p_{33}(1230)$ and to a less well-determined extent in $f_{37}(1950)$. These are the first two resonances on the well-known Δ Regge trajectory, and defining, by Eq. (2.32), the multipole resonance couplings analogously to the helicity couplings one finds

$$p_{33}(1230), \quad |E_{1+}/M_{1+}| < 0.02$$

$$f_{37}(1950), \quad |E_{3+}/M_{3+}| < 0.35$$

The number 0.35 is the upper limit possible from any of columns 4–7 of Table 1; an average value from those columns is

$$|E_{3+}/M_{3+}| = 0.15$$

(ii) Dominance of the isovector amplitude in helicity $\frac{3}{2}$ of the $d_{13}(1520)$. From Eq. (2.15b) one finds $A^S/A^{V1} = (A^n + A^p)/(A^n - A^p)$, giving for the $d_{13}(1520)$

$$|A_{3/2}^S/A_{3/2}^{V1}| < 0.16$$

(iii) $A_{1/2}^p$ both for the $d_{13}(1520)$ and $f_{15}(1690)$, small and compatible with zero.

The above are the sort of results that suggest the operation of selection rules, and this aspect is discussed in detail in the next section, along with other possibilities and selection rules.

In addition to the above, we should note some more facts, which turn out to be critically important theoretically. We note that the signs of the resonance couplings in Table 1 are determined not just relative to each other, but relative to the sign of the Born terms. The following moderate or large couplings have well-determined signs (and, usually, *approximate magnitudes*):

$$p_{33}(1230): A_{1/2}^\Delta, A_{3/2}^\Delta \qquad\qquad s_{11}(1535): A_{1/2}^p, A_{1/2}^n$$

$$d_{13}(1520): A_{3/2}^p, A_{1/2}^n, A_{3/2}^n \qquad s_{31}(1650): A_{1/2}^\Delta$$

$$d_{33}(1670): A_{1/2}^\Delta, A_{3/2}^\Delta \qquad\qquad s_{11}(1700): A_{1/2}^p$$

$$d_{15}(1670): A_{3/2}^n \qquad\qquad\qquad f_{15}(1690): A_{3/2}^p, A_{1/2}^n$$

$$f_{37}(1950): A_{1/2}^\Delta, A_{3/2}^\Delta \qquad\qquad p_{11}(1470): A_{1/2}^p$$

On the $d_{33}(1670)$, listed above as well-determined, there is not a universal consensus, the two isobar model analyses being totally uncertain as to the sign. The relatively good determination of these amplitudes appears to us to be one of the strong points of the fixed-t dispersion relation approach.

3. Theory of Radiative Baryon Decays

It is well known that the naive quark model gives a first-order agreement with the observed radiative transition properties of baryons, the greater part of the knowledge of which we have reviewed in Section 2. By the naive quark model of baryons we mean, in the first place, three quarks in a potential, since this is the only form yet in which radiative transition properties have been calculated, though more sophisticated forms of the constituent quark model are under development, such as relativistic bag models (Chodos *et al.*, 1974*a*, 1974*b*) or infrared confinement gauge theories (Kogut and Susskind, 1975a). In Section 3.1 we outline the two versions of the naive quark model in which calculations have been mainly carried out. These are the original so-called nonrelativistic quark model (Becchi and Morpurgo, 1965*a*; Copley *et al.*, 1969*a*, 1969*b*; Dalitz and Sutherland, 1966, Moorhouse, 1966; Faiman and Hendry, 1968) and the four-dimensional oscillator model of Feynman *et al.* (1971), which gives results very close to those of the harmonic oscillator version of the non-relativistic quark model.

3.1. Electromagnetic Interactions of Three Quarks in a Potential

3.1.1. The "Nonrelativistic" Quark Model

We outline here the simplest version of electromagnetic interactions in this model, leaving refinements for later discussion in Section 3.3.

The quarks are supposed bound or contained by an effective potential. The original version of this model (Becchi and Morpurgo, 1965*a*) had heavy quarks (explicitly, 5 GeV/c^2 mass) deeply bound in a potential well broad enough so that the heavy quarks had nonrelativistic motion; hence, the name "nonrelativistic" quark model. The theory soon diverged into a more illogical picture in which the effective quark mass is rather light; hence, the quotation marks on "nonrelativistic."

The most easily calculable, and much the most used, version of the "nonrelativistic" model, which also fits in quite well with the average energy levels of the known resonance multiplets, is the harmonic oscillator shell model. The Hamiltonian for three particles at positions $\mathbf{r}_j$, momenta $\mathbf{p}_j$ and mass m_j ($j = 1, 2, 3$) interacting via harmonic oscillator forces is

$$H = \sum_j \frac{\mathbf{p}_j^2}{2m_j} + \frac{1}{2} \sum_{i<j} \frac{1}{3} \beta_{ij} (\mathbf{r}_i - \mathbf{r}_j)^2 \tag{3.1}$$

where $\frac{1}{3}\beta_{ij}$ is a constant giving the strength of the forces between quarks i and j. Simplifying to the nucleon case with mean equality of masses and strong forces of the nonstrange quarks, so that $m_j = m$, $\beta_{ij} = \beta$, Eq. (3.1) becomes

$$H = \sum_j \frac{\mathbf{p}_j^2}{2m} + \frac{1}{2}\left(\frac{\beta}{3}\right) \sum_{i<j} (\mathbf{r}_i - \mathbf{r}_j)^2 \tag{3.2}$$

the corresponding shell-model Hamiltonian being

$$H_{sm} = \sum_j \frac{\mathbf{p}_j^2}{2m} + \frac{1}{2} \beta \sum_j \mathbf{r}_j^2 \qquad (3.3)$$

Defining the center-of-mass position vector **R** and relative coordinates $\boldsymbol{\lambda}$, $\boldsymbol{\rho}$ by

$$\mathbf{R} = \tfrac{1}{3}(\mathbf{r}_1 + \mathbf{r}_2 + \mathbf{r}_3), \qquad \boldsymbol{\lambda} = \tfrac{1}{6}(\mathbf{r}_1 + \mathbf{r}_2 - 2\mathbf{r}_3), \qquad \boldsymbol{\rho} = \tfrac{1}{2}(\mathbf{r}_2 - \mathbf{r}_3) \qquad (3.4)$$

we can rewrite the above Hamiltonians as

$$H = \frac{1}{2} \frac{\mathbf{P}^2}{3m} + \left[\frac{1}{2m} (\mathbf{p}_\lambda^2 + \mathbf{p}_\rho^2) + \frac{1}{2}\beta(\boldsymbol{\lambda}^2 + \boldsymbol{\rho}^2) \right] \qquad (3.5)$$

$$H_{sm} = \frac{1}{2} \frac{\mathbf{P}^2}{3m} + \frac{1}{2}\beta \mathbf{R}^2 + \left[\frac{1}{2m} (\mathbf{p}_\lambda^2 + \mathbf{p}_\rho^2) + \frac{1}{2}\beta(\boldsymbol{\lambda}^2 + \boldsymbol{\rho}^2) \right] \qquad (3.6)$$

where **P**, $\mathbf{p}_\lambda$, $\mathbf{p}_\rho$ are the momenta canonically conjugate to **R**, $\boldsymbol{\lambda}$, $\boldsymbol{\rho}$, respectively.

The spectrum of excitation due to the internal coordinates is the same in either of the Hamiltonians (3.5) or (3.6). However, if we use the shell-model Hamiltonian [either (3.6) or, equivalently, (3.3)], which is calculationally somewhat simpler, we have to take care to eliminate the false center-of-mass oscillations due to the second term in Eq. (3.6). The other operators in the Hamiltonian [$SU(3)$, spin] are left implicit and assumed to combine with the spatial part (which may include a confining force additional to the Hamiltonian given above) in such a way as to give only totally symmetric states [in $SU(3)$, spin and spatial parts combined] as the physical states of the system. The Hamiltonian above is that effective in such totally symmetric states. A natural way of implementing totally symmetric wave functions of fermion quarks is to have the quarks colored.

Many authors, including Dalitz (1965) and Faiman and Hendry (1965) early pointed out the utility of the quark harmonic oscillator shell model for energy level and decay calculations, and an extensive recent study by Horgan and Dalitz (1973) and Horgan (1973) has been made. Though there are many other possible potentials, we can illustrate our points most simply in this section by using the harmonic oscillator. Also, this three-dimensional harmonic oscillator is closely related to the four-dimensional model of Feynman, Kislinger, and Ravndal, discussed below, which is explicitly committed to a four-dimensional oscillator.

The symmetric oscillator shell model leads to the well-known spectrum of quark model supermultiplets (Faiman and Hendry, 1965; Horgan and Dalitz, 1973) with oscillator wave functions with radial and orbital angular momentum excitation: $\{\mathbf{56}, L = 0^+\}_0$, $\{\mathbf{70}, L = 1^-\}_1$, $\{\mathbf{56}, L = 2^+\}_2$, $\{\mathbf{56}, L = 0^+\}_2, \ldots$. The $\mathbf{70}, L = 1^-$ and $\mathbf{56}, L = 2^+$ supermultiplets are shown in Fig. 5.

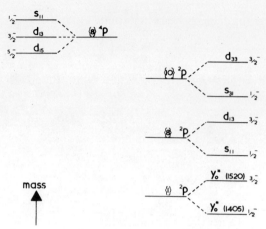

Fig. 5a. The $\{70, 1^-\}$ supermultiplet of the L-excitation quark model. The $\{SU(3)\}^{2S+1}L$ multiplets are shown, split into the J_P submultiplets. The nucleonic state corresponding to each submultiplet is indicated, in pion–nucleon scattering notation. [Candidates for the SU(3) singlet states are also indicated.] There is an evident mass ordering in the $\{SU(3)\}^{2S+1}L$ multiplets, which is that shown in the figure. The ordering of the J^P submultiplets, within the $\{SU(3)\}^{2S+1}L$ multiplets, is subject to experimental uncertainty.

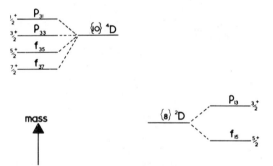

Fig. 5b. The $\{56, 2^+\}$ supermultiplet of the L-excitation quark model. As in Fig. 5(a), the ordering of the $\{SU(3)\}^{2S+1}L$ multiplets is taken from the known experimental masses. The ordering of the J^P submultiplets within $\{10\}^4\ d$ is not experimentally significant.

3.1.2. Photon Interaction in the "Nonrelativistic" Model

The nonrelativistic form of the electromagnetic interaction with real photons of the quarks in the potential (derived, for example, by using the Foldy–Wouthuyson transformation) is given by the Hamiltonian

$$H_{\text{e.m.}}^{\text{NR}} = \sum_{j=1}^{3} \left[-i\mu_j \boldsymbol{\sigma}_j \cdot (\mathbf{k} \times \mathbf{A}) + 2\mathbf{p}^j \cdot \mathbf{A} \frac{eq_j}{2m_j} \right] \tag{3.7}$$

where $\mathbf{A}$ is the electromagnetic vector potential, $\mathbf{k}$ is the photon momentum, p^j the jth quark momentum, m_j is the jth quark mass, $\boldsymbol{\sigma}_j$ the jth quark

Pauli spin operator, μ_j the jth quark magnetic moment, and q_j the jth quark charge in units of e (so $q_j = \frac{2}{3}$ for a proton quark, and $-\frac{1}{3}$ for a neutron or strange quark).

From the structure of the electromagnetic current as a U-spin scalar we know that $\mu_j \propto q_j$; in order to have the correct proton magnetic moment, using Eq. (3.7), we must have

$$\mu_j = q_j \mu_p = q_j \left(\frac{2.79e}{2M} \right) \tag{3.8}$$

where μ_p is the proton magnetic moment, and M is the proton mass. Then the quark model predicts, nearly correctly [as does $SU(6)$]

$$\mu_n = -\tfrac{2}{3}\mu_p \tag{3.9}$$

where μ_n is the neutron magnetic moment.

The only undetermined parameters in Eq. (3.7) are the quark masses, $m_j = m$ for proton or neutron quarks. If the quarks are point (structureless) particles, as they seem to be in high-energy electromagnetic and neutrino interactions (Perkins, 1976; Mo, 1975; Taylor, 1976; Llewellyn-Smith, 1976), then they have only the Dirac magnetic moment given from Eq. (3.8)

$$\frac{e}{2m} = 2.79 \frac{e}{2M} \tag{3.10}$$

or

$$m \sim 340 \, \text{MeV}/c^2 \tag{3.11}$$

Another way of estimating the effective quark mass has been given by Faiman and Hendry (1965). The spatial wave functions of the nucleons contain a factor $\exp\left[-\frac{1}{2}\alpha^2(\mathbf{r}_1^2 + \mathbf{r}_2^2 + \mathbf{r}_3^2)\right]$ where [from Eq. (3.3) or Eq. (3.6)]

$$\alpha^2 = (m\beta)^{1/2} \tag{3.12}$$

can be used to measure the interaction radius in strong decay processes such as

$$N^*[d_{13}(1520)] \to N + \pi \tag{3.13}$$

They find by fitting the decays such as (3.13) a value of

$$\alpha^2 = 0.10 \, (\text{GeV}/c)^2 \tag{3.14}$$

corresponding to an interaction radius of about 0.8 fm, which seems in itself quite reasonable and serves to verify the estimate of α^2 from Eq. (3.12). But also in the harmonic oscillator model, the level spacing is given by $(\beta/m)^{1/2}$, and the separation of the mean masses of the different multiplets is about 400 MeV, so

$$(\beta/m)^{1/2} = 4.4 \text{ GeV} \tag{3.15}$$

and dividing Eq. (3.12) by Eq. (3.15) we get

$$m \sim 250 \text{ MeV}/c^2$$

This is somewhat smaller than the previous estimate of 340 MeV; preserving the same level spacing, and requiring a mass of 340 MeV, would lead to a value of α^2 equal to $0.136 \,(\text{GeV}/c)^2$ and a correspondingly smaller interaction radius. We will adopt the value $m = 340 \text{ MeV}/c^2$, so that as far as photon interactions with *nucleons* (including nucleon resonances) are concerned, the Hamiltonian (3.7) becomes

$$H_{\text{e.m.}}^{\text{N.R.}} = \sum_{j=1}^{3} q_j \left[-i\mu_p \boldsymbol{\sigma}_j \cdot (\mathbf{k} \times \mathbf{A}) + 2\mathbf{p}^j \cdot \mathbf{A} \frac{e}{2m} \right] \tag{3.16}$$

where $m = 340 \text{ MeV}/c^2$.

The model thus formulated is paradoxical in two respects: (a) With such a mass the quarks in the Δ, for example, are unbound; this paradox may be resolved by quark confinement theories (Section 3.3). (b) The basis of derivation of Eq. (3.16) is nonrelativistic, but with such a mass and a usual hadronic radius the quarks are relativistic; such problems are discussed in Section 3.3 below. Here we accept Eq. (3.16) and review its consequences. The review is mainly carried out in Section 3.2; here we just discuss a few desirable consequences of a light quark mass in Eqs. (3.7) and (3.16).

(i) We note that the first term in Eq. (3.16) is an interaction with the quark magnetic moment; the last term is an interaction with the quark charge. Let us consider the photoexcitation (or photon decay) of the $d_{13}(1520)$ resonance, which, as shown in Table 1, is classified into an $[8, 2]$ of $SU(3) \otimes SU(2)_{\text{spin}}$, so that it has quark spin $S = \frac{1}{2}$ (doublet state). Let us consider in particular the (s-channel) helicity $\frac{3}{2}$ couplings $A_{3/2}^p$ and $A_{3/2}^n$, this helicity being made up of z-component of spins, $S_z(\gamma) = 1$ and $S_z(N) = \frac{1}{2}$, as shown in the diagram:

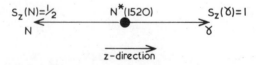

The nucleon has no quark orbital angular momentum, so all its spin comes from the quark spin: $S_z(N) = S_{Qz}(n) = \frac{1}{2}$. Now the *magnetic moment* term of Eq. (3.16) must flip this quark spin, so for interactions *by the magnetic moment* $S_{Qz}(N^*) = -\frac{1}{2}$, which is impossible since $S_z(N^*) = \frac{3}{2}$ and the N^* has only quark orbital angular momentum $L = 1$. So all the $A^p_{3/2}$ and $A^n_{3/2}$ coupling must come from the last term in Eq. (3.16), that is, the interaction by the quark charge. As listed below, in Section 3.2, with a quark mass of 340 MeV/c^2 the calculated couplings are about 20%–40% less than the experimental couplings; with a larger quark mass the agreement would be much worse.

(ii) Both the quark magnetic moment interaction and the quark charge interaction can contribute to the helicity $\frac{1}{2}$ amplitudes of the d_{13} (1520). The $A^p_{1/2}$ coupling is very small experimentally, and calculation in the "nonrelativistic" harmonic oscillator model gives

$$\tilde{A}^p_{1/2}[d_{13}(1520)] = -\frac{2}{3}\left(\frac{\pi}{k}\right)^{1/2} \alpha e^{-k^2/6\alpha^2}\left\{\mu_p\frac{k^2}{\alpha^2} - \frac{e}{2m}\right\} \qquad (3.17)$$

where k is the photon momentum in the center-of-mass system, so that $k^2 = 0.22$ (GeV/c)^2. Since $\mu_p \simeq e/2m$ we can see that there is a strong cancellation in the term within the braces for our value of $\alpha^2 = 0.136$ (GeV/c)^2 given above. This is, of course, only a semiquantitative observation because the values of the constants are not determined with certainty in such models.

Similarly,

$$\tilde{A}^p_{1/2}[f_{15}(1690)] = -\frac{2}{3}\left(\frac{2}{5}\right)^{1/2} (\pi k)^{1/2} e^{-k^2/6\alpha^2}\left\{\mu_p\frac{k^2}{2\alpha^2} - \frac{e}{2m}\right\} \qquad (3.18)$$

and here k is larger, so that $\frac{1}{2}k^2 = 0.17$ (GeV/c)^2; thus, we have simultaneous strong cancellation in (3.17) and (3.18), both in accord with experiment. These rather remarkable cancellations in agreement with experiment were first pointed out by Copley *et al.* (1969*a*, 1969*b*).

As in (i) above, they would not be possible without a considerable contribution from the interaction with the quark charge, due to a small quark mass in the last term of Eq. (3.16).

(iii) A rather different point is that to explain the observed mass splittings, for instance those in the decuplet, one requires a mass difference between the λ quark and the proton or neutron quark:

$$m_\lambda - m \simeq 130 \text{ MeV}/c^2 \qquad (3.19)$$

giving

$$m_\lambda \simeq 476 \text{ MeV}/c^2 \qquad (3.20)$$

and a λ quark magnetic moment of

$$\frac{e}{2m_\lambda} \cdot q_\lambda \simeq \frac{2}{3}\mu_p q_\lambda \qquad \left(q_\lambda = -\frac{1}{3}\right) \qquad (3.21)$$

With this value and the previous values of the proton quark and neutron quark magnetic moments $(e/2m) \cdot (\frac{2}{3})$, and $(e/2m) \cdot (-\frac{1}{3})$, respectively, one calculates the magnetic moment of the $\Lambda(1115)$ as

$$\mu(\Lambda) = -0.66 \text{ nuclear magnetons} \qquad (3.22)$$

This is in better agreement with the measured magnetic moment of -0.67 ± 0.06, than the $SU(3)$ prediction that $\mu(\Lambda) = -0.93$.

3.1.3. The Four-Dimensional Oscillator Model

Some authors, most notably Feynman, et al. (1971), have propounded four-dimensional oscillator models. We will briefly describe the model of Feynman et al., which has some relativistic features, even though it is not a fully covariant theory.

The hadron spectrum is described by a hadron propagator

$$K = P^2 - \pi \qquad (3.23)$$

where P is the hadron four-momentum and π is a mass-square operator depending only on the internal motion and given by

$$-\pi = \frac{1}{2}\xi^2 + \frac{1}{2}\eta^2 + \frac{1}{2}\Omega^2 x^2 + \frac{1}{2}\Omega^2 y^2 \qquad (3.24)$$

where x and y are relative coordinates given in terms of the four-vectors of position and time of particles 1, 2, 3 by*

$$x = \frac{1}{6}(r_1 + r_2 - 2r_3), \qquad y = \frac{1}{2(3)^{1/2}}(r_1 - r_2) \qquad (3.25)$$

and are thus, apart from a factor $[1/2(3)^{1/2}]$ the four-dimensional versions of $\boldsymbol{\lambda}$ and $\boldsymbol{\rho}$ defined in Section 3.1.1 above. ξ and η in Eq. (3.24) are the canonical momenta conjugate to x and y.

The ground-state wave function is in momentum space

$$h_0(\xi, \eta) = \exp[(\xi^2 + \eta^2)/2\Omega] \qquad (3.26a)$$

and in configuration space

$$h_0(x, y) = \exp[(x^2 + y^2)\Omega/2] \qquad (3.26b)$$

where $x^2 = x_0^2 - \mathbf{x}^2$, etc. If one allowed the timelike modes to be excited one would have time-states of negative norm and positive energy. Feynman et al. arbitrarily exclude such states.

A value of $\Omega = 1.05 \ (\text{GeV}/c)^2$ is taken as the correct level spacing between the supermultiplets $\mathbf{56}, L = 0^+$, $\mathbf{70}, L = 1^-$, and $\mathbf{56}, L = 2^+$. Since $\Omega/12$ corresponds to α^2 in the nonrelativistic model [Eq. (3.12)], this value

*Feynman et al. (1971) describe particles 1, 2, 3 by c, b, a, respectively.

of Ω corresponds to $\alpha^2 = 0.09$. We can note that the quark mass does not appear explicitly in this four-dimensional oscillator model.

The electromagnetic interaction is obtained as the perturbation to the propagator K of Eq. (3.23):

$$\delta K = 3 \sum_i q_i e [\not{p}_i(r_i) + \not{A}(r_i)\not{p}_i] \qquad (3.27)$$

To obtain the magnetic moment we note that this is the perturbation to the mass squared of the hadron: $K = 2M\delta M$, where M is the nucleon mass if the hadron is the nucleon. So that sandwiching Eq. (3.27) between proton and neutron states, respectively, gives, as the magnetic moment contribution

$$M = 3 \cdot \frac{1}{2M} i\boldsymbol{\sigma} \cdot (\mathbf{k} \times \mathbf{A}) \qquad \text{(proton state)}$$

$$M = -2 \cdot \frac{1}{2M} i\boldsymbol{\sigma} \cdot (\mathbf{k} \times \mathbf{A}) \qquad \text{(neutron state)}$$

corresponding to proton and neutron magnetic moments of 3 and -2 nuclear magnetons compared with the experimental values of 2.79 and -1.91. Contrary to the "nonrelativistic" quark model, there is no input assumption on the quark mass, which simply does not enter the calculation of Feynman *et al.* (1971).

Transition matrix elements, $N^* \to N + \gamma$, are obtained by sandwiching Eq. (3.27) between the N^* and the N quark states, some assumptions or approximations being made on the quark Dirac spinors (the results of which approximations, however, probably involve mostly correction terms to the main contribution, that main contribution being very similar to the "nonrelativistic" model). We refer the reader to the original paper for details.

3.2. Comparison with Experimental Results

In this section we compare with experimental results—notably those of Table 1 on radiative decays of N^* resonances—the quark models discussed above. In the above discussion the nature of the electromagnetic current implied by the models was emphasized. However, as we shall see, there are great variations in results, depending on the different processes and resonances involved. These depend on the nature of the quark model wave functions and form a critical check on the first-order validity of these wave functions. We have already seen this in the discussion of the $d_{13}(1520)$ decays in points (i) and (ii) of Section 3.1.2 above, where, for example, the inhibition of magnetic coupling in the helicity $\frac{3}{2}$ decay is a result of the spin $J = \frac{3}{2}$ of the $d_{13}(1520)$ being formed from quark spin $S = \frac{1}{2}$, quark orbital angular momentum $L = 1$.

Some resonance couplings from partial wave analysis (including, but not exclusively, all the better-known couplings) are compared with naive quark model calculations in Table 2. The analysis results include all the couplings whose signs, at least, are well determined; some others that do not satisfy that criterion have been included, so as to complete the **70**, $L = 1^-$ and **56**, $L = 2^+$ supermultiplets (except for the p_{33} member of the latter, for reasons explained in the discussion of Table 1). The partial wave analysis couplings are an average given by the Particle Data Group (1976), and thus differ, but not significantly, from the average of couplings of Table 1. The quark model calculations are those from the model of Feynman *et al.* (1971) and the "nonrelativistic" model, as discussed in the text. The calculations in the "nonrelativistic" model are as described in Section 3.1.2, with a quark mass of 340 MeV/c^2 and a pure Dirac moment for the quark. In the multiplet column the resonances are ascribed to their quark model supermultiplet $\{SU(6), L^p\}_n$, where L is the quark orbital angular momentum and n is the excitation level (in a harmonic oscillator model); the resonances are also ascribed to their $SU(3) \otimes SU(2)$ quark spin multiplets, within the supermultiplets.

The reader may note the near identity between the Feynman, Kislinger, and Ravndal four-dimensional oscillator model and the "nonrelativistic" quark model with oscillator wave functions and a quark mass of 340 MeV. We list some successful comparisons of theory and experiment.

(i) Signs. The reader may count 16 well-determined signs, of which 15 agree with the quark model predictions, the 16th being the proton coupling of the $p_{11}(1470)$. This resonance is usually assigned to a radial excitation of the **56**, $L = 0^+$; such resonances might be strongly mixed.

The sign agreement is nontrivial because the photon belongs to a different $SU(3)$ multiplet from the pion and the process $\gamma + N \to \pi + N$ is "$SU(3)$ inelastic," so that none of the numbers are predicted by $SU(3)$, (Moorhouse and Oberlack, 1973; Moorhouse and Parsons, 1973). Of the higher symmetries $SU(6)_W$ is an obvious candidate (Lipkin and Meshkov, 1965; Barnes, *et al.*, 1965), but that predicts, for example, the helicity $\frac{3}{2}$ couplings of the $d_{13}(1520)$ to be zero, in plain contradiction with experiment. The quark model agreement is a consequence of the wave functions and the assumed form of the electromagnetic interaction, discussed above. We note that it is more significant than the agreement obtained by various forms of broken symmetry, such as "l-broken" $SU(6)_W$ (Petersen and Rosner, 1972, 1973; Faiman and Plane, 1972) or the similar phenomenology inspired by the Melosh transformation (discussed in detail by Hey in this volume) in two respects:

Table 2. A Comparison of Some Resonance Helicity Couplings from Partial Wave Analysis with Naive Quark Model Calculations

Multiplet	State	λ	Analyses average $\tilde{A}_\lambda^p$ or A_λ^Δ	Analyses average $\tilde{A}_\lambda^n$	Quarks (FKR) $\tilde{A}_\lambda^p$	Quarks (FKR) $\tilde{A}_\lambda^n$ or A_λ^Δ	Quarks (NR) $\tilde{A}_\lambda^p$ or A_λ^Δ	Quarks (NR) $\tilde{A}_\lambda^n$
$\{56, 0^+\}_0$ [10, 4]	$p'_{33}(1230)$	$\tfrac{1}{2}$	-139 ± 5			-108	-103	
		$\tfrac{3}{2}$	-256 ± 5			-187	-178	
[8, 2]	$d'_{13}(1520)$	$\tfrac{1}{2}$	-10 ± 15	-75 ± 15	-34	-31	-29	-30
		$\tfrac{3}{2}$	171 ± 15	-129 ± 10	109	-109	112	-112
$\{70, 1^-\}_1$ [8, 4]	$s'_{11}(1535)$	$\tfrac{1}{2}$	63 ± 25	-49 ± 35	156	-108	160	-109
	$d''_{11}(1700)$	$\tfrac{1}{2}$	43 ± 30	-37 ± 40	0	30		
	$d''_{13}(1700)$	$\tfrac{1}{2}$	-20 ± 45	18 ± 55	0	-10		
		$\tfrac{3}{2}$	19 ± 45	18 ± 80	0	-40		
	$d'_{15}(1670)$	$\tfrac{1}{2}$	16 ± 10	-32 ± 36	0	-38	0	-38
		$\tfrac{3}{2}$	21 ± 12	-62 ± 40	0	-53	0	-53
[10, 2]	$s'_{31}(1650)$	$\tfrac{1}{2}$	46 ± 36			47	47	
	$d_{33}(1670)$	$\tfrac{1}{2}$	72 ± 26			88	92	
		$\tfrac{3}{2}$	72 ± 45			84	91	
[8, 2]	$f_{15}(1690)$	$\tfrac{1}{2}$	-5 ± 30	25 ± 10	-10	30	-15	$+41$
		$\tfrac{3}{2}$	127 ± 35	-16 ± 20	60	0	70	0
	$p'_{13}(1810)$	$\tfrac{1}{2}$	-25 ± 50	24 ± 70	100	-30		
		$\tfrac{3}{2}$	-31 ± 50	-4 ± 60	-30	0		
$\{56, 2^+\}_2$ [10, 4]	$p_{31}(1910)$	$\tfrac{1}{2}$	-11 ± 20			-30		
	$f_{35}(1890)$	$\tfrac{1}{2}$	21 ± 30			-20		
		$\tfrac{3}{2}$	-10 ± 60			-90		
	$f_{37}(1950)$	$\tfrac{1}{2}$	-69 ± 16			-50		
		$\tfrac{3}{2}$	-76 ± 20			-70		
$\{56, 0^+\}_2$ [8, 2]	$p'_{11}(1470)$	$\tfrac{1}{2}$	-74 ± 15	34 ± 35	27	-18	32	-20
	$p''_{11}(1780)$	$\tfrac{1}{2}$	18 ± 40	18 ± 50	-40	10		

^{a}(GeV)$^{-1} \times 10^{-3}$.

(a) The broken symmetries or "Melosh" phenomenologies contain arbitrary parameters; we have seen that the quark models do not, at the level of significance used in this comparison.

(b) The quark models give an absolute prediction on the signs (more precisely the sign is predicted relative to the Born terms); the symmetry schemes predict generally just signs within a multiplet.

(ii) Selection Rules.

(a) We noted in Section 2.5.2 that the $p_{33}(1230)$ [and to a lesser extent the $f_{37}(1950)$] had magnetic coupling dominance (Becchi and Morpurgo, 1965a). This is very simply explained in the quark model. The photoexcitation or decay of the p_{33}, which is in the same spatial state as the nucleon, involves a $0^+ \rightarrow 0^+$ transition, which is forbidden to an electric multipole transition by angular momentum and parity considerations. A similar argument holds for the f_{37}.

(b) We have noted in our introduction to the nonrelativistic quark model in Section 3.1.2 that the $d_{13}(1520)$, proton state helicity $\frac{1}{2}$ couplings are predicted to be small, because of a cancellation (Copley et al., 1969a, 1969b); they are even smaller than the explicit calculation finds, but this difference is undoubtedly within the errors of the model!

(c) There is also a selection rule due to Copley et al. (1969a, 1969b) with the elementary electromagnetic interaction considered above, that the helicity $\frac{3}{2}$ coupling $A_{3/2}^n$, for the photo-excitation of $I = \frac{1}{2}$ resonances belonging to 56-plets (and hence with quark spin $\frac{1}{2}$) vanishes identically.

This selection rule is reflected in the quark model zero appearing in the appropriate place $(A_{3/2}^n)$ for the $f_{15}(1690)$ and $p_{13}(1800)$ couplings. This selection rule cannot be said to be verified, though zero is certainly within the range of experimental error!

(d) Another selection rule (Moorhouse, 1966) that is a result of the quark model wave functions is that the *proton* couplings of [8, 4] members of a 70-plet vanish. This rule may well be violated for the $s_{11}(1700)$, but there is independent evidence of strong mixing between this resonance and the $s_{11}(1535)$, which belongs to an [8, 2] multiplet. The $d_{15}(1670)$ is almost certainly unmixed, and we can say that the evidence inclines towards these couplings being rather small.

(iii) Orders of Magnitude. In general, experimentally large couplings are predicted large and experimentally small couplings are predicted small.

After these successes we can also list some failures.

(iv) Magnitude. The calculation of well-determined large couplings, is, in the case of the $p_{33}(1232)$, $d_{13}(1520)$, and $f_{15}(1690)$, up to about 50% too small.

(v) Failures of Detail. For example, the helicity $\frac{3}{2}$ excitation of the $d_{13}(1520)$ is not pure isovector (Section 2.5.2), as predicted, but contains a small, but significant, isoscalar component.

3.3. Modifications of the Naive Quark Model

In the last sections we have seen a success of the naive quark model in predicting signs and relative magnitudes that is so marked that this itself serves as a validation of the naive quark model and its spectral assignments; at the same time we have seen a failure in the absolute magnitudes of matrix elements of 20%–70%, and some failures of detail. It is natural to ask whether some nonfundamental modifications can correct this situation, with the most obvious such modifications being the v/c, $(v/c)^2 \cdots$ corrections to the nonrelativistic Hamiltonian of Eq. (3.7). These were considered some years ago by Bowler (1970) and by Copley *et al.* (1969*a*, 1969*b*) and from a more fundamental point of view by Close and Osborn (1971). Such modifications give rise inter alia to a spin-orbit coupling term in the electromagnetic interaction of Eq. (3.7); this spin-orbit coupling invalidates the selection rule (ii)(c) of Section 3.1 above but leaves the other selection rules of (ii) intact. The relevant matrix elements thus provide a possible experimental test of the occurrence or nonoccurrence of spin-orbit terms; however, more accurate photoproduction data off neutrons ($\gamma n \to \pi^- p$) are needed.

Kubota and Ohta (1976) have considered $(v/c)^2$ corrections so that Eq. (3.16) becomes

$$H_{\text{e.m.}}^{\text{N.R.}} = \sum_{j=1} q_j \left\{ \left[-i\mu_p \boldsymbol{\sigma}_j \cdot (\mathbf{k} \times \mathbf{A}) + 2\mathbf{p}^j \cdot \mathbf{A} \frac{e}{2m} \right] \right.$$
$$\left. + \frac{k}{4m} \left[2i\mu_p \boldsymbol{\sigma}^j \times \mathbf{p}^j \cdot \mathbf{A} + \frac{e}{2m} \mathbf{k} \cdot \mathbf{A} \right] \right\}$$

This form omits Wigner spin rotation terms of the same order considered by Close and Copley (1970). Kubota and Ohta show that this modification of Eq. (3.16) leads to a marked improvement of the agreement with experiment, correcting some of the failures referred to in the last section, including that in the $d_{13}(1520)$.

The Hamiltonian of Eq. (3.5) for the interquark interaction generates a spatial wave function, which can be regarded as approximately correct in the rest-frame of the hadron. Fujimura, *et al.* (1970); Licht and Pagnamenta (1970*a*, 1970*b*); Le Yaouanc *et al.* (1972); Lipes (1972); Gonzales and Watson (1972); and Kellett (1974) have emphasized the importance of the Lorentz contraction of the spatial wave function in the direction of recoil in, for example, $N^* \to N + \gamma$, where N^* is at rest but N is recoiling. Again, such modifications, in so far as they have been considered in detail, do not seem to help much in the problem of the magnitude of the matrix elements, at any rate in the harmonic oscillator model, where the effect of such modifications on the matrix elements is easiest to work out. It might well be that while the spectrum of the baryons is approximately that of a harmonic oscillator, the effective Hamiltonian is such that the wave functions are considerably different; certainly, a nonoscillator effective Hamiltonian is suggested by the combination of infrared confinement (Kogut and Susskind, 1975a) and asymptotic freedom, and different basic wave functions could given important corrections to matrix elements.

The validity of expansions in powers of v/c [including the zeroth-order expression in Eq. (3.7)] is in doubt when the effective quark mass is $\sim 0.30\ \text{GeV}/c^2$ and hadron dimensions are of the order of less than 1 fm, so that $v/c \sim 0.8$. These considerations have led Le Yaouanc et al. (1972) to propose the retention of the spin-$SU(3)$ structure of the quark wave functions of hadrons, while replacing the nonrelativistic two-component quark spinors χ_s with four-component Dirac spinors

$$u_s(p_j) = \left(\frac{E_j + m}{2E_j}\right)^{1/2} \left(\begin{array}{c} \chi_s \\ \dfrac{\boldsymbol{\sigma} \cdot p_j}{E_j + m}\chi_s \end{array}\right) \tag{3.28}$$

where p_j is the internal quark momentum of quark j, and $E_j = (m^2 + \mathbf{p}_j^2)^{1/2}$. So far as the interaction with real photons is concerned, one obtains matrix elements by sandwiching the untransformed relativistic Hamiltonian, (in the radiation gauge)

$$H_{\text{e.m.}} = \sum_{j=1}^{3} \boldsymbol{\alpha}^j \cdot \mathbf{A}(\mathbf{x}_j) \tag{3.29}$$

between the usual quark wave functions with Eq. (3.28) substituted for the Pauli spinors χ_s. For example, the wave function for the positively charged state is given, in momentum space, by

$$\Delta^+ = u_{1/2}(p_1)u_{1/2}(p_2)u_{1/2}(p_3) \cdot \tfrac{1}{3}(p_1 p_2 n_3 + p_1 n_2 p_3 + n_1 p_2 p_3)F(p_1, p_2, p_3) \tag{3.30}$$

where $\tfrac{1}{3}(p_1 p_2 n_3 + p_1 n_2 p_3 + n_1 p_2 p_3)$ is the $SU(3)$ wave function corresponding to two proton quarks (p_j) and one neutron quark (n_j) and $F(p_1, p_2, p_3)$ is the momentum space wave function.

The recipe above is just to undo the Foldy–Wouthuysen transformation and to revert to what would be an accurate calculation to all orders in v/c in a free field theory. However, the quarks have the same spatial, or momentum space, distribution as in the "nonrelativistic" quark model, given by $F(p_1, p_2, p_3)$ in Eq. (3.30).

It is immediately seen from Eqs. (3.28) and (3.29) that for free photon matrix elements, an important part of the change has the same effect as replacing the constant $1/2m$, where m is the quark mass, which appears in

Eq. (3.16):

$$H_{e.m.}^{NR} = \sum_{j=1}^{3} q_j \left[-i\frac{e}{2m}\boldsymbol{\sigma}_j \cdot (\mathbf{k} \times \mathbf{A}) + 2\mathbf{p}^j \cdot \mathbf{A}\frac{e}{2m} \right] \qquad (3.16)$$

by the operator $1/2F_j$, where $F_j = (m^2 + \mathbf{p}_j^2)^{1/2}$. Now in the harmonic oscillator quark model the distribution of a single quark, momentum $\mathbf{p}$, is given by

$$|\psi|^2 \propto e^{-\mathbf{p}^2/\gamma^2} \qquad (3.31)$$

where $\gamma \sim 0.3$ GeV$/c$. Consequently, in the modified theory given by Eqs. (3.28) and (3.29), any quark mass m, with value less than or of the order of the canonical value 0.34 GeV$/c^2$ of Eq. (3.11), would give the same order of results as the "nonrelativistic" quark model. For example, we could take $m = 0.01$ GeV$/c$ without qualitative change.

By this change in mass one might have hoped to attain enough change in the modified theory to better account for some of the absolute magnitudes and details of the photonic transition matrix elements. However, some trial calculations show that this is probably not so, using oscillator spatial (momentum space) wave functions. For instance, one can obtain an isoscalar component as required in the $d_{13}^+(1520) \to N + \gamma$ transition by using a radius of the d_{13} different from that of the nucleon. But for likely differences in radius the isoscalar component thus generated is too small.

4. Radiative Transitions of Mesons

4.1. Old Mesons

We have considerable information about the magnetic dipole transitions in which a vector meson emits a pseudoscalar meson and a photon: $1^- \to 0^- + \gamma$, the paradigm example, an early triumph for the quark model first calculated by Becchi and Morpurgo (1965b), being $\omega \to \pi\gamma$. Two new results were presented at the 1975 SLAC conference, and we reproduce in Table 3 below the experimental results, mainly those given by Bemporad (1976) at that conference. In the quark model these magnetic dipole transitions go by the *quark magnetic moment*:

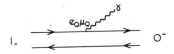

Before we present the quark model results, let us very briefly outline how they are arrived at. The calculation given by Becchi and Morpurgo is as follows. They note that the only covariant vertex for $\omega \to \pi\gamma$ is

$$f\varepsilon^{\alpha\beta\gamma\delta}\left(\frac{\partial A_\beta}{\partial x^\alpha}\right)\left(\frac{\partial \omega_\delta}{\partial x^\gamma}\right)\pi$$

and that for an ω decaying at rest this gives rise to a matrix element

$$\frac{f}{2(2)^{1/2}}\left[\frac{M_\omega}{(m_\pi^2 + k^2)^{1/2}}\right]^{1/2}(\mathbf{k} \times \boldsymbol{\varepsilon}) \cdot \mathbf{n} \tag{4.1}$$

where $\mathbf{n}$ is the direction of the polarization.

A nonrelativistic quark model calculation, proceeding via the quark magnetic moment $e\mu_p$ ($e\mu_p$ being equal to the proton moment from the baryon quark model) gives

$$\frac{\mu_p e}{2^{1/2}}(\mathbf{k} \times \boldsymbol{\varepsilon}) \cdot \mathbf{n} \tag{4.2}$$

and this latter expression (4.2) gives a width of $0.6\,MeV$ compared with the experimental value of $0.9\,MeV$, which is certainly not in exact agreement, but is perhaps as good as one could expect when applying a nonrelativistic model to a strongly recoiling system.

However, Becchi and Morpurgo remarked that such a quark model calculation would be accurate in nonrecoil circumstances, that is if $m_\pi \simeq M_\omega$, $k \simeq 0$, and then one could equate Eqs. (4.1) and (4.2) and find f:

$$\frac{f}{2(2)^{1/2}}\left(\frac{M_\omega}{\sim M_\omega}\right)^{1/2} = \frac{\mu_p e}{2^{1/2}} \quad \text{or} \quad f = 2e\mu_p \tag{4.3}$$

They then *assume* that f is only weakly dependent on mass and use the matrix element (4.1) with the value (4.3) for f, obtaining

$$\Gamma_{\text{calc}} = 1.17\ \text{MeV} \tag{4.4}$$

to be compared with $\Gamma_{\text{exp}} = 0.9$.

What now is the situation on the other transitions? To answer this we must know the magnetic moments of the different quarks. Now the level spacing in baryons, plus the baryon transition rates (which give information on the radius of the quark wave functions), suggest $m_p = 0.3\,\text{GeV}/c^2$ and indeed, taking $m_p = 336\,\text{MeV}/c^2$ gives, for the Dirac moment of the proton quark,

$$\mu_p = \frac{1}{2m_p} = \mu_{\text{proton}} \tag{4.5}$$

Decuplet level spacing and other considerations give

$$m_\Lambda = 476\,\text{MeV}/c^2$$

leading to

$$\mu_\Lambda = \frac{1}{2m_\lambda} = 0.7\mu_{\text{proton}} \qquad (4.6)$$

giving closer agreement with experiment on the magnetic moment of the $\Lambda(1115)$ than the $SU(3)$ prediction, which puts $\mu_\Lambda = \mu_p$. The new predicted value is -0.66, the $SU(3)$ prediction is -0.93, and experiment is -0.67 ± 0.06 nuclear magnetons.

The other information needed to calculate the rates are the wave functions of the ϕ and the η, that is, the proportions of the strange and nonstrange parts of the wave functions

$$S = \bar\lambda\lambda, \qquad N = (1/2^{1/2})(\bar pp + \bar nn) \qquad (4.7)$$

since by the Zweig rule, or quark diagrams, there are to first approximation no transitions between N and S. Tests, related to the new mesons and gluon exchange, are discussed by Harari (1976b). As indicated by the small branching ratio into three pions, and the mass formula, the ϕ is nearly all $\bar\lambda\lambda$

$$\phi \simeq S + (0.08)N \qquad (4.8)$$

while "ideal mixing" is S only. The mass-mixing from $SU(3)$-breaking formulas gives (Morpurgo, 1973)

$$\eta = (\cos\theta)S + (\sin\theta)N \qquad (4.9)$$

where $45° \leqslant \theta \leqslant 58°$. Since in $\phi \to \eta\gamma$ only the $(\cos\theta)S$ part of η contributes there is a factor $\cos^2\theta$ in the rate, which lies between $\frac{1}{2}$ and $\sim\frac{1}{4}$. The experimental comparison is shown in Table 3 for the mesonic radiative

Table 3. *Comparison of Experimental Results of the Width* Γ *(in keV) with Quark Model Calculations for the Mesonic Radiative Transitions,* $1^- \to 0^- + \gamma$, *Proceeding by the Quark Magnetic Moment*

Process	Quark model		Experiment
	(1)	(2)	
$\omega \to \pi\gamma$	600	1170	~900
$\phi \to \eta\gamma$	36–71	55–110	65 ± 15 (NEW)
$\phi \to \pi^0\gamma$	9?[a]	18?[a]	5.9 ± 2.1 (NEW)
$\rho^- \to \pi\gamma$	66	130	35 ± 10
$\rho^0 \to \eta\gamma$	62–42	80–55	<160
$K^{*+} \to K^+\gamma$	75	112	<80[b]
$K^{*0} \to K^0\gamma$	133	200	75 ± 25

[a]These numbers are sensitive to the small admixture of nonstrange quarks in ϕ, which we have assumed to be 0.08. If we were to replace this by 0.06, for example, the numbers given would be 10 for (2) and 6 for (1).

[b]Bemporad *et al.* (1973).

transitions, $1^- \to 0^- + \gamma$, proceeding by the quark magnetic moment. Column 1 of the calculations is the straightforward "nonrelativistic" calculation using expression (4.1) as matrix element; column 2 calculates using the quasirelativistic prescription of Becchi and Morpurgo (1965b) explained in the text. Where a theoretical range is given, it corresponds to uncertainty in the mixing angle. In terms of the octet–singlet mixing angle for η defined by $\eta = -\eta_1 \sin \theta_p + \eta_8 \cos \theta_p$, 45° and 58° correspond, respectively, to $\theta_p = -10°$ and $\theta_p = -23°$; $-10°$ comes from a quadratic mixing formula, and $-23°$ from a linear mixing formula. Other theoretical arguments—for example, that of Chanowitz (1975) using the $\eta \to \gamma\gamma$ and $\eta \to \pi^+\pi^-\gamma$ rates—favor $\theta_p \sim -10°$ or (numerically) less. This smaller mixing angle gives the worse agreement with the experimental width of $\phi \to \eta\gamma$, as is seen from Table 3.

If we had used the $SU(3)$ prescription for the quark moments, $\mu_\Lambda = \mu_p$, the K^{*0} transitions calculated would have been increased by a factor 1.4 and the $\phi \to \eta\gamma$ by a factor of 2, in both cases giving a worse agreement with experiments. On the other hand, $K^{*+} \to K^+\gamma$ would have been decreased by a factor of 0.6, giving better agreement with experiment.

We see a qualitative agreement with experiment, but with variations of a factor of about 3 in the rate or 1.7 in the matrix element for column 2 and less for column 1. This qualitative agreement, with experimentally large rates predicted large, and small rates predicted small, but with 40%–70% disagreement in the magnitude of the matrix elements, is what we also get in the radiative transitions of baryons.

The uncertainties in the ϕ and η wave functions prevent us from assessing the extent of the discrepancy in the decays $\phi \to \eta\gamma$, $\phi \to \pi^0\gamma$, but the apparently large discrepancy between theory and experiment (Gobbi et al., 1974) in the decay $\rho^- \to \pi^-\gamma$ is surprising; one would have expected an experimental result ~ 100 keV rather than 35 ± 10 keV, since the kinematics of $\rho \to \pi\gamma$ and $\omega \to \pi\gamma$ are nearly the same and the predicted ratio—say R—of these two decays only involved the isospin Clebsch–Gordan coefficients of the quark model and the U-spin singlet nature of the photon. In support of this expectation one should note that the "vector dominance" prediction (Gell-Mann et al., 1962; Morpurgo, 1973) of the ratio $\Gamma(\pi^0 \to \gamma\gamma)/\Gamma(\omega \to \pi^0\gamma)$ agrees with experiment to within about 10% (Morpurgo, 1973; Chanowitz, 1975) and the vector dominance argument involves the ratio R. Further, it should be noted that the experimental result for $\rho^- \to \pi^-\gamma$ is obtained by assuming that the coefficient of the nuclear amplitude in the Primakoff effect is independent of A. If this assumption is relaxed then one can only put a lower limit of 30 ± 10 keV and an upper limit of 80 ± 10 keV on the $\rho \to \pi\gamma$ width (Gobbi et al., 1974).

A full discussion of the radiative decays of the old mesons has been given recently by Etim-Etim and Greco (1976).

4.2. New Mesons

4.2.1. The J/ψ, and Associated, Mesons

Since particles that have quantum numbers in the family 0^{++}, 1^{++}, ..., are observed in radiative decays of ψ' (3.68), at energies intermediate between J/ψ (3.1) and ψ(3.68) the most natural interpretation of this system of particles is that they are two-fermion systems of spin $\frac{1}{2}$ fermions with J/ψ (3.1) being 1^3s_1, ψ' (3.68) being a radial excitation 2^3s_1, and the intermediate levels being ^3p_J states, $J = 0, 1, 2$. This interpretation is shown in Fig. 6 and the detailed level assignment is discussed in Section 4.2.3 below.

Even without the hypothesis of charmed quarks (Glashow *et al.*, 1970), a natural interpretation of this new spectrum would have been in terms of a fourth quark. In fact, the ψ particles were predicted on the basis of the charm hypothesis (Appelquist and Politzer, 1975). We shall adopt the point of view that the ψ spectrum is formed from a fourth quark, c, though nothing in our discussion will be specific to the weak interaction properties of the GIM charmed quark.

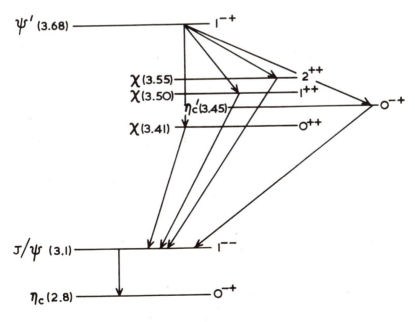

Fig. 6. Energy level diagram of charmonium.

4.2.2. Magnetic Dipole Transition

There has been reported from DESY a resonance at $2.83\,\text{GeV}/c$ observed in the modes

$$\psi(3.1) \rightarrow \eta(2.83) + \gamma$$
$$\llcorner \gamma + \gamma \qquad (4.10)$$

It should be emphasized that this state awaits further experimental confirmation. The state at 2.83 is the candidate for the predicted η_c resonance, the pseudoscalar 0^{-+} bound $\bar{c}c$ state. If so, Eq. (4.10) is a $1^- \rightarrow 0^- + \gamma$ decay of the kind just considered, proceeding by the *charmed* quark magnetic moment. If we take the canonical mass of $\sim 2\,\text{GeV}/c^2$ for the charmed quark, then we obtain $\mu_c = 1/2m_c = \mu_p/6$. The factor of $1/6$ is fortunate from the point of view of experimental comparison because it leads to a width

$$(3.1 \rightarrow 2.83 + \gamma) = 16\,\text{keV} \qquad (4.11)$$

which is still too large a proportion of the total width $(3.1 \rightarrow \text{all}) = 70\,\text{keV}$, since the search for monochromatic γ rays (Simpson *et al.*, 1975) sets an upper limit of about 7 keV on the width. Without the factor of $1/6$ in the amplitude coming from the assumption of point quarks (or at least zero anomalous magnetic moment), the discrepancy would have been a factor of 10 in the amplitude or 100 in the width. The discrepancy would be further reduced if the mass of the charmed quark were, in fact, greater than $2\,\text{GeV}/c^2$, as discussed below, or if the assumed η_c had a considerable admixture of noncharmed quarks from mixing with η, η'. While some such mixing seems plausible, it is most unlikely to be large enough to account for a factor of 2 in the rate. A general discussion of the η, η', η_c mass mixing formalism is given by Gaillard *et al.* (1975).

4.2.3. Electric Dipole Transitions

The χ (3.41) and two states at 3.50 and 3.55 GeV, respectively, have been discovered (Braunschweig *et al.*, 1975; Feldman *et al.*, 1975; Tanenbaum *et al.*, 1975; Luth, 1976; Goldhaber, 1976) in radiative decays of the ψ' (3.68) and have been interpreted as three of the four even charge conjugation states of charmonium predicted to lie between J/ψ (3.1) and ψ' (3.68). A fourth state at 3.45 GeV has also been observed (Daboul and Kraseman, 1975) in radiative decay of the ψ' (3.68), though unlike the other three this 3.45 GeV state has not yet any visible hadronic decay modes. The predicted states in question are 0^{++}, 1^{++}, and 2^{++}, being ^3P_J states of charmonium; and also 0^{-+}, a $^1S_0$ state, being the first radially excited state of paracharmonium commonly known as n'_c, whose ground state η_c was discussed in Section 4.2.2.

Of these observed states the χ (3.41) is established from observed hadronic decays to be one of the series $0^{++}, 2^{++}, 4^{++}, \ldots$. The angular distribution in $\psi' \to \gamma + \chi$ (3.41) is in accord with that appropriate to the spin-0 assignment—namely, $(1 + \cos^2 \theta)$—and this spin-0 angular distribution is not observed in the 3.50 and 3.55 region (Goldhaber, 1976); these observations suggest that the 0^{++} state is at 3.41 GeV and the 0^{-+} state at 3.45 GeV. In accord with these assignments Chanowitz and Gilman (1976) have given arguments for assigning χ (3.41), χ (3.50), and χ (3.55) to the 0^{++}, 1^{++}, and 2^{++} states, respectively, and the 3.45 to the 0^{-+} state. Alternatively Harari (1976a) has suggested assigning the 3.45 level to be a 2^{-+} level, a $^1D_2$ state which would be the first Regge recurrence of the lowest 0^-, $^1S_0$ state. We adopt the orthodox point of view on the observed states, which is illustrated in the level diagram (Fig. 6). We do not include in the level diagram the broader 1^{--} states observed at higher energies, since our purpose is mainly with the radiative transitions, which have not been observed for these higher, broader levels where they will have a small branching fraction.

Precise experimental information on the radiative decay widths $\psi' \to \gamma + \chi$ is not available at the time of writing, but there are some indications. An experiment searching for any monochromatic γ rays from ψ' decay sets an upper limit of 5% on such decays (Simpson *et al.*, 1975):

$$\frac{\Gamma(\psi' \to \gamma + \chi)}{\Gamma(\psi' \to \text{all})} < 5\% \qquad (4.12)$$

On the other hand, later experiments give the value of about 8% for the $\chi = {}^3P_0$ ratio (Whitaker *et al.*, 1976). A branching ratio of 5% corresponds to a width of 11 keV and one of 8% to 18 keV.

A question that presents itself is whether these observed levels and the radiative transition rates can be accommodated within a potential theory with constituent quarks. In this connection it is worth noting that the ascending level ordering 0^{++}, 1^{++}, 2^{++} is that given by the spin-orbit force

$$\frac{1}{m^2 r} \frac{\partial V}{\partial r} \mathbf{L} \cdot \mathbf{S} \qquad (4.13)$$

from a short-range Coulomb type potential such as is suggested by the asymptotic freedom calculations, though the observed relative *splitting* is different, suggesting the presence of, for example, tensor forces.

4.2.4. Qualitative Calculations

The transitions are all electric dipole, and as a first guide to their magnitude one can make the usual nonrelativistic calculation. For example,

$$\Gamma[\psi' \to \chi(3.41) + \gamma] = \frac{16}{27} \cdot \frac{1}{137} \cdot \frac{(2J+1)}{9}[(3.68 - 3.41)\rho]^2 \frac{3.41}{3.68} k$$

(4.14)

where J is the spin of the $\chi(3.41)$ (so $2J + 1 = 1$), k is the photon energy (0.26 GeV), and ρ is approximately the interquark distance in the wave functions, the exact relationship of ρ to the root mean square interquark distance depending on the wave functions used. The expression (4.14) reduces to

$$\Gamma[\psi' \to \chi(3.41) + \gamma] = 8.66\rho^2 \text{ keV}$$

(4.15)

and the corresponding expression for a χ in the neighborhood of 3.52 GeV/c is

$$\Gamma[\psi' \to \chi + \gamma] = 2.16(2J + 1)\rho^2 \text{ keV}$$

(4.16)

where J is the spin of the χ and ρ is inserted in these formulas in units of GeV^{-1}.

We see that the search for monochromatic γ rays experiment (Simpson et al., 1975) sets an upper limit of about 1.1 GeV^{-1} on ρ, though later experiments are somewhat less restrictive (Whitaker et al., 1976). We conclude that within the charmonium picture, either (a) the radius of the charmed states is smaller than that of the old mesons, *or* (b) the wave functions are of an extraordinary type [or (c) both together, in some milder degree]. By its nature (b) requires complicated considerations and we shall not pursue it here, but we will consider a little further the consequences of possibility (a).

Firstly, we note that if $\phi(\mathbf{r})$ is the spatial wave function of the decaying vector meson, J/ψ, ψ', ψ'', ..., and M is its mass, then

$$\Gamma_{e^+e^-} \simeq \Gamma_{\mu^+\mu^-} \simeq 3\left(\frac{1}{137}\right)^2 \frac{16\pi}{3} \frac{4}{9}\left[\frac{\phi(0)}{M}\right]^2$$

(4.17)

where the extra factor of 3 over the van Royen–Weiskopf expression arises from color (Barbieri et al., 1975, 1976); the corresponding formulas for the ρ, ω, ϕ decays would have $\frac{1}{2}$, $\frac{1}{18}$, $\frac{1}{9}$, respectively, instead of the factor $\frac{4}{9}$. Bearing in mind the decay widths 6.5, 0.75, 1.3 keV for the ρ, ω, ϕ, respectively, and the widths (Luth et al., 1975) 5 keV, 2 keV, ..., for the J/ψ, ψ', ψ'', ..., we see that the formula (4.17) implies that the value of the spatial wave function at the origin is greater for the new mesons than for the old mesons, because of the $1/M$ factor. And, indeed, a smaller radius, as in (a), implies an increase in $\phi(0)$.

Another consequence, in some respects less desirable, is that a smaller radius for a given level spacing implies a relatively large mass m_c of the charmed quark component with stronger binding. [A larger mass than the previously assumed $2\,\text{GeV}/c$ most desirably gives a smaller rate when substituted in our previous calculation for $J/\psi(3.1) \to \eta_c(2.83) + \gamma$; at the same time a mass larger than $2\,\text{GeV}/c$ is somewhat less desirable for the $K_L - K_S$ mass difference calculations (Gaillard and Lee, 1974), and definitely less desirable in the quantitative calculations described in Section 4.2.5 below.]

If the quark p-wave levels, χ, are approximately intermediate between the lowest s-state $J/\psi(3.1)$ and the first radially excited s-state ψ' (3.68), it may not be unreasonable for the purpose of illustration, to approximate the wave functions by harmonic oscillator wave functions (Schnitzer, 1975), with a spring constant appropriate to the particlar energy levels being considered. Apart from normalization constants, the ground state J/ψ, P wave (χ), and first radially excited state (ψ') have radial wave functions

$$\exp\left(-\frac{1}{2}\frac{r^2}{\rho^2}\right), \qquad r\exp\left(-\frac{1}{2}\frac{r^2}{\rho^2}\right), \qquad \left(3-2\frac{r^2}{\rho^2}\right)\exp\left(-\frac{1}{2}\frac{r^2}{\rho^2}\right)$$

$$(4.18)$$

corresponding to root mean square quark separation

$$\bar{r} = (\tfrac{3}{2})^{1/2}\rho, \qquad (\tfrac{5}{2})^{1/2}\rho, \qquad (\tfrac{7}{2})^{1/2}\rho \qquad (4.19)$$

The oscillator formulas for the e^+e^- widths are, from Eq. (4.17),

$$J/\psi: \qquad \Gamma_{e^+e^-} = 22.2/\rho^3 \text{ keV} \qquad (4.20)$$

$$\psi': \qquad \Gamma_{e^+e^-} = 23.6/\rho^3 \text{ keV} \qquad (4.21)$$

where ρ is the parameter of Eqs. (4.14)–(4.16) and is to be inserted in these formulas in units of GeV^{-1} (the corresponding formula for the ρ-meson is $\Gamma_{e^+e^-} = 420/\rho^3 \text{ keV}$). From Eqs. (4.20) and (4.21) we see that values of ρ of about $1.7\,\text{GeV}^{-1}$ and $2.2\,\text{GeV}^{-1}$ for the J/ψ and ψ', respectively, are needed to obtain agreement with the observed rates and that the values correspond to radii

$$\bar{r}(J/\psi) = (\tfrac{3}{2})^{1/2}\rho = 2.0 \text{ GeV}^{-1} = 0.40 \text{ fm} \qquad (4.22)$$

$$\bar{r}(\psi') = (\tfrac{3}{2})^{1/2}\rho = 4.1 \text{ GeV}^{-1} = 0.82 \text{ fm} \qquad (4.23)$$

The value of $\rho = 1.1$ deduced from the $\psi' \to \chi + \gamma$ transition would give an interquark distance

$$\bar{r}(\psi') = (\tfrac{7}{2})^{1/2}\rho = 0.41 \text{ fm} \qquad (4.24)$$

The inconsistency between (4.23) and (4.25) points to the need for somewhat more sophisticated potentials and wave functions (Section 4.2.5 below). For comparison, the value of ρ required to give the ρ-meson e^+e^- width is $\rho = 4\,\text{GeV}^{-1}$ corresponding to

$$\bar{r}(\rho\text{-meson}) = 0.84\,\text{fm}$$

and this is to be compared with Eq. (4.22); the corresponding value for the ρ'-meson being a radial excitation would be expected to be larger—say, about 1.5 fm—and this should be compared with Eq. (4.23).

The oscillator level spacing formula is

$$2/\rho^2 = m_c \Delta E \tag{4.25}$$

and 0.3 GeV is an approximate value for the level spacing ΔE; the value $\rho = 1.1$, from the $\psi' \to \chi + \gamma$ transition rates gives $m_c = 5.5\,\text{GeV}/c^2$, $\rho = 2.2$ from the $\psi' \to e^+e^-$ gives $m_c = 1.4\,\text{GeV}/c^2$ and $\rho = 1.7$ from $J/\psi \to e^+e^-$ gives $m_c = 2.4\,\text{GeV}/c^2$. These varying values signal an inadequacy in the harmonic oscillator treatment; fortunately theoretical guidance is available to form a basis for a more qualitative calculation described in the following sections.

4.2.5. Quarks in an Instantaneous Potential

Already before the discovery of the J/ψ (3.1), Appelquist and Politzer (1975) had begun to formulate the idea of charmonium as a consequence of the Glashow *et al.* (1970) charmed quarks interacting by massless colored gluons to give rise to an asymptotically free theory. The asymptotically free aspect of the theory leads to a Coulomb-like interaction at short distances giving rise to orthocharmonium states ($^3S_1$ states such as the $J/\psi, \psi'$) and paracharmonium ($^1S_0$ states—the η_c). On the same basis Appelquist *et al.* (1975) predicted p-wave states of charmonium, lying between the then known J/ψ (3.1) and ψ' (3.68), which have subsequently been found as the χ (3.41) etc., and which have been described in Section 4.2.3.

Another aspect of the same non-Abelian color gauge theory is strong forces at larger distances, which may be confining forces. The lattice gauge theories (Kogut and Susskind, 1975a) suggest a static-spin and flavor independent linear confining force. Taken with the asymptotically free Coulomb-like interaction this suggests an effective instantaneous potential of the form

$$V(r) = \lambda r - \tfrac{4}{3}\alpha_s \gamma_\mu a(1/r)\gamma^{\mu b} + C \tag{4.26}$$

where C is a constant addition to the potential representing effects (such as the effect of higher open channels (Eichten *et al.*, 1975b) not explicitly included in Eq. (4.26); λ is the confinement coupling constant, and α_s is the coupling constant derived from asymptotically free theories. In principle α_s is dependent on the momenta at which the particles are interacting and, of course, it is quite possible to have such a momentum-dependent potential. More simply, it can be taken to have some average value appropriate to the mass of the meson under calculation. For example, values that have been considered (De Rujula *et al.*, 1975) are

$$\alpha_s[\phi(1.0)] = 0.5, \qquad \alpha_s[J/\psi(3.1)] = 0.28 \qquad (4.27)$$

the connection between these two values being given by the asymptotic freedom formula appropriate to four flavors and three colors:

$$\alpha_s[\psi] = \left\{ 1 - \frac{25}{12\pi}\alpha_s[\phi]\ln\left(\frac{1.019}{3.1}\right)^2 \right\}^{-1} \alpha_s[\phi] \qquad (4.28)$$

In nonrelativistic kinematics the potential of Eq. (4.26) is

$$V = \lambda r - \tfrac{4}{3}\alpha_s(1/r) + C \qquad (4.29)$$

and is illustrated by the solid line in Fig. 7. As suggested by Kogut and Susskind (1975*b*), the confining potential may be effectively softened at large distances by the effect of quark–antiquark pair creation, as shown by the dotted line, the escape of probability represented by that dotted potential being in the form of quark–antiquark pairs and *not* of single quarks, which remain confined.

Various authors (Eichten *et al.*, 1975*a*; Harrington *et al.*, 1975; Gunion and Wiley, 1975) have shown, with various representations of a binding or confining potential, that an effective quark mass in the region 1.5–2.0 GeV/c will give the J/ψ, ψ' at their observed energies and the center-of-gravity p wave in the region 3.4–3.5 GeV. In particular, Eichten *et al.* (1975*a*) have investigated a potential corresponding to Eq. (4.29), finding the center of gravity of the p-wave multiplet [Eq. (4.29) does not give a p-wave splitting] at 3.45 and too large $2S \to {}^3P_J$ transitions compared with the experimental values (Whitaker *et al.*, 1976). In a later paper (Eichten *et al.*, 1976) they attempt to correct the radiative transition values by taking the charmed meson decay channel into the wave function. The radiative transitions have been calculated in approximate agreement with

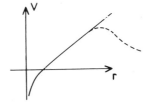

Fig. 7. The potential of Eq. (4.26) (solid line), softened at large distances (dotted line).

experiments by Henriques *et al.* (1976). Jackson (1976) has used dipole sum rules to obtain upper and lower bounds on the radiative transitions $^3P_J \rightarrow \gamma + J/\psi$.

4.2.6. Level Splitting in a Potential

The asymptotically free part of (4.26) has been represented for convenience in the Feynman gauge. In relativistic kinematics both the confining and the asymptotically free part give rise to a spin-orbit force. Expanding in powers of $1/m$ (equivalently v/c) where m is the quark mass, one finds if λr in Eq. (4.26) is a Lorentz scalar (Henriques *et al.*, 1976)

$$\frac{1}{2m^2} \mathbf{L} \cdot \mathbf{S} \left[-\frac{1}{r}\frac{\partial}{\partial r}(\lambda r) + \frac{3}{r}\frac{\partial}{\partial r}\left(-\frac{4}{3}\alpha_s\frac{1}{r}\right) \right]$$

$$= \left(-\frac{\lambda}{r} + \frac{4\alpha_s}{r^3}\right)\frac{1}{4m^2}[J(J+1) - L(L+1) - S(S+1)] \qquad (4.30)$$

the opposite sign and different magnitude of the λ and α_s parts coming from the scalar and vector nature of the original potentials in Eq. (4.26) [the same result would hold using the radiation gauge instead of the Feynman gauge in Eq. (4.26)].

In Eq. (4.30), J is the meson spin, S is the quark spin, and L is the quark orbital angular momentum. The p-wave triplet with $J = 0, 1, 2$ and $L = 1$, $S = 1$ would be split by Eq. (4.30) *alone* in the usual ratio $(^3P_2 - {}^3P_1)/(^3P_1 - {}^3P_0) = 2/1$, which would only be in accord with experiment if the $^3P_2$ level were near to 3.70. It is interesting to note that though the effect of the asymptotically free part is to give the levels in the order $^3P_0$, $^3P_1$, $^3P_2$ the scalar confining potential alone would give the opposite order, so it diminishes the splitting. (A *vector* confining potential would increase the splitting.)

Expanding in powers of $1/m$ also gives a tensor force

$$\frac{4}{3}\alpha_s\frac{1}{4m^2}\left[\frac{3(\boldsymbol{\sigma}^a \cdot \mathbf{r})(\boldsymbol{\sigma}^b \cdot \mathbf{r})}{r^2} - \boldsymbol{\sigma}^a \cdot \boldsymbol{\sigma}^b\right]\frac{1}{r^3} \qquad (4.31)$$

giving a contribution to the energy levels of $^3P_0$, $^3P_1$, $^3P_2$ proportional to $(\alpha_s/3m^2)(-4, 2, -4/15)$, respectively. Thus, the tensor force tends to decrease or reverse the $(^3P_2 - {}^3P_1)/(^3P_1 - {}^3P_0)$ ratio from the spin-orbit coupling, which gives contributions proportional to $(\alpha_s/m^2)(-4, -2, 2)$ together with a (reversed) contribution from the λ term. Because of the λ term and possible second-order effects, careful calculation is appropriate, provided one is prepared to grant some validity to the instantaneous potential model.

Henriques *et al.* (1976) have shown that the correct *p*-wave levels can be attained by a potential of the form (4.26) (softened at large distances) together with the J/ψ and ψ' masses, and the approximate radiative widths $\Gamma(J/\psi \to e^+e^-)$, $\Gamma(\psi' \to e^+e^-)$, $\Gamma(\psi' \to \gamma + {}^3P_J)$. The value of λ obtained is about 0.7 GeV/fm, while in GeV units the value of the running coupling constant α_s in the neighborhood of ψ (3.1) is in the region ~0.25 to ~0.4 (where values of α_s near 0.25 correspond to the use of a momentum-dependent asymptotic freedom formula for α_s inside the equation of motion). These values of α_s are larger than those obtained by quantum chromodynamic calculation of the ratio $\Gamma(J/\psi\text{-hadrons})/\Gamma(J/\psi \to e^+e^-)$ which is about 0.2.

The *p*-wave splitting, and other features such as the *S–D* splitting, and the $0^- - 1^-$ splitting, can be critically dependent on the Lorentz-covariant characteristics of the long-range potential. Schnitzer (1976*a*) has shown that, if the Lorentz scalar–scalar form $1^a \lambda r 1^b$ of Henriques *et al.* (1976) is replaced by a form $\gamma_\mu^a \lambda r \gamma^{\mu b}$, then it is not possible to attain the *p*-wave splitting without extra terms in the potential. Schnitzer (1976*b*) has suggested that extra terms such as an effective quark–gluon magnetic moment may be required for correcting the tensor force and the spin–spin interaction.

If the α_s term predominates in Eq. (4.30) for *s* and *d* levels, then the spin-orbit coupling splits corresponding *s* and *d* levels, with the *d* wave higher. The tensor force gives no contribution to the *s*-wave energy but gives a negative contribution to the *d*-wave energy. Also, the tensor force directly couples the *s* and *d* levels, thus increasing any splitting already induced.

Appendix

To express experimental quantities in terms of the helicity amplitudes H_i we write the differential cross section in the center-of-mass frame as [see Eq. (2.4)]

$$\frac{d\sigma}{d\Omega} = \frac{q}{k} \sum |\langle \chi_f | \mathscr{F} | \chi_i \rangle|^2 \tag{A.1}$$

where χ_i, χ_f are two-component spinors appropriate to the polarization of the initial and final nucleon, and $\mathscr{F}$ contains the polarization of the γ according to Eq. (2.3); the summation (or average where appropriate) is taken over unobserved spins. The matrix elements in Eq. (A.1) can be trivially expressed as superpositions of the helicity matrix elements of Eq. (2.6) and via Eq. (2.9) in terms of the $H_{\text{flip}}(\theta)$ and ϕ. One readily obtains the following expressions, which are needed for the data fitting and comparison:

(i) Differential cross section $\sigma(\theta)$ [see Eq. (2.10)]:

$$\sigma(\theta) = \frac{q}{k}[|H_N(\theta)|^2 + |H_D(\theta)|^2 + |H_{SP}(\theta)|^2 + |H_{SA}(\theta)|^2] \qquad (A.2)$$

(ii) Differential cross sections for photons polarized perpendicular and parallel to the production plane:

$$\sigma_\perp(\theta) = \tfrac{1}{2}(q/k)[|H_{SP}(\theta) + H_{SA}(\theta)|^2 + |H_N(\theta) - H_D(\theta)|^2] \qquad (A.3)$$

$$\sigma_\parallel(\theta) = \tfrac{1}{2}(q/k)[|H_{SP}(\theta) - H_{SA}(\theta)|^2 + |H_N(\theta) + H_D(\theta)|^2] \qquad (A.4)$$

(iii) Polarized photon asymmetry [from (ii)], $\Sigma(\theta)$:

$$\Sigma(\theta)\sigma(\theta) \equiv \sigma_\perp - \sigma_\parallel$$

$$= 2(q/k)\mathrm{Re}[H_{SP}(\theta)H_{SA}^*(\theta) - H_N(\theta)H_D^*(\theta)] \qquad (A.5)$$

(iv) Polarization of the final nucleon in the direction $\mathbf{k} \times \mathbf{q}$, $P(\theta)$:

$$P(\theta)\sigma(\theta) = (2q/k)\mathrm{Im}[H_{SP}(\theta)H_D^*(\theta) + H_N(\theta)H_{SA}^*(\theta)] \qquad (A.6)$$

(v) Polarized target asymmetry, where σ_+ and σ_- are the cross sections for nucleons polarized parallel and antiparallel, respectively, to $\mathbf{k} \times \mathbf{q}$, $T(\theta)$:

$$T(\theta)\sigma(\theta) \equiv (\sigma_+ - \sigma_-)$$

$$= 2(q/k)\,\mathrm{Im}[H_{SP}(\theta)H_N^*(\theta) + H_D(\theta)H_{SA}^*(\theta)] \qquad (A.7)$$

There are also some intrinsically double polarization quantities which are the subject of experimental measurement, involving a polarized beam and a polarized target. The double polarization quantities E, F, G, H are defined in terms of experimental observations of a differential cross section $d\sigma/d\Omega$ with a polarized target and polarized beam:

$$d\sigma/d\Omega = \sigma(\theta)[1 - (P_T \cos 2\phi)\Sigma - (P_x P_T \sin 2\phi)H + (P_x P_\odot)F + (P_y)T$$

$$- (P_y P_T \cos 2\phi)P + (P_z P_T \sin 2\phi)G - (P_z P_\odot)E] \qquad (A.8)$$

where (P_x, P_y, P_z) is the polarization of the target (the z direction being the beam direction and the y direction the normal to the reaction plane), P_T is the transverse polarization of the beam at an angle ϕ to the reaction plane and $P_\odot$ is the degree of right circular polarization of the beam.

The quantities G, H, E, and F are given in terms of the helicity amplitudes by

$$G(\theta)\sigma(\theta) = -2(q/k)\operatorname{Im}[H_{SP}(\theta)H^*_{SA}(\theta) + H_N(\theta)H^*_D(\theta)] \qquad \text{(A.9)}$$

$$H(\theta)\sigma(\theta) = -2(q/k)\operatorname{Im}[H_{SP}(\theta)H^*_D(\theta) + H_{SA}(\theta)H^*_N(\theta)] \qquad \text{(A.10)}$$

$$E(\theta)\sigma(\theta) = (q/k)[|H_{SA}(\theta)|^2 - |H_{SP}(\theta)|^2 + |H_D(\theta)|^2 - |H_N(\theta)|^2] \qquad \text{(A.11)}$$

$$F(\theta)\sigma(\theta) = 2(q/k)\operatorname{Re}[H_{SA}(\theta)H^*_D(\theta) + H_{SP}(\theta)H^*_N(\theta)] \qquad \text{(A.12)}$$

References

Appelquist, T., and Politzer, H. D. (1975), *Phys. Rev. Lett.* **34**, 43.
Appelquist, T., De Rujula, A., Politzer, H. D., and Glashow, S. L. (1975), *Phys. Rev. Lett.* **34**, 365.
Argyres, E. N., Contogouris, A. P., Holden, J. P., and Svec, M. (1973), *Phys. Rev. D* **8**, 2068.
Barbieri, R., Gatto, R., Kögerler, R., and Kunski, Z. (1975), *Phys. Lett.* **57B**, 455.
Barbieri, R., Gatto, R., and Kögerler, R. (1976), *Phys. Lett.* **60B**, 183.
Barbour, I. M., and Crawford, R. L. (1976), *Nucl. Phys.* **B111**, 358.
Barbour, I. M., and Moorhouse, R. G. (1974), *Nucl. Phys.* **B69**, 637.
Barnes, K. J., Carruthers, P., and von Hippel, F. (1965), *Phys. Rev. Lett.* **14**, 82.
Becchi, C., and Morpurgo, G. (1965*a*), *Phys. Rev.* **140B**, 687.
Becchi, C., and Morpurgo, G. (1965*b*), *Phys. Lett.* **17**, 352.
Bemporad, C. (1976), *Proceedings of the 1975 International Symposium on Lepton and Photon Interactions at High Energies* Stanford University, Aug. 21-27, 1975. Ed. W. T. Kirk (Stanford: Stanford Linear Accelerator Center).
Bemporad, C., Beusch, W., Duffey, J. P., Polgar, E., Websdale, D., Wilson, J. D., Zaimidoroga, O., Frendenreich, K., Frosch, R., Gentit, F. X., Muhlemann, P., Astbury, P., Codling, J., Lee, J. G., and Letheren, M. (1973), *Nucl. Phys.* **B51**, 1.
Berends, F. A., and Donnachie, A. (1975*a*), *Nucl. Phys.* **B84**, 342.
Berends, F. A., and Donnachie, A. (1975*b*), private communication.
Berends, F. A., and Weaver, D. L. (1971), *Nucl. Phys.* **B30**, 575.
Bowcock, J. E., and Burkhardt, H. (1975), *Rep. Prog. Phys.* **38**, 1099.
Bowler, K. C. (1970), *Phys. Rev. D* **1**, 926.
Braunschweig, W., Martin, H. U., Sander, H. G., Schmitz, D., Sturm, W., Wallraff, W., Berkelman, K., Cords, D., Felst, R., Gadermann, E., Grindhammer, G., Hultschig, H., Joos, P., Koch, W., Kotz, U., Krehbiel, H., Kreinick, D., Ludwig, J., Mess, K. H., Moffeit, K. C., Petersen, A., Poelz, G., Ringel, J., Sauerberg, K., Schmuser, P., Vogel, G., Wiik, B. H., Wolf, G., Buschorn, G., Kotthaus, R., Kruse, U. E., Lierl, H., Oberlack, H., Pretzl, K., Schliwa, M., Orito, S., Suda, T., Totsuka, Y., and Yamada, S. (1975), *Phys. Lett.* **57B**, 407.
Cashmore, K. J., Hey, A. J. G., and Litchfield, P. (1975), *Nucl. Phys.* **98**, 237.
Chanowitz, M. S. (1975), *Phys. Rev. Lett.* **35**, 977.
Chanowitz, M. S., and Gilman, F. J. (1976), SLAC-PUB-1746, LBL-4864.
Chau, Y. C., Dombey, N., and Moorhouse, R. G. (1967), *Phys. Rev.* **163**, 1632.
Chew, G. F., Goldberger, M. L., Low, F. E., and Nambu, Y. (1957), *Phys. Rev.* **106**, 1345.
Chodos, A., Jaffe, R. L., Johnson, K., Thorn, C. B., and Weisskopf, V. F. (1974*a*), *Phys. Rev. D* **9**, 3471.

Chodos, A., Jaffe, R. L., Johnson, K., and Thorn, C. B. (1974b), *Phys. Rev. D* **10**, 2599.
Close, F. E., and Copley, L. A. (1970), *Nucl. Phys.* **B19**, 477.
Close, F. E., and Osborn, H. (1971), *Phys. Lett.* **34B**, 400.
Copley, L. A., Karl, G., and Obryk, E. (1969a), *Phys. Lett.* **29B**, 117.
Copley, L. A., Karl, G., and Obryk, E. (1969b), *Nucl. Phys.* **B13**, 303.
Crawford, R. L. (1975), *Nucl. Phys.* **B97**, 125.
Daboul, J., and Krasemann, H. (1975), Desy preprint No. 75/46.
Dalitz, R. H. 1965, *High Energy Physics*: Proceedings of the 1965 Les Houches School, Ed. C. DeWitt and M. Jacob (New York: Gordon and Breach).
Dalitz, R. H. (1968), *Proceedings of the Second Hawaii Topical Conference in Particle Physics*, 1967, Ed. S. Pakvasa and S. F. Tuan (Honolulu: Univ. of Hawaii).
Dalitz, R. H., and Sutherland, D. G. (1966), *Phys. Rev.* **146**, 1180.
De Rujula, A., Georgi, H., and Glashow, S. L. (1975), *Phys. Rev. D* **12**, 147.
Devenish, R. C. E., Leigh, W. J., Lyth, D. H., and Rankin, W. A. (1971), *Nuovo Cimento* **1A**, 155.
Devenish, R. C. E., Lyth, D. H., and Rankin, W. A. (1972), Danesbury Laboratory report No. DNPL/P109.
Devenish, R. C. E., Lyth, D. H., and Rankin, W. A. (1973), *Phys. Lett.* **43B**, 44.
Devenish, R. C. E., Lyth, D. H., and Rankin, W. A. (1974), *Phys. Lett.* **52B**, 277.
Eichten, E., Gottfried, K., Kinoshita, T., Lane, K. D., and Yan, T. M. (1975a), *Phys. Rev. Lett.* **34**, 369.
Eichten, E., Gottfried, K., Kinoshito, T., Lane, K. D., and Yan, T. M. (1975b), *Phys. Rev. Lett.* **36**, 500.
Etim-Etim and Greco, M., (1976), CERN preprint No. TH-2174.
Faiman, D., and Hendry, A. W. (1965), *Phys. Rev.* **173**, 1720.
Faiman, D., and Hendry, A. W. (1968), *Phys. Rev.* **180**, 1572.
Faiman, D., and Plane, D. E. (1972), *Nucl. Phys.* **B50**, 379.
Feldman, G. J., Jean-Marie, B., Dadoulet, B., Vanucci, F., Abrams, G. S., Boyarski, A. M., Breidenbach, M., Bulos, F., Chinowsky, W., Friedberg, C. E., Goldhaber, G., Hanson, G., Hartill, D. L., Johnson, A. D., Kadyk, J. A., Larsen, R. R., Litke, A. M., Luke, D., Lulu, E. A., Luth, C., Lynch, H. L., Morehouse, C. C., Paterson, J. M., Perl, M. L., Pierre, F. M., Pun, T. P., Rapidis, P., Richter, B., Schwitters, R. F., Tanenbaum, W., Trilling, G. H., Whitaker, J. S., Winkelman, R. C., and Wiss, J. E. (1975), *Phys. Rev. Lett.* **35**, 821.
Feynman, R. P., Kislinger, M., and Ravndal, F. (1971), *Phys. Rev. D* **3**, 1706.
Fujimura, K., Kobayasi, T., and Namiki, M. (1970), *Prog. Theor. Phys.* **43**, 73.
Gaillard, M. K., and Lee, B. W. (1974), *Phys. Rev. D* **10**, 897.
Gaillard, M. K., Lee, B. W., and Rosner, J. L. (1975), *Rev. Mod. Phys.* **47**, 277.
Gell-Mann, M., Sharp, D., and Wagner, W. G. (1962), *Phys. Rev. Lett.* **8**, 261.
Gilman, F. J., and Karliner, I. (1973), *Phys. Lett.* **46B**, 426.
Gilman, F. J., and Karliner, I. (1974), *Phys. Rev. D* **10**, 2194.
Gilman, F. J., Kugler, M., and Meshkov, S. (1973), *Phys. Lett.* **45B**, 481.
Glashow, S. L., Iliopoulos, J., and Maiani, L. (1970), *Phys. Rev. D* **2**, 1258.
Gobbi, B., Rosen, J. L., Scott, H. A., Shapiro, S. L., Strawczynski, L., and Meltzer, C. M. (1974), *Phys. Rev. Lett.* **33**, 1450.
Goldhaber, G. (1976), LBL report No. LBL-4884 (1976).
Gonzales, M. A., and Watson, P. J. S. (1972), *Nuovo Cimento* **12A**, 889.
Gourdin, M., and Salin, Ph. (1963), *Nuovo Cimento* **27**, 193.
Greenberg, D. W. (1964), *Phys. Rev. Lett.* **13**, 598.
Gunion, J. F., and Willey, R. S. (1975), *Phys. Rev. D* **12**, 174.
Harari, H. (1976a), *Phys. Lett.* **64B**, 469.

Harari, H. (1976*b*) *Proceedings of the 1975 International Symposium on Lepton and Photon Interactions at High Energies*, Stanford University, 21–27 August 1975. Ed. W. T. Kirk. (Stanford, Stanford Linear Accelerator Center), p. 317.

Harrington, B. J., Park, S. Y., and Yildiz, A. (1975), *Phys. Rev. Lett.* **34**, 706.

Henriques, A. B., Kellett, B. H., and Moorhouse, R. G. (1976), *Phys. Lett.* **64B**, 85.

Hey, A. J. G., Rosner, J., and Weyers, J. (1973), *Nucl. Phys.* **B61**, 205.

Hontebeyrie, M., Procureur, J., and Salin, Ph. (1973), *Nucl. Phys.* **B55**, 83.

Horgan, R. (1973), *Nucl. Phys.* **B71**, 514 (1973).

Horgan, R., and Dalitz, R. H. (1973), *Nucl. Phys.* **B66**, 135.

Jackson, J. D. (1976), *Phys. Rev. Lett.* **37**, 1107.

Kellett, B. H. (1974), *Ann. Phys. N.Y.* **87**, 60.

Knies, G., Moorhouse, R. G., and Oberlack, H. (1974), *Phys. Rev. D* **9**, 2680.

Kogut, J., and Susskind, L. (1975*a*), *Phys. Rev. D* **11**, 395.

Kogut, J., and Susskind, L. (1975*b*), *Phys. Rev. Lett.* **34**, 767.

Kubota, T., and Ohta, K. (1976), *Phys. Lett.* **65B**, 374.

Lehmann, H. (1954), *Nuovo Cimento Suppl.* **14**, 153.

Le Yaouanc, A., Oliver, L., Pene, O., and Raynal, O. C. (1972), *Nucl. Phys.* **B37**, 552.

Licht, D., and Pagnamenta, A. (1970*a*), *Phys. Rev. D* **2**, 1150.

Licht, D., and Pagnamenta, A. (1970*b*), *Phys. Rev. D* **2**, 1156.

Lipes, R. G. (1972), *Phys. Rev. D* **5**, 2849.

Lipkin, H. J., and Meshkov, S. (1965), *Phys. Rev. Lett.* **14**, 670.

Llewellyn-Smith, C. (1976), *Proceedings of the 1975 International Symposium on Lepton and Photon Interactions at High Energies*, Stanford University, 21–27 August 1975. Ed. W. T. Kirk (Stanford, Stanford Linear Accelerator Center, 1976), p. 709.

Luth, V. (1976), *Bull. Am. Phys. Soc.* **21**, 679.

Luth, V., Boyarski, A. M., Lynch, H. L., Breidenbach, M., Bulos, F., Feldman, G. J., Fryberger, D., Hanson, G., Hartill, D. L., Jean-Marie, B., Larsen, R. R., Luke, D., Morehouse, C. C., Paterson, J. M., Perl, M. L., Pun, T. P., Rapidis, P., Richter, B., Schwitters, R. F., Tanenbaum, W., Vannucci, F., Abrams, G. S., Chinowsky, W., Friedberg, C. E., Goldhaber, G., Kadyk, J. A., Litke, A. M., Lulu, B. A., Pierre, F. M., Sadoulet, B., Trilling, G. H., Whitaker, J. S., Winkelmann, F. C., and Wiss, J. E. (1975), *Phys. Rev. Lett.* **35**, 1124.

Lyth, D. H. (1974), *Proceedings of the XVII International Conference on High Energy Physics*, London, July 1974, Ed. J. R. Smith (Chilton, SRC Rutherford Laboratory), p. II–147.

Melosh, H. J., (1974), *Phys. Rev. D* **9**, 1095.

Metcalf, W. J., and Walker, R. L. (1974), *Nucl. Phys.* **B76**, 253.

Mo, L. W. (1975), *Proceedings of the 1975 International Symposium on Lepton and Photon Interactions at High Energies*, Stanford University, 21–27 August 1975. Ed. W. T. Kirk (Stanford, Stanford Linear Accelerator Center), p. 651.

Moorhouse, R. G. (1966), *Phys. Rev. Lett.* **16**, 771.

Moorhouse, R. G., and Oberlack, H. (1973), *Phys. Lett.* **43B**, 44.

Moorhouse, R. G., and Parsons, N. H. (1973), *Nucl. Phys.* **B62**, 109.

Moorhouse, R. G., Oberlack, H., and Rosenfeld, A. (1974), *Phys. Rev. D* **9**, 1.

Morpurgo, G. (1965), Physics **2**, 95.

Morpurgo, G. (1973), *Proceedings of the 9th International School of Sub-Nuclear Physics "Ettore Majorana": Properties of the Fundamental Interactions*, Ed. A. Zichichi (Editrice Compositori, Bologna).

Noelle, P., Pfeil, W., and Schwela, D. (1971), *Nucl. Phys.* **B26**, 461. Particle Data Group (1976), Data tables, *Phys. Lett.*, April.

Peierls, R. F. (1958), *Phys. Rev. Lett.* **1**, 175.

Perkins, D. H. (1976), *Proceedings of the 1975 International Symposium on Lepton and Photon Interactions at High Energies*, Stanford University, 21–27 August 1975. Ed. W. T. Kirk (Stanford, Stanford Linear Accelerator Center), p. 571.

Petersen, W. P., and Rosner, J. L. (1972), *Phys. Rev. D* **6**, 820.

Petersen, W. P., and Rosner, J. L. (1973), *Phys. Rev. D* **7**, 747.

Pfeil, W., and Schwela, D. (1972), *Nucl. Phys.* **B45**, 379.

Rutherglen, J. G. (1974), *Proceedings of the XVII International Conference on High Energy Physics*, London, July 1974. Ed. J. R. Smith (Chilton, SRC Rutherford Laboratory), p. II-151.

Sakurai, J. J. (1958), *Phys. Rev. Lett.* **1**, 259.

Salin, Ph. (1963), *Nuovo Cimento* **28**, 1295.

Schnitzer, H. J. (1976*a*), *Phys. Rev. D* **13**, 74.

Schnitzer, H. J. (1976*b*), *Phys. Lett.* **65B**, 239.

Simpson, W., Beron, B. L., Ford, R. L., Hilger, E., Hofstadter, R., Howell, R. L., Hughes, E. B., Liberman, A. D., Martin, T. W., O'Neill, L. H., and Resvanis, L. K. (1975), *Phys. Rev. Lett.* **35**, 699.

Tanenbaum, W., Whitaker, J. S., Abrams, G. S., Boyarski, A. M., Breidenbach, M., Bulos, F., Chinowsky, W., Feldman, G. J., Friedberg, C. E., Goldhaber, G., Hanson, G., Hartill, D. L., Jean-Marie, B., Kadyk, J. A., Larsen, R. R., Litke, M. M., Luke, D., Lulu, B. A., Luth, V., Lynch, H. L., Morehouse, C. C., Paterson, J. M., Perl, M. L., Pierre, F. M., Pun, T. P., Rapidis, P., Richter, B., Sadoulet, B., Schwitters, R. F., Trilling, G. H., Vannucci, F., Winkelmann, F. C., and Wiss, J. E. (1975), *Phys. Rev. Lett.* **35**, 1323.

Taylor, R. E. (1976), *Proceedings of the 1975 International Symposium on Lepton and Photon Interactions at High Energies*, Stanford University, 21–27 August 1975. Ed. W. T. Kirk (Standford, Stanford Linear Accelerator Center), p. 571.

Walker, R. L. (1969), *Phys. Rev.* **182**, 1729.

Walker, R. L. (1970), *Proceedings of the 4th International Symposium on Electron and Photon Interactions at High Energies* Liverpool, 14–20 September 1969. Ed. D. W. Braben and R. E. Rand. (Daresbury Laboratory).

Whitaker, J. S., *et al.* (1976), *Phys. Rev. Lett.* **37**, 1596.

Wilson, R. R. (1958), *Phys. Rev.* **110**, 1212.

_____ *3*

Low-Energy Pion
Photoproduction

A. Donnachie
and
G. Shaw

1. Introduction

While pion photoproduction has been discussed generally in the preceding
article, the low-energy region ($E_\gamma \leq 450$ MeV) offers additional features of
particular interest. Over this region the Watson theorem [cf. Moorhouse,
Eq. (2.27)] is valid to a good approximation for all multipoles, so that a
more thorough phenomenological analysis is possible than at higher energies. There are also fewer multipoles contributing significantly to the
reaction. Theoretically, the process is well understood, the dominant features arising from the contributions of the Born terms [Fig. 1(a)] and the

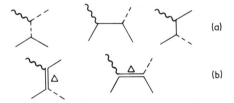

Fig. 1. The dominant contributions to low-energy pion photoproduction.

A. Donnachie and G. Shaw • Department of Theoretical Physics, University of Manchester,
Manchester M13 9PL, England

$\Delta(1236)$ resonance [Fig. 1(b)]. Qualitatively, the π^0 reactions

$$\gamma + p \to \pi^0 + p \tag{1.1a}$$

$$\gamma + n \to \pi^0 + n \tag{1.1b}$$

are completely resonance dominated, while the $\pi^\pm$ reactions

$$\gamma + p \to \pi^+ + n \tag{1.2a}$$

$$\gamma + n \to \pi^- + p \tag{1.2b}$$

have, in addition, a large background, predominantly in the s-wave multipole, coming mainly from the Born terms. The Born terms themselves dominate at the lowest energies ($E_\gamma \leqslant 200$ MeV) in charged pion production in agreement with threshold theorems. The resonance is of particular interest because of the dynamics of its excitation and because it provides an excellent means of testing the fundamental isospin and T-invariance properties of the electromagnetic current.

In discussions of this energy region, the partial wave decomposition has been almost invariably carried out in terms of the electric and magnetic multipole amplitudes $\mathcal{M}_{l\pm} \equiv E_{l\pm}, M_{l\pm}$ [Moorhouse, Eq. (2.32)], which have a simpler threshold behavior than the partial wave helicity amplitudes [Moorhouse, Eq. (2.12)]. In terms of these, the Watson theorem—assuming unitarity, T-invariance, and neglecting inelastic production—takes the form

$$\mathcal{M}^I_{l\pm} = \pm |\mathcal{M}^I_{l\pm}| e^{i\delta^I_{l\pm}} \tag{1.3}$$

where $\delta^I_{l\pm}$ is the πN phase shift corresponding to angular momentum $J = l \pm \frac{1}{2}$, and isospin $2I$.

2. Dispersion Relations

Theoretical discussion has usually been based on the fixed momentum transfer dispersion relations [cf. Moorhouse, Eq. (2.23)]. From these, a set of coupled equations for the multipole amplitudes $\mathcal{M}_\alpha \equiv \mathcal{M}^I_{l\pm}$ can be projected, of the form

$$\mathcal{M}_\alpha(W) = B_\alpha(W) + \frac{1}{\pi} \int_{W_0}^\infty dW' \frac{\operatorname{Im} M_\alpha(W')}{W' - W}$$
$$+ \frac{1}{\pi} \sum_\beta \int_{W_0}^\infty dW' K_{\alpha\beta}(W, W') \operatorname{Im} \mathcal{M}_\beta(W') \tag{2.1}$$

Explicit expressions for the Born terms B_α, and the kernels $K_{\alpha\beta}$, can be found in Berends et al. (1967). In the derivation, the partial wave expansion is used outside the physical region, and as the energy increases, this region exceeds the Lehman ellipse bounding the region of convergence of the partial wave expansion. Because of this, the equations are only formally valid for $E_\gamma \leqslant 450$ MeV, although they are probably a reasonable approximation to much higher energies (see, e.g., Shaw, 1966; Devenish *et al.*, 1972; Moorhouse, Chapter 1 of this volume, Section 2.3.5). The dispersion relations were first studied in the static limit (Chew *et al.*, 1957), in which an expansion is made in the ratio of pion to nucleon mass, $\mu = m_\pi/M$, and all but leading-order terms neglected. In this approximation, the equation for the magnetic dipole resonance excitation amplitude M_{1+}^3, and the corresponding equation for the πN scattering amplitude f_{1+}^3, differ only by a multiplicative factor. This leads directly to the so-called CGLN solution

$$M_{1+}^3 = \frac{(\mu_p - \mu_n)kf_{1+}^3}{2fq} = \frac{\mu_v k}{2fq^2} e^{i\delta_{1+}^3} \sin \delta_{1+}^3 \qquad (2.2)$$

where f is the pseudovector pion–nucleon coupling constant,* and μ_p and μ_n are the magnetic moments of the proton and neutron, respectively. In contrast, the electric quadrupole resonance excitation amplitude E_{1+}^3 satisfies a homogenous equation in the static limit, the Born term being zero. This equation has the obvious solution

$$E_{1+}^3 = 0 \qquad (2.3)$$

These solutions are formally unique if the static model equations, and the Watson theorem, are assumed at all energies, provided that $\delta_{1+}^3(\infty) \leqslant \pi$ and $E_{1+}^3(\infty)$, $M_{1+}^3(\infty)$ are zero.

While the static model is obviously a crude approximation, the prediction of dominantly magnetic dipole excitation of the $\Delta(1236)$ resonance is striking, and is confirmed by experimental data. Indeed, a semiquantitative description of the data is immediately obtained by saturating the absorptive parts in the fixed-t dispersion relations by the M_{1+}^3 term alone, using Eq. (2.2) (Höhler and Schmidt, 1964). In particular, resonance dominance of the π^0 reactions is correctly predicted, as is the existence of the large, predominantly s-wave background in the $\pi^\pm$ reactions. Empirically, the resonance peak itself is shifted below the 90° point, $E_\gamma \approx 350$ MeV, as shown in Fig. 2.

Subsequently, all these qualitative features have been confirmed in calculations using the exact form of the coupled equations (2.2) (Donnachie and Shaw, 1966; Korth *et al.*, 1967; Berends *et al.*, 1967; Engels *et al.*, 1968; Schwela and Weizel, 1969). In principle, the solutions for M_{1+}^3, E_{1+}^3 would still be formally unique, under the same conditions as before provided the inhomogeneous terms in Eq. (2.1) could be evaluated exactly at

---

*Related to the usual pseudoscalar coupling by $4M^2 f^2 = g^2/4\pi \approx 14.5$.

large s. In practice, this is not possible, and the solutions are no longer unique. (The way in which this comes about is discussed by Lyth, Section 4.5.) This problem is dealt with either by introducing cutoff parameters, which are then adjusted to give a good fit to the data, or by constructing solutions in the limited energy range of interest (by iterating from the static model solutions or by using conformal mapping techniques), assuming reasonably high energy tails. The resulting solutions for M_{1+}^3 are similar to the CGLN solution of Eq. (2.2). More specifically, the magnitude at resonance is almost unchanged, but the shape is somewhat modified. Similarly, the electric quadrupole excitation E_{1+}^3 is again predicted to be small at resonance, typically $\leqslant 3\%$ of M_{1+}^3. However, below resonance it is no longer negligible, the ratio $-E_{1+}^3/M_{1+}^3$ rising to 20%–30% near threshold. A full discussion, including results for other waves, is given by Berends *et al.* (1967), who estimate the uncertainties in the calculation arising from errors in the input πN phases, the πN coupling constant, and the small but uncertain contributions from the second resonance region. On the whole, the agreement with experiment was found to be excellent, although some discrepancies were apparent, especially in low-energy π^0 production, where the multipoles are small and sensitive to corrections. Some differences are to be expected, of course, arising from the uncertain higher-energy contributions to the integrals in Eq. (2.1).

More recently, extensive and accurate data on the proton reactions, and measurements of the π^-/π^+, and to a lesser extent the $\pi^0 n/\pi^0 p$, production ratios on deuterium, have enabled discrete energy multipole analyses of the type discussed in Moorhouse (Section 2.4) to be carried out with impressive precision over this energy region (Noelle *et al.*, 1971; Berends and Weaver, 1971; Pfeil and Schwela, 1972; Aleksandrov *et al.*,

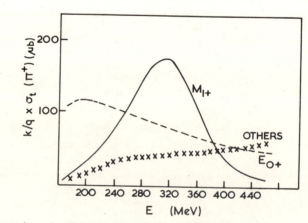

Fig. 2. Approximate breakdown of the total cross section for π^+ production into its various components.

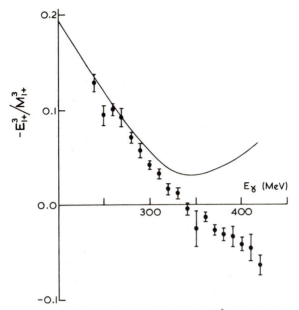

Fig. 3. The relative magnitude of the electric quadrupole (E^3_{1+}) and magnetic dipole (M^3_{1+}) resonance excitation amplitudes. The curve is a typical dispersion relation prediction of the type discussed in the text (Berends *et al.*, 1967). The data points are from the multipole analysis of Berends and Donnachie (1975). Statistical errors only are shown.

1972; Berends and Donnachie, 1975).* While the resulting amplitudes are similar to the predicted ones, there are differences in detail, which are usually plausibly associated with neglected contributions. For example, E^3_{0+} shows a steeper energy dependence than predicted, probably owing to the neglect of the $s_{31}(1650)$ resonance in the theoretical calculations summarized above (cf. Berends and Weaver, 1971). In general, however, all the most important features are confirmed, including the smallness of the electric quadrupole amplitude E^3_{1+} relative to the magnetic dipole amplitude M^3_{1+} in the resonance region. Indeed, the most recent analysis (Berends and Donnachie, 1975) suggests that E^3_{1+} vanishes very close to resonance (in exact agreement with the naive quark model), as shown in Fig. 3.

3. Low-Energy Theorems and Threshold Amplitudes

Standard soft-photon arguments lead directly to the Kroll–Ruderman theorem for the threshold amplitudes, giving

$$E^{(-)}_{0+} = \frac{eg}{2M} + O(\mu), \qquad E^{(+,\,0)}_{0+} = 0 + O(\mu^2) \tag{3.1}$$

*These authors give an extensive list of the relevant experimental papers.

A. Donnachie and G. Shaw

Table 1. Values of the Threshold Multipoles E_{0+} in Units of $10^{-2}\hbar/m_\pi c$

Method	$\pi^0 p$	$\pi^+ n$	$\pi^- p$
Low-energy theorems			
De Baenst (1970)	−0.19	2.55	−3.15
Dispersion relation results			
Born poles only	−0.81	2.80	−3.16
Born $+\Delta(1232)$ terms	−0.40	2.81	−3.18
Donnachie and Shaw (1966)	−0.09	2.88	−3.29
Berends et al. (1967)	−0.01	2.93	−3.29
Phenomenological analysis			
Donnachie and Shaw (1967)		2.83 ± 0.04	−3.19 ± 0.11
Adamovich (1970)		2.83 ± 0.05	−3.18 ± 0.20
Mullensiefen (1968)	−0.17		

where $\mu = (m_\pi/M)$ and g is the πN coupling constant, $g^2/4\pi \approx 14.5$ (Kroll and Ruderman, 1954). With the development of current algebra and soft-pion techniques, it was realized (Gaffney, 1967) that combining these with gauge invariance would allow these results to be extended to higher order in μ. This is thoroughly discussed by de Baenst* (1970), who derives the expressions

$$E_{0+}^{(-)} = \frac{eg}{2M} [1 - \mu + O(\mu^2)] \tag{3.2}$$

$$E_{0+}^{(+,0)} = \frac{eg}{2M}\left[-\mu + \frac{3\mu^2}{2} + \kappa^{(+,0)}\mu^2 + O(\mu^3) \right] \tag{3.3}$$

where

$$\kappa^{(+,0)} \equiv \kappa^{(s,v)} = \tfrac{1}{2}(\kappa_p \pm \kappa_n), \qquad \kappa_p = 1.79, \qquad \kappa_n = -1.91 \tag{3.4}$$

are the appropriate combinations of anomalous magnetic moments. The result for $E_{0+}^{(-)}$ is identical to that obtained from the Born poles of Fig. 1(a), whereas that for $E_{0+}^{(+,0)}$ differs by a term which is proportional to $\kappa^{s,v}$. This correction is therefore important for π^0, but not for $\pi^\pm$ production (see Table 1). This term is automatically incorporated if pseudovector, rather than pseudoscalar, coupling is used to evaluate the Born approximation, when in addition to the pole terms a contact term is present (Dombey and Read, 1973). This term does not appear in the unsubtracted dispersion relations described above. In these calculations, the low-energy theorem is

*References to earlier work can be found in this paper.

approximately realized for π^0 by the large, negative Born poles being almost completely canceled by the dispersion integral contributions, particularly that of the $\Delta(1236)$ resonance. As can be seen from Table 1, the low-energy theorems, dispersion relations, and phenomenological data analyses agree on the threshold multipoles within about $3 \times 10^{-3} \, \hbar/m_\pi c$, i.e., within about 10% of the largest multipole $E_{0+}^{\pi^-}$.

Independent information on this latter amplitude is obtained from measurements of the Panofsky ratio for stopped π^-:

$$P = \frac{\sigma(\pi^- p \to \pi^0 n)}{\sigma(\pi^- p \to \gamma n)} \tag{3.5}$$

Application of detailed balance leads to the relation

$$P|E_{0+}^{\pi^-}|^2 = \frac{2(M+1)}{9(2M+1)} v_0 (a_1 - a_3)^2$$

$$= 0.0242 \, (a_1 - a_3)^2 \tag{3.6}$$

where $(a_1 - a_3)$ is the charge exchange combination of πN scattering lengths and v_0 is the threshold velocity of the neutral pion in the reaction $\pi^- p \to \pi^0 n$. The Panofsky ratio is well known from experiment, the most accurate results being 1.56 ± 0.05 (Jones *et al.*, 1961), 1.53 ± 0.02 (Cocconi *et al.*, 1961) and 1.51 ± 0.04 (Ryan, 1963). A weighted mean of all experimental results gives 1.54 ± 0.02 (Ryan, 1963). Using the latter value, together with the result of Adamovich *et al.* (1970) for $E_{0+}^{\pi^-}$ (Table 1) gives

$$(a_1 - a_3) = 0.254 \pm 0.018 \tag{3.7}$$

Results from πN scattering itself lie in the range 0.24–0.30 depending upon the method and data used (for a compilation, see Nagels *et al.*, 1976).

4. Isospin Selection Rules and C, T Conservation

In the preceding discussion we have made the conventional assumption that the electromagnetic current transforms under symmetry operations in the same way as the electric charge operator

$$Q = I_3 + Y/2 \tag{4.1}$$

That is, we have assumed that it transforms under isospin as a scalar and the third component of a vector, leading to the $\Delta I = 0, 1$ rule for lowest-order transitions; and that C and T invariance hold to all orders. Terms that do not obey these rules are often referred to as "exotic," and general conditions, even in their most restrictive form of minimal interactions, do not exclude such terms (Lee, 1965). For spin $0, \frac{1}{2}$ fields, the principle of minimal electromagnetic interactions uniquely specifies the form of the interaction, the current depending only on the charge. However, for higher-spin particles, this is no longer true, and parameters other than the charge are needed to fix the interaction. For example, for a single spin-1 particle, the gyromagnetic ratio g must be specified in addition to the charge, in order to determine the current. In the absence of electromagnetic interactions g does not appear in the equations of motion, although it appears in the Lagrangian density. There is thus no reason to restrict its isospin transformation properties, and the current need not transform as the charge. Similarly, minimal interactions are perfectly compatible with C and T violation (Lee, 1965). (Throughout this article, we assume the validity of both the PCT theorem and P invariance, for strong and electromagnetic interactions. C and T conservation are thus equivalent assumptions.) In view of this, it is important to test both the $\Delta I = 0, 1$ rule, and T conservation as directly as possible.

Low-energy pion photoproduction provides a suitable testing ground for both. While discussing these tests, it is important to keep in mind the context provided by our knowledge of other reactions. In the case of T violation, experimental study of a wide variety of processes has led to no evidence for such effects in electromagnetic interactions (see below). On the other hand, in the case of $\Delta I \geq 2$ terms, we have almost no knowledge at all from experiments in other processes. (Grishin *et al.*, 1966; Dombey and Kabir, 1966). We shall therefore begin our discussion by considering tests for $\Delta I = 2$ terms on the assumption that T is conserved, turning to the question of its possible violation later.

4.1. The $|\Delta I| \leq 1$ Rule

If this rule is violated, then in addition to the usual isoscalar amplitudes A^0 leading to the $I = \frac{1}{2}\pi N$ final state, and the isovector amplitudes $A^{1,3}$ leading to the $I = \frac{1}{2}, \frac{3}{2}$ final states, there can be an isotensor amplitude A^2 leading to the $I = \frac{3}{2}$ final state. It is then convenient to introduce amplitudes for photoexcitation of the $I = \frac{3}{2}$ state on protons and neutrons:

$$
\begin{aligned}
_pA^3 &= (\tfrac{2}{3})^{1/2}[A^3 - (\tfrac{3}{5})^{1/2}A^2] \\
_nA^3 &= (\tfrac{2}{3})^{1/2}[A^3 + (\tfrac{3}{5})^{1/2}A^2]
\end{aligned}
\tag{4.2}
$$

which become equal in the absence of isotensor terms.

Most discussions concentrate on a possible isotensor excitation of the $\Delta(1236)$ resonance in a magnetic dipole transition M_{1+}^2, i.e., they try to detect a difference in the dominant resonance excitation amplitudes $_pM_{1+}^3$, $_nM_{1+}^3$. The most direct and model-independent way of doing this (Shaw,

1967) is by a comparison of the resonance-dominated reactions $\gamma + p \rightarrow \pi^0 + p$, $\gamma + n \rightarrow \pi^0 + n$. In contrast, the large nonresonant backgrounds in the charged pion photoproduction reactions $\gamma + n \rightarrow \pi^- + p$, $\gamma + p \rightarrow \pi^+ + n$ make extraction of the resonant multipoles more difficult. One possible approach (Sanda and Shaw, 1970a, b; Shaw, 1972) is to study the quantity

$$\Delta(W) = (k/q)[\sigma_t(\pi^-) - \sigma_t(\pi^+)] \qquad (4.3)$$

which should vary smoothly over the resonance region in the absence of an appreciable M_{1+}^2 excitation. Unfortunately, this test is rather insensitive compared to that using the π^0 reactions, and it is more usual to analyze the π^- data in the context of the dispersion relation approach outlined previously. This is easily extended to include possible isotensor terms (Sanda and Shaw, 1970a, b). Assuming C invariance, the isotensor amplitudes A_i^2 have the same crossing properties as the isoscalar amplitudes A_i^0, and obey the same fixed-t dispersion relations, except that the Born terms are zero. Assuming T-invariance, the Watson theorem holds for the corresponding multipole amplitudes $M_{l\pm}^2$, $E_{l\pm}^2$, which must therefore have the same phase (modulo π) as the corresponding isovector amplitudes $M_{l\pm}^3$, $E_{l\pm}^3$. Further, to the extent that the diagrams of Fig. 1 dominate the absorptive parts, isotensor terms can only enter via the resonant terms Im M_{1+}^2, Im E_{1+}^2. On substituting these into the dispersion integrals, contributions will occur in other multipoles; however, unless the isotensor terms are very large, these will be small, and can reasonably be neglected. Within the context of dispersion theory, it is therefore sensible to adopt a model in which isotensor terms occur in the resonant multipoles only, and to characterize them by the single real parameter

$$t = M_{1+}^2 / M_{1+}^3 \qquad (4.4)$$

which is taken to be constant over the resonance region. A possible isotensor electric quadrupole excitation is usually neglected because of the very small electric quadrupole excitation of the resonance. The assumption of constancy for t corresponds to pure Breit–Wigner forms for the resonant amplitudes M_{1+}^2, M_{1+}^3. More generally, it would be expected to be slowly varying. For a full discussion of such models, and their extension to include T-violation, see Donnachie and Shaw (1972, 1973), and references therein.

Turning to experiment, the bubble chamber measurements of the PRFN group (Rossi et al., 1973) on $\gamma n \rightarrow \pi^- p$ indicate a dip in $\Delta(W)$ corresponding to $t \approx -0.26$. However, the data of the ABBHHM collaboration (Benz et al., 1973) disagree with this, indicating a smooth behavior, and this is confirmed by accurate measurements of the π^- / π^+

ratio (Fujii *et al.*, 1972; von Holtey *et al.*, 1972), which is much less sensitive to uncertainties arising from deuteron corrections. (For a discussion of this point, see Benz and Soding, 1973.) Thus the data, in this respect at least, are consistent with no $I = 2$ term. However, the "dip test" on $\Delta(W)$, while rather model independent, is not very sensitive, a clear dip only emerging if $|t| \geq 0.15$. In order to obtain a more precise limit, Donnachie and Shaw (1973) performed a fit to the π^+ data and the π^-/π^+ ratio data, based on the simple model described above, in order to determine both $_pM_{1+}^3$ and $_nM_{1+}^3$. For other multipoles, the predictions of Berends *et al.* (1967) were used, except for the E_{0+}, M_{1-} multipoles. In the latter cases, slowly varying background terms were added and determined, together with $_pM_{1+}^3$, $_nM_{1+}^3$, from fits over the energy range 200 MeV $< E_\gamma <$ 400 MeV. The best fit corresponded to no isotensor term, $t = 0.00$, and on comparing with the best fits possible for $t = \pm 0.05$, the authors estimated an upper limit of about 3% on the isotensor term (see Fig. 4).

This upper limit rests on our understanding of the large nonresonant backgrounds in the $\pi^\pm$ reactions. Much less model-dependent information

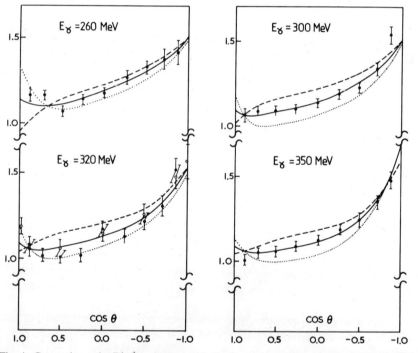

Fig. 4. Comparison of π^-/π^+ ratio data with the fits of Donnachie and Shaw (1973). The data points are from Fujii *et al.* (1972) (●) and von Holtey *et al.* (1972) (○). The solid line corresponds to $t = 0.00$ (best fit), the dashed line to $t = 0.05$, and the dotted line to $t = -0.05$.

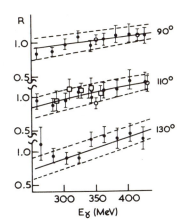

Fig. 5. Comparison of the predictions of Donnachie and Shaw (1973) for the ratio $\gamma n \to \pi^0 n / \gamma p \to \pi^0 p$ with the data of Di Capua *et al.* (1973) ($\square$), Clifft *et al.* (1974) ($\bullet$), and Christ *et al.* (1976) ($\bigcirc$). The solid line corresponds to $t = 0.00$ and the dotted lines to $t = \pm 0.05$.

is provided by more recent measurements of the ratio

$$R = \frac{\sigma(\gamma n \to \pi^0 n)}{\sigma(\gamma p \to \pi^0 p)} \tag{4.5}$$

(Di Capua *et al.*, 1973; Clifft *et al.*, 1974; Christ *et al.*, 1976), since for the π^0 reactions, the nonresonant background is very small. The data on this quantity are shown in Fig. 5, compared to the predictions of Donnachie and Shaw (1973) based on their charged pion analysis. Again, an upper limit of about 2–3% is obtained, confirming the more model-dependent $\pi^\pm$ result.

Thus it can be concluded that the $|\Delta I| \leqslant 1$ rule is satisfied to high precision, at least for the $\Delta(1236)N\gamma$ vertex. At present no tests in other reactions have been carried out, although several have been suggested (Grishin *et al.*, 1966; Dombey and Kabir, 1966; Bergmann, 1968). Perhaps the most promising is to test for possible $\pi^+ \pi^0$ asymmetries in

$$e^+ e^- \to \pi^+ \pi^- \pi^0 \tag{4.6}$$

above the $\rho\pi$ threshold (Bergmann, 1968).

4.2. Time Reversal Invariance

If we consider the possibility of T-violation in the electromagnetic interactions of hadrons (Barshay, 1965; Bernstein *et al.*, 1965), then an obvious test (Christ and Lee, 1966) is to compare the processes

$$\gamma + n \to \pi^- + p \tag{4.7a}$$

$$\pi^- + p \to \gamma + n \tag{4.7b}$$

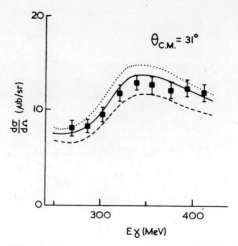

Fig. 6. Comparison of the radiative capture data of Schinzel (1970) with the predictions of Donnachie and Shaw (1973) based on fits to photoproduction data. The solid line corresponds to T conservation, while the dashed and dotted lines correspond to T-violating phases in M^3_{1+} of $+10°$ and $-5°$, respectively.

If T is violated, the Watson theorem (1.3) becomes

$$\mathcal{M}^I_{l\pm} = \pm |\mathcal{M}^I_{l\pm}| \exp[i(\delta^I_{l\pm} \pm \Delta^I_{l\pm})] \tag{4.8}$$

the sign of the T-violating phases $\Delta^I_{l\pm}$ being opposite for reactions (4.7a) and (4.7b). At threshold, where only one multipole (E_{0+}) contributes with $\delta^I_{0+} = 0$, this is already sufficient to give detailed balance between these reactions. Hence the Panofsky ratio relation (3.6), which rests on detailed balance, is not a test for T-violation. However, at higher energies, and especially in the first resonance region, Christ and Lee (1966) showed that quite modest T-violating phases could lead to appreciable deviations from detailed balance in reactions (4.7a) and (4.7b). This is illustrated in Fig. 6, which shows the effect of introducing a small T-violating phase into the resonant M^3_{1+} multipole in the model of Donnachie and Shaw (1973) discussed earlier.* The data shown (Schinzel, 1970; Favier et al., 1970) are consistent with no T-violation. This is confirmed by later experiments (Guex et al., 1975; Comiso et al., 1975) on the radiative capture reaction, which can now be compared to the results of detailed multipole analysis for the photoproduction reactions. This comparison is made in Fig. 7 for data in the first resonance region. As can be seen, there is at present no evidence

*For each value of the T-violating phase, the parameters are adjusted to fit the photoproduction data, so that all the curves give a reasonable fit to the $\gamma n \to \pi^- p$ data used in this analysis.

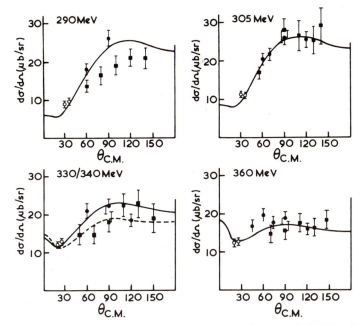

Fig. 7. Comparison of the predictions of the fit of Berends and Donnachie (1975) to the direct reaction $\gamma n \to \pi^- p$ with data on the inverse reaction $\pi^- p \to \gamma n$ from Schinzel (1970) (O), Guex *et al.* (1975) (●), and Comiso *et al.* (1975) (■).

for deviations from detailed balance, except at $E_\gamma = 285$ MeV, where there are inconsistencies in the data, and we can conclude that there is no evidence for T-violation in this process.

Finally, we note that tests for T-violation in electromagnetic inter-actions have been carried out in a variety of other reactions (see Baird *et al.*, 1969; Rock *et al.*, 1970; Schrock *et al.*, 1971; Layter *et al.*, 1972; Thaler *et al.*, 1972; Jane *et al.*, 1974, 1975), and there is at present no evidence to suggest that T (or C) is violated in the strong or electromag-netic interactions.

References

Adamovich, M. I., Larionova, V. G., Kharlamov, S. P., and Yaguda, F. R. (1970), *Sov. J. Nucl. Phys.* **7**, 360.

Aleksandrov, Yu. M., Grushin, V. F., Leikin, E. M., and Rotvain, A. Ya (1972), *Nucl. Phys.* **B45**, 589.

Baird, J. K., Miller, P. D., Dress, W. B., and Ramsey, N. F. (1969), *Phys. Rev.* **179**, 1285.

Barshay, S. (1965), *Phys. Rev. Lett.* **17**, 78.

Benz, P., and Söding, P. (1973), Daresbury Laboratory report No. DL/R32.

Benz, P., Braun, O., Butenschon, H., Finger, H., Gall, D., Idschok, U., Kiesling, C., Knies, G., Kowalski, H., Muller, K., Nellen, B., Schiffer, R., Schlamp, P., Schnackers, H. J., Schulz, V., Söding, P., Spitzer, H., Stiewe, P., Storim, F., and Weigl, J. (1973), *Nucl. Phys.* **B65**, 158.

Berends, F. A., and Donnachie, A. (1975), *Nucl. Phys.* **B84**, 342.

Berends, F. A., and Weaver, D. L. (1971), *Nucl. Phys.* **B30**, 575.

Berends, F. A., Donnachie, A., and Weaver, D. L. (1967), *Nucl. Phys.* **B4**, 1, 54, 103.

Bergmann, E. E. (1968), *Phys. Rev.* **172**, 1441.

Bernstein, J., Feinberg, G., and Lee, T. D. (1965), *Phys. Rev.* **139**, 1650.

Chew, G. F., Goldberger, M. L., Low, F. E., and Nambu, Y. (1957), *Phys. Rev.* **106**, 1345.

Christ, N., and Lee, T. D. (1966), *Phys. Rev.* **148**, 1650.

Christ, A., Noldecke, G., Nuding, W., Reichelt, T., and Stanek, H. (1976), *Nucl. Phys.* **B102**, 429.

Clifft, R. W., Gabathuler, E., Littenberg, L. S., Marshall, R., Rock, S. E., Thompson, J. C., Ward, D. L., and Brookes, G. R. (1974), *Phys. Rev. Lett.* **33**, 1500.

Cocconi, V. T., Fazzini, T., Fidecaro, G., Legros, M., Lipman, N., and Merrison, W. (1961), *Nuovo Cimento* **22**, 494.

Comiso, J., Blasberg, D., Haddock, R., Nefkens, B., Truoel, P., and Verhey, L. (1975), *Phys. Rev. D* **12**, 719.

de Baenst, P. (1970), *Nucl. Phys.* **B24**, 633.

Devenish, R. C. E., Lyth, D. H., and Rankin, W. A. (1972), Daresbury Laboratory report No. DNPL/P109.

Di Capua, E., Pocci, P., Severi, M., Tau, L., Fiorentino, E., Palmonari, F., Reale, A., Satta, L., and Ubaldini, G. (1973), *Nuovo Cimento Lett.* **8**, 692.

Dombey, N., and Kabir, P. K. (1966), *Phys. Rev. Lett.* **17**, 730.

Dombey, N., and Read, B. J. (1973), *Nucl. Phys.* **B60**, 65.

Donnachie, A., and Shaw, G. (1966), *Ann. Phys. (N.Y.)* **37**, 333.

Donnachie, A., and Shaw, G. (1967), *Nucl. Phys.* **87**, 556.

Donnachie, A., and Shaw, G. (1972), *Phys. Rev. D* **5**, 1117.

Donnachie, A., and Shaw, G. (1973), *Phys. Rev. D* **8**, 4198.

Engels, J., Mullensiefen, A., and Schmidt, W. (1968), *Phys. Rev.* **175**, 1991.

Favier, J., Alder, J. C., Joseph, C., Vaucher, B., Schinzel, D., Zupancic, C., Bressani, T., and Chiavassa, E. (1970), *Phys. Lett.* **31B**, 609.

Fujii, T., Homma, S., Huke, K., Kato, S., Okuno, H., Takasaki, F., Kondo, T., Yamada, S., Endo, I., and Fujii, H. (1972), *Phys. Rev. Lett.* **28**, 1672.

Gaffney, G. W. (1967), *Phys. Rev.* **161**, 1599.

Grishin, V. G., Lyuboshitz, V. L., Ogievetskii, V. I., and Podgoretski (1966), *Yadern. Fiz.* **4**, 126 [*Sov. J. Nucl. Phys.* **4**, 90].

Guex, L. H., Joseph, C., Tran, M. T., Vaucher, B., Winkelmann, E., Bayer, W., Hilscher, H., Schmitt, H., Zupancic, C., and Truoel, P. (1975), *Phys. Lett.* **55B**, 101.

Höhler, G., and Schmidt, W. (1964), *Ann. Phys. (N.Y.)* **28**, 34.

Jane, M. R., Jones, B. D., Lipman, N. H., Owen, D. P., Penney, B. K., Walker, T. G., Gettner, M., Grannis, P., Uto, H., Anderson, J., Bellamy, E. H., Green, M. G., Kirkby, J., Osmon, P. E., Strong, J. A., Thomas, D. H., and Solomonides, C. (1974), *Phys. Lett.* **48B**, 260, 265.

Jane, M. R., Grannis, P., Jones, B. D., Lipman, N. H., Owen, D. P., Peterson, V. Z., Toner, W. T., Bellamy, E. H., Green, M. G., Kirkby, J., Solomonides, C., Strong, J. A., and Thomas, D. H. (1975), *Phys. Lett.* **59B**, 99.

Jones, D. P., Murphy, P. G., O'Neill, P. L., and Wormald, J. R. (1961), *Proc. Phys. Soc.* **77**, 77.

Korth, W., Rollnik, H., Schwela, D., and Weizel, R. (1967), *Z. Phys.* **202**, 451.

Kroll, N. M., and Ruderman, M. A. (1954), *Phys. Rev.* **93**, 233.

Layter, J. G., Appel, J. A., Kotlewski, A., Lee, W., Stein, S., and Thaler, J. J. (1972), *Phys. Rev. Lett.* **29**, 316.

Lee, T. D. (1965), *Phys. Rev. B* **140**, 959, 967.

Mullensiefen, A. (1968), *Z. Phys.* **211**, 360.

Nagels, M. M., de Swart, J. J., Nielsen, H., Oades, G. C., Petersen, J. L., Tromborg, B., Gustafson, G., Irving, A. C., Jarlskog, C., Pfeil, W., Pilkuhn, H., Steiner, F., and Tauscher, L. (1976), *Nucl. Phys.* **B109**.

Noelle, P., Pfeil, W., and Schwela, D. (1971), *Nucl. Phys.* **B26**, 461.

Pfeil, W., and Schwela, D. (1972), *Nucl. Phys.* **B45**, 379.

Rock, S., Borghini, M., Chamberlain, O., Fuzesy, R. Z., Morehouse, C. C., Powell, T., Shapiro, G., Weisberg, H., Cottrell, R. L. A., Litt, J., Mo, L. W., and Taylor, R. E. (1970), *Phys. Rev. Lett.* **24**, 748.

Rossi, V., Piazza, A., Susinno, G., Carbonara, F., Gialanella, G., Napolitano, M., Rinzivillo, R., Votano, L., Mantovani, G. C., Piazzoli, A., and Lodi-Rizzini, E. (1973), *Nuovo Cimento* **13A**, 59.

Ryan, J. W. (1963), *Phys. Rev.* **130**, 1554.

Sanda, A. I., and Shaw, G. (1970*a*), *Phys. Rev. Lett.* **24**, 1310.

Sanda, A. I., and Shaw, G. (1970*b*), *Phys. Rev. D* **3**, 520.

Schinzel, D. (1970), Ph.D. thesis, University of Karlsruhe.

Schrock, B. L., Haddock, R. P., Helland, J. A., Longo, M. J., Wilson, S. S., Young, K. K., Cheng, D., and Perez-Mendez, V. (1971), *Phys. Rev. Lett.* **26**, 1659.

Schwela, D., and Weizel, R. (1969), *Z. Phys.* **221**, 71.

Shaw, G. (1966), *Nuovo Cimento*, **44A**, 1276.

Shaw, G. (1967), *Nucl. Phys.* **B3**, 338.

Shaw, G. (1972), Proceedings of the Informal Meeting on Electromagnetic Interactions, Frascati, May 1972.

Thaler, J. J., Appel, J. A., Kotlewski, A., Layter, J. G., Lee, W., and Stein, S. (1972), *Phys. Rev. Lett.* **29**, 313.

von Holtey, G., Knop, G., Stein, H., Stumpfig, J., and Wahlen, H. (1972), *Phys. Lett.* **40B**, 589.

4

Exclusive Electroproduction Processes

D. H. Lyth

1. Introduction

In this article we are concerned with the formalism for the electroproduction process $eN \rightarrow e +$ hadrons, where the invariant mass of the hadrons is in the resonance region, i.e., not greater than about 2 GeV. Since the general formalism is not much more difficult for an arbitrary hadronic target, we treat this case in the early stage of our discussion.

To a good approximation, the hadrons may be taken to interact with a single virtual photon generated by the accelerating electron. However, because of the relativistic nature of the problem as well as our lack of understanding of hadronic structure, one cannot write the transition amplitude as the matrix element of an appropriate electromagnetic field taken between the initial and final wave functions, as is the case for nuclear transitions. In particular, the dependence of the amplitude upon the momentum transfer between the electrons cannot be interpreted as the Fourier transform of some charge and current distribution within the hadrons.

Instead, a treatment of the electroproduction amplitude has to be like that of hadronic scattering amplitudes. The electroproduction amplitude is, apart from known factors, just the matrix element of the electromagnetic

D. H. Lyth • Department of Physics, University of Lancaster, Lancaster, England

current between the initial and final hadronic states. This quantity presumably has various theoretical properties: the Lorentz covariance, analytic, crossing, and unitarity properties shared by hadronic amplitudes; similar $SU(6)$ properties perhaps related to the quark model; relationships with weak amplitudes coming from pion pole dominance and from the commutation relations between the electromagnetic and weak currents. These properties provide the immediate motivation for measuring the electroproduction amplitude, but it should be remembered that the systematics revealed by new measurements often include previously unsuspected features. In particular, it is quite unclear what trends to expect when a large number of transition form factors for the processes $N \to N^*$ have been measured as functions of the virtual photon mass squared k^2.

References are given for the first published work in the case of topics of direct concern or for useful reference sources for other topics (e.g., manipulation with the Dirac equation) and where stated. In the case of the dispersion theory of single-pion electroproduction, which provides the backbone of current analysis of experiments and constitutes a significant portion of this chapter, a fairly complete list of references is included.

2. The Photon Polarization Vector

2.1. One-Photon Exchange

The reaction with which we are dealing is

$$e + A \to e + B$$

where A is a hadron and B is one or more hadrons, and e is either an electron or a muon. To first order in the electromagnetic coupling constant α, this process is given by one-photon exchange, as illustrated in Fig. 1. Although higher-order processes are not necessarily negligible (particularly the bremsstrahlung process, where an electron emits an almost-real photon), they can be calculated to sufficient accuracy in all the experimental situations so far encountered (Mo and Tsai, 1969; Bartl and Urban, 1966; Urban, 1970), and once these "radiative corrections" have been

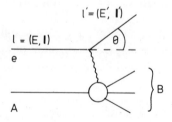

Fig. 1. Single-photon exchange in $eA \to eB$.

applied during the experimental analysis they can be ignored thereafter. The Hamiltonian for electromagnetic interactions is

$$H = \int d^3\mathbf{x} J_\mu(x) A^\mu(x) \tag{2.1}$$

where J is the electromagnetic current and A is the electromagnetic field. It is assumed that the electron and muon have pointlike interactions, i.e., that the dependence of J on the electron or muon field is

$$J_\mu = e\bar\psi(x)\gamma_\mu\psi(x) \tag{2.2}$$

We want the T-matrix element, defined in terms of the S matrix by

$$S = 1 + i(2\pi)^4\delta^4(P_i - P_f)T \tag{2.3}$$

with state normalization

$$\langle E|E'\rangle = 2E(2\pi)^3\delta^3(\mathbf{p} - \mathbf{p}') \tag{2.4}$$

Second-order perturbation theory gives the one-photon exchange expression (Dalitz and Yennie, 1957; Fubini *et al.*, 1958)

$$T = e\varepsilon_\mu T^\mu/k^2 \tag{2.5}$$

where

$$k \equiv p - p' \tag{2.6}$$

is the momentum of the virtual photon and

$$e\varepsilon_\mu \equiv \left\langle \begin{array}{c}\text{final}\\\text{electron}\end{array}\middle| J_\mu(0) \middle| \begin{array}{c}\text{initial}\\\text{electron}\end{array}\right\rangle$$

$$= e\bar u_f\gamma_\mu u_i \tag{2.7}$$

$$T_\mu \equiv \langle B|J_\mu(0)|A\rangle \tag{2.8}$$

Use has been made of the translation-invariance relation

$$\exp[i(p_2 - p_1)x]\langle p_2|J_\mu(x)|p_1\rangle = \langle p_2|J_\mu(0)|p_1\rangle \tag{2.9}$$

but the T-matrix element is independent of the choice of origin because the extra phases imparted to ε_μ and T_μ by a different choice would be equal and opposite as a result of energy-momentum conservation. Time reversal invariance (together with the Hermiticity of J_μ) requires that T should be real for particles in definite helicity states, and in accordance with the usual convention we shall choose ε_μ and T_μ to be separately real.

The matrix element of $J_\mu(0)$ can also be expressed as a matrix element of the spatial Fourier transform of $J_\mu(x)$. Integration of Eq. (2.9) gives

$$\int d^3\mathbf{x} \exp(-i\mathbf{K}\cdot\mathbf{x})\langle\mathbf{p}'|J_\mu(x)|\mathbf{p}\rangle$$
$$= (2\pi)^3\delta^3(\mathbf{k} - \mathbf{K})\exp(-ik_0t)\langle\mathbf{p}'|J_\mu(0)|\mathbf{p}\rangle \tag{2.10}$$

The virtual photon mass squared is related to the electron scattering angle θ, initial energy E, and final energy E' by

$$-k^2 = |k^2| = 2EE'(1 - \cos \theta) + |k^2|_{\min} \qquad (2.11)$$

$$|k^2|_{\min} = m_e^2 k_0^2/(EE') \qquad (2.12)$$

where m_e is the electron (or muon) mass and the photon energy k_0 is given by

$$k_0 = E - E' \qquad (2.13)$$

The amplitude (2.5) may be compared with the amplitude for the photoproduction process

$$\gamma + A \to B$$

which according to first-order perturbation theory is

$$T_{\text{phot}} = e_\mu T^\mu \qquad (2.14)$$

where T_μ is still the matrix element of J_μ between states A and B. The four-momentum transfer squared between the states is now of course $k^2 = 0$. The (real) photon polarization vector e_μ has zero time component, the vector $\mathbf{e}$ lying along the direction of the electric field for a plane-polarized photon. The phase of the amplitude for circularly polarized photons is conventionally defined by

$$e_\mu(\pm 1) = 2^{-1/2}(0, \mp 1, -i, 0) \qquad (2.15)$$

where the z axis is of course along the photon direction. The x axis for a two-body hadron state is taken to lie in the plane of the hadrons, on the same side of the beam as the meson in the case of a baryon target (cf. Fig. 2).

Because of the similarity between the electroproduction and photoproduction amplitudes, the quantity ε_μ is termed the "virtual photon polarization vector." For zero angle scattering the limit $k^2 \to 0$ is almost attained [Eq. (2.12)], and is actually possible for elastic scattering $(A = B)$, where one can have $k_0 \to 0$.

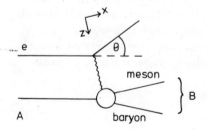

Fig. 2. The $\phi = 0$ configuration for single-meson production.

2.2. Current Conservation

Current conservation in the form

$$\partial J_\mu(x)/\partial x^\mu = 0 \qquad (2.16)$$

leads to the conditions

$$k_\mu \varepsilon^\mu = 0 \qquad (2.17)$$

$$k_\mu T^\mu = 0 \qquad (2.18)$$

but it is easy to see that there is no restriction on the scattering amplitude T except for $k^2 = 0$. First observe that the condition on ε_μ is automatically satisfied in virtue of the Dirac equation. This condition says that ε_μ has no component parallel with the four-vector k_μ. The amplitude T depends only on $\varepsilon_\mu T^\mu$, so it is independent of any component of T_μ that is parallel to k_μ. But the condition on T_μ says only that this component vanishes!

When $k^2 = 0$ this geometrical argument breaks down. To see what happens then we write

$$k = (k_0, 0, 0, k_z) \qquad (2.19)$$

$$k_z \varepsilon_z = k_0 \varepsilon_0 \qquad (2.20)$$

$$k_z T_z = k_0 T_0 \qquad (2.21)$$

and hence

$$T = e\{[1 - (k_0/k_z)^2]\varepsilon_0 T_0 - \varepsilon_x T_x - \varepsilon_y T_y\}/k^2 \qquad (2.22)$$

In the limit $k^2 \to 0$, current conservation says tht T depends only on the transverse components T_x and T_y, not on the longitudinal components. It therefore becomes a multiple of the photoproduction amplitude, as expected. This fact is exceedingly useful in the analysis of electroproduction data, since it enables the amplitudes as a function of k^2 to be tied down at the point $k^2 = 0$.

2.3. Helicity Amplitudes

Because ε_μ satisfies the condition $k \cdot \varepsilon = 0$, it has only three independent components. It is therefore useful to expand ε in the form (Manweiler and Schmidt, 1971)

$$\varepsilon_\mu = \sum_r e_\mu(r) a_r \qquad (2.23)$$

where the three vectors $e_\mu(r)$ each satisfy the condition $k \cdot e = 0$. In view of the photoproduction amplitude discussed earlier, it is convenient to take two of them to be

$$e_\mu(\pm 1) = 2^{-1/2}(0, \mp 1, -i, 0) \tag{2.24}$$

where the choice of axes is the same as in photoproduction (Fig. 2). The third vector is chosen to be orthonormal:

$$e_\mu(0) = (k_z, 0, 0, k_0)/|(k^2)^{1/2}| \tag{2.25}$$

Its phase is such that when k^2 is timelike then in the photon rest frame one has $e_\mu = (0, 0, 0, 1)$ giving the helicity amplitudes their standard phase (Section 4), but there is no particular motivation for the choice when $k^2 = 0$. The inverse of our expansion is (for $k^2 < 0$)

$$a_r = (-1)^r e_\mu^*(r)\varepsilon^\mu \tag{2.26}$$

The expansion decomposes the quantity ε_μ into pieces that transform like states of definite angular momentum when the electron momenta are rotated around the azimuthal angle of the electron plane, with $\phi = 0$ corresponding to the configuration shown in Fig. 2 and positive ϕ corresponding to clockwise rotation of the electron plane around the direction of the virtual photon. Since ε transforms as a four-vector, ε_0 and ε_z are independent of ϕ and

$$\varepsilon_x(\phi) = \cos \phi \, \varepsilon_x(0) - \sin \phi \, \varepsilon_y(0) \tag{2.27}$$

$$\varepsilon_y(\phi) = \sin \phi \, \varepsilon_x(0) + \cos \phi \, \varepsilon_y(0) \tag{2.28}$$

It follows that

$$a_r(\phi) = \exp(-ir\phi)a_r(0) \tag{2.29}$$

which is the transformation law for a state whose z component of angular momentum is r.

In addition, the components a_r are unchanged if the electron states are subject to a boost along the direction of the virtual photon. This is obvious for $r = \pm 1$, and for $r = 0$ follows from the fact that $(k_z T_0 - k_0 T_z)$ is the (xy) component of the tensor $\varepsilon_{\mu\nu\gamma\delta}k^\delta T^\gamma$.

The electroproduction matrix element T may be expressed in terms of the components a_r as

$$T = (e/k^2)\sum_r a_r f_r \tag{2.30}$$

where

$$f_r = e_\mu(r)T^\mu \tag{2.31}$$

Because of the transformation properties just described, this constitutes a decomposition of T into "helicity amplitudes" f_r, where r is the helicity of the virtual photon.

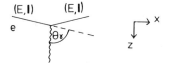

Fig. 3. The Breit frame.

The current conservation conditions on ε_μ and T_μ lead to

$$a_0 = |(k^2)^{1/2}|\varepsilon_0/k_z \qquad (2.32)$$

$$f_0 = \pm|(k^2)^{1/2}|T_0/k_z \qquad (2.33)$$

so that a_0 and f_0 both vanish when $k^2 \to 0$. (The extension to $k^2 > 0$ for f_0 is needed later; we shall never need a_0 for $k^2 > 0$.)

The decomposition of T into helicity amplitudes is not a Lorentz-invariant procedure, except for boosts along the direction of the virtual photon. However, if the target is at rest in the laboratory frame, one can reach the c.m. frame for the virtual process

$$"\gamma" + A \to B$$

by just such a boost and the lack of invariance can be ignored. In contrast, for colliding beams one has to take it into account (Brown and Lyth, 1973).

2.4. Explicit Computation of the Polarization Vector

For computational purposes it is convenient to boost the laboratory frame along the direction of the virtual photon so as to reach a frame where $k = (0, 0, 0, k_z)$. This is called the Breit frame, and in it the electron suffers no energy loss (Fig. 3).

We calculate the polarization vector in the Breit frame, for electrons with initial helicity λ and final helicity λ'. In terms of Pauli spinors the defining equation gives (Pilkuhn, 1967)

$$\varepsilon_0(\lambda', \lambda) = \begin{cases} 2E\chi_2^\dagger(\lambda')\chi_1(\lambda) \\ 2m_e\chi_2^\dagger(\lambda')\chi_1(\lambda) \end{cases} \qquad (2.34)$$

$$\varepsilon_i(\lambda', \lambda) = \begin{cases} \pm 2|\mathbf{l}|\chi_2^\dagger(\lambda')\sigma_i\chi_1(\lambda) \\ 0 \end{cases} \qquad (2.35)$$

where the first expression in each case is for $\lambda' = \lambda = \pm\frac{1}{2}$ and the second is for $\lambda' = -\lambda$. The spinors are given in the $\phi = 0$ configuration of Fig. 2 by

$$\chi_1(\tfrac{1}{2}) = \begin{pmatrix} \cos(\theta_\gamma/2) \\ -\sin(\theta_\gamma/2) \end{pmatrix}, \qquad \chi_1(-\tfrac{1}{2}) = \begin{pmatrix} \sin(\theta_\gamma/2) \\ \cos(\theta_\gamma/2) \end{pmatrix}$$

$$\chi_2(\tfrac{1}{2}) = \begin{pmatrix} \sin(\theta_\gamma/2) \\ -\cos(\theta_\gamma/2) \end{pmatrix}, \qquad \chi_2(-\tfrac{1}{2}) = \begin{pmatrix} \cos(\theta_\gamma/2) \\ \sin(\theta_\gamma/2) \end{pmatrix}$$

(we have fixed the phases by rotating the usual $\theta_\gamma = 0$ spinors around the y axis, the generators of rotation being the Pauli spin matrices). It is convenient to introduce the parameter

$$\varepsilon = (1 - \cos^2\theta_\gamma)/(1 + \cos^2\theta_\gamma) \tag{2.36}$$

which satisfies a number of useful relations, e.g.,

$$\varepsilon/(1 - \varepsilon) = \tfrac{1}{2}\tan^2\theta_\gamma \tag{2.37}$$

$$(1 - \varepsilon)/(1 + \varepsilon) = \cos^2\theta_\gamma \tag{2.38}$$

Using Eq. (2.26), the coefficients a_r are given by

$$a_0(\lambda', \lambda) = \begin{cases} 2E[2\varepsilon/(1 + \varepsilon)]^{1/2} \\ \mp 2m_e[(1 - \varepsilon)/(1 + \varepsilon)]^{1/2} \end{cases} \tag{2.39}$$

$$\exp(-i\phi)a_{-1}(\lambda', \lambda) = -\exp(i\phi)a_1(-\lambda', -\lambda)$$

$$= \begin{cases} 2^{1/2}|\mathbf{l}|[(1 + \varepsilon)^{1/2} \pm (1 - \varepsilon)^{1/2}]/(1 + \varepsilon)^{1/2} \\ 0 \end{cases} \tag{2.40}$$

where in each case the first line is for $\lambda' = \lambda = \pm\tfrac{1}{2}$ and the second line is for $\lambda' = -\lambda = \pm\tfrac{1}{2}$.

The quantity ε relates to the probability of finding the photon in a plane-polarized state. We have

$$\frac{N_\| - N_\perp}{N_\| + N_\perp} = \frac{|\varepsilon_x|^2 - |\varepsilon_y|^2}{|\varepsilon_x|^2 + |\varepsilon_y|^2}$$

$$= \frac{-2a_1a_{-1}}{|a_1|^2 + |a_{-1}|^2} = \varepsilon \tag{2.41}$$

where the a_r are understood to have their $\phi = 0$ values. When $\varepsilon = 1$ the photon is plane polarized in a direction parallel with the electron plane, whereas when $\varepsilon = 0$ there is an equal probability of parallel and perpendicular polarization.

If the electron mass is ignored, these formulas may be simplified using momentum conservation to give

$$a_r = 0 \qquad (\lambda = -\lambda') \tag{2.42}$$

$$a_0 = |(k^2)^{1/2}|[2\varepsilon/(1 - \varepsilon)]^{1/2} \qquad (\lambda = \lambda') \tag{2.43}$$

$$2^{1/2}\exp(-i\phi)a_{-1}(\lambda', \lambda) = -2^{1/2}\exp(i\phi)a_1(-\lambda', -\lambda)$$

$$= |k^2|[(1 + \varepsilon)^{1/2} \pm (1 - \varepsilon)^{1/2}](1 - \varepsilon)^{1/2} \tag{2.44}$$

where the last equation is for $\lambda = \lambda' = \pm\frac{1}{2}$. We see that in this approximation ε also gives the fraction of helicity zero photons in the sense that

$$\varepsilon = |a_0|^2/(|a_1|^2 + |a_{-1}|^2) \qquad (2.45)$$

So far we have been working in the Breit frame. We have seen that once the electron states have been specified the quantities a_r are the same in the Breit frame and in the laboratory frame. However, the electron helicities have been specified in the Breit frame, and in general the helicity states of a particle are mixed by any Lorentz boost not along the direction of the particle's momentum. On the other hand, the states of a *massless* particle are not mixed. For highly relativistic electrons the only kinematic region where the electron mass is nonnegligible is the extreme forward region, defined by

$$|k^2| \sim m_e^2$$

We therefore deduce that for highly relativistic electrons away from the extreme forward direction the coefficients a_r are given by Eqs. (2.42)–(2.44) in which the helicities can be taken to refer to the laboratory frame.

It is worth noting also that any result involving a spin sum can be obtained using the exact equations (2.39), (2.40), since the sum can be performed in the Breit frame. In particular the interpretation of ε given by Eq. (2.41) is exactly valid in the laboratory frame for unpolarized electrons.

An explicit expression for ε in terms of laboratory frame quantities is obtained by writing in the Breit frame

$$\varepsilon^{-1} = 1 + 2l_{1z}^2/l_{1x}^2 = 1 + 2|k^2|/l_{1x}^2 \qquad (2.46)$$

The final expression is obviously valid also in the laboratory frame and evaluating l_x with the aid of energy and momentum conservation gives ε in terms of laboratory frame quantities:

$$\varepsilon^{-1} = 1 + |k^2||\mathbf{k}|^2/(2E^2E'^2\sin^2\theta) \qquad (2.47)$$

This expression is exact, but neglecting the electron mass it can be simplified to give

$$\varepsilon^{-1} = 1 + 2(|\mathbf{k}|^2/|k^2|)\tan^2(\theta/2) \qquad (2.48)$$

The approximate expression is valid for highly relativistic electrons except in the extreme forward direction, where $|k^2| \sim m_e^2$. We see that ε vanishes in the backward direction. As θ decreases with E and E' fixed it rises, and in the "near-forward" region defined by

$$m_e^2 \ll |k^2| \ll k_0^2$$

it achieves a maximum value given by

$$\varepsilon = 2EE'/(E^2 + E'^2) \qquad (2.49)$$

which is close to unity when $E \simeq E'$. As θ moves into the extreme forward region ε decreases again, vanishing when $\theta = 0$.

2.5. Photon Density Matrix

The cross section will involve the quantity $|T|^2$, which is given by

$$|T|^2 = \sum_{r,s} (a_r^* a_s)(f_r^* f_s) \qquad (2.50)$$

It is necessary to sum over unmeasured final spins and to average over unmeasured initial spins, so for unpolarized electrons scattering from an unpolarized spin-$\frac{1}{2}$ target we will require the quantity

$$\frac{1}{4} \sum_{\text{spins}} |T|^2 = \left(\frac{e^2}{k^2}\right) \sum_{rs} \rho_{rs} \rho_{rs}^H \qquad (2.51)$$

where

$$\rho_{rs} = \frac{1}{4} \sum_{\text{spins}} a_r^* a_s \qquad (2.52)$$

$$\rho_{rs}^H = \sum_{\text{spins}} f_r^* f_s \qquad (2.53)$$

(the inclusion of the factor $\frac{1}{4}$ in ρ_{rs} is just convention). If any of the final spins are observed, it is necessary to omit them from the sum. If the initial electron is polarized it is convenient to define

$$\rho_{rs}(\pm\tfrac{1}{2}) = \frac{1}{2} \sum_{\text{final spin}} a_r^* a_s$$

$$= \rho_{rs} \pm \rho_{rs}^A \qquad (2.54)$$

where ρ_{rs} is the unpolarized quantity and the quantity (ρ_{rs}^A/ρ_{rs}) is a measure of the polarization asymmetry. (If the final electron spin is measured, one simply omits the spin sum. If the initial hadron is polarized one can treat ρ^H in a similar fashion.)

The quantity ρ_{rs} may be regarded as the virtual photon density matrix. The virtual photon is in a definite state if there exists a basis in which ρ factorizes. In general, in the basis where ρ is diagonal the photon corresponds to an incoherent mixture of the basis states, the proportions being given by the diagonal elements of ρ.

We have introduced the corresponding hadron quantity ρ^H mainly for compactness. It could also be regarded as the virtual photon density matrix if we regarded the photon as having being emitted rather than absorbed at the hadron vertex (either point of view is of course equally valid in covariant perturbation theory).

The notation acquires more significance if we consider the related processes with a timelike photon. For example, the analytic continuation of the ρ^H describing single-pion electroproduction

$$e + N \to e + \pi + N$$

is the density matrix of the rho meson in the reaction

$$\pi + N \to \rho + N$$

when k^2 is equal to the rho mass squared (Manweiler and Schmidt, 1971).

2.6. Explicit Evaluations of the Photon Density Matrix

In all cases the ϕ dependence of ρ_{rs} is given by

$$\rho_{rs}(\phi) = e^{-i(r-s)\phi} \rho_{rs}(0) \tag{2.55}$$

In this section we calculate $\rho(0)$, which will simply be denoted by ρ.

If the initial and final electrons are unpolarized, the spin sum may be performed over Breit frame helicity states and Eqs. (2.39) and (2.40) yield the exact formulas

$$\rho_{++} = \rho_{--} = \frac{\frac{1}{2}|k^2|}{1 - \varepsilon} \tag{2.56}$$

$$\rho_{+-} = \rho_{-+} = -\varepsilon\rho_{++} \tag{2.57}$$

$$\rho_{00} = 2\varepsilon\rho_{++} + [2m_e^2] \tag{2.58}$$

$$\rho_{+0} = \rho_{0+} = -\rho_{-0} = -\rho_{0-}$$
$$= \{\varepsilon(1+\varepsilon)\}^{1/2}\rho_{++}\left[1 + \frac{1-\varepsilon}{1+\varepsilon}\frac{4m_e^2}{|k^2|}\right]^{1/2} \tag{2.59}$$

We see that when $\varepsilon = 0$, i.e., when $\theta = 0$ and π, ρ is diagonal. For practical purposes ρ_{00} can be ignored in these cases. When $\theta = \pi$ this is simply because $\rho_{00} \ll \rho_{++}$; when $\theta = 0$, $\rho_{00} \sim \rho_{++}$ but then the helicity amplitude f_0 will be negligible compared with $f_{\pm 1}$ because of the current conservation condition Eq. (2.33). The remaining elements ρ_{++} and ρ_{--} are equal in accordance with the fact that $\varepsilon = 0$ corresponds to zero polarization of the virtual photon.

When $\varepsilon \simeq 1$ all the elements of ρ are the same order of magnitude, but since this happens in the near-forward direction (with $E \simeq E'$) the elements with r, $s = 0$ will again not usually be relevant. The other elements correspond to the photon being plane polarized in the plane of the electrons.

The terms in square brackets are relevant only when m_e cannot be ignored; i.e., for highly relativistic electrons they are relevant only in the extreme forward direction. If the electron mass is ignored, Eqs. (2.42)–(2.44) enable us also to calculate ρ when the initial electron has a well-defined helicity in the laboratory frame. The final electron also has a well-defined helicity and

$$\rho_{rs}(\lambda) = \tfrac{1}{2}a_r^*(\lambda, \lambda)a_s(\lambda, \lambda) \tag{2.60}$$

which gives for the asymmetry

$$\rho_{++}^A = -\rho_{--}^A = (1 - \varepsilon^2)^{1/2}\rho_{++} \tag{2.61}$$

$$\rho_{+-}^A = \rho_{-+}^A = \rho_{00}^A = 0 \tag{2.62}$$

$$\rho_{+0}^A = -\rho_{0+}^A = \rho_{-0}^A = -\rho_{0-}^A = [(1 - \varepsilon)/(1 + \varepsilon)]^{1/2}\rho_{+0} \tag{2.63}$$

We see that the asymmetry is maximal when $\varepsilon = 0$, i.e., near the backward direction, and is small when $\varepsilon \simeq 1$, i.e., in the near-forward direction with $E \simeq E'$. We emphasize again that the formulas do not apply in the extreme forward direction, where Eqs. (2.75)–(2.77) below are valid.

Electrons in a storage ring may also be polarized along an axis perpendicular to their direction of motion. The spin states are then sums and differences of helicity states. It is easy to see that in this case ρ has no spin dependence when m_e is negligible, from the following argument. Consider a final electron of definite helicity; ignoring m_e the amplitude vanishes when the initial electron has one or other value for its helicity and is therefore the same apart from a sign whether we add or subtract the two helicity states.

Instead of calculating ρ by explicit calculation of ε_μ one can employ trace techniques (Jones, 1965; Dombey, 1969). This method is particularly convenient when dealing with the extreme forward direction. In this approach, we define the quantity

$$L_{\mu\nu} = \tfrac{1}{4} \sum_{\text{spins}} \varepsilon_\mu^* \varepsilon_\nu$$

$$= \tfrac{1}{4}\text{Tr}[\gamma_\mu(\gamma \cdot l' + m_e)\gamma_\nu(\gamma \cdot l + m_e)]$$

$$= l_\mu l_\nu' + l_\nu l_\mu' + \tfrac{1}{2}k^2 g_{\mu\nu} \tag{2.64}$$

Using Eqs. (2.26) and (2.20) we can write for r, $s \neq 0$

$$\rho_{rs} = \tfrac{1}{2}[L_{yy} + rsL_{xx} + ir(L_{xy} - rsL_{yx})] \tag{2.65}$$

$$\rho_{0s} = [|(k^2)^{1/2}|/(2^{1/2}k_z)][sL_{0x} + iL_{0y}] \tag{2.66}$$

$$\rho_{r0} = [|(k^2)^{1/2}|/(2^{1/2}k_z)][rL_{x0} + iL_{y0}] \tag{2.67}$$

$$\rho_{00} = [|k^2|/k_z^2]L_{00} \tag{2.68}$$

It is easily verified that these equations lead to the expressions for ρ that we have already obtained.

If the initial electron has a component of spin equal to $\pm\frac{1}{2}$, along the direction of a vector **s**, then we have to include the projection matrix $(1 \pm \gamma_5\gamma \cdot s)$, where the normalization and fourth component of s are defined by $s^2 = 1$ and $s \cdot l = 0$. The additional contribution is

$$L_{\mu\nu}^A = \tfrac{1}{4}m_e \, \text{Tr}[(\gamma_\mu\gamma \cdot l\gamma_\nu + \gamma_\mu\gamma_\nu\gamma \cdot l')\gamma_5\gamma \cdot s]$$

$$= im_e\varepsilon_{\alpha\mu\nu\gamma}s^\alpha(l - l')^\gamma \tag{2.69}$$

For a highly relativistic electron with a definite helicity, $m_e s \simeq l$ and away from the extreme forward direction

$$L_{\mu\nu}^A \simeq -i\varepsilon_{\alpha\mu\nu\gamma}l^\alpha l'^\gamma \tag{2.70}$$

which leads to our expressions for ρ^A.

In the extreme forward direction we have to keep m_e and with the axes in Fig. 2

$$m_e s = (l, E \sin\theta_\gamma, 0, E \cos\theta_\gamma) \tag{2.71}$$

Using this expression, together with the small-θ expressions

$$|k^2| \simeq |k^2|_{\min}\left(1 + \frac{\theta^2}{\theta_0^2}\right) \tag{2.72}$$

$$\varepsilon^{-1} = 1 + \frac{k_0^2}{2EE'}\frac{\theta^2 + \theta_0^2}{\theta^2} \tag{2.73}$$

$$\theta_0 \equiv \frac{m_e k_0}{EE'} \tag{2.74}$$

we obtain

$$\rho_{++}^A = -\rho_{--}^A = \frac{1}{2}|k^2|_{\min} + \frac{EE'(E + E')}{k_0}\theta^2 \tag{2.75}$$

$$\rho_{+-}^A = \rho_{-+}^A = \rho_{00}^A = 0 \tag{2.76}$$

$$\rho_{+0}^A = -\rho_{0+}^A = \rho_{-0}^A = -\rho_{0-}^A = -\frac{|(k^2)^{1/2}|}{2^{1/2}}\frac{EE'\theta}{k_0} \tag{2.77}$$

These expressions differ from those obtained earlier when θ is in the extreme forward region (not of course in the remainder of the small-θ

172   D. H. Lyth

region). When $\theta = 0$

$$\rho_{++}(-\tfrac{1}{2}) = \rho_{++} - \rho^A_{++} = 0 \tag{2.78}$$

$$\rho_{--}(\tfrac{1}{2}) = \rho_{--} + \rho^A_{--} = 0 \tag{2.79}$$

so that electrons with helicity $\pm\tfrac{1}{2}$ do not produce photons with helicity ∓ 1. This is in accordance with angular momentum conservation. However, when θ is nonzero there is no such relationship, even when θ is in the extreme forward region. In other words, an electron with a definite helicity has a comparable probability of producing photons with both helicity states ± 1, even in the extreme forward region.*

If the initial electron is polarized perpendicularly to its direction of motion, the fourth component of s vanishes and s is a unit vector (pointing along the direction of polarization). We see again that when m_e is negligible the asymmetry vanishes in this case. In the extreme forward direction we use the axes in Fig. 2 to write

$$s = (0, \cos\theta_\gamma \cos\phi_p, \sin\theta_\gamma \sin\phi_p, -\sin\theta_\gamma \cos\phi_p) \tag{2.80}$$

where ϕ_p specifies the direction of polarization ($\phi_p = 0$ when the polarization is in the electron-scattering plane, the spin pointing toward the same side of the beam as the final electron). This gives

$$\rho^A_{++} = -\rho^A_{--} = -m_e E' \theta \cos\phi_p \tag{2.81}$$

$$\rho^A_{\pm 0} = -\rho^A_{0+} = -2^{1/2} m_e |(k^2)^{1/2}| e^{\pm i\phi_p} \tag{2.82}$$

The quantity (ρ^A_{++}/ρ_{++}) is zero when $\theta = 0$ and rises to a maximum when

$$\theta = \frac{m_e k_0^2}{EE'(E^2 + E'^2)^{1/2}} \tag{2.83}$$

the maximum value being

$$\left(\frac{\rho^A_{++}}{\rho_{++}}\right)_{max} = \frac{-E' \cos\phi_p}{(E^2 + E'^2)^{1/2}} \tag{2.84}$$

These results keeping m_e could also have been obtained by our earlier method of calculating a_r, using an appropriate generalization of Eqs. (2.34) and (2.35) valid in the laboratory frame (Pilkuhn, 1967).

3. Form Factors

3.1. Introduction

In the previous section we decomposed the electroproduction amplitude into a product $a_r f_r$, and discussed the first term, which describes the

*Contrary to what is implied in the article of Dombey (1969).

electron vertex and which can be calculated explicitly. This leaves the quantity f_r, which describes the hadron vertex.

In this section we consider the theoretical description of the hadron vertex in the case where there is only a single hadron in the final state, i.e., we are dealing with the real or virtual transition

$$\gamma M \to M'$$

The hadron M' may be unstable, and more generally our considerations also apply when M' is any multihadron system possessing a well-defined spin; in other words our description also applies to partial wave amplitudes for the real or virtual scattering process

$$\gamma M \to \text{hadrons}$$

With a suitable description of the hadron states, the amplitudes f_r are functions of the single variable k^2. For nonrelativistic systems they are the form factors describing the shape of the hadrons and the terminology is retained for elementary particles. Because they are functions of only one variable they should provide a convenient meeting point between theory and experiment. They have an intermediate status between coupling constants (single numbers) and scattering amplitudes (functions of at least two variables).

The physical region of k^2 for the electroproduction process $eM \to eM'$ is

$$k^2 < 0$$

With the usual postulate of crossing symmetry the amplitudes $f_r(k^2)$ may be continued from this region to describe three related processes. First there is the real photonic decay and transition $M' \leftrightarrow M\gamma$ for which

$$k^2 = 0$$

Secondly there is the corresponding virtual photonic decay $M' \to Me\bar{e}$, for which

$$4m_e^2 < k^2 < (M' - M)^2$$

Finally there is the electron–positron annihilation process $e\bar{e} \to M\bar{M}'$, for which

$$k^2 > (M' + M)^2$$

3.2. Helicity Amplitudes

Consider a virtual photon moving along the z axis with helicity r, which interacts with a target M at rest with z component of spin $-\lambda$. The

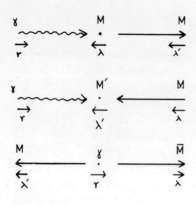

Fig. 4. Helicity amplitudes for the $\gamma MM'$ vertex. The first part of the figure indicates the laboratory frame process $\gamma M \to M'$ and the second the same process in the c.m. frame. The third part indicates the c.m. frame for the crossed process $\gamma \to \bar{M}M'$.

hadron M' that is produced will have helicity $-\lambda' = -(\lambda - r)$ and will be moving along the z axis (Fig. 4).

The amplitude for this process is unchanged by the Lorentz boost along the $-z$ axis which brings the final hadron to rest (Fig. 4). We are then in the theoretically convenient center-of-mass frame for the virtual process $\gamma M \to M'$. The helicity of the initial hadron is now λ, and the final hadron has z component of spin $-\lambda' = -(\lambda - r)$. In this frame, the momentum of the virtual photon is given by

$$|\mathbf{k}|^2 = [k^2 - (M' + M)^2][k^2 - (M' - M)^2]/4M'^2 \qquad (3.1)$$

and its energy k_0 is given by

$$k_0 = (k^2 + |\mathbf{k}|^2)^{1/2} = [M'^2 - M^2 - k^2]/2M' \qquad (3.2)$$

Unless otherwise stated, $|\mathbf{k}|$ and k_0 will denote these quantities from now on. If $M = M'$,

$$|k|^2 = -k^2(1 - k^2/4M^2) \qquad (3.3)$$

$$k_0 = -k^2/2M \qquad (3.4)$$

The analytic continuation of $|\mathbf{k}|$ into the timelike region of k^2 is needed in theoretical work. The cut is taken to lie between the branch points $(M' \pm M)^2$ and therefore the continuation of $|\mathbf{k}|$ is negative in the region $k^2 > (M + M')^2$.

The amplitude will be denoted by $f_{\lambda r}$ and is defined (Eq. 2.31) by

$$f_{\lambda r} \equiv \langle M' | e_\mu(r) J^\mu(0) | M \rangle \qquad (3.5)$$

It can also be interpreted as the amplitude which describes the interaction of a (γM) state whose total spin is S' (the final hadron spin), the photon and hadron helicities still being r and λ. Denote the state vector for that case by $|S'S'_z \lambda r\rangle$, and the state vector for the definite-momentum (γM) state simply by $|\lambda r\rangle$. Then from the overlap (Jacob and Wick, 1959),

$$\langle S', S'_z = r - \lambda, \lambda, r|\lambda r\rangle = [(2S' + 1)/4\pi]^{1/2} \qquad (3.6)$$

we deduce that the amplitude describing a transition from the definite-angular-momentum state is just

$$[4\pi/(2S' + 1)]^{1/2}f_{\lambda r} \qquad (3.7)$$

Clearly this same amplitude also describes the (real or virtual) decay $M' \rightarrow \gamma M$, where either the γM system has definite angular momentum S' or the hadron M moves along the direction of quantization of M', and where r and λ are still the helicities.

Parity invariance implies (Jacob and Wick, 1959)

$$f_{\lambda r} = \eta\eta'(-1)^{S'-S}f_{-\lambda,-r} \qquad (3.8)$$

where η and η' are the intrinsic parities of the hadrons and S and S' are their spins.

We may consider the analytic continuation of k^2 to the timelike region $k^2 > 0$. As already implied, the amplitude $f_{\lambda r}$ will still be an helicity amplitude for the virtual process $\gamma M \leftrightarrow M'$ since only the electron lines are crossed in making the continuation. In order to have the standard relative phase (Jacob and Wick, 1959) we keep the positive value of $|(k^2)^{1/2}|$ in the definition (2.25) for $f_{\lambda 0}$. For the usual reasons (perturbation theory, axiomatic field theory) we expect that apart from a possible phase $f_{\lambda r}$ will also be the helicity amplitude for the colliding beam reaction $\gamma \rightarrow \bar{M}M'$. In say the rest frame of the photon, angular momentum conservation $r = \lambda - \lambda'$ implies that the helicity λ of M is unchanged by crossing (Fig. 4). It is easy to show using the method of Trueman and Wick (1964) that there will be no helicity-dependent phase in our case, provided that the state vectors of the (always collinear) hadrons are constructed by Lorentz boosting state vectors from rest. [The state vector corresponding to a particle moving along the $-z$ axis then differs by a factor $d^s_{\lambda\lambda}(\pi)$ from that obtained by rotating the state vector corresponding to a particle moving along the $+z$ axis.]

A helicity-independent phase can be ignored for practical purposes (it relates to interference between different hadronic states) but could be obtained once the various conventions have been defined, by expanding the helicity amplitudes in terms of invariant amplitudes as described in Section 3.6 (Cohen-Tannoudji *et al.*, 1968).

We have already noted that at the point $k^2 = 0$ the amplitudes corresponding to photon helicity ± 1 will be equal to the amplitudes for the real photon transition $\gamma M \leftrightarrow M'$. If the masses of the hadrons are different, current conservation yields for the helicity zero amplitude [Eq. (2.33)] when $k^2 < 0$

$$f_0 = (|(k^2)^{1/2}|/|\mathbf{k}|)\langle M'|J_0(0)|M\rangle \qquad (3.9)$$

which vanishes at $k^2 = 0$ unless the masses M and M' are the same. To investigate the case $M = M'$ let the hadrons have equal and opposite momenta so that $k_0 = 0$. Then the definition (3.5) gives

$$f_{\lambda 0} = \langle M\lambda \ -\mathbf{p}|J_0(0)|M\lambda \mathbf{p}\rangle \qquad (3.10)$$

and taking $\mathbf{p} \to 0$ we deduce that between rest states

$$f_{\lambda 0}(k^2 = 0) = \langle M|J_0(0)|M\rangle \qquad (3.11)$$

This matrix element is proportional to the electric charge Q of the hadrons. We have

$$Q\langle 0|\mathbf{p}\rangle = \int d^3\mathbf{x}\langle 0|J_0(\mathbf{x}, t)|\mathbf{p}\rangle$$

$$= \int d^3\mathbf{x} \exp(i\mathbf{p} \cdot \mathbf{x})\langle 0|J_0(0)|\mathbf{p}\rangle \exp[-i(p_0 - M)t]$$

$$= (2\pi)^3\delta^3(\mathbf{p})f_{0\lambda} \qquad (3.12)$$

Recalling the factor $2M$ in the state normalization [Eq. (2.4)] we deduce that

$$f_{\lambda 0}(k^2 = 0) = 2MQ \qquad (3.13)$$

If the initial and final hadrons are not identical (even with the same mass) $f_{\lambda 0}$ still vanishes owing to the orthogonality of the states. Incidentally, one sees that the time independence of Q implies the vanishing of $f_{\lambda 0}$ for $M \neq M'$, previously derived from the differential form of current conservation.

3.3. Multipole Amplitudes

The helicity amplitudes introduced above are a natural choice within the context of elementary particle physics. However, when radiative transitions of the form $\lambda M \leftrightarrow M'$ are studied in classical or nuclear physics they are most conveniently described in terms of multipole amplitudes for which the photon is in a state corresponding to an electromagnetic field of definite multipolarity (Rose, 1957). These amplitudes are also useful in elementary particle physics because they have simple properties near the threshold $k^2 = (M' - M)^2$, and in the case of identical hadrons because their values at $k^2 = 0$ are equal to the static electric and magnetic multipole moments of the hadron (Section 3.5).

The simplest definition of multipole amplitudes is via the angular momentum subtotal

$$\mathbf{j} = \mathbf{L} + \mathbf{S}_\gamma \tag{3.14}$$

where L is the total orbital angular momentum of the γM system and S_γ is the spin of the photon. Clearly the total angular momentum is

$$\mathbf{S}' = \mathbf{j} + \mathbf{S} \tag{3.15}$$

where S and S' are the hadron spins.

We shall show later that 2^j is the multipolarity. This is well known for classical and nuclear physics, where M can be taken to be at rest so that j is the total angular momentum of the photon.

In addition to the quantum number j, we continue to use the photon helicity r and the total angular momentum S'. The amplitude corresponding to a γM state with these quantum numbers will be denoted by

$$[4\pi/(2j+1)]^{1/2} g_{jr} \tag{3.16}$$

We then define multipole amplitudes by

$$2^{1/2} E_j \equiv g_{j1} + g_{j-1} \tag{3.17}$$

$$2^{1/2} M_j \equiv g_{j1} - g_{j-1} \tag{3.18}$$

It will be shown below that aside from normalization factors and a Clebsch–Gordan coefficient these are the multipole amplitudes of classical and nuclear physics (Jackson, 1962; Blatt and Weiskopf, 1952), and that aside from different normalization factors they are also those commonly used in single-pion photo- and electroproduction (Chew *et al.*, 1957; Dennery, 1961).

When the photon helicity is zero it is convenient to define "scalar" (Zagury, 1966) or "longitudinal" (Dennery, 1961) multipoles by pulling out the current conservation factor $(k^2)^{1/2}$, provided the masses M and M' are not equal. One defines

$$\pm g_{j0}/|(k^2)^{1/2}| \equiv L_j/k_0 \equiv S_j/|\mathbf{k}| \qquad (k^2 \lesssim 0) \tag{3.19}$$

From Eqs. (2.21) and (2.31) we see that L and S are related to the matrix elements of, respectively, J_z and J_0. They have no branch point at $k^2 = 0$ but instead vanish when, respectively, $k_0 = 0$ and $|\mathbf{k}| = 0$. Since the multipole amplitudes already vanish when $|\mathbf{k}| = 0$ the scalar amplitude S_j is the most convenient object (Zagury, 1966). Note that $|\mathbf{k}|$ is defined by analytic continuation for $k^2 > (M - M')^2$ as we explained after Eq. (3.4) above.

When the masses M and M' are equal, g_{j0} has no branch point at $k^2 = 0$ but it is still convenient to define

$$g_{j0} \equiv S_j/(1 - k^2/4M^2)^{1/2} \tag{3.19a}$$

which is what the unequal mass expression reduces to. The square root is defined to be positive for $k^2 < 4M^2$.

Parity invariance implies $g_{j-r} = \eta\eta'(-1)^j g_{jr}$ [e.g., by combining Eqs. (3.8) and (3.25)], which leads to the well-known fact that E and S vanish for odd j between states of the same parity and for even j between states of opposite parity, and vice versa for M.

In order to relate the multipole amplitudes to the helicity amplitudes, we need the overlap between the relevant (γM) states. Denote the momentum state used for defining helicity amplitudes by $|r\lambda\rangle$ as before, and denote the angular-momentum state used for defining multipole amplitudes by $|S'S'_z jr\rangle$. We decouple the j from the spin S of the hadron

$$|S'S'_z jr\rangle = \sum_{j_z} (S'S'_z|jj_z SS_z)|jj_z r\rangle|S_z\rangle \tag{3.20}$$

where the parentheses denote the usual Clebsch–Gordan coefficient. In the same spirit we decompose the momentum state

$$|r\lambda\rangle = |r\rangle|S_z = -\lambda\rangle \tag{3.21}$$

where the first factor denotes the entire state vector apart from the hadron spin-state vector. The overlap of the states $|jj_z r\rangle$ and $|r\rangle$ vanishes unless $j_z = r$ when it is $[(2j+1)/4\pi]^{1/2}$ by the same argument (Jacob and Wick, 1959) that gives Eq. (3.6). The desired overlap is therefore

$$\langle S', S'_z = r - \lambda, j, r|r\lambda\rangle = (S', r - \lambda|jrS - \lambda)[(2j+1)/4\pi]^{1/2} \tag{3.22}$$

and recalling the normalization (3.16) we obtain with $\lambda' \equiv \lambda - r$

$$f_{\lambda r} = \sum_j (S', -\lambda'|jrS - \lambda)g_{jr} \tag{3.23}$$

Using the relation (Rose, 1957)

$$(S' - \lambda'|jrS - \lambda) = (-1)^{S-\lambda}[(2S'+1)/(2j+1)](jr|S\lambda S' - \lambda') \tag{3.24}$$

together with the orthogonality of the Clebsch–Gordan coefficients we obtain the inverse relation

$$g_{jr} = [(2j+1)/(2S'+1)]^{1/2} \sum_\lambda (-1)^{S-\lambda}(jr|S\lambda S' - \lambda')f_{\lambda r} \tag{3.25}$$

The relationship between helicity and multipole amplitudes (with a different normalization for the multipoles) first seems to have been written down by De Celles et al. (1962) and has been derived more recently (with the same normalization and notation for the multipoles) by Close and Cottingham (1975). These papers do not, however, define the multipoles, from the same viewpoint as we (or as each other), and consequently the reasoning employed in their derivations of the equation is different.

For a spin-$\frac{1}{2}$ target the relation between helicity amplitudes and multipoles for $r = \pm 1$ is (with $\lambda = \pm\frac{1}{2}$)

$$\pm(2S'+1)(2j+1)^{-1/2}g_{jr} = (S'+\tfrac{3}{2})^{1/2}f_{\lambda r} - (S'-\tfrac{1}{2})^{1/2}f_{-\lambda r} \qquad (j = S'+\tfrac{1}{2})$$
$$(3.26)$$

$$(2S'+1)(2j+1)^{-1/2}g_{jr} = (S'-\tfrac{1}{2})^{1/2}f_{\lambda r} + (S'+\tfrac{3}{2})^{1/2}f_{-\lambda r} \qquad (j = S'-\tfrac{1}{2})$$
$$(3.27)$$

and for $r = 0$ it is

$$2^{1/2}(2S'+1)^{1/2}(2J+1)^{-1/2}g_{j0} = f_{\frac{1}{2}0} \mp f_{-\frac{1}{2}0} \qquad (j = S' \pm \tfrac{1}{2}) \qquad (3.28)$$

Von Gehlen (1969) has defined multipoles for single-pion electroproduction by means of these equations.

3.4. Threshold Behavior

The main reason for considering multipoles for elementary particle reactions is their simple behavior at the threshold $k^2 = (M' - M)^2$ for the transition $\gamma M \leftrightarrow M'$. The behavior can easily be established on the assumption that it corresponds to the usual orbital angular momentum factor $|\mathbf{k}|^L$ (a different derivation will be given in Section 3.5). Recall that j is a combination of the orbital angular momentum L and the photon spin $S_\gamma = 1$. We need the overlap between $|S'S'_zjr\rangle$ and $|S'S'_zjL\rangle$. Decouple j from the hadron spin in both cases:

$$|S'S'_zjr\rangle = \sum (S'\lambda'|jj_zSS_z)|j_zjr\rangle|S_z\rangle \qquad (3.29)$$

$$|S'S'_zjL\rangle = \sum (S'\lambda'|jj_zSS_z)|j_zjL\rangle|S_z\rangle \qquad (3.30)$$

The overlap between the nontrivial factors is (Jacob and Wick, 1959)

$$\langle j_zjr|j_zjL\rangle = [(2L+1)/(2j+1)]^{1/2}(jr|L01r) \qquad (3.31)$$

and since this is independent of j_z it is also the desired overlap. Evaluating the Clebsch–Gordan coefficient we find the following threshold behavior for multipoles:

$$(j+1)^{1/2}E_j + j^{1/2}g_{j0} \sim |\mathbf{k}|^{j-1} \qquad (3.32)$$

$$M_j \sim |\mathbf{k}|^{j} \qquad (3.33)$$

$$j^{1/2}E_j - (j+1)^{1/2}g_{j0} \sim |\mathbf{k}|^{j+1} \qquad (3.34)$$

[recall the definitions (3.17) and (3.18) of E and M as sums and differences of g_{j+} and g_{j-}].

Assuming that the amplitudes have the usual property that they are analytic in k^2 (and hence in $|\mathbf{k}|^2$) near threshold except for the threshold factor, and introducing the scalar amplitude S_j [Eq. (3.19)] this gives when the masses M and M' are not equal

$$M_j \sim |\mathbf{k}|^j \tag{3.35}$$

$$S_j \sim |\mathbf{k}|^j \tag{3.36}$$

$$E_j \sim |\mathbf{k}|^{j-1} \tag{3.37}$$

$$|\mathbf{k}|^{-(j-1)}[j^{1/2}E_j + (j+1)^{1/2}(k_0 S_j/|\mathbf{k}|)] \to 0 \tag{3.38}$$

For $j = 0$, E and M do not exist, while from Eq. (3.31) S has the exceptional behavior

$$S_0 \sim |\mathbf{k}|^2 \tag{3.39}$$

When the hadron masses M and M' are equal the physical region $0 < k^2 < (M' - M)^2$ which we have been considering no longer exists. Nevertheless we shall see in Section 3.5 that our behavior holds, with the modification demanded by the definition (3.19a) of S_j, viz.,

$$S_j \sim |\mathbf{k}|^{j-1} \tag{3.36a}$$

$$|\mathbf{k}|^{-(j-1)}[j^{1/2}E_j + (j+1)^{1/2}S_j] \to 0 \tag{3.38a}$$

The only exception is the behavior of S_0 when M and M' are identical. From Eq. (3.13) this is given by

$$S_0(k^2 = 0) = 2MQ \tag{3.40}$$

For the helicity amplitudes these conditions yield a large number of constraints. When M and M' are not too different the threshold lies not too far from the physical region for electroproduction. A parameterization of the k^2 dependence of the amplitudes is therefore better made for multipole amplitudes than for helicity amplitudes.

There is another threshold at the point $k^2 = (M + M')^2$, corresponding to the process $\bar{e}e \to \bar{M}M'$. In order to determine the behavior of the helicity amplitudes for this process, we replace the hadron helicities λ and λ' by the orbital angular momentum L and the total hadronic spin J'. In terms of intermediate amplitudes h defined by

$$h_{j'r} = \sum_{\lambda\lambda'} (j'r|S\lambda S' - \lambda')f_{\lambda r} \tag{3.41}$$

the relevant amplitudes are (Jacob and Wick, 1959)

$$\sum_r (j'r|L01r)h_{j'r} \sim |\mathbf{k}|^L \tag{3.42}$$

Since the multipoles already have simple behavior at the point $k^2 = (M' - M)^2$ it is desirable to rewrite these formulas in terms of them. This may be done in a compact fashion owing to the similarity between the definitions of h and g. We have

$$h_{j'r} = \sum_j A_{j'j}^A [(2S' + 1)/(2j + 1)]^{1/2} g_{jr} \qquad (3.43)$$

where

$$A_{j'j}^r \equiv \sum_{\lambda\lambda'} (-1)^{S-\lambda} (j'r|S\lambda S' - \lambda')(jr|S\lambda S' - \lambda') \qquad (3.44)$$

The threshold behavior is explicitly [cf. Eqs. (3.32)–(3.34)]

$$h_{j0} \sim |\mathbf{k}|^{j-1}$$

$$h_{j1} \sim |\mathbf{k}|^{j-1}$$

$$h_{j-1} \sim |\mathbf{k}|^{j-1}$$

$$h_{j1} - h_{j-1} \sim |\mathbf{k}|^{j}$$

$$(j/2)^{1/2}(h_{j1} + j_{j-1}) - (j + 1)^{1/2} h_{j0} \sim |\mathbf{k}|^{j+1} \qquad (3.45)$$

These equations give the constraints on the multipoles at the point $k^2 = (M' + M)^2$. For example, in the case of a spin-$\frac{1}{2}$ target one has

$$A_{j'j}^r = \begin{cases} \pm r(S' + \tfrac{1}{2})^{-1} & (j = j' = S' \pm \tfrac{1}{2}) \\ (S' + \tfrac{3}{2})^{1/2}(S' - \tfrac{1}{2})^{1/2}(S' + \tfrac{1}{2})^{-1} & (j \neq j', \quad r \neq 0) \\ 1 & (j \neq j', \quad r = 0) \end{cases} \qquad (3.46)$$

and Eq. (3.45) gives, after some rearrangement, the following behavior for the multipoles at $k^2 = (M' + M)^2$:

$$(S' + 1)^{-1/2} M_+ + S'^{-1/2}(S' + \tfrac{3}{2})^{1/2}[(S' - \tfrac{1}{2})^{1/2} E_-$$

$$+ (S' + \tfrac{1}{2})^{1/2}(k_0/|\mathbf{k}|)S_-] \sim |\mathbf{k}|^{S'+3/2}$$

$$M_+ \sim |\mathbf{k}|^{S'-1/2}$$

$$E_- \sim |\mathbf{k}|^{S'-1/2} \qquad (3.47)$$

$$(S' + \tfrac{1}{2})^{1/2} E_+ + (S' + \tfrac{3}{2})^{1/2}(k_0/|\mathbf{k}|)S_+ \sim |\mathbf{k}|^{S'+1/2}$$

$$(S' + 1)^{-1/2} E_+ + S'^{-1/2}(S' + \tfrac{3}{2})^{1/2}(S' - \tfrac{1}{2})^{1/2} M_- \sim |\mathbf{k}|^{S'+1/2}$$

$$M_- \text{ and } E_+ \sim |\mathbf{k}|^{S'-1/2}$$

The $\pm$ subscript denotes $j = S' \pm \frac{1}{2}$. Parity invariance will cause either the first three amplitudes or the last three to vanish. Note that $|\mathbf{k}|$ [Eq. (3.1)] is defined to be positive in the spacelike region and hence negative for $k^2 > (M' + M)^2$.

The rearrangement referred to has been done so that (i) the helicity zero multipoles appear only once in each set of amplitudes (ii) the amplitudes are still unconstrained at the point $k^2 = (M' - M)^2$. One sees from Eqs. (3.35)–(3.38) that there they have the behavior $|\mathbf{k}|^{S'+n}$, where, respectively, $n = \frac{1}{2}, \frac{1}{2}, -\frac{3}{2}, \frac{3}{2}, -\frac{1}{2}, -\frac{1}{2}$.

A different method of discussing the behavior of the helicity and multipole amplitudes at the thresholds will be described in Section 3.6.

3.5. Multipole Moments

The only property of multipole amplitudes that makes them particularly useful in elementary particle physics is their behavior at the threshold $k^2 = (M' - M)^2$, which we have discussed. This is related to the reason for their use in the classical and nuclear physics description of radiative transitions (Jackson, 1962; Blatt and Weiskopf, 1952; Rose, 1957). In that case the point is that they correspond to a transition for which the electromagnetic field is given by a single term in its multipole expansion, and for which the low-energy amplitude is therefore the matrix element of a single multipole moment of the electric charge or of the magnetic term $\mathrm{div}(\mathbf{r} \times \mathbf{J})$. For identical hadrons this becomes simply the static multipole moment, which is of course measurable by other methods than electron scattering.

This last connection also holds for sufficiently long-lived elementary particles, but in practice one is dealing with at most spin-$\frac{1}{2}$ so that the elaborate general formalism of multipoles is irrelevant. Nevertheless it is interesting to verify that formally the connection between multipole amplitudes and multipole moments can still be established for the case of elementary particles. This will also provide a justification for our nomenclature "multipole," which has been missing in the discussion so far.

In classical and nuclear physics one considers the angular dependence of the photon emitted in the transition $M' \to \gamma M$ between *fixed* hadron states. We therefore take M' to be at rest but take M to be an integral over states with a fixed energy,

$$\|\mathbf{p}\rangle = \int d\Omega |\mathbf{p}\rangle \tag{3.48}$$

The spin of the hadrons is quantized along a fixed direction and has components m and m'. The multipole moment of the state $\|\mathbf{p}\rangle$ will depend only on the spin component m, since no angular dependence was introduced in forming the state.

Following Gottfried (1966) we define certain moments of the electromagnetic current operator

$$T^r_{jj_z}(|\mathbf{k}|) \equiv \int d^3 x f_\mu(|\mathbf{k}|, \mathbf{x}) J^\mu(x) \tag{3.49}$$

$$f_\mu \equiv \int d\Omega_k \, e^{-i\mathbf{k}\cdot\mathbf{x}} \, e^{ij_z\phi} \, d^j_{j_z - r}(\theta) \, e_\mu(-r) \tag{3.50}$$

where θ and ϕ are the polar coordinates of $\mathbf{k}$ and d is the usual rotation function (Rose, 1957). In the definition (2.24), (2.25) of the polarization vector the axes employed have to be rotated into the $\mathbf{k}$ direction.

There are the moments of the electromagnetic current which produce an electromagnetic field of definite multipolarity (Rose, 1957) in the case of an oscillating classical current (Jackson, 1962) or of a nuclear matrix elements (Blatt and Weisskopf, 1952). In the limit $\mathbf{k} \to 0$, one can calculate (Gottfried, 1966)

$$T_{jj_z}^1 + T_{jj_z}^{-1} = \text{const } \dot{Q}_{jj_z} |\mathbf{k}|^{j-1} \tag{3.51}$$

$$T_{jj_z}^1 - T_{jj_z}^{-1} = \text{const } M_{jj_z} |\mathbf{k}|^{j} \tag{3.52}$$

$$T_{jj_z}^0 = ik_0 \left(\frac{2j}{j+1}\right)^{1/2} \text{const } Q_{jj_z} |\mathbf{k}|^{j-1} \tag{3.53}$$

where Q and M are the electric and magnetic moments

$$Q_{jj_z} \equiv \int d^3\mathbf{r}\, r^j Y_{jj_z}(\theta\phi) J_0 \tag{3.54}$$

$$-(j+1)M_{jj_z} \equiv \int d^3\mathbf{r}\, r^j Y_{jj_z}(\theta\phi) \nabla \cdot (\mathbf{r} \times \mathbf{J}) \tag{3.55}$$

Using translational invariance Eq. (2.9) and the definition (2.31) of the amplitude f_r, the matrix element of T between the states which we are considering is found to be (with k_0 equal to the energy of the virtual photon)

$$\langle M' | T_{jj_z}^{-r}(|\mathbf{k}|) | |\mathbf{p}| \rangle = (2\pi)^3 |\mathbf{p}|^{-2}\, \delta(|\mathbf{p}| - |\mathbf{k}|) \exp(ik_0 t) A \tag{3.56}$$

where

$$A = \int d\Omega \, \exp(ij_z\phi) d_{j_z r}^j(\theta) f_r(\theta\phi) \tag{3.57}$$

The amplitude f_r denotes the transition where the photon is moving along the θ, ϕ direction with helicity r and where the initial and final hadrons have spin components m and m' along the $\theta = 0$ direction. The integral is therefore equal (Jacob and Wick, 1959) to $[4\pi/(2j + 1)]^{1/2}$ times the transition amplitude from an initial state whose total angular momentum apart from the hadron spin [Eq. (3.14)] has a value j, and z component

$$j_z = m' - m \tag{3.58}$$

(The photon still has helicity r and the hadrons still have spin components m and m' along the z axis.) This transition amplitude is $[4\pi(2j + 1)]^{1/2}$ times our multipole amplitude g_{jr} [Eq. (3.16)] except for the overlap $(S'm'|jj_zSm)$ between the initial state appearing here and the one used in defining g.

Inserting the known value of the constant in Eqs. (3.51)–(3.53) (Gottfried, 1966) we deduce a relationship between the threshold behavior of multipoles and the matrix element of electric and magnetic moments between states at rest. Besides verifying the relation (3.38) between E_j and S_j we find with $\bar{M} \equiv (MM')^{1/2}$

$$2\bar{M}\langle m'|Q_{jj_z}|m\rangle = KN_j(S'm'|jj_zSm)\lim[|\mathbf{k}|^{-j}S_j] \qquad (3.59)$$

$$2\bar{M}\langle m'|M_{jj_z}|m\rangle = KN'_j(S'm'|jj_zSm)\lim[|\mathbf{k}|^{-j}M_j] \qquad (3.60)$$

The factor

$$K = (2\pi)^3|\mathbf{p}|^{-2}\,\delta(\mathbf{p})2\bar{M}\,\exp[i(M'-M)t] \qquad (3.61)$$

contains the expected time dependence but would otherwise be equal to unity for states of unit norm. The nontrivial factors are given by

$$N_j = i^j(2j+1)!!\,(2j+1)^{-1/2}(4\pi)^{-1/2} \qquad (3.62)$$

$$N'_j = i[j/(j+1)]^{1/2}N_j \qquad (3.63)$$

All dependence upon the spin components m, m', and j_z is contained in the Clebsch–Gordan coefficient, in accordance with the Wigner–Eckart theorem (Rose, 1957).

If the initial and final hadrons are the same, these expressions give the electric and magnetic moments of the hadron. The electric and magnetic fields generated by the hadron are determined entirely by these moments (Jackson, 1962). The fields and moments do not depend on the precise state $\int d^3\mathbf{p}\,g(\mathbf{p})|\mathbf{p}\rangle$ of the particle, provided that g depends only on $|\mathbf{p}|$ and is approximately a delta function. (The best known example of this statement is the fact that a uniformly charged sphere generates a field that depends only on the total charge, not on the radius of the sphere.)

As a special case we find that the charge of a hadron is given by

$$Q \equiv (4\pi)^{1/2}Q_{00} = S_0(k^2=0)/2M \qquad (3.64)$$

where S_0 is the $j=0$ scalar multipole for the transition $\gamma M \to M$. We have already derived this equation [Eqs. (3.13), (3.39a)]. since Q is a conserved quantity we also deduce that for other transitions $\gamma M' \to M$,

$$S_0(\mathbf{k}=0) = 0 \qquad (3.65)$$

in accordance with Eq. (3.39).

For a spin-$\frac{1}{2}$ particle the magnetic moment in Bohr magnetons is given by (Blatt and Weisskopf, 1952)

$$e\mu \equiv (4\pi/3)^{1/2}2MM_{10} = 6^{-1/2}\lim[M_1|(k^2)^{1/2}|] \qquad (3.66)$$

where M_{10} denotes the expectation value of that quantity for a hadron with spin component $m = +\frac{1}{2}$, and M_1 is the magnetic multipole for the transition $\gamma M \to M$.

It may be shown by generalizing Eqs. (3.51)–(3.53) that higher-order terms in the expansion of multipoles near $|\mathbf{k}|^2 = 0$ are given by matrix elements of operators similar to Q_{jj_z} and M_{jj_z}, except that the r dependence involves higher powers of r. However, the value of the matrix element now depends on the details of how the state is formed. For nonrelativistic particles it is possible to define a localized state by writing $\int d^3 \mathbf{p} | \mathbf{p} \rangle$ and one can then show for identical hadrons that the higher derivatives of the multipoles near $k^2 = 0$ explore the higher moments of the charge current distributions. For relativistic particles this localized state is not covariant and no obvious generalization suggests itself (Schweber, 1961). In this context it is clear that a nonrelativistic particle is one whose size (in units $\hbar = c = 1$) is large compared with its inverse mass, i.e., whose multipoles are negligible except when $|\mathbf{p}| \ll M$. This is true of nucleii but not of elementary particles.

Nevertheless, one speaks loosely of, say, the first derivative of S_0 at $k^2 = 0$ as defining the "charge radius" of an elementary particle.

3.6. Invariant Amplitudes

The helicity amplitudes for $f_{\lambda r}$ and the multipole amplitudes g_{ir} are constrained at the thresholds $k^2 = (M' \pm M)^2$ and the helicity zero amplitudes $f_{\lambda 0}$ and g_{j0} also vanish at $k^2 = 0$.

These kinematic constraints may be avoided by the use of a suitable set of Lorentz-invariant amplitudes, as may the corresponding kinematic singularities (Section 3.7). Such a procedure has not been necessary in the theoretical investigations that have been undertaken to date, but nevertheless we shall discuss the invariant amplitude approach briefly.

The procedure is to write the matrix element T_μ [Eq. (2.8)] in an explicitly covariant form, using Dirac spinors (Omnes, 1971) or their higher-spin generalization (Joos, 1962; Theis and Hertl, 1970; Devenish *et al.*, 1975). One expects, e.g., from perturbation theory, that the invariant amplitudes defined in this way are analytic functions of k^2, which are everywhere free of kinematic constraints. In order to calculate cross sections (and for certain theoretical purposes such as the imposition of unitarity) one still needs the helicity amplitudes. The connection is quite difficult to make, so that in many cases (including, as we said, all those treated so far) it is simpler to learn to live with the constrained helicity or multipole amplitudes.

In the cases of spin 0 and $\frac{1}{2}$ there are, of course, no problems in either approach. For example, the most general form of T_μ that satisfies current conservation for identical spinless particles is

$$T_\mu = e(p + p')_\mu F(k^2) \tag{3.67}$$

The helicity amplitude [Eq. (3.10)] is given by

$$f_{00} = (4M^2 - k^2)^{1/2} eF(k^2) \tag{3.68}$$

which has the expected behavior at $k^2 = 4M^2$ (Section 3.4). The normalization [Eq. (3.13)] is given by

$$F(0) = 1 \qquad (3.69)$$

This form factor, defined by Eq. (3.68) for $k^2 < 0$ may be analytically continued to $k^2 > 4M^2$, where it describes the process $e\bar{e} \rightarrow \pi^+\pi^-$ (Section 3.7).

For identical spin-$\frac{1}{2}$ particles the most general matrix element satisfying current conservation is

$$T_\mu = e\bar{u}_2[\gamma_\mu F_1(k^2) + (i/2M)\sigma_{\mu\nu}k^\nu F_2(k^2)]u_1 \qquad (3.70)$$

where

$$\sigma_{\mu\nu} \equiv (i/2)(\gamma_\mu\gamma_\nu - \gamma_\nu\gamma_\mu)$$

Writing this expression in terms of Pauli spinors (Pilkuhn, 1967) in the Breit frame we find for the nonvanishing helicity amplitudes

$$f_{\frac{1}{2}1} = f_{-\frac{1}{2}-1} = e2^{1/2}|(k^2)^{1/2}|G_M \qquad (3.71)$$

$$f_{\frac{1}{2}0} = f_{-\frac{1}{2}0} = e2MG_E \qquad (3.72)$$

where we have introduced the Sachs form factors

$$G_M \equiv F_1 + F_2 \qquad (3.73)$$

$$G_E \equiv F_1 + (k^2/4M^2)F_2 \qquad (3.74)$$

The relations (3.26)–(3.28), (3.18), (3.19), (3.64), and (3.66) then give

$$F_1(0) = G_E(0) = 1 \qquad (3.75)$$

$$1 + F_2(0) = G_M(0) = \mu \qquad (3.76)$$

where μ is the magnetic moment of the particle in Bohr magnetons. We have defined $G_{E,M}$ for $k^2 < 0$ but they may be continued to $k^2 > 4M^2$, where they describe the process $e\bar{e} \rightarrow N\bar{N}$.

The first of relations (3.47) for the behavior of the multipoles at the threshold $k^2 = 4M^2$ gives when expressed in terms of G_E and G_M the relation

$$G_M(4M^2) = G_E(4M^2) \qquad (3.77)$$

which also follows directly from the definitions of these quantities in terms of the invariant amplitudes.

If the hadron spins S and S' are higher than $\frac{1}{2}$ the formulas become more complicated. The case $S = \frac{1}{2}$ and $S' = \frac{3}{2}$ (equal parities) has been worked out by Jones and Scadron (1973) and the case $S = \frac{1}{2}$ with S' arbitrary by Devenish *et al.* (1975). An expression for T_μ in terms of invariant amplitudes for arbitrary spins has been written down by Theis and Hertl (1970), but the relationship with helicity or multipole amplitudes has not been given in this case.

3.7. Analyticity and Unitarity

As soon as analytic properties were formulated for scattering amplitudes it was realized that they could be expected also to apply to form factors (Chew *et al.*, 1958). The invariant amplitudes of the last section (Section 3.6) are expected to be real-analytic functions of k^2 except for the cuts corresponding to intermediate states in the timelike process $e\bar{e} \to \bar{M}M$. The discontinuity across such a cut may be obtained from the relevant Feynman graph by the usual rules (Landau, 1959; Cutkosky, 1960).

These analytic properties will also be shared by the helicity and multipole amplitudes, provided that they are divided by

$$[k^2 - (M' - M)^2]^{n/2}[k^2 - (M' + M)^2]^{m/2}$$

where the integers n and m are the powers of $|\mathbf{k}|$ appearing in the threshold behavior. When the photon helicity is zero there is in general an extra factor $(k^2)^{1/2}$ coming from current conservation. Strictly speaking the division is only necessary for odd n and m, i.e., we only need to eliminate square root zeros to obtain good analytic behavior.

In the absence of "anomalous" thresholds (Bjorken and Drell, 1964; Bjorken and Walecka, 1966), the cut runs over values of k^2 equal to the masses squared of systems that couple to the $\bar{M}M'$ system. The amplitude just above the cut satisfies the (extended) unitarity relation (Pilkuhn, 1967), which for a two-body intermediate state reads

$$\text{Im}(f_{\lambda\lambda}/\rho) = \sum_{\mu\mu'} (\rho')^2 [f'_{\mu\mu'}/\rho'][T_{\mu\mu',\lambda\lambda'}/(\rho'\rho)]^* \qquad (3.78)$$

Here f denotes the helicity form factor of Section 3.2 with $\lambda' = \lambda - r$ and f' denotes the corresponding form factor for the intermediate state. The quantity T is the $j = 1$ partial wave T-matrix element for the hadronic transition from the intermediate to the final state (Jacob and Wick, 1959) normalized so that the unitarity limit for $|T|$ is unity. The quantities ρ and ρ' represent any threshold factors that have to be divided out of f and f', but need not be included when they are real.

For example, the contribution of the two-pion intermediate state for the pion form factor is

$$\text{Im}\, F_\pi = F_\pi \exp(-i\delta) \sin \delta \qquad (3.79)$$

where δ is the $I = J = 1$ phase shift for $\pi\pi$ scattering. Its contribution for the proton and neutron form factors is (using isospin and C invariance)

$$-\text{Im } G_E^n = \text{Im } G_E^p = [\rho/(2^{1/2}M)]F_\pi f_+^{1*} \qquad (3.80)$$

$$-\text{Im } G_M^n = \text{Im } G_M^p = [\rho/2]F_\pi f_-^{1*} \qquad (3.81)$$

where

$$\rho \equiv (k^2/4 - \mu^2)[(k^2 - 4\mu^2)/k^2]^{1/2} \qquad (3.82)$$

and $f_\pm^1$ are the partial wave amplitudes for $\pi\pi \to N\bar{N}$ with the normalization of Frazer and Fulco (1960). The kinematic factors are in agreement with our general Eq. (3.58) bearing in mind the normalization of $G_{E,M}$, F_π, and $f_\pm^1$.

If one of the hadrons M or M' can couple to an intermediate state $(m_1 m_2)$ such that $m_1^2 + m_2^2$ is smaller than M^2 or M'^2, there will be an "anomalous" threshold below the normal threshold at $k^2 = 4\mu^2$. The discontinuity across the associated "anomalous" cut is not given by extended unitarity. If the external hadron is almost unstable so that $M = m_1 + m_2$ the anomalous cut begins at $k^2 \simeq 2\mu^2$, and if the hadron is actually unstable the cut leaves the real axis. If both external hadrons are unstable the anomalous threshold will be in the physical region because the associated triangle graph represents an actually possible process (Bjorken and Drell, 1964; Bjorken and Walecka, 1966). These anomalous cuts have not yet been studied in detail. From the mass condition we deduce that they will be absent for pion, kaon, and nucleon form factors but present for, say, Λ and Σ form factors as well as for nuclear form factors where the position of the anomalous threshold is related to the radius of the nucleus and determines the slope of the form factor at threshold (cf. the discussion at the end of Section 3.5).

The mesons with $J^{PC} = (-1)^{--}$ such as the ρ, ω, ϕ, and ψ give delta function contributions to $\text{Im } f$ when their widths are neglected. These give pole contributions to the form factors of the form

$$f/\rho = cM_r^2/(M_r^2 - k^2 - iM_r\Gamma_r) \qquad (3.83)$$

where ρ again represents any necessary threshold factor. The constant may be calculated by comparing the integrated $e\bar{e}$ cross-section formula (Cabibbo and Gatto, 1961)

$$\sigma(e\bar{e} \to \bar{M}M') = \frac{\alpha}{6} \frac{[k^2 - (M' + M)^2]^{1/2}[k^2 - (M' - M)^2]^{1/2}}{k^6} |f_{\lambda\lambda'}|^2 \qquad (3.84)$$

with the general formula (Pilkuhn, 1967)

$$\sigma = \frac{16\pi}{k^2}(2J + 1)|T_J|^2 \qquad (3.85)$$

where T_J is the partial wave amplitude normalized so that the unitarity limit of $|T_J|$ is unity. This gives the result for the $e\bar{e} \rightarrow \bar{M}M'$ process

$$ef_{\lambda\lambda'} = 2^{1/2}24\pi k^2[k^2 - (M' + M)^2]^{-1/4}[k^2 - (M' - M)^2]^{-1/4}T_J \qquad (3.86)$$

If the resonance can decay into $(\bar{M}M')$, the unitarity relation $\text{Im}(T_J^{-1}) = 1$ now gives

$$|c| = 24\pi e^{-1}M_r[M_r^2 - (M' + M)^2]^{-1/4}$$
$$\times [M_r^2 - (M' - M)^2]^{-1/4}\Gamma_{ee}^{1/2}\Gamma_{\bar{M}M'}^{1/2}/\rho(M_r^2) \qquad (3.87)$$

where Γ_{ee} is the partial width for decay into an $e\bar{e}$ pair and $\Gamma_{\bar{M}M'}$ is the partial width for decay into an $(\bar{M}M')$ pair with helicities λ and λ'. For example, the constant for the pion form factor defined by Eq. (3.68) is

$$c = [6/(\alpha M_r)][1 - M^2/M_r^2]^{-3/4}\Gamma_{ee}^{1/2}\Gamma_{\pi\pi}^{1/2} \qquad (3.88)$$

One commonly defines a γ-meson coupling f_v by

$$f_v^2/4\pi \equiv (\alpha^2/3)(M_r/\Gamma_{ee}) \qquad (3.89)$$

If the resonance lies below the $\bar{M}M'$ threshold one has to replace $\Gamma_{\bar{M}M'}^{1/2}$ (combined with any necessary kinematic factors) by a (meson-M-M') coupling constant. This may be defined by means of a suitable Lorentz-covariant expression for the coupling (cf. Section 3.6 above) or may simply be defined by writing $c = g/f_v$. In the latter case the coupling may be related to that measured in other reactions by noting that the residue of the pole in the partial wave amplitude T_J will factorize.

In order that analyticity should be a useful constraint it is necessary to make some assumption about the asymptotic behavior of the form factor. In the case of the pion form factor, whose modulus is measurable along the entire cut [being related to the cross section for $e\bar{e} \rightarrow \pi^+\pi^-$ by Eq. (3.81)] one can derive interesting constraints from the very weak assumption that the quantity

$$(k^2 - 4\mu^2)^{-1/2}\log F_\pi(k^2) \qquad (3.90)$$

vanishes as $k^2 \rightarrow \infty$ (μ is the pion mass). By writing a dispersion relation for this quantity one deduces (Truong and Vinh-Mau, 1969) for $k^2 < 0$

$$\log|F(k^2)| \leq \exp\left[\frac{(4\mu^2 - k^2)^{1/2}}{\pi} \int_{4\mu^2}^{\infty} dx \frac{\log|F(x)|}{(x - 4\mu^2)^{1/2}(x - k^2)}\right] \qquad (3.91)$$

and for $k^2 \geq 0$

$$\arg F(k^2) \geq -\frac{(k^2 - 4\mu^2)^{1/2}}{\pi}P \int_{4\mu^2}^{\infty} dx \frac{\log|F(x)|}{(x - 4\mu^2)^{1/2}(x - k^2)} \qquad (3.92)$$

as well as the inequality [assuming that $\text{Im } F_\pi(4\mu^2) = 0$ and that $|F| \neq 0$ for $k^2 > 4\mu^2$]

$$0 \leq \int_{4\mu^2}^{\infty} dx \, \frac{(d/dx) \log |F(x)|}{(x - 4\mu^2)^{1/2}} \tag{3.93}$$

These relations become equalities if F has no zeros in the complex plane (Bowcock and Kannelopoulos, 1968). Experimentally they are satisfied and are perhaps equalities (Ecker and Schwela, 1974).

To derive generally interesting results one needs the usual assumption that the form factor vanishes at infinity, or at any rate that this is true after division by a known power of k^2. Nonrelativistic form factors are of course related to the Fourier transforms of the charge and current distributions (Section 3.5) and will therefore vanish at infinity. For elementary particles there are no convincing theoretical arguments about the asymptotic behavior. However, a naive dimensional analysis relates the rate of fall-off of form factors to the number of constituents (Matveef et al., 1973; Brodsky and Farrar, 1973). This gives, with the usual quark model,

$$F_\pi \sim k^{-2} \tag{3.94}$$

$$G_M \sim k^{-4} \tag{3.95}$$

$$G_E \sim k^{-4} \tag{3.96}$$

which agrees with what is observed at finite k^2, as we shall see.

If a form factor vanishes at infinity [faster than $\log^{-1}(k^2)$] it will satisfy a dispersion relation. Neglecting anomalous cuts for simplicity the dispersion relation reads

$$F(k^2) = \frac{1}{\pi} \int_{4\mu^2}^{\infty} dx \, \frac{\text{Im } F(x)}{x - k^2} \tag{3.97}$$

A known value of F at (say) $k^2 = 0$ can be exploited to give a more convergent integral

$$F(k^2) - F(0) = \frac{k^2}{\pi} \int_{4\mu^2}^{\infty} dx \, \frac{\text{Im } F(x)}{x(x - k^2)} \tag{3.98}$$

and threshold zeros may be exploited in the same way.

In the case of the pion and nucleon form factors unitarity and analyticity (together with the assumption that the form factors vanish asymptotically) have considerable predictive power. They determine the shape of the pion form factor for $k^2 \lesssim 1$ $(\text{GeV}/c)^2$ given the mass and width of the ρ meson, as follows.

First one verifies by inspection that a function satisfying analyticity and elastic unitarity together with the requirement $F_\pi(0) = 1$ is (Omnes, 1958)

$$F_\pi(k^2) = D^{-1}(k^2) \equiv \exp\left[\frac{k^2}{\pi} \int_{4\mu^2}^{\infty} dx \frac{\delta(x)}{x(x - k^2)}\right] \qquad (3.99)$$

The asymptotic behavior of this solution is $(k^2)^{-\delta(\infty)/\pi}$ and it is therefore unique provided that F is required to vanish asymptotically like $k^{-2\varepsilon}$ and that

$$\delta(\infty) \leqslant \pi + \varepsilon \qquad (3.100)$$

[If $\delta(\infty) > N\pi + \varepsilon$ it can be multiplied by an arbitrary Nth-order polynomial normalized to unity at $k^2 = 0$]. This is a reasonable requirement in the presence of a single resonance. The pion scattering amplitude is given under similar assumptions by (Chew and Mandlestam, 1960)

$$\exp(-i\delta)\sin\delta = \rho N D^{-1} \qquad (3.101)$$

where ρ is given by Eq. (3.82). D^{-1} is the expression we have just given for F_π, and N satisfies an unsubtracted dispersion relation with a cut along the negative k^2 axis. On both theoretical and experimental grounds it appears that N is approximately a constant, and if this is so one can deduce from Eq. (3.96) the expression (Frazer and Fulco, 1960)

$$F_\pi^{-1} \equiv D = A + Bk^2 + CX(k^2) \qquad (3.102)$$

where for $k^2 \leqslant 0$

$$X(k^2) \equiv -\frac{1}{\pi}|\rho(k^2)| \log[(1 - k^2/4\mu^2)^{1/2} + (-k^2/4\mu^2)^{1/2}] \qquad (3.103)$$

(elsewhere X is defined by analytic continuation). The three constants A, B, and C are determined by the condition $F_\pi(0) = 1$, together with a knowledge of the mass and width of the ρ meson.

Except for $|k^2| \gg 1$ $(\text{Gev}/c)^2$ this formula is almost the same as the ρ dominance expression

$$F_\pi = M_\rho^2/(M_\rho^2 - k^2) \qquad (k^2 \leqslant 4\mu^2) \qquad (3.104)$$

$$F_\pi = \frac{M_\rho^2}{M_\rho^2 - k^2 - iM_\rho\Gamma_\rho[(k^2 - 4\mu^2)/(M_\rho^2 - 4\mu^2)]^{3/2}} \qquad (k^2 > 4\mu^2)$$
$$(3.105)$$

It can be shown (Lyth, 1972) that the inclusion of the variation of N and of the neglected terms in the elastic unitarity equation does not substantially affect the result for $|k^2| \leqslant 1$ $(\text{GeV}/c)^2$. With the inclusion of the ω contribution near $k^2 = M_\omega^2$ the formula gives a good description of the presently available data in the interval $4\mu^2 < k^2 \leqslant 1$ $(\text{GeV}/c)^2$ (Benaksas, 1972). We shall see later that the ρ-pole expression seems to be valid in the spacelike region out to $k^2 \sim -4$ $(\text{GeV}/c)^2$. Unitarity and analyticity also determine the shape of the two-pion contribution to the nucleon form factor provided that the $\pi\pi \to N\bar{N}$ amplitude is known in the unphysical region $4\mu^2 < k^2 < 4M^2$. In the ρ region this quantity may be estimated from unitarity and analyticity in terms of the ρNN and πNN coupling constants (Frazer and Fulco, 1960), or calculated from these requirements quite accurately if πN scattering data are also included (Hohler et al., 1976). The resulting quantities $\text{Im}\,G_{E,M}$ deviate substantially from ρ dominance because of a strong N exchange contribution to the $\pi\pi \to N\bar{N}$ amplitudes.

In principle the solution (3.99) of the unitarity and analyticity equations can be generalized to the case where several intermediate states contribute (Bjorken, 1960) but convenient asymptotic requirements sufficient to ensure uniqueness have not yet been formulated for this case.

4. The Single-Pion Electroproduction Amplitude

4.1. Preliminary Remarks and Kinematics

In the last section we studied the amplitude f_r which describes the virtual transition $\gamma M \to M'$, where M' was a single hadron or at least a hadronic system of definite spin and parity. If M' consists of two or more hadrons with definite momenta the amplitude f_r is more complicated. For a fixed virtual photon mass it has the same kinematic properties as an ordinary scattering amplitude, and of course when the virtual photon mass is zero $f_{\pm 1}$ are the scattering amplitudes for real photoproduction.

In this section we make a detailed study of the amplitude for the virtual transition $\gamma N \to \pi N$, including its photoproduction limit. This is the only electroproduction amplitude that has so far received detailed attention. It will be clear that the methods we describe apply equally to any two-body state.

It will be convenient to write down the kinematic quantities that we shall need. In the usual notation these are as follows:

nucleon mass	M	initial nucleon c.m. energy	E_1
pion mass	μ	final nucleon c.m. energy	E_2
c.m. energy	W	pion c.m. energy	E_π
c.m. energy squared	s	photon c.m. energy	k_0
virtual photon mass squared	k^2	pion c.m. momentum	$|\mathbf{q}|$
four-momentum transfer squared		photon c.m. momentum	$|\mathbf{k}|$
between the nucleons	t	nucleon c.m. scattering angle	θ
four-momentum transfer squared			
between the photon and nucleon	u		

Useful kinematic relations are

$$s + t + u = 2M^2 + \mu^2 + k^2 \tag{4.1}$$

$$E_1 = (s + M^2 - k^2)/2W \tag{4.2}$$

$$E_2 = (s + M^2 - \mu^2)/2W \tag{4.3}$$

$$E_\pi = (s + \mu^2 - M^2)/2W \tag{4.4}$$

$$k_0 = (s + k^2 - M^2)/2W \tag{4.5}$$

$$|\mathbf{k}|^2 = [k^2 - (W + M)^2][k^2 - (W - M)^2]/4s \tag{4.6}$$

$$|\mathbf{q}|^2 = [s - (M + \mu)^2][s - (M - \mu)^2]/4s \tag{4.7}$$

$$t = -2|\mathbf{k}||\mathbf{q}|(1 - \cos \theta) - |t|_{\min} \tag{4.8}$$

$$|t|_{\min} = 2|\mathbf{k}||\mathbf{q}| - [(s - k^2 - M^2)(s - k^2 - \mu^2) - 2M^2(k^2 + \mu^2)] \tag{4.9}$$

When $s \gg M^2$,

$$|t|_{\min} \simeq [M^2(\mu^2 - k^2)^2]/[s(s - k^2)] \tag{4.10}$$

4.2. Helicity Amplitudes and Multipoles

The helicity formalism (Jacob and Wick, 1959) gives a simple description of the process $\gamma N \to \pi N$ both for real photons (Walker, 1969) and virtual photons (Jones, 1965). The particles have well-defined helicities in the c.m. frame and one uses the axes shown in Fig. 2. The phase of the final-state vector is defined by

$$|\theta\rangle = \exp(-iJ_y\theta)|0\rangle \tag{4.11}$$

with the usual Pauli representation for the spin operators and $\theta = 0$ spin states (Jacob and Wick, 1959). We denote f_r for this case by $f_{\mu\lambda r}$, where λ and μ are the initial and final nucleon helicities. In photoproduction one commonly defines (Walker, 1969)

$$A_{-\mu,1-\lambda} = f_{-\mu,\lambda,1}/(8\pi W) \tag{4.12}$$

We expand f into partial waves (Jacob and Wick, 1959)

$$f_{\mu\lambda r}(\theta) = \sum_J f_{\mu\lambda r}^J [(2J + 1)/4\pi]^{1/2} d_{r-\lambda,-\mu}^J(\theta) \tag{4.13}$$

It is useful to define amplitudes

$$2^{1/2} f_{\lambda r}^{n+} \equiv f_{\frac{1}{2}\lambda r}^J + f_{-\frac{1}{2}\lambda r}^J$$
$$2^{1/2} f_{\lambda r}^{(n+1)-} \equiv f_{\frac{1}{2}\lambda r}^J - f_{-\frac{1}{2}\lambda r}^J \tag{4.14}$$

where

$$n \equiv J - \tfrac{1}{2} \tag{4.15}$$

These are the amplitudes (Section 3.2) describing transitions to a πN state with total angular momentum J and parity $(-1)^{l+1}$, where l is the superscript n or $n+1$. Clearly l is the orbital angular momentum of the πN system.

The commonly defined multipoles (Walker, 1969) differ from those defined in our general discussion. By comparing our definitions (3.26) and (3.27) with the definitions in Walker's paper we deduce expressions for our multipoles,

$$E_{J+1/2} = -c(J + \tfrac{3}{2})^{1/2}(J+1)^{1/2} E_{n+} \tag{4.16}$$

$$M_{J-1/2} = -c(J - \tfrac{1}{2})^{1/2} J^{1/2} M_{n+} \tag{4.17}$$

$$E_{J-1/2} = +c(J - \tfrac{1}{2})^{1/2} J^{1/2} E_{(n+1)-} \tag{4.18}$$

$$M_{J+1/2} = -c(J + \tfrac{3}{2})^{1/2}(J+1)^{1/2} M_{(n+1)-} \tag{4.19}$$

$$c = 2W/(2J+1) \tag{4.20}$$

Similarly the commonly defined scalar multipoles are related to those of Eq. (3.28) by

$$S_{J+1/2} = c(J + \tfrac{1}{2})^{1/2}(J+1)^{1/2} S_{n+} \tag{4.21}$$

$$S_{J-1/2} = c(J + \tfrac{1}{2})^{1/2} J^{1/2} S_{(n+1)-} \tag{4.22}$$

[Note, however, that the first paper to introduce scalar multipoles (Zagury, 1966) had a definition different by a factor $(n + 1)$.]

In terms of these multipoles the quantities that are unconstrained at the thresholds $k^2 = (W \pm M)^2$ are [Eq. (3.47)]

$$[k_0/|\mathbf{k}|]S_{l-} + [(l-1)/l]E_{l-} - [1/l]M_{l-} \sim [\phi_+^{l+1}\phi_-^{l}] \tag{4.23}$$

$$M_{l-} \sim [\phi_+^{l-1}\phi_-^{l}] \tag{4.24}$$

$$E_{l-} \sim [\phi_+^{l-1}\phi_-^{l-2}] \tag{4.25}$$

$$[k_0/|\mathbf{k}|]S_{l+} - E_{l+} \sim [\phi_+^{l+1}\phi_-^{l+2}] \tag{4.26}$$

$$lM_{l+} + E_{l+} \sim [\phi_+^{l-1}\phi_-^{l}] \tag{4.27}$$

where

$$\phi_\pm \equiv k^2 - (W \pm M)^2 \tag{4.28}$$

These thresholds are of course at the physically unattainable γN threshold energies $W = [M \pm (k^2)^{1/2}]$. In addition there is the πN

threshold at $W = (M + \mu)$ where all of the multipoles have the same behavior:

$$M_{l\pm}, E_{l\pm}, S_{l\pm} \sim |\mathbf{q}|^l \tag{4.29}$$

4.3. Invariant Amplitudes

The helicity amplitudes have kinematic singularities and constraints that may be located by an expansion into a suitable set of invariant amplitudes (Cohen-Tannoudji *et al.*, 1968). Such a set is defined by a suitable covariant expansion of the current matrix element T_μ (Ball, 1961)

$$T_\mu = \bar{u}_2 \gamma_5 \{ \gamma_\mu [-B_5 + (\gamma \cdot k)B_1] + \tfrac{1}{2}(p_1 + p_2)_\mu [2B_2 + (\gamma \cdot k)B_6]$$
$$+ q_\mu [2B_3 + (\gamma \cdot k)B_8] + k_\mu [2B_4 + (\gamma \cdot k)B_7] \} u_1 \tag{4.30}$$

where k, p_1, q, and p_2 are the four momenta of the photon, initial nucleon, pion, and final nucleon.

The current conservation condition (2.18) on T_μ yields the constraints

$$2k^2(B_4 + \tfrac{1}{2}B_1) = (t - k^2 - \mu^2)B_3 - \tfrac{1}{2}(s - u)B_2 \tag{4.31}$$

$$k^2 B_7 = B_5 - \tfrac{1}{4}(s - u)B_6 + \tfrac{1}{2}(t - k^2 - \mu^2)B_3 \tag{4.32}$$

As a result of the current conservation condition (2.17) on ε_μ the electroproduction matrix element (2.5) does not depend upon the amplitudes B_4 and B_7. In accordance with the general discussion of Section 2.2 we see that the remaining amplitudes are restricted only at $k^2 = 0$ when

$$(t - \mu^2)B_3 - \tfrac{1}{2}(s - u)B_2 = 0 \tag{4.33}$$

$$B_5 - \tfrac{1}{4}(s - u)B_6 + \tfrac{1}{2}(t - \mu^2)B_8 = 0 \tag{4.34}$$

The four independent amplitudes commonly used for photoproduction (Chew *et al.*, 1957) are

$$A_1 = B_1 - MB_6 \tag{4.35}$$

$$A_2 = 2B_2/(t - \mu^2) \tag{4.36}$$

$$A_3 = -B_8 \tag{4.37}$$

$$A_4 = -\tfrac{1}{2}B_6 \tag{4.38}$$

In order to relate the invariant and helicity amplitudes we evaluate T_μ in the c.m. frame and obtain (Pilkuhn, 1967)

$$T_0 = \chi_2^\dagger [-\boldsymbol{\sigma} \cdot \hat{\mathbf{q}}\phi_7 - \boldsymbol{\sigma} \cdot \mathbf{k}\phi_8]\chi_1 \tag{4.39}$$

$$\mathbf{T} = \chi_2^\dagger [\boldsymbol{\sigma}\phi_1 - i(\boldsymbol{\sigma} \cdot \hat{\mathbf{q}})(\boldsymbol{\sigma} \times \mathbf{k})\phi_2 + (\boldsymbol{\sigma} \cdot \hat{\mathbf{k}})\hat{\mathbf{q}}\phi_3 + (\boldsymbol{\sigma} \cdot \hat{\mathbf{q}})\hat{\mathbf{q}}\phi_4$$
$$+ (\boldsymbol{\sigma} \cdot \hat{\mathbf{k}})\hat{\mathbf{k}}\phi_5 + (\boldsymbol{\sigma} \cdot \hat{\mathbf{q}})\hat{\mathbf{k}}\phi_6]\chi_1 \tag{4.40}$$

where

$$\phi_1 = (E_1^+ E_2^+)^{1/2}[W^- B_1 - B_5] \tag{4.41}$$

$$\phi_2 = (E_1^- E_2^-)^{1/2}[-W^+ B_1 - B_5] \tag{4.42}$$

$$\phi_3 = (E_1^- E_2^-)^{1/2}[(2B_3 - B_2) + W^+(\tfrac{1}{2}B_6 - B_8)] \tag{4.43}$$

$$\phi_4 = (E_1^+ E_2^+)^{1/2} E_2^-[-(2B_3 - B_2) + W^-(\tfrac{1}{2}B_6 - B_8)] \tag{4.44}$$

$$\phi_7 = (E_1^+ E_2^-)^{1/2} k^{-2}\{-E_1^-(k^2 B_1 + W^+ B_5) + 2W|\mathbf{k}|^2(B_2 + \tfrac{1}{2}W^- B_6)$$
$$+ [\tfrac{1}{2}k_0(\mu^2 + k^2 - t) - k^2 E_\pi][2B_3 - B_2 - W^-(\tfrac{1}{2}B_6 - B_8)]\} \tag{4.45}$$

$$\phi_8 = (E_1^- E_2^+)^{1/2} k^{-2}\{E_1^+(k^2 B_1 - W^- B_5) - 2W|\mathbf{k}|^2(B_2 - \tfrac{1}{2}W^+ B_6)$$
$$- [\tfrac{1}{2}k_0(\mu^2 + k^2 - t) - k^2 E_\pi][2B_3 - B_2 + W^+(\tfrac{1}{2}B_6 - B_8)]\} \tag{4.46}$$

(ϕ_5 and ϕ_6 will not be needed but follow from the current conservation condition on T_μ.) We have used the notation $\hat{\mathbf{q}} \equiv \mathbf{q}/|\mathbf{q}|$, $\hat{\mathbf{k}} \equiv \mathbf{k}/|\mathbf{k}|$, $E_i^\pm \equiv E_i \pm M$ and $W^\pm \equiv W \pm M$ and we have eliminated B_4 and B_7 using Eqs. (4.31), (4.32). Using Eq. (4.11) from the θ dependence of the Pauli spinors we deduce that the helicity amplitudes as given by Eq. (2.30) are

$$f_{+++} = f_{---} = 2^{1/2} \sin \tfrac{1}{2}\theta[\phi_1 + \phi_2 + \cos^2 \tfrac{1}{2}\theta(\phi_3 + \phi_4)] \tag{4.47}$$

$$f_{-++} = -f_{+--} = -2^{1/2} \cos \tfrac{1}{2}\theta[\phi_1 - \phi_2 - \sin^2 \tfrac{1}{2}\theta(\phi_3 - \phi_4)] \tag{4.48}$$

$$f_{+-+} = f_{-+-} = 2^{-1/2} \sin \tfrac{1}{2}\theta \sin \theta(\phi_3 - \phi_4) \tag{4.49}$$

$$f_{--+} = f_{++-} = -2^{-1/2} \cos \tfrac{1}{2}\theta \sin \theta(\phi_3 + \phi_4) \tag{4.50}$$

$$f_{++0} = -f_{--0} = (|(k^2)^{1/2}|/|\mathbf{k}|) \cos \tfrac{1}{2}\theta(\phi_7 + \phi_8) \tag{4.51}$$

$$f_{-+0} = f_{+-0} = -(|(k^2)^{1/2}|/|\mathbf{k}|) \sin \tfrac{1}{2}\theta(\phi_7 + \phi_8) \tag{4.52}$$

From Eqs. (3.26)–(3.28) and (4.13)–(4.22) we deduce an expansion of the ϕ amplitudes into the conventionally defined multipoles which was first derived (Chew *et al.*, 1957; Dennery, 1961) by somewhat different methods. The argument of the Legendre polynomial P_l is $z = \cos \theta$:

$$\phi_1 = \sum (lM_{l+} + E_{l+})P'_{l+1} + [(l+1)M_{l-} + E_{l-}]P'_{l-1} \tag{4.53}$$

$$\phi_2 = \sum [(l+1)M_{l+} + lM_{l-}]P'_l \tag{4.54}$$

$$\phi_3 = \sum (E_{l+} - M_{l+})P''_{l+1} + (E_{l-} + M_{l-})P''_{l-1} \tag{4.55}$$

$$\phi_4 = \sum (M_{l+} - E_{l+} - M_{l-} - E_{l-})P''_l \tag{4.56}$$

$$\phi_7 = \sum (lS_{l-} - (l+1)S_{l+})P'_l \tag{4.57}$$

$$\phi_8 = \sum [(l+1)S_{l+}P'_{l+1} - lS_{l-}P'_{l-1}] \tag{4.58}$$

The inverse of this expansion is

$$
E_{l+} = \frac{1}{2(l+1)} \int_{-1}^{1} dz \left[\phi_1 P_l - \phi_2 P_{l+1} - \phi_3 \frac{l}{2l+1} (P_{l-1} - P_{l+1}) \right.
$$

$$
\left. + \phi_4 \frac{l+1}{2l+3} (P_l - P_{l+2}) \right] \tag{4.59}
$$

$$
M_{l+} = \frac{1}{2(l+1)} \int_{-1}^{1} dz \left[\phi_1 P_l - \phi_2 P_{l+1} - \phi_3 \frac{1}{2l+1} (P_{l-1} - P_{l+1}) \right] \tag{4.60}
$$

$$
E_{(l+1)-} = \frac{1}{2(l+1)} \int_{-1}^{1} dz \left[\phi_1 P_{l+1} - \phi_2 P_l - \phi_3 \frac{l+2}{2l+3} (P_l - P_{l+2}) \right.
$$

$$
\left. - \phi_4 \frac{l+1}{2l+1} (P_{l-1} - P_{l+1}) \right] \tag{4.61}
$$

$$
M_{(l+1)-} = \frac{1}{2l+1} \int_{-1}^{1} dz \left[-\phi_1 P_{l+1} + \phi_2 P_l + \phi_3 \frac{1}{2l+3} (P_l - P_{l+2}) \right] \tag{4.62}
$$

$$
S_{l+} = \frac{1}{2(l+1)} \int_{-1}^{1} dz [\phi_7 P_{l+1} + \phi_8 P_l] \tag{4.63}
$$

$$
S_{(l+1)-} = \frac{1}{2(l+1)} \int_{-1}^{1} dz [\phi_7 P_{l+1} + \phi_8 P_{l+1}] \tag{4.64}
$$

The threshold constraints (4.23)–(4.28) on the multipoles are equivalent to the requirement that the invariant amplitudes are finite and nonzero at the points $k^2 = (W \pm M)^2$ (Devenish and Lyth, 1975).

In a similar fashion the helicity amplitudes for the t-channel process $\gamma\pi \rightarrow \bar{N}N$ may also be expressed in terms of the invariant amplitudes. The result with the Jacob–Wick relative phase conventions (labeling the γ, π, $\bar{N}$, N as particles 1, 2, 3, and 4) is (Devenish, 1971)

$$
g_{++1} + g_{--1} = K(t/4)^{1/2} B_1 \tag{4.65}
$$

$$
g_{++1} - g_{--1} = -P[K_0 B_1 + t^{1/2} B_2 - M K_0 B_6] \tag{4.66}
$$

$$
g_{-+1} + g_{+-1} = -K[MB_1 + P^2 B_6] \tag{4.67}
$$

$$
g_{-+1} - g_{+-1} = P[B_5 - yB_6] \tag{4.68}
$$

$$
K|(k^2)^{1/2}|g_{++0} = k^2 yB_1 + t^{1/2} K_0 yB_2 - tK^2 B_3 + Mk^2 B_5
$$

$$
- MK_0^2 yB_6 + MK^2 K_0 t^{1/2} B_8 \tag{4.69}
$$

$$
-P^{-1}|(k^2)^{1/2}|g_{-+0} = K_0 B_5 - K_0 yB_6 + K^2 t^{1/2} B_8 \tag{4.70}
$$

The g's are related to the t-channel helicity amplitudes apart from an overall phase by

$$2(2)^{1/2}g_{\pm\pm1} = f_{\pm\pm1}/\sin \theta_t \tag{4.71}$$

$$2(2)^{1/2}g_{+-1} = f_{+-1}/(2 \cos^2 \tfrac{1}{2}\theta_t) \tag{4.72}$$

$$2(2)^{1/2}g_{-+1} = f_{-+1}/(2 \sin^2 \tfrac{1}{2}\theta_t) \tag{4.73}$$

$$2g_{++0} = f_{++0} \tag{4.74}$$

$$2g_{-+0} = f_{-+0}/\sin \theta_t \tag{4.75}$$

The kinematic quantities are

$$K_0 = (t + k^2 - \mu^2)(4t)^{-1/2} \tag{4.76}$$

$$K^2 = (K_0^2 - k^2)^{1/2} \tag{4.77}$$

$$P = (\tfrac{1}{4}t - M^2)^{1/2} \tag{4.78}$$

$$y = \tfrac{1}{4}(s - u) \tag{4.79}$$

$$\cos \theta_t = t^{-1/2}P^{-1}(s - u) \tag{4.80}$$

4.4. Analyticity and Unitarity

When expressed in terms of the invariant amplitudes B_n, the helicity amplitudes have the expected kinematic singularities and constraints, and this implies that the amplitudes B_n are related to the Joos amplitudes (Joos, 1962) by an expansion whose coefficients possess no singularities in s or t at fixed k^2 (Cohen-Tanoudjii *et al.*, 1968). On the basis of perturbation theory one therefore expects that the amplitudes B_n are analytic in s and t for fixed spacelike k^2 except for the pion and nucleon poles and cuts along $s > (M + \mu)^2$, $u > (M + \mu)^2$ and $t > 4\mu^2$. For timelike values of k^2 greater than $2\mu^2$ one will encounter also anomalous thresholds (Bjorken and Walecka, 1966; Bjorken and Drell, 1964).

It is possible to replace the amplitudes B_n by six amplitudes which automatically satisfy the current conservation condition $k_\mu T^\mu = 0$ (Fubini *et al.*, 1958), and such amplitudes were used in most early theoretical calculations (Barbour, 1963; Salin, 1964; Loubaton, 1965; Zagury, 1966, 1967, 1968; Walecka and Zucker, 1968; Adler, 1968; Von Gehlen, 1969, 1970; Berends, 1970; Crawford, 1971). When these six amplitudes are expressed in terms of the amplitudes B_n one finds kinematic singularities and constraints which have to be taken into account in theoretical work. There is no advantage in using them, and the amplitudes B_n have been used in most recent work (Manweiler and Schmidt, 1971; Devenish and Lyth, 1972*a, b*, 1973, 1975) (see also Vik, 1967).

In order to utilize the analytic properties it is necessary to make an assumption concerning the asymptotic behavior at fixed k^2. In particular the widely used fixed-t dispersion relations require a knowledge of the large s behavior for fixed negative k^2 and t. To this end, suppose that the t-channel helicity amplitudes are bounded by s^α with $\alpha < 1$, which corresponds to normal Regge behavior and also to the assumption that the virtual photon cross section vanishes. One then finds from Eqs. (4.65)–(4.70)

$$B_3 \sim s^\alpha \tag{4.81}$$

$$B_1, B_2, B_5', B_6, \text{ and } B_8 \sim s^{\alpha-1} \tag{4.82}$$

where

$$B_5' \equiv B_5 - \tfrac{1}{4}(s-u)B_6 \tag{4.83}$$

Because of its better asymptotic behavior one replaces B_5 by B_5' (Manweiler and Schmidt, 1971; Devenish and Lyth, 1973).

It is convenient to write dispersion relations for amplitudes that correspond to photons of isospin 0 and 1, viz.,

$$
\begin{aligned}
B_n(\gamma p \to \pi^+ n) &= (2)^{1/2}[B_n^{(0)} + B_n^{(-)}] \\
B_n(\gamma n \to \pi^- p) &= (2)^{1/2}[B_n^{(0)} - B_n^{(-)}] \\
B_n(\gamma p \to \pi^0 p) &= B_n^{(0)} + B_n^{(+)} \\
B_n(\gamma n \to \pi^0 n) &= -B_n^{(0)} + B_n^{(+)}
\end{aligned}
\tag{4.84}
$$

The ($\pm$) amplitudes correspond to photons of isospin 1 and the (0) amplitude to photon of isospin 0. Except for $B_3^{(-)}$, which will be dealt with shortly, the fixed-t dispersion relations are

$$B_n(s) = R_n\left(\frac{1}{M^2-s} + \frac{\eta_n}{M^2-u}\right) + \frac{1}{\pi}\int_{s_0}^\infty ds' \operatorname{Im} B_n(s')\left(\frac{1}{s'-s} + \frac{\eta_n}{s'-u}\right) \tag{4.85}$$

where $s_0 \equiv (M+\mu)^2$, and $\eta_1 = \eta_2 = \eta_6 = -\eta_3 = -\eta_5' = -\eta_6 = 1$ except for $B_n^{(-)}$ when they are -1. The residues of the poles may be obtained from perturbation theory (in any gauge) or from extended unitarity [Eq. (4.97)] if we give the nucleon an infinitesimally small width, and they have been given by numerous authors. They are

$$R_1 = -\tfrac{1}{2}eg[F_1(k^2) + 2MF_2(k^2)] \tag{4.86}$$

$$R_2 = \tfrac{1}{2}egF_1(k^2) \tag{4.87}$$

$$R_3 = \tfrac{1}{4}egF_1(k^2) \tag{4.88}$$

$$R_5' = -\tfrac{1}{4}egF_2(k^2)(t-k^2-\mu^2) \tag{4.89}$$

$$R_6 = -egF_2(k^2) \tag{4.90}$$

$$R_8 = -\tfrac{1}{2}egF_2(k^2) \tag{4.91}$$

where F_1 and F_2 are the nucleon form factors [Eq. (3.70) and following] and g is the πNN coupling constant. The appropriate form factors for $B^{(0)}$ and $B^{(\pm)}$ are, respectively, F^s and F^v, where

$$F^s = F^p + F^n \qquad (4.92)$$

$$F^v = F^p - F^n \qquad (4.93)$$

The amplitude $B_3^{(-)}$ requires a subtraction unless $\alpha < 0$ in Eq. (4.81). This amplitude is the only one that has the pion pole at $t = \mu^2$, and from the appropriate Feynman graph (in any gauge) or from extended unitarity one deduces

$$\lim_{t \to \mu^2} [(\mu^2 - t) B_3^{(-)}(s, t, k^2)] = egF_\pi(k^2) \qquad (4.94)$$

where F_π is the pion form factor. We deduce that for $t = \mu^2$, $\alpha = 0$. For $t < 0$ Regge theory suggests that α may be negative but still fairly close to zero when cuts are taken into account and in the photoproduction limit where data exist one finds $\alpha \sim 0$ experimentally.

Accordingly it is desirable if not absolutely necessary to make a subtraction for $B_3^{(-)}$ in order to ensure good convergence of the dispersion integral. This gives

$$B_3^{(-)}(s) = \tfrac{1}{4} egF_1 \left(\frac{1}{M^2 - s} + \frac{1}{M^2 - u} \right) + \left[\frac{egF_\pi}{\mu^2 - t} + \Delta B(s_1) \right]$$

$$+ \frac{1}{\pi} \int_{s_0}^{\infty} ds' \operatorname{Im} B_3^{(-)}(s') \left(\frac{1}{s' - s} + \frac{1}{s' - u} - \frac{1}{s' - s_1} - \frac{1}{s' - u_1} \right)$$

$$(4.95)$$

The subtraction point s_1 is arbitrary and $s_1 + u_1 + t = 2M^2 + \mu^2 + k^2$. The term in square brackets is just $B_3^{(-)}(s_1)$ minus its nucleon pole contribution and we have exhibited the pion pole before denoting the remainder by ΔB.

We note at this point that the current conservation condition (4.33) at $k^2 = 0$ when evaluated at the pole positions is equivalent to $F_1^n = 0$ and $F_1^p = F_\pi$, i.e., the neutron charge has to vanish and the p and π^+ charges have to be equal. This is of course just charge conservation for the reactions $\gamma n \to \pi^- p$ and $\gamma p \to \pi^+ n$.

Returning to ΔB this quantity is determined at $k^2 = 0$ by the same current conservation condition (for example if the subtraction point is chosen so that $s_1 = u_1$, then ΔB vanishes). Away from $k^2 = 0$ one may estimate ΔB from a fixed-s dispersion relation if $B_3^{(-)}$ vanishes as $t \to \infty$ with fixed k^2 and $s = s_1$. When $s_1 = 0$ and $k^2 = 0$ this seems to be the case since the backward photoproduction cross section decreases quite rapidly at high energies. On the basis of Regge theory we may expect the same behavior to persist for $k^2 \neq 0$. Accordingly we choose $s_1 = 0$ and write a fixed-s dispersion relation for ΔB (Devenish and Lyth, 1973). This gives

$$
\begin{aligned}
B_3^{(-)}(st) = {}&\tfrac{1}{4}egF_1\left(\frac{1}{M^2 - s} + \frac{1}{M^2 - u}\right) + \frac{egF_\pi}{\mu^2 - t} \\
&+ \frac{1}{\pi}\int_{4\mu^2}^{\infty} dt' \frac{\operatorname{Im} B_3^{(-)}(s_1, t')}{t' - t} - \frac{\tfrac{1}{4}egF_1}{M^2 - s_1} \\
&- \frac{1}{\pi}\int_{s_0}^{\infty} ds' \frac{\operatorname{Im} B_3^{(-)}(s', t)}{s' - s_1} \\
&+ \frac{1}{\pi}\int_{s_0}^{\infty} ds' \operatorname{Im} B_3^{(-)}(s't)\left[\frac{1}{s' - s} + \frac{1}{s' - u}\right] \\
&+ \frac{1}{\pi}\int_{s_0}^{\infty} du' \frac{\operatorname{Im} B_3^{(-)}(u', t(u', s_1)) - \operatorname{Im} B_3^{(-)}(u', t)}{u' - u_1}
\end{aligned}
\tag{4.96}
$$

In the last integral, $t(u, s)$ is the function defined by $2M^2 + \mu^2 + k^2 = s + u + t$. It is understood that convergence of the s' and u' integrals is only guaranteed if the upper limit is taken to infinity simultaneously in all of them.

We see that the pion pole is now accompanied by a contribution from higher-mass intermediate states in the channel $\gamma\pi \to \bar{N}N$, for example, meson resonance contributions. When $k^2 = 0$ the current conservation condition (4.33) determines this extra contribution in terms of the other quantities appearing in the dispersion relation. An estimate with $s_1 = 0$ shows that for $-1 \ (\mathrm{GeV}/c)^2 \lesssim t < 0$ (which includes the full angular range for $k^2 = 0$ in the first resonance region) the extra contribution is negligible compared with the pion pole, but that for more negative t values it is probably comparable (Devenish and Lyth, 1973). It decreases only slowly as t goes negative, which suggests that it is generated mostly by fairly high-mass states. Its estimate for $k^2 < 0$ will therefore be difficult, but it seems reasonable to suppose that it remains negligible compared with the pion pole for fairly small t.

The indeterminacy of the subtraction constant was not recognized until recently (Manweiler and Schmidt, 1971; Devenish and Lyth, 1973, 1975). Previous authors had used a subtraction point $s_1 = \infty$ (which presumably causes divergent or only slowly convergent integrals) together with some arbitrary (and implicit) prescription for continuing ΔB away from $k^2 = 0$. In some cases their $k^2 = 0$ limit did not satisfy the current conservation condition.

The central problem in utilizing the dispersion relations is the theoretical or phenomenological determination of the direct channel imaginary parts. Unitarity is of considerable assistance here. For a πN state of isospin $I = \frac{1}{2}$ or $\frac{3}{2}$ the multipoles (or partial wave helicity amplitudes) satisfy the unitarity requirement

$$\text{Im } M_{l\pm}^I = M_{l\pm}^I \exp(-i\delta_{l\pm}^I) \sin \delta_{l\pm}^I + (\text{Im } M_{l\pm}^I)^{\text{in}} \qquad (4.97)$$

and similarly for $E_{l\pm}$ and $S_{l\pm}$, where δ is the πN phase shift. The second term vanishes for $s < (M + 2\mu)^2$ and is negligible in practice up to considerably higher energies for most of the partial waves. When the second term is not negligible the πN partial wave is no longer described by a single parameter δ and the separation of the first term from the second is not a unique procedure. In practice we shall see that it is most convenient to take δ to be a continuation of the elastic phase shift which has the property that $\sin \delta$ falls smoothly to zero with increasing energy, i.e., to make δ go to 0, or π (in principle $n\pi$), whichever is nearest. If this is done the first term falls smoothly to zero as one moves out of the elastic region while the second term takes over (Lyth, 1971, 1972).

In the region of a narrow resonance the partial wave helicity amplitudes [Eq. (4.14)] are given by

$$f_{\lambda r}^{l\pm} = \frac{g_{\lambda r}(k^2)(\Gamma_{\pi N}/|\mathbf{q}|_r)^{1/2}}{M_r^2 - s - iM_r\Gamma} \qquad (4.98)$$

where apart from normalization $g_{\lambda r}$ is the form factor for the transition $\gamma N \to N^*$. Here M_r is the resonance mass, Γ its full width, and $\Gamma_{\pi N}$ its partial width into the πN channel. The amplitude for any other two-body final state is obtained by replacing $\Gamma_{\pi N}$ by the appropriate partial width.

When $r = \pm 1$ and $k^2 = 0$, $g_{\lambda r}$ is related to the γN partial width by unitarity (Pilkuhn, 1967)

$$g_{\lambda r}(0) = 8\pi M_r^2 (\Gamma_{\gamma N}^{\lambda r}/|\mathbf{k}|_r)^{1/2} \qquad (4.99)$$

The sign of the square root here and in the preceding equation is physically significant and can be determined relatively to, say, that of a pole term by phenomenological analysis of experiments.

In practice the shape of these extreme narrow resonance approximation requires modification particularly to take into account the threshold behavior $f^{l\pm} \sim |\mathbf{q}|^l$ [Eq. (4.29)] and the stronger behavior $\text{Im } f^{l\pm} \sim |\mathbf{q}|^{2l+1} f^{l\pm}$ following from unitarity and the behavior of the πN phase shift. It is also desirable at any rate for small k^2 to take into account the behavior at the threshold $k^2 = (W - M)^2$ and for this purpose the multipole amplitudes are most appropriate. A simple procedure satisfying these requirements is to multiply the resonant amplitudes by a factor $[\text{const}|\mathbf{k}|^n|\mathbf{q}|^l/(q^2 + X^2)^l]$, where the constant is chosen to preserve the normalization at $s = M_r^2$ and also to multiply the width by a similar factor $[\text{const}|q|^{2l+1}/(|\mathbf{q}|^2 + X^2)^l]$. The denominators are necessary to avoid unphysical increases as $s \to \infty$ and X

is an arbitrary constant. The power of $|\mathbf{k}|$ is $n = l$ for $M_{l\pm}$, E_{l+} and S_{l+} and $n = l - 2$ for E_{l-} and S_{l-} [Eqs. (4.23)–(4.28)].

Only the imaginary part of f is required for input into the dispersion relations, and since this quantity possesses the resonance peak one may hope that it is not sensitive to details of the shape of f away from resonance. The real part of f is hopefully obtainable as output from the dispersion relation. Its shape will not be directly related to that of Im f and will automatically take into account the full threshold constraints (4.23)–(4.28).

In this account we have discussed only the commonly used fixed-t dispersion relations. In principle one may consider dispersion integrals over other contours in the s–t plane, e.g., fixed s or fixed θ, and one may also write down dispersion relations for the partial wave amplitudes themselves. In such cases one will require a knowledge of the imaginary part for $t > 4\mu^2$ in all the amplitudes, or conversely may hope to learn something about the amplitude in this region. Such considerations have proved very powerful for the similar (but much easier) case of πN scattering but have not been undertaken so far for electroproduction or even to any serious extent for photoproduction.

For any dispersion relation one has to check whether the partial wave expansion for Im B which one truncates for practical calculations does in fact converge. The interval of convergence can be computed by calculating the boundary of the double spectral functions. For the fixed-t dispersion relation it is such [at least for $-k^2 \lesssim 1 \, (\text{GeV}/c)^2$] that one can use the relation for t values that cover the whole of the angular range in the first resonance region but only for a gradually decreasing angular range as the energy is increased (Devenish and Lyth, 1972a, b).

4.5. Construction of Unitary and Analytic Amplitudes

In this section we show how to construct amplitudes that satisfy both the fixed-t dispersion relations (4.85) and the unitarity relation (4.97) (for any choice of the inelastic term). Our discussion proceeds along standard lines except that we take particular care to establish what free parameters enter into the solution and also allow a nonzero inelastic term. The practical application of our mathematical results will be described in the next section.

One works with multipoles

$$\overline{M}_{l\pm} = (|\mathbf{k}| \, |\mathbf{q}|)^{-l} M_{l\pm} \qquad (4.100)$$

and similarly for $E_{l\pm}$ and $(S_{l\pm}/|\mathbf{k}|)$. These barred multipoles are nonzero at the πN threshold $W = M + \mu$, and at the γN thresholds $W = M \pm (k^2)^{1/2}$. Actually E_{l-} and $(S_{l-}/|\mathbf{k}|)$ have poles at the latter point. These latter multipoles only exist for the cases $l \geqslant 2$, which are not our primary concern, as will become clear; also we shall see that the behavior of the barred multipoles at the γN thresholds is not crucial; the same applies to their behavior at the negative energy points $W = -(M + \mu)$ and $W = -[M \pm (k^2)^{1/2}]$, where it follows from McDowell symmetry (Dennery, 1961) [also from Eqs. (4.23)–(4.28) for the latter points] that they have square root branch points.

One may project out the multipoles from invariant amplitudes that have been calculated from fixed-t dispersion relations, whose input includes the imaginary parts of the multipoles themselves. This gives an expression of the form

$$\overline{M}(s) = L(s) + \frac{1}{\pi} \int_{s_0}^{\infty} ds' \, \frac{\exp[-i\delta(s')] \sin \delta(s') \overline{M}(s' + i\varepsilon)}{s' - s} \quad (4.101)$$

where $\overline{M}$ is any of the barred multipoles for a definite isospin πN state and δ is the corresponding πN phase shift suitably continued above the elastic region [recall the discussion following Eq. (4.97)]. The first term receives four contributions,

$$L(s) = (\text{pole contribution}) + \left(\int dt' \, \text{Im} \, B_3^{(-)} \text{contribution} \right)$$

$$+ \frac{1}{\pi} \int_{s_0}^{\infty} ds' \, \frac{\text{Im} \, \overline{M}_{in}(s')}{s' - s} + \sum_{\substack{n = \text{all} \\ \text{multipoles}}} \int_{s_0}^{\infty} ds' K_n(ss') \, \text{Im} \, \overline{M}_n(s') \quad (4.102)$$

The first two come from the indicated parts of the fixed-t dispersion relations and the others come from the input multipoles to these relations. The kinematic kernel K_n may be calculated explicitly (von Gehlen, 1969).

If one had calculated $\overline{M}$ directly from a dispersion relation, L would have been (apart from its second term) the contribution of the unphysical cuts including that arising from the kinematic branch point at $s = 0$ (Dennery, 1961). In the fixed-t approach the analytic properties of $\overline{M}$ are not strictly relevant, but they provide a guide to what one may expect for the convergence and nature of the kinematic factor K_n. If the factor relating the barred multipoles to the unbarred ones were multiplied by a function with singularities in s the barred multipoles would acquire singularities that would manifest themselves as a strong dependence of L on $\text{Im} \, \overline{M}$; conversely, if it were multiplied by a polynomial the resultant zeros of $\overline{M}$ could come about as a cancellation between the two terms of Eq. (4.101). Neither of these features is desirable for practical calculations, as we shall see in Section 4.6, but still the precise choice of $\overline{M}$ is not so crucial as it would be if dispersion relations were to be evaluated for this quantity.

If L is temporarily assumed to be known, Eq. (4.101) for $\overline{M}$ has the well-known solution

$$\overline{M} = N/D \qquad (4.103)$$

where

$$D \equiv \exp\left[-\frac{1}{\pi} \int_{s_0}^{\infty} ds' \frac{s}{s'} \frac{\delta(s')}{s' - s} \right] \qquad (4.104)$$

and provided that $\delta(\infty) \leqq 0$ and $\overline{M}(\infty) = 0$

$$N - LD = -\frac{1}{\pi} \int_{s_0}^{\infty} ds' \frac{L(s') \operatorname{Im} D(s')}{s' - s} \qquad (4.105)$$

[see Lyth (1972) for a proof that handles the inelastic contribution to L].

If $\delta(\infty)$ is positive, the asymptotic behavior $D \sim s^{\delta(\infty)/\pi}$ means that subtractions may be needed in the equation for N with the consequent introduction of free parameters. If $\overline{M}$ is known to vanish at infinity faster than s^{-n}, where n is an integer, then the number of parameters one needs is $[\delta(\infty)/\pi] - n$. [We are taking $\delta(\infty)/\pi$ to be integral, recalling the discussion following Eq. (4.97).]

A slightly subtle point (but one of practical importance, as we shall see) is that L will not in general vanish as rapidly as $\overline{M}$ but will rather cancel with the second term in Eq. (4.101). For example, if $n = 1$

$$L(s) \xrightarrow[s \to \infty]{} \Lambda/s \qquad (4.106)$$

$$\Lambda = \frac{1}{\pi} \int_{s_0}^{\infty} ds \operatorname{Im} \overline{M}(s) \qquad (4.107)$$

Then the expression for N will have an additional contribution

$$-\Lambda \lim_{s \to \infty} [s^{-1} D(s)] \qquad (4.108)$$

In principle Λ is known if L is known but in practice one's estimate of L may only be valid for small s, and then Λ is an effective free parameter. Note that the contribution of a resonance that dominates $\operatorname{Im} \overline{M}$ is

$$\frac{1}{\pi} \int ds' \frac{\operatorname{Im} \overline{M}(s')}{s' - s} \approx \frac{\Lambda}{M^2 - s - iM\Gamma} \qquad (4.109)$$

so that Λ determines the resonance coupling strength.

The asymptotic behavior postulated in writing down the fixed-t dispersion relations suggests (with reasonable large-angle behavior) that the multipoles vanish asymptotically and that therefore the barred quantities vanish faster than s^{-1}.

Of the πN phase shifts only $\delta_{1+}^{3/2}$ is resonant in the elastic region, and we may therefore take $\delta_{l\pm}^{I}(\infty) = 0$ except for $\delta_{1+}^{3/2}(\infty) = \pi$ [recall the remarks after Eq. (4.97)]. Then all the multipoles are given by the parameter-free equations (4.103)–(4.105) except for $M_{1+}^{3/2} E_{1+}^{3/2}$ and $S_{1+}^{3/2}$, which require the additional term (4.108) involving Λ.

So far we have assumed that L is known. The t-channel contribution is difficult to calculate, but we have noted earlier that it is probably small in the first resonance region for small k^2. The contribution from the inelastic region of the s-channel may be estimated phenomenologically, as we shall see. This leaves the contribution from the elastic region, which involves the multipoles obtained from the unitary solution (4.103). One has here a consistency problem for which no closed solution has been written down, but we shall see in Section 4.6 that a numerical solution is fairly easy to obtain.

4.6. Practical Calculations

Apart from the current algebra results to be discussed in Section 4.7 and the symmetry ideas that were described in Chapter 2, we have now listed all the known theoretical properties of the single-pion electro-production amplitude and incidentally of the photoproduction amplitude, which is its $k^2 = 0$ limit. We shall next discuss an approximation scheme that allows the amplitude to be estimated in terms of relatively few parameters. Corresponding roughly to the historical development we start with the simplest approximation and gradually elaborate it.

The simplest possible approximation concerning the fixed-t dispersion relation is to keep only the pole terms, which is consistent with current conservation provided that the subtraction point s_1 for $B_3^{(-)}$ is taken at infinity. As we shall see, this approximation is quite good for charged pions at low energies and near the forward direction, at least for $-k^2 \lesssim 1$ $(\text{GeV}/c)^2$. One may seek to satisfy elastic unitarity by keeping only the pole terms for L in the unitary expression (4.103) for the multipoles. The πN phase shifts are all small in the elastic region except for the resonant $\delta_{1+}^{3/2}$, and nonresonant δ_{0+}^{I}, which latter rise to around 30° in the resonance region. These affect the resonant multipoles $M_{1+}^{3/2}$, $E_{1+}^{3/2}$, and $S_{1+}^{3/2}$ and the nonresonant multipoles E_{0+}^{I} and S_{0+}^{I}.

Consider first the multipole $M_{1+}^{3/2}$. The pole contribution to L in the unitary expression (4.103) may be approximated when $-k^2 \lesssim 1 \,(\text{GeV}/c)^2$ by (Fubini *et al.*, 1958)

$$L(k^2, s) \simeq \frac{1}{4\pi} \frac{2}{3} \frac{g G_M^v(k^2)}{s - M^2} \tag{4.110}$$

Similar approximations for elastic πN scattering (Chew *et al.*, 1957) show that the πN partial wave $F_{1+}^{3/2} \equiv \exp(i\delta)\sin\delta/|\mathbf{q}|^3$ is also given by the unitary expression (4.103) but with an L that may be approximated by

$$L_{\pi N}(s) \simeq \frac{2}{3M}\frac{g^2}{4\pi}\frac{1}{s - M^2} \qquad (4.111)$$

We therefore deduce that

$$\overline{M}_{1+}^{3/2}(s, k^2) \simeq \frac{M}{g}G_M^v(k^2)F_{1+}^{3/2}(s) \qquad (4.112)$$

The normalization of $\overline{M}$ and F are given by Eqs. (4.106) and (4.107). The detailed shape of $\overline{M}$ and F is given by the unitary expression (4.103) but may be written within the accuracy of the approximations in the form discussed after Eq. (4.99) above, which reproduces the important singularities (Alcock *et al.*, 1968),

$$\overline{M}_{1+} = \frac{cX(s)G_M^*(k^2)}{M_r^2 - s - iM\Gamma Y(s)X(s)} \qquad (4.113)$$

where M_r and Γ_r are the mass and width of the $\Delta(1232)$ resonance,

$$X(s) = (|\mathbf{q}|_r^2 + a^2)/(|\mathbf{q}|^2 + a^2) \qquad (4.114)$$

$$a \simeq \mu \quad \text{(pion mass)} \qquad (4.115)$$

$$y(s) = |\mathbf{q}|^3/|\mathbf{q}|_r^3 \qquad (4.116)$$

(note that $X = y = 1$ when $s = M_r^2$) and

$$c = \frac{2}{3}\frac{1}{4\pi}gG_M^v(0) \qquad (4.117)$$

The function $G_M^*(k^2)$ is proportional to the $\gamma N \to N^*$ transition form factor and is normalized to unity at $k^2 = 0$ so that from (4.112) it is given by

$$G_M^*(k^2) = G_M(k^2)/G_M(0) \qquad (4.118)$$

We note incidentally that the normalization condition (4.106), (4.107) for F, combined with the unitarity requirement $F(M_r^2) = 1/|\mathbf{q}|_r^3$ leads to an expression for the resonance width

$$\Gamma = \frac{2}{3}\frac{q_r^3}{MM_r}\frac{g^2}{4\pi} \qquad (4.119)$$

which is in $\sim 20\%$ agreement with experiment.

As we shall see, this theoretical prediction for the coupling strength of the $\Delta(1232)$ resonance in the process $\gamma N \to N^* \to \pi N$ with real and virtual photons is in similar agreement with experiment, and the prediction for the shape is in good agreement except on the top side of the resonance for $-k^2 \lesssim 1$ $(\text{GeV}/c)^2$.

It has been necessary to assume that our estimate (4.110) of $L(s)$ is valid at energies well above the resonance position so as to allow an estimate of the constant Λ. There is no particular reason *a priori* why the neglected contributions should (or should not) be small, but the success of the predictions justifies their neglect [i.e., Eq. (4.107) is satisfied quite well with the estimated value of Λ].

In order to be consistent, we should have included the contribution to $L(s)$ coming from $\text{Im}\, M_{1+}$ itself. When this contribution is included [e.g., using the approximation (4.113) or experiment] the change in L is not very great, but one encounters a problem common to any calculation going beyond the pole approximation. This is that the separate contributions to L and in particular that of M_{1+} do not generally have $1/s$ behavior and do not therefore permit an estimate of the coupling strength Λ. In other words if we wish to go beyond the pole approximation the normalization of M_{1+} cannot be predicted theoretically.

To a good approximation the contribution of $\text{Im}\, M_{1+}$ to L depends only on the coupling strength Λ (besides the resonance mass and width), i.e., the shape of the resonance is not of prime importance. Thus one can evaluate the unitary solution (4.103) quite accurately for any desired value of Λ, taking the shape from Eq. (4.113).

For the other resonant multipoles E_{1+} and S_{1+} the contributions to L from both the poles and from $\text{Im}\, M_{1+}$ are small. Although they are not $\sim 1/s$ at large s, it seems reasonable to take their smallness as an indication that the coupling strength Λ is small for them compared with the M_{1+} coupling, and hence that M_{1+} is the dominant multipole. This is verified experimentally.

The approximation presented so far has three parameters, the coupling strengths $\Lambda(k^2)$ for $M_{1+}^{3/2}$, $E_{1+}^{3/2}$ and $S_{1+}^{3/2}$. In addition one requires the πNN coupling constant g, the nucleon form factors F_1 and F_2 and the pion form factor F_π, which may be taken from experiment, as well as the t-channel contribution, which as discussed earlier should not matter in the first resonance region for $-k^2 \lesssim 1$ $(\text{GeV}/c)^2$. The contributions of the multipoles to L may be handled in practice by first keeping only $\text{Im}\, M_{1+}$ to obtain a first approximation to the multipoles, then evaluating the contributions using this approximation; one can check that the contributions would not be significantly altered by using the resultant second approximations for the multipoles instead.

Most of the published calculations of resonance region electroproduction effectively employ some version of the approximation outlined so far. As we shall see it gives a good description of the photoproduction amplitude in the first resonance region, and as far as one can tell of the electroproduction amplitude also for $-k^2 \lesssim 1$ (Gev/c)^2. The imaginary part of the dominant $M_{1+}^{3/2}$ is given well by the resonance formula (4.113) in the region $-k^2 \lesssim 1$ (GeV/c)^2 with parameters (Devenish and Lyth, 1972)

$$M_r = 1.232 \text{ GeV}$$

$$\Gamma_r = 0.114 \text{ GeV}$$

$$cM^{-1}\Gamma^{-1} = 3.52 \; (\mu\text{b})^{1/2}$$

In order to evaluate the dispersion relation at higher energies it is obviously necessary to include the contribution of higher-mass resonances. In principal one may perhaps employ multichannel unitarity to determine their contribution (Walecka and Zucker, 1968), but in practice the situation is very complicated and in any case the number of free parameters has not yet been determined. In practice therefore one has to parameterize their contributions, e.g., by Eqs. (4.98) and following. At energies above the resonance region one may parameterize the imaginary parts by smooth functions of s and t.

These extra contributions to the L terms in the unitary expression (4.103) for the first resonance region multipoles are small but not totally negligible (Devenish and Lyth, 1972). Note that even the contributions of resonances in the multipoles themselves (e.g., from the s_{11} resonance in S_{0+} or from the p_{33}, resonance in M_{1+}) may be included without double counting provided that the prescription $\sin \delta \to 0$ which we mentioned earlier is used to continue the elastic phase shift to higher energies.

The scheme outlined above provides a description of the electroproduction and photoproduction amplitudes in terms of a number of parameters which can be quite small compared with the number and complexity of the amplitudes one is discussing (Devenish and Lyth, 1975; Devenish et al., 1974). Its big limitation, as we have noted, is that it only works over a limited interval of t corresponding in general to some forward range of angles. For large values of $-t$ the partial wave expansion may fail to converge, as we have noted, and in any case severe cancellations between the separate contributions will take place. The backward region may perhaps be handled by means of a fixed-u dispersion relation so as to provide a complete angular coverage in the resonance region, but the central region at high energies (s, t, u all large) is quite inaccessible through the dispersion relation approach.

4.7. Current Algebra and the Weak Nucleon Form Factor

We have seen that the amplitudes

$$T_\mu^{(\pm)} \equiv \langle \pi N' | J_\mu(0) | N \rangle$$

describe the electroproduction process

$$eN \rightarrow e\pi N'$$

where the superscript denotes the isospin state [Eqs. (4.84)]. In a similar fashion, the amplitudes

$$[e^{-1}\mu^2 f_\pi/(\mu^2 - q^2)]T_\mu^{(\pm)}(q) \equiv \int d^4x \, \exp(iqx)\langle N'|T(\partial^\nu A_\nu^{(\pm)}(x), J_\mu(0))|N\rangle \tag{4.120}$$

describes, respectively, the weak processes

$$ep \rightarrow e(\bar{e}\nu)n$$

$$en \rightarrow e(e\bar{\nu})p$$

where the $(\bar{e}\nu)$ or $(e\bar{\nu})$ system has momentum q and $J^P = 0^-$. Here $A_\nu^{(\pm)}$ is the weak axial vector current.

These amplitudes are related, because the latter has a pole when $q^2 = \mu^2$, whose residue is proportional to the former. In fact the square bracket which we have pulled out before defining $T_\mu(q)$ is such that $T_\mu^{(\pm)}(q^2 = \mu^2) = T_\mu^{(\pm)}$, when f_π is the $(e\nu\pi)$ coupling. The current algebra hypothesis (Gell-Mann, 1962) allows one to define also a current $A_\nu^{(0)}$ and hence an amplitude $T_\mu^{(0)}(q)$, which reduces to the electroproduction amplitude $T_\mu^{(0)}$.

When $q = 0$, the current algebra hypothesis allows one to calculate $T_\mu(q)$ in terms of the electromagnetic and weak axial vector form factors of the nucleon. To see this one defines in addition to $T_\mu(q)$ the weak amplitude

$$T_{\nu\mu}(q) = \int d^4x \, \exp(iqx)\langle N'|T(A_\nu(x)J_\mu(0))|N\rangle \tag{4.121}$$

which describes the production of a $(\nu\bar{e})$ or (νe) pair which may have $J^P 1^-$ as well as 0^-. One also needs the weak nucleon form factor

$$S_\mu = \langle N'|A_\mu(0)|N\rangle \tag{4.122}$$

which for the $(+)$ isospin state is given by

$$S_\mu = \bar{u}'\gamma_5[\gamma_\mu g_A(k^2) + k_\mu h_A(k^2)]u \tag{4.123}$$

where k is the four-momentum transfer between the nucleons (other isospin states are trivially related). After integrating by parts, setting $q = 0$ and using the current algebra relations

$$[Q^{(\pm)}(0), J_\mu(0)] = \mp iA_\mu^{(\pm)}(0)$$
$$[Q^{(0)}(0), J_\mu(0)] = 0 \tag{4.124}$$

where

$$Q(t) \equiv \int d^3x A_0(x) \tag{4.125}$$

one obtains the identities

$$T_\mu^{(\pm)}(0) = \lim_{q \to 0} (iq^\nu T_{\nu\mu}^{(\pm)}) \pm iS_\mu^{(\pm)}$$

$$T_\mu^{(0)}(0) = \lim_{q \to 0} (iq^\nu T_{\nu\mu}^{(0)})$$

(4.126)

(Riazuddin and Lee, 1966; Adler and Gilman, 1966). The first term on the right-hand side can be calculated in terms of the nucleon electromagnetic form factors.

If $T_\mu(q)$ is expanded in terms of invariants $B_n(q^2)$ by means of Eq. (4.30), $B_n(0)$ is related to the nucleon form factors. Using for convenience the Goldberger–Trieman relation $f_\pi \simeq Mg_A(0)/g$ and defining a normalized form factor by $G_A(q^2) \equiv g_A(q^2)/g_A(0)$ one obtains (Nambu and Schrauner, 1962; Riazuddin and Lee, 1966; Adler and Gilman, 1966; Furlan *et al.*, 1966; Benfatto *et al.*, 1972)

$$B_1^{(+)} - MB_0^{(+)} + B_2^{(+)} = (\tfrac{1}{2}eg/M)F_2^{(v)}(k^2)$$

(4.127)

$$B_5^{(-)} = -(eg/M)[G_A(k^2) - F_1^v(k^2) - 2MF_2^v(k^2)]$$

(4.128)

$$B_1^{(-)} + 2B_4^{(-)} - 2MB_7^{(-)} = -ef_\pi^{-1}h_A(k^2)$$

(4.129)

The left-hand sides of these relations involve the amplitudes B_n at the point $q = 0$, whereas in electroproduction they can only be measured on the surface $q^2 = \mu^2$. This surface does not include the point $q = 0$ unless the pion mass vanishes, when $q = 0$ is the πN threshold. The current algebra relations are therefore only useful when the amplitudes have a small variation between the point $q = 0$ and a nearby point (e.g., the πN threshold) on the surface $q^2 = \mu^2$.

Insight into this question may be obtained by considering the fixed-t dispersion relations (4.85). They may be expected to hold for the weak amplitudes $B_n(q)$ at least for $q^2 \lesssim \mu^2$, and Im $B_n(q)$ will still be given by extended unitarity relations like (4.97). Insofar as these relations can be solved in the fashion that we have discussed, the variation of Im B_n with q^2 is given essentially by that of the weak N–N^* transition form factors and may therefore be expected to be small in the small interval $0 < q^2 < \mu^2$ (e.g., from experiment, from analytic properties in q^2, or from the analogy with electromagnetic form factors).

The hypothesis that the imaginary parts are slowly varying leads to the same statement for the full amplitudes provided that the dispersion relations are unsubtracted. This is the case for the fixed-t dispersion relations giving the left-hand sides of the first two current algebra relations. One may therefore obtain sum rules giving $F_2^{v,s}$ and $[G_A - F_1^v]$ as integrals over the imaginary parts of single-pion electroproduction amplitudes (Fubini *et al.*, 1965; Riazuddin and Lee, 1966; Adler and Gilman, 1966). Alternatively one may regard the current algebra relations as giving the electroproduction amplitude at threshold with $q^2 = 0$ and merely use the dispersion integral to estimate the small correction in going to the physical point $q^2 = \mu^2$ (Furlan *et al.*, 1966; Benfatto *et al.* 1972, 1973; Dombey and Read, 1973). Obviously, these two approaches will be equivalent provided that the dispersion relation gives the correct electroproduction amplitude at threshold. As we shall see, this seems to be the case experimentally.

The third relation which gives the form factor h_A is on a different footing because the amplitudes B_4 and B_7 do not contribute to the electroproduction cross section, as we have seen. One may attempt to evaluate them in terms of the measurable amplitudes by using current conservation, but for $q^2 \neq \mu^2$ the condition $\partial_\mu J^\mu(x) = 0$ leads to a nonzero result for $k_\mu T^\mu$ (Nauenberg, 1966) which gives the expression (Benfatto *et al.*, 1972, 1973)

$$k^2(B_1^{(-)} + 2B_4^{(-)}) = (t - k^2 - q^2)B_3^{(-)} - \tfrac{1}{2}(s - u)B_2^{(-)}$$
$$+ ef_\pi^{-1}\mu^{-2}(q^2 - \mu^2)[2Mg_A(t) + th_A(t)] \qquad (4.130)$$

as well as the expected expression [cf. Eqs. (4.31) and (4.32)]

$$k^2 B_7^{(-)} = B_5^{(-)} - \tfrac{1}{2}(s - u)B_6^{(-)} + \tfrac{1}{2}(t - k^2 - q^2)B_8^{(-)} \qquad (4.131)$$

When these expressions are evaluated at the point $q = 0$ (i.e., $s = u = M^2$, $t = k^2$, $q^2 = 0$) and substituted into the current algebra relation for h_A, the two sides of the relation become identically equal.

The current algebra relation for h_A cannot therefore be tested with electroproduction amplitudes. One could have anticipated this result simply by observing (Adler and Gilman, 1966) that the contribution of h_A to S_μ is parallel to the four-vector k_μ, whereas the electroproduction amplitude involves only those components of T_μ that are perpendicular to k_μ (Section 2.2).

5. Cross Sections

5.1. Phase Space

It is necessary to convert the above amplitudes into electro- and photoproduction cross sections. To this end we define a Lorentz-invariant reduced phase space element (Pilkuhn, 1967)

$$d \operatorname{Lips}(P^2, p_1, \ldots, p_n) = (2\pi)^4 \delta(P - \textstyle\sum p_i) \prod_i [(2\pi)^3 2E_i]^{-1} d^3 p_i \qquad (5.1)$$

The cross section for the production of particles with momentum $p_1 \cdots p_n$ is given in terms of the T-matrix element defined in Eqs. (2.3) and (2.4) by

$$\frac{d\sigma}{d \text{ Lips}} = \frac{1}{2Mp_L} |T|^2 \tag{5.2}$$

where M is the mass of the target and p_L is the beam momentum in the rest frame of the target. Unmeasured initial/final spins must be averaged/summed over.

For $n = 2$, useful expressions are

$$\begin{aligned}
d \text{ Lips}(s; p_1, p_2) &= [16\pi^2 p_L]^{-1} dE'_L \, d\phi \\
&= [32\pi^2 p_c s^{1/2}]^{-1} \, dt \, d\phi \\
&= p'_c [16\pi^2 s^{1/2}]^{-1} d(\cos\theta_c) \, d\phi
\end{aligned} \tag{5.3}$$

In the first line the final energy E'_L of one of the particles and the beam momentum p_L are to be evaluated in the laboratory frame (target rest frame). In the second and third lines the initial momentum p_c, the final momentum p'_c, and the scattering angle θ_c are to be evaluated in the c.m. frame. The quantities s and t are as usual the total four-momentum squared and the four-momentum transfer squared. The azimuthal angle ϕ is the same in the laboratory and c.m. frames.

5.2. Single Hadron in the Final State

First consider the case where there is only a single hadron in the final state. Let the initial electron be unpolarized and let the initial hadron be an unpolarized spin-$\frac{1}{2}$ object. Then including the spin average our various formulas give (neglecting m_e)

$$\frac{d\sigma}{d \cos\theta_c d\phi} = \frac{\alpha}{2^6 \pi} \frac{1}{k^2} \frac{1}{1-\varepsilon} \frac{E'}{EM} \sum_{rs} \frac{\rho_{rs}}{\rho_{++}} \rho_{rs}^H \tag{5.4}$$

The ϕ dependence is given by Eq. (2.54):

$$\rho_{rs} = e^{-i(r-s)\phi} \rho_{rs}|_{\phi=0} \tag{5.5}$$

However there can be no ϕ dependence for two-body scattering unless some direction is picked out by a particle polarization. When there is no ϕ dependence

$$\sum_{rs} \frac{\rho_{rs}}{\rho_{++}} \rho_{rs}^H = \rho_{++}^H + \rho_{--}^H + 2\varepsilon\rho_{00}^H \tag{5.6}$$

The energy and mass of the virtual photon are related in the laboratory frame by

$$|k^2| = 2Mk_0 + M^2 - M'^2 \tag{5.7}$$

where M' is the mass of the final hadron. For elastic scattering $M = M'$ and (neglecting m_e)

$$\varepsilon^{-1} = 1 + 2\left(1 + \frac{1}{4}\frac{|k^2|}{M^2}\right)\tan^2\frac{\theta}{2} \tag{5.8}$$

For unpolarized elastic electron–nucleon scattering these formulas when combined with Eqs. (2.52) and (3.71)–(3.74) give the Rosenbluth formula (Rosenbluth, 1950)

$$\frac{d\sigma}{d\Omega_L} = \left(\frac{d\sigma}{d\Omega_L}\right)_{\text{N.S.}}\left[\frac{G_E^2 + \tau G_M^2}{1 + \tau} + 2\tau G_M^2 \tan^2\frac{\theta}{2}\right] \tag{5.9}$$

where the first factor is the cross section for a nucleon with no structure,

$$\left(\frac{d\sigma}{d\Omega_L}\right)_{\text{N.S.}} = \frac{\alpha^2}{4E^2}\frac{\cos^2(\theta/2)}{\sin^4(\theta/2)}\left[1 + \frac{2E}{M}\sin^2\frac{\theta}{2}\right]^{-1} \tag{5.10}$$

We have written the cross sections in terms of laboratory frame variables and used the notation $\tau \equiv -k^2/(4M^2)$.

5.3. More than One Hadron in the Final State

If there is more than one hadron in the final state, we can use a recurrence relation for d Lips (Pilkuhn, 1967) to write (neglecting m_e)

$$\frac{d\sigma}{d \text{ Lips } dE' d\cos\theta \, d\phi} = \frac{\alpha}{32\pi^2}\frac{1}{|k^2|}\frac{1}{1-\varepsilon}\frac{E'}{EM}\sum_{rs}\frac{\rho_{rs}}{\rho_{++}}\rho_{rs}^H \tag{5.11}$$

where d Lips now refers only to the hadrons.

Sometimes the variables E' and θ are replaced by the Lorentz-invariant quantities k^2 and s. The appropriate Jacobian gives

$$ds \, dk^2 = 4MEE' \, dE' d\cos\theta \tag{5.12}$$

It is customary to define a virtual photon cross section by

$$\frac{d\sigma_\gamma}{d \text{ Lips}} = \frac{1}{8(s - M^2)}\sum_{rs}\frac{\rho_{rs}}{\rho_{++}}\rho_{rs}^H \tag{5.13}$$

In the limit $k^2 = 0$ this becomes

$$\frac{d\sigma_\gamma}{d \text{ Lips}} = \frac{1}{8(s - M^2)}(\rho_{++}^H + \rho_{--}^H - \varepsilon\rho_{+-}^H - \varepsilon\rho_{-+}^H), \tag{5.14}$$

which according to our discussion of the last section is a real photoproduction cross section [the normalization for a spin $-\frac{1}{2}$ target is correct in virtue of Eqs. (2.14), (2.30), (2.52), (5.2), (5.3), and the $k^2 = 0$ relation $k_0 = (s - M^2)/2M$]. The photon is an incoherent mixture of components with plane polarization parallel to and perpendicular to the electron scattering plane, since the square bracket is diagonal with these photon states as a basis.

In terms of $d\sigma_\gamma$, the electroproduction cross section with two or more hadrons in the final state for an unpolarized electron scattering from an unpolarized spin-$\frac{1}{2}$ target is

$$\frac{d\sigma}{dE'd\cos\theta\,d\phi} = \Gamma\,d\sigma_\gamma \qquad (5.15)$$

(omitting the common factor d Lips), where

$$\Gamma = \frac{\alpha}{2\pi^2}\frac{E'}{E}\frac{s-M^2}{2M}\frac{1}{|k^2|}\frac{1}{1-\varepsilon} \qquad (5.16)$$

The quantity Γ can be regarded as the flux of virtual photons per unit element of electron phase space. (If the phase space is taken to be $dk^2\,ds$ then Γ has to be divided by the factor $4MEE'$.) Away from $k^2 = 0$ the definitions of $d\sigma_\gamma$ and Γ are pure convention (Hand, 1963) and as a matter of fact authors concerned with the extreme forward direction have used the different photon flux

$$N = \frac{\alpha}{2\pi^2}\frac{E'}{E}\frac{k_0}{|k^2|}\frac{1}{1-\varepsilon} \qquad (5.17)$$

The definitions are equivalent when $k^2 = 0$ because

$$k_0 = \frac{s-M^2-k^2}{2M} \qquad (5.18)$$

If there are only two hadrons in the final state and all the particles are unpolarized, then parity invariance requires that the cross section should be an even function of the azimuthal angle ϕ. One can then write (Driver *et. al.*, 1971)

$$d\sigma_\gamma = d\sigma_u + \varepsilon\,d\sigma_s + \varepsilon\cos 2\phi\,d\sigma_T + [2\varepsilon(1+\varepsilon)]^{1/2}\cos\phi\,d\sigma_I \qquad (5.19)$$

where

$$\frac{d\sigma_u}{dt} = c(\rho_{++}^H + \rho_{--}^H) \qquad (5.20)$$

$$\frac{d\sigma_s}{dt} = 2c\rho_{00}^H \qquad (5.21)$$

$$\frac{d\sigma_T}{dt} = -c(\rho_{+-}^H + \rho_{-+}^H) \qquad (5.22)$$

$$\frac{d\sigma_I}{dt} = \frac{c}{2^{1/2}}(\rho_{-0}^H + \rho_{0-}^H - \rho_{+0}^H - \rho_{0+}^H) \qquad (5.23)$$

$$c^{-1} \equiv 32\pi s^{1/2}|\mathbf{k}_{c.m.}|(s-M^2) \qquad (5.24)$$

We see that the dependence of the cross section upon the azimuthal angle ϕ and the parameter ε picks out an "unpolarized" cross section σ_u, a "scalar" cross section σ_s, a "transverse interference" term σ_T, and a "scalar-transverse interference" term σ_I.

If there is a direction of polarization that breaks the symmetry with respect to ϕ, or if there are more than two hadrons in the final state (which also breaks the symmetry), then there will be terms involving $\sin \phi$ and $\sin 2\phi$ in the cross section. The modified formula for polarized electrons can be calculated in any particular case using, for example, our Eqs. (2.60)–(2.62), (2.74)–(2.76), or (2.80), and (2.81).

If the hadrons are not observed there can be no ϕ dependence and the virtual total γp cross section may be written (Hand, 1963)

$$\sigma_\gamma = \sigma_u + \varepsilon \sigma_s \tag{5.25}$$

where σ_u comes from helicity ± 1 photons and σ_s from helicity 0 photons.

For the case of πN scattering, the use of Eqs. (3.17)–(3.19), (3.26)–(3.28), (4.16)–(4.22), (2.53), and (5.19)–(5.21) gives the virtual integrated $\gamma N \to \pi N$ in terms of multipoles

$$\sigma_u + \varepsilon \sigma_s = \frac{8\pi s^{1/2}|\mathbf{q}|}{s - M^2} \sum_l (l+1)^2 \left[\tfrac{1}{2}(l+2)(|E_{l+}|^2 + |M_{(l+1)-}|^2) \right.$$

$$\left. + \tfrac{1}{2}l(|M_{l+}|^2 + |E_{(l+1)-}|^2) + \frac{|k^2|\varepsilon}{|\mathbf{k}|^2}(l+1)(|S_{l+}|^2 + |S_{(l+1)-}|^2) \right] \tag{5.26}$$

In the region of a resonance, the contribution of the resonant multipoles is also the resonant contribution to the total cross section, multiplied by the decay fraction of the resonance into the relevant πN channel.

5.4. Equivalent Photon Approximation

When the photon is almost real in the sense that its mass squared is small compared with its energy, it is emitted in the near-forward direction since Eq. (2.11) together with momentum conservation gives

$$\sin^2\theta_\gamma = \left[\frac{(|k^2| - |k^2|_{min})}{k_0^2} \right] \left[\frac{E_2}{E_1} \right] \cos^2\frac{\theta}{2}$$

$$< \frac{|k^2|}{k_0^2} \tag{5.27}$$

Provided that $|k^2|$ is also small on any relevant hadron mass scale, i.e., that

$$|k^2| \ll M_H^2 \tag{5.28}$$

the helicity amplitudes $f_{\pm 1}$ will be close to their photoproduction values and f_0 will be negligible.

Under these circumstances it will be a good approximation to regard the electron beam as equivalent to a polarized real photon beam. The number of photons per electron per unit of electron phase space is given by N or Γ, which are equivalent since Eq. (5.26) certainly implies $|k^2| \ll s - M^2$. This number peaks strongly in the forward direction.

If only a crude approximation to the integrated electron cross section is required one can write

$$\frac{d\sigma}{dE'd\,\text{Lips}} \simeq L(E, E')\frac{d\sigma_\gamma}{d\,\text{Lips}} \qquad (5.29)$$

where

$$L(E, E') = 2\pi \int_{-1}^{1} d(\cos\theta)N(E, E', \cos\theta)$$

$$= [\alpha/(\pi E^2 k_0)](E^2 + E'^2)\log(2EE'm_e^{-1}k_0^{-1})$$

$$- EE' - \tfrac{1}{2}(E + E')^2 \log[E(E + E')/k_0]$$

$$\simeq [\alpha/(\pi E^2 k_0)][\log(E/m_e) - \tfrac{1}{2}] \qquad (k_0 \ll E)$$

$$\simeq [\alpha/(\pi E^2 k_0)]\log(E/m_e) \qquad [\log(E/m_e) \gg 1] \quad (5.30)$$

The first expression was first derived by Dalitz and Yennie (1957), but in precise detail it has no compelling motivation because (i) it is not Lorentz invariant, (ii) in any particular frame the expression for N is a matter of convention away from $k^2 = 0$, and (iii) for any particular N the approximations used are only valid near $\theta = 0$. However, all derivations (Fermi, 1924; Weizsacker, 1934; Landau and Lifshitz, 1934; Curtis, 1956) lead to substantially the same result when $k_0 \ll E$ and $\log(E/m_e) \gg 1$.

References

Adler, S. L. (1968), *Ann. Phys. N.Y.* **50**, 189.

Adler, S. L., and Gilman, F. (1966), *Phys. Rev.* **152**, 1460.

Adler, S. L., and Weisberger, W. I. (1968), *Phys. Rev.* **169**, 1392.

Alcock, J. W., Burkhardt, H., McCauley, G. P., and Lyth, D. H. (1968), *Nuovo Cimento* **54A** 957.

Ball, J. S. (1961), *Phys. Rev.* **124**, 2014.

Barbour, I. M. (1963), *Nuovo Cimento* **27**, 1382.

Bartl, A., and Urban, P. (1966), *Acta Phys. Austriaca* **24**, 139.

Benaksas, D., Cosme, G., Jean-Marie, B., Julian, S., Laplanche, F., Lefrançois, J., Liberman, A. D., Parrout, G., Repellin, J. P., and Sauvage, G. (1972), *Phys. Lett.* **39B**, 289.

Benfatto, G., Nicolo, F., and Rossi, G. C. (1972), *Nucl. Phys.* **B50**, 205.

Benfatto, G., Nicolo, F., and Rossi, G. C. (1973), *Nuovo Cimento* **14A**, 425.
Berends, F. A. (1970), *Phys. Rev.* **D1**, 2590.
Bjorken, J. D. (1960), *Phys. Rev. Lett.* **4**, 473.
Bjorken, J. D., and Drell, S. (1965), *Relativistic Quantum Fields* (New York, McGraw-Hill).
Bjorken, J. D., and Walecka, J. D. (1966), *Ann. Phys. N.Y.* **38**, 35.
Blatt, J. M., and Weisskopf, V. F. (1962), *Theoretical Nuclear Physics* (New York, Wiley).
Bowcock, J. E., and Kannelopoulos, T. (1968), *Nucl. Phys.* **B3**, 417.
Brodsky, S. J., and Farrar, G. R. (1973), *Phys. Rev. Lett.* **31**, 1153.
Brown, C. J., and Lyth, D. H. (1973), *Nucl. Phys.* **B53**, 323.
Cabibbo, N., and Gatto, R. (1961), *Phys. Rev.* **124**, 1577.
Chew, G. F., and Mandelstam, S. (1960), *Phys. Rev.* **119**, 467.
Chew, G. F., Goldberger, M. L., Low, F. E., and Nambu, Y. (1957), *Phys. Rev.* **101**, 1570 and 1579.
Chew, G. F., Karplus, R., Gasiorowicz, S., and Zachariasen, F. (1958), *Phys. Rev.* **111**, 265.
Close, F. E., and Cottingham, W. N. (1975), *Nucl. Phys.* **B99**, 61.
Cohen-Tannoudji, G., Morel, A., and Navelet, H. (1968), *Ann. Phys. N.Y.* **46**, 239.
Crawford, R. L. (1971), *Nucl. Phys.* **B28**, 573.
Curtis, R. B. (1956), *Phys. Rev.* **104**, 211.
Cutkosky, R. E. (1960), *J. Math. Phys.* **1**, 429.
Dalitz, R. H., and Yennie, D. R. (1957), *Phys. Rev.* **105**, 1958.
DeCelles, P. C., Durand, L., and Marr, R. B. (1962), *Phys. Rev.*, **126**, 1882.
Dennery, P. (1961), *Phys. Rev.* **124**, 2000.
Devenish, R. C. E. (1971), private communication.
Devenish, R. C. E., and Lyth, D. H. (1972*a*), *Phys. Rev.* **D5**, 47.
Devenish, R. C. E., and Lyth, D. H. (1972*b*), *Nucl. Phys.* **B43**, 228.
Devenish, R. C. E., and Lyth, D. H. (1973), *Nucl. Phys.* **B59**, 256.
Devenish, R. C. E., and Lyth, D. H. (1975), *Nucl. Phys.* **B93**, 109.
Devenish, R. C. E., Lyth, D. H., and Rankin, W. A. (1974), *Phys. Lett.* **52B**, 227.
Devenish, R. C. E., Eisenschitz, T. S., and Korner, J. G. (1975), DESY preprint No. 75/48.
Dombey, N. (1969), *Rev. Mod. Phys.* **41**, 236.
Dombey, N., and Read, B. J. (1973), *Nucl. Phys.* **B60**, 65.
Driver, C., Heinloth, K., Höhne, K., Hofmann, G., Karow, P., Schmidt, D., and Rathje, J. (1971), *Phys. Lett.* **35B**, 770.
Ecker, G., and Schwela, D. (1974), *Nuovo Cimento Lett.* **9**, 587.
Fermi, E. (1924), *Z. Phys.* **29**, 315.
Frazer, W. R., and Fulco, J. R. (1960), *Phys. Rev.* **117**, 1603.
Fubini, S., Nambu, Y., and Wataghin, V. (1958), *Phys. Rev.* **111**, 329.
Fubini, S., Farland, G., and Rossetti, C. (1965), *Nuovo Cimento* **40**,1171.
Furlan, G., Jengo, R., and Remiddi, E. (1966), *Nuovo Cimento* **44A**, 427.
Gell-Mann, M. (1962), *Phys. Rev.* **125**, 1067.
Gottfried, K. (1966), *Preludes in Theoretical Physics* (Amsterdam, North-Holland).
Hand, L. N. (1963), *Phys. Rev.* **183**, 1834.
Hohler, G., Pietarinen, E., Sabba-Stefanescu, I., Borkowski, F., Simon, G. G., Walther, V. H., and Wendling, R. D. (1976), *Nucl. Phys.* **B114**, 505.
Jackson, J. D. (1962), *Classical Electrodynamics* (New York, Wiley).
Jacob, M., and Wick, G. C. (1959), *Ann. Phys. N.Y.* **7**, 404.
Jones, H. F. (1965), *Nuovo Cimento* **40A**, 1018.
Jones, H. F., and Scadron, M. D. (1973), *Ann. Phys. N.Y.* **81**, 1.
Joos, H. (1962), *Fortschr. Phys.* **10**, 65.
Landau, L. D. (1959), *Nucl. Phys.* **13**, 181.
Landau, L., and Lifschitz, E. (1934), *Phys. Z. Sowjetunion* **6**, 244.

Loubaton, J. P. (1965), *Nuovo Cimento* **39**, 591.
Lyth, D. H. (1971), *Nucl. Phys.* **B30**, 173.
Lyth, D. H. (1972), *Nucl. Phys.* **B45**, 512.
Manweiler, R., and Schmidt, W. (1971), *Phys. Rev. D* **3**, 2752.
Matveef, V. A., Muradyan, R. M., and Tavkelidze, A. N. (1973), *Nuovo Cimento Lett.* **7**, 719.
Mo, L. W., and Tsai, Y. S. (1969), *Rev. Mod. Phys.* **41**, 205.
Nambu, Y., and Schrauner, E. (1962), *Phys. Rev.* **128**, 862.
Nauenberg, M. (1966), *Phys. Lett.* **22**, 201.
Omnes, R. (1958), *Nuovo Cimento* **8**, 316.
Omnes, R. (1971), *Introduction to Particle Physics* (New York, Wiley).
Pilkuhn, H. (1967), *The Interactions of Hadrons* (Amsterdam, North-Holland).
Riazuddin, and Lee, B. W. (1966), *Phys. Rev.* **146**, 1202.
Rose, M. (1957), *Elementary Theory of Angular Momentum* (New York, Wiley).
Rosenbluth, M. N. (1950), *Phys. Rev.* **79**, 615.
Salin, Ph. (1964), *Nuovo Cimento* **32**, 521.
Schweber, S. (1961), *An Introduction to Relativistic Quantum Field Theory* (New York, Harper and Row).
Theis, W. R., and Hertl, P. (1970), *Nuovo Cimento* **66**, 152.
Trueman, L. T., and Wick, G. C. (1964), *Ann. Phys. N.Y.* **26**, 332.
Truong, T. N., and Vinh-Mau, R. (1969), *Phys. Rev.* **177**, 2494.
Urban, P. (1970). *Topics in Applied QED* (Berlin, Springer).
Vik, R. C. (1967), *Phys. Rev.* **163**, 1535.
Von Gehlen, G. (1969), *Nucl. Phys.* **B9**, 17.
Von Gehlen, G. (1970), *Nucl. Phys.* **B20**, 102.
Walecka, J. D., and Zucker, P. A. (1968), *Phys. Rev.* **167**, 1479.
Walker, R. L. (1969), *Phys. Rev.* **182**, 1729.
Zagury, N. (1966), *Phys. Rev.* **145**, 1112 erratum; **150**, 1406.
Zagury, N. (1967), *Nuovo Cimento* **52A**, 506.
Zagury, N. (1968), *Phys. Rev.* **165**, 1934.

5

Form Factors and Electroproduction

A. Donnachie, G. Shaw, and D. H. Lyth

1. Introduction

In this chapter, we discuss the phenomenology of electron scattering

$$e^- + N \rightarrow e^- + X$$

where the produced hadron mass lies in the "resonance region" $W = M_x \lesssim 3 - 4 \text{ GeV}/c^2$. The main interest in this region lies in the Q^2 dependence of the elastic $(X \equiv N)$ and transition $(X \equiv N^*, \Delta)$ form factors, which may be extracted from such experiments. These have important implications for any model of the internal structure of baryons, in particular the quark model, and their behavior may also be linked, via duality arguments, to the deep inelastic structure functions at small $x = Q^2/2M\nu$. In addition, there are interesting features in the nonresonant backgrounds, particularly for the $\pi^\pm$ electroproduction reactions. At threshold, this background is linked via current-algebra arguments to the nucleon axial vector form factor $G_A(Q^2)$, whereas in the higher-mass range $M_x \sim 3-4 \text{ GeV}/c^2$, the one-pion exchange term may be separated out, leading to estimates of the pion form factor $F_\pi(Q^2)$. We discuss these various topics in turn.

A. Donnachie and G. Shaw • Department of Theoretical Physics, University of Manchester, Manchester M13 9PL, England
D. H. Lyth • Department of Physics, University of Lancaster, Lancaster, England

2. Nucleon Elastic Form Factors

The measurement of the proton charge radius by Hofstadter and McAllister (1955) initiated a period of intense activity in elastic electron scattering, leading in turn to important developments in dispersion theory. In particular, the existence of the ρ meson was predicted (Frazer and Fulco, 1959, 1960a, b) prior to its experimental observation. While subsequent developments have been less spectacular, there have been interesting advances in both experiment and theory, and many detailed reviews and data compilations have appeared over the years (see, for example, Drell and Zachariasen, 1961; Hofstadter, 1963; Wilson, 1971; Bartoli et al., 1972; Bartel et al., 1973; Gourdin, 1966, 1974; Höhler, 1976; Höhler et al., 1976a). In this account, we shall summarize the main features, especially with regard to more recent developments, referring to other sources where appropriate for extensive documentation.

2.1. Proton Form Factors

Unpolarized elastic electron–proton scattering has been extensively measured over the range $0 \lesssim Q^2 \lesssim 35$ $(\text{GeV}/c)^2$ (see, e.g., the recent compilation of Höhler et al., 1976a). After appropriate radiative corrections (Bartl and Urban, 1966; Mo and Tsai, 1969) the results are invariably interpreted in the one-photon approximation of Fig. 1(a). In this approximation, the cross section is given by the Rosenbluth formula [cf. Chapter 4 of this volume, Eq. (5.9)]:

$$R(Q^2, \theta) = \left(\frac{d\sigma}{d\Omega}\right)_L \Big/ \left(\frac{d\sigma}{d\Omega}\right)_{\text{NS}} = \left(\frac{G_E^2 + \tau G_M^2}{1 + \tau}\right) + 2\tau G_M^2 \tan^2\left(\frac{\theta}{2}\right) \quad (2.1)$$

where the NS (no structure) cross section is given by Lyth, Eq. (5.10), and

$$\tau = \frac{Q^2}{4M^2} = -\frac{k^2}{4M^2} \quad (2.2)$$

The form factors can thus be determined from the intercept and slope of a "Rosenbluth plot" of R against $\tan^2(\theta/2)$ at fixed Q^2. The linearity of

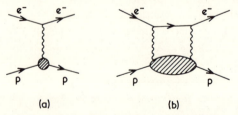

(a)   (b)

Fig. 1. One-photon and two-photon exchange in elastic electron scattering.

the plot is a consistency test for the neglect of the 2γ contributions of Fig. 1(b), which are estimated theoretically, albeit with considerable uncertainty, to be about 1% (Greenhut, 1969, and references therein). However, while the observed linearity of such plots is consistent with pure 1γ exchange, 2γ contributions are not necessarily excluded. A more direct test is the comparison of the $e^{\pm}$ ($\mu^{\pm}$) cross sections $\sigma^{\pm}$, which gives

$$R_{2\gamma} \equiv \frac{\mathrm{Re}(A_{2\gamma})}{A_{1\gamma}} = \frac{1}{2}\left(\frac{\sigma^+ - \sigma^-}{\sigma^+ + \sigma^-}\right) \tag{2.3}$$

in an obvious notation, with an implied spin average. The experimental data have been reviewed by Bartoli *et al.* (1972), and are summarized by

$$\begin{aligned} R_{2\gamma} &\lesssim 1\%, \qquad Q^2 \lesssim 1\ (\mathrm{GeV}/c)^2 \\ &\lesssim 2\%, \qquad Q^2 \lesssim 5\ (\mathrm{GeV}/c)^2 \end{aligned} \tag{2.4}$$

A nonzero imaginary part of the 2γ amplitude, which cannot contribute to the differential cross section by interference, can be detected via a nonzero recoil proton polarization, or target asymmetry. For $Q^2 \lesssim 1.0\ (\mathrm{GeV}/c)^2$, the data have been summarized by Bartoli *et al.* (1972) and are consistent with zero polarization with errors typically of order 1–3%. Somewhat less precise data (Kirkman *et al.*, 1970) extend to 1.9 $(\mathrm{GeV}/c)^2$, and are again consistent with zero.

Ignoring 2γ exchange, G_E and G_M can be determined separately out to $Q^2 \simeq 3\ (\mathrm{GeV}/c)^2$, and are approximately proportional. At larger Q^2, the contribution of G_E to the unpolarized cross section is kinematically suppressed, so that it is essentially undetermined. In contrast, assuming G_E/G_M does not increase dramatically at higher Q^2, G_M may be extracted out to $Q^2 \simeq 34\ (\mathrm{GeV}/c)^2$. A determination of G_E in this larger range—both for its own sake, and to check the above assumption used in the extraction of G_M—must await double polarization experiments, where the $G_E G_M$ interference term can be measured directly (Dombey, 1969).

Compilations of the extensive data available, and the resulting form factor values, have been made by several authors (Bartel *et al.*, 1973; Höhler *et al.*, 1976*a*) and the results are usually summarized in terms of the *approximate* scaling law

$$G_M^p(Q^2) \simeq \mu_P G_E^p(Q^2), \qquad 0 \leqslant Q^2 \leqslant 3\ (\mathrm{GeV}/c)^2 \tag{2.5}$$

and the *approximate* "dipole form" (Hofstadter, 1956)

$$G_M^p(Q^2) \simeq G_D \equiv \left(\frac{0.71}{0.71 + Q^2}\right)^2, \qquad 0 \leqslant Q^2 \leqslant 34\ (\mathrm{GeV}/c)^2 \tag{2.6}$$

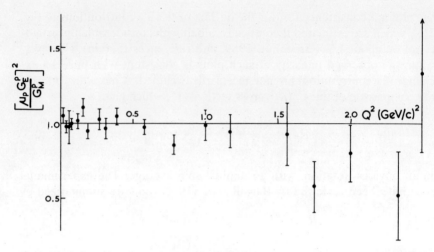

Fig. 2. Deviations from the scaling law, Eq. (2.5). Data points are taken from the compilation of Bartel *et al.* (1973) and usually combine the results of several experiments at each Q^2.

Deviations from both these rules can be exhibited in a rather elegant way by plotting the discrepancy function

$$\Delta(Q^2, X) = \frac{d\sigma_L}{d\sigma_{DS}} - 1 \qquad (2.7)$$

against

$$X = \left[1 + \frac{1+\tau}{1+\mu_p^2\tau} 2\tau\mu_p^2 \tan^2\left(\frac{\theta}{2}\right)\right]^{-1} \qquad (2.8)$$

at fixed Q^2, where DS means the calculated values using Eqs. (2.5) and (2.6). In this "modified Rosenbluth plot" (Höhler *et al.*, 1976a, b) the intercept gives directly deviations from the dipole law (2.6), whereas the slope is proportional to the violation of the scaling law (2.5). Examples of such plots are given by the authors cited. Here (Figs. 2 and 3) we merely show typical data on deviations from the scaling law and the dipole form, the latter falling by more than three orders of magnitude over the region shown. A parameter of particular interest is

$$r_E^2 = -6\left(\frac{dG_E^p(Q^2)}{dQ^2}\right)_{Q^2=0} = r_1^2 + \frac{6(\mu_p - 1)}{4M^2} \qquad (2.9)$$

where

$$r_1^2 = -6\left(\frac{dF_1^p(Q^2)}{dQ^2}\right)_{Q^2=0} \qquad (2.10)$$

The quantity $r_E(r_1)$ can be loosely interpreted as the rms charge (Dirac) radius of the proton. [Nonrelativistically, the form factor is given by the Fourier transform of the charge density, leading directly to (2.9).] A purely empirical fit to data (Borkowski *et al.*, 1975) gives $r_E^p = 0.87 \pm 0.02$ fm, $r_1^p = 0.71 \pm 0.02$ fm, whereas a careful analysis based on dispersion relations gives (Höhler *et al.*, 1976*b*) $r_E^p = 0.84 \pm 0.01$ fm, $r_1^p = 0.76 \pm 0.01$ fm. (Note $r_D \simeq 0.81$ fm.)

In the timelike region $k^2 = -Q^2 \geqslant 4M^2$, information can be obtained from the processes $e^+e^- \to p\bar{p}$, for which the cross-section formula (Cabibbo and Gatto, 1961) is

$$\frac{d\sigma}{d\Omega} = \frac{\alpha^2}{4M^2}(1 - \tau^{-1})^{3/2}[\sin^2\theta G_E^2 + (1 + \cos^2\theta)G_M^2] \qquad (2.11)$$

and the threshold condition [Chapter 4, Eq. (3.77)] implies a breakdown of the scaling law (2.5) unless both G_E, G_M vanish at $k^2 = 4M^2 = (3.54 \text{ GeV}/c)^2$. That this is not the case is shown by the recent detection of $p\bar{p} \to e^+e^-$ at threshold (Bassompiere *et al.*, 1977), giving $G_E = G_M = 0.51 \pm 0.08$. Somewhat above threshold $[k^2 = 4.3 \text{ (GeV}/c)^2]$ the cross section for $e^+e^- \to p\bar{p}$ has been measured (Castellano *et al.*, 1973) giving $\sigma = 0.91 \pm 0.22$ nb. If the threshold relation $G_E = G_M$ remains valid,

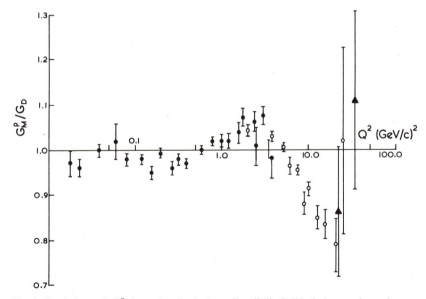

Fig. 3. Deviations of G_M^p from the dipole form Eq. (2.6). Solid circles are from the compilation of Bartel *et al.* (1973), open circles from Kirk *et al.* (1973), and closed triangles from Taylor (1976).

this corresponds to $|G_E| = |G_M| = 0.27 \pm 0.04$. Both of these results are noticeably larger than the prediction of the dipole ($G_D \simeq 0.06$ at the latter value of k^2). Finally, for higher values of k^2 [5.1, 6.6, 6.8 (GeV/c)^2] only upper limits on $p\bar{p} \rightarrow e^+ e^-$ are available (Conversi et al., 1965; Hartill et al., 1969).

2.2. Neutron Form Factors

The first information on neutron form factors was obtained by scattering thermal neutrons on atomic electrons (Havens et al., 1947; Fermi and Marshall, 1947), yielding a value of the rms charge radius

$$(r_E^n)^2 = -6\left(\frac{dG_E^n}{dQ^2}\right)_{Q^2=0} \tag{2.12}$$

If the neutron's interaction arises solely from the Pauli (anomalous moment) term (Foldy, 1958), then

$$F_1^n(Q^2) = 0 \tag{2.13a}$$

$$G_E^n(Q^2) = -\tau G_M^n(Q^2) \tag{2.13b}$$

and

$$\left(\frac{dG_E^n(Q^2)}{dQ^2}\right)_{Q^2=0} = \frac{\mu_n}{4M^2} = 2.12 \times 10^{-2}\,\text{fm}^2 \tag{2.14}$$

The most recent and precise experiments yield $(1.89 \pm 0.04) \times 10^{-2}\,\text{fm}^2$ (Krohn and Ringo, 1973) and $(1.99 \pm 0.03) \times 10^{-2}\,\text{fm}^2$ (Koester et al., 1975), suggesting that the Foldy term is dominant, F_1^n being small and probably negative at small Q^2.

Away from $Q^2 = 0$, information must be derived from elastic and quasielastic electron–deuteron scattering, with the uncertainties this entails. (For a full discussion, see Gourdin, 1966, 1974, and references therein.) Elastic scattering is most useful at the lower Q^2 values and determines G_E^S (and hence G_E^n), the contributions from G_M^S being negligible. A thorough discussion covering the range $0 \leqslant Q^2 \leqslant 1$ (GeV/c)^2, and summarizing earlier work, is given by Galster et al. (1971). The results depend somewhat on the deuteron wave function used, but are inconsistent with $G_E^n = 0$ in all cases. Further, for the more realistic wave functions* (Feschbach and Lomon, 1967; Hamada and Johnson, 1962; McGee, 1966) the results are smaller than the Foldy term, Eq. (2.13b), indicating that F_1^n

---

*We exclude the Hulthen wave functions (Hulthen and Sugawara, 1957), with or without core. Further, if the results in this case are fitted with a smooth form consistent with the radius value (2.14), the chi-squared is unacceptable (cf. Galster et al., 1971).

is small and negative. Results for the Feschbach–Lomon wave function are shown in Fig. 4. Subsequently, measurements have been made at very low Q^2 [0.05–0.5 fm^2 ≲ 0.02 (GeV/c)^2] (Berard *et al.*, 1973) allowing a precise evaluation of the slope, Eq. (2.12). Results are obtained using Feschbach–Lomon wave functions (Feschbach and Lomon, 1968) corresponding to D-state probabilities of 4.57%, 5.20%, and 7.53%, and subsequently fitting the results with either linear or quadratic forms. The slope values obtained vary from $(1.77 \pm 0.40) \times 10^{-2}$ fm^2 to $(2.32 \pm 0.22) \times 10^{-2}$ fm^2, in excellent agreement with the more precise values from thermal neutron scattering quoted earlier. Thus we can conclude that our knowledge of $G_E^n(Q^2)$ for $0 \le Q^2 \le 1$ (GeV/c)^2 is in an encouragingly satisfactory state.

Measurements of quasielastic electron–deuteron scattering enlarge the accessible range up to $Q^2 \simeq 2$ (GeV/c)^2, and allow the determination of the magnetic form factor G_M^n. The most recent experiments are those of Hanson *et al.* (1973) and Bartel *et al.* (1973), who give a useful compilation of earlier results. Over this extended range G_E^n remains small, and consistent with either of the curves shown in Fig. 4. Results on G_M^n are usually

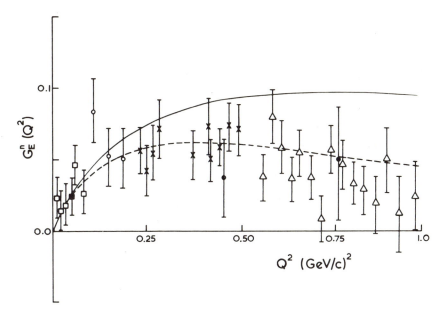

Fig. 4. Results from elastic electron–deuteron scattering on G_E^n, from the analysis of Galster *et al.* (1971), using the Feshback–Lomon wave function (see text). The different data sets used are as follows: (□) Drickey and Hand (1962); (■) Grossetete, *et al.* (1966); (○) Benaksas *et al.* (1966); (●) Buchanan and Yearian (1965); (△) Elias *et al.* (1969); (×) Galster *et al.* (1971). The solid curve is the Foldy term, Eq. (2.13b), where we have taken $G_M^n = -\mu_n G_D$. The dashed curve is $G_E^n = -\mu_n \tau G_D/(1 + 4\tau)$.

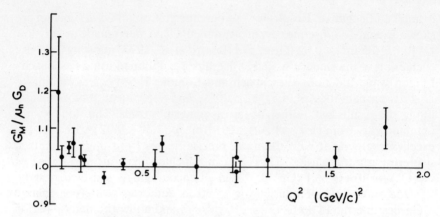

Fig. 5. Results on G_M^n obtained from quasielastic deuteron scattering. The data points are taken from the compilation of Bartel *et al.* (1973) and usually combine the results of more than one experiment at each Q^2.

expressed in terms of the approximate scaling law

$$G_M^n(Q^2)/\mu_n \simeq G_M^p(Q^2)/\mu_p \simeq G_D, \qquad Q^2 \leqslant 2 \,(\text{GeV}/c)^2 \qquad (2.15)$$

and are shown in Fig. 5.

2.3. Theoretical Interpretation

2.3.1. Asymptotic Behavior

Nonrelativistically, form factors are the Fourier transforms of charge distributions,* so that composite models with nonsingular potentials (and hence finite wave functions at the origin) lead naturally to falloffs like Q^{-4}, or faster. (See, e.g., Drell, 1967.) Early attempts to construct relativistic models (Amati *et al.*, 1969; Ball and Zachariasen, 1968; Ciafoloni and Menotti, 1968) led to similar results if logarithmic factors are ignored, i.e., $F_\pi, F_1, F_2 \simeq Q^{-4}$ (so that $G_E \sim Q^{-2}$). However, these results are based on the assumed dominance of particular ladder diagrams, which may well not be valid. Perhaps more promising are the results obtained by naive dimensional arguments, relating the asymptotic falloff to the number of constituents (Matveev *et al.*, 1973; Brodsky and Farrar, 1973). In the usual quark model, this gives† $F_\pi \sim Q^{-2}$, $G_E, G_M \sim Q^{-4}$. This agrees with experiment for G_M^p, and while no high-Q^2 information is available for $F_\pi, G_E,$

---

*The dipole form corresponds to an exponential wave function. For other examples, see Hofstadter (1956).

†Such behavior has been found in a specific quark model calculation utilizing vector gluon exchange, by Alabiso and Schierholz (1975). See Section 5.

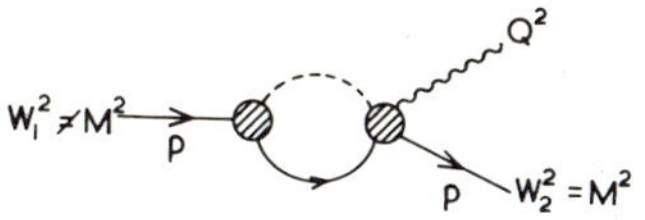

Fig. 6. Contribution of the πN intermediate state to sidewise dispersion relations.

extrapolation of the lower Q^2 trends [Eq. (2.5) and Section 3 below] is encouraging. In addition, similar arguments lead to interpretations of purely hadronic processes at large momentum transfer, which are in at least qualitative agreement with experiment (see, e.g., Landshof, 1974). It would be very interesting to test the prediction for G_E more directly, by means of the polarization measurements discussed above.

2.3.2. Anomalous Magnetic Moments

Attempts to calculate these have usually been based on "sidewise" dispersion relations (Bincer, 1960), in which one disperses (at fixed Q^2) in the mass of one of the nucleon legs, the other remaining on-shell. The lowest intermediate state is πN (s_{11}, p_{11} waves only), the discontinuity of Fig. 6 being related by unitarity to the pion photo- ($Q^2 = 0$) and electroproduction amplitudes. Early attempts (Drell and Pagels, 1965; Drell and Silverman, 1968) included only the threshold region contribution (estimated using low-energy theorems) in analogy to the discussion of the electron anomalous moment, retaining only the $e\gamma$ intermediate state (Drell and Pagels, 1965) and obtained encouraging qualitative agreement. Subsequent work has extended this to include higher-mass contributions both elastic (Bluvstein *et al.*, 1973; Deo and Singh, 1974; Love and Rankin, 1970, and references therein) and inelastic (Love and Rankin, 1970; Lee *et al.*, 1974). While these contributions are somewhat uncertain, some are clearly large, so that the success of the early threshold calculations is presumably accidental.

2.3.3. Phenomenological Analyses

Most analyses of form-factor data have been based on the dispersion relations in $Q^2 = -k^2$ summarized in Chapter 4 of this volume (Section 3), an early success being the prediction of the ρ meson (Frazer and Fulco, 1959, 1960). Subsequent analyses have often been framed in simple pole approximations of the form*

$$G_{E,M}(Q^2) = \sum_i \frac{\alpha_i m_i^2}{m_i^2 + Q^2} \qquad (2.16)$$

*Many such analyses have been published. In what follows, only a small sample is cited to illustrate the main features. Many more references may be found in Höhler and Pietarinen (1975a) and Höhler (1976).

with the vector meson couplings, and in some cases masses, being adjusted to fit the data. Such fits quickly led to the prediction of an isovector, 1^- meson in addition to the ρ (765). Thus, for example, Chan *et al.* (1966) found that a rather light mass $m = 975 \text{ MeV}/c^2$ was required with m_ρ fixed at its experimental value. Subsequently, a whole series of such states were predicted from duality considerations and incorporated into form-factor fits using formulas like

$$\mu_p G_E^p(Q^2) = G_M^p(Q^2) = \frac{\gamma \Gamma(1 - \alpha(Q^2))}{\Gamma(r + 1 - \alpha(Q^2))} \tag{2.17}$$

(Frampton, 1970; di Vecchia and Drago, 1969; Jengo and Remiddi, 1969) where

$$\alpha(Q^2) = \frac{1}{2} + \frac{Q^2}{2m_\rho^2} \tag{2.18}$$

Thus one has the spectrum $m_n^2 = m_0^2(1 + 2n)$, $n = 0$, 1, etc., so that the first ρ' occurs at $m_1 \simeq 1300 \text{ MeV}/c^2$. An interesting sidelight on this calculation (Gounaris, 1971) is that it predicts appreciable couplings of the ρ''s in $F_1^v(Q^2)$, the well-known resemblance to ρ dominance in this case arising from "accidental" cancellations. In contrast, a similar formula for the pion form factor indicates that they do decouple in this case, since a value $r \simeq 1$ is required to fit the data.

Alternatively (Shaw, 1972), one may work directly in terms of F_1^v, F_2^v (or equivalently F_1^v, G_m^v), taking the resemblance of F_1^v to ρ dominance more or less literally. That is, the low-Q^2 data are interpreted in a simple two-pole model ($m_0 \sim 660 \text{ MeV}/c^2$, $m_1 \sim 1200 \text{ MeV}/c^2$) in which the ρ' decouples from the charge form factors F_1^v, F_π, F_K. The light ρ mass required to fit the nucleon data is ascribed to the inadequacy of the narrow-width approximation, and $\omega \pi^0$ is suggested as the most probable ρ' decay mode.

Of course, the ρ contributions need not be approximated by a simple pole, but can be evaluated explicitly using the unitarity relations [see Chapter 4, Eqs. (3.80) and (3.81)]. Such a calculation has recently been carried out by Höhler and Pietarinen (1975*a,b*), using information on the pion form factor and the πN scattering amplitudes as input. References to earlier calculations may be found in Höhler (1976). In this way, they are able to calculate the ρ contributions to within about 5%–10%, and in the spacelike region they may be parameterized by

$$2F_1^\rho(Q^2) = \frac{0.955 + 0.090(1 + Q^2/0.355)^{-2}}{(1 + Q^2/0.536)} \tag{2.19}$$

$$2F_2^\rho(Q^2) = \frac{5.355 + 0.962(1 + Q^2/0.268)^{-1}}{(1 + Q^2/0.603)} \tag{2.20}$$

Contributions from the low-mass wing of the ρ are indeed enhanced. This had been noted earlier (Frazer and Fulco, 1960) and arises from the N exchange contribution to the πN amplitude. Such an effect is not present in the pion form factor.

Subsequently Höhler *et al.* (1976b) have exploited these results to carry out a careful analysis of the data, assuming the forms

$$\tfrac{1}{2}F_n^v = \tfrac{1}{2}F_n^\rho + \sum_i \frac{a_n^i}{m_i^2 + Q^2} \qquad (n = 1, 2) \tag{2.21}$$

$$\tfrac{1}{2}F_n^s = \frac{a_n^\omega}{m_\omega^2 + Q^2} + \frac{a_n^\phi}{m_\phi^2 + Q^2} + \sum_j \frac{a_n^j}{m_j^2 + Q^2} \tag{2.22}$$

The main conclusions are as follows.

(*i*) *Isovector Form Factors.* Over the region $0 \leqslant Q^2 \leqslant 1 \,(\mathrm{GeV}/c)^2$, F_i^v is dominated by the ρ contribution alone, whereas F_2^v receives an important contribution from a ρ' meson of mass about $1200 \,\mathrm{MeV}/c^2$. The contributions are of opposite sign, giving a dipolelike structure for F_2^v, and attempts to increase the ρ' mass to $1600 \,\mathrm{MeV}/c^2$ result in a poor fit to the data. These results are similar to the much more qualitative picture noted earlier, and it is tempting to identify the ρ' with the state possibly seen in $e^+ e^- \to \omega \pi^0$ at Frascati (Conversi *et al.*, 1974). However, there is not really any guarantee that this pole represents a resonance rather than a broad continuum contribution centered in the same mass range.

(*ii*) *Isoscalar Form Factors.* Here it is F_1^s that has the pronounced dipolelike structure, F_2^s being of course very small compared to F_2^v. The most startling feature of the simplest fit is the large ϕNN coupling, violating Zweig's rule. However, this coupling can be greatly reduced, if an appreciable coupling to an ω' (1250) is assumed. Clearly, if a ρ' (1200) exists, an ω' and a ϕ' are to be expected. This implies at least four (ω, ϕ, ω', ϕ') low-lying poles in the scalar form factors, so that separation of their couplings without additional assumptions will be extremely difficult.

For a fuller discussion on the ρNN, ωNN, and ϕNN couplings implied by the different fits, and their comparison with simple symmetry schemes, we refer to the original authors.

3. The Pion Form Factors

As discussed in Chapter 4 (Section 4.4), the pion pole contribution to $\pi^\pm$ electroproduction occurs in the amplitude $B_3^{(-)}$, and has the form [Chapter 4, Eq. (4.95)]

$$B_3^{(-)} = \frac{egF_\pi(Q^2)}{m_\pi^2 - t} \tag{3.1}$$

232

It was suggested by Frazer (1959) that its characteristically sharp variation with t might allow the pion form factor $F_\pi(Q^2)$ to be extracted from data on forward charged pion production. However, this contribution to the cross section is distorted by a rapidly changing kinematic factor, and the quantity $(m_\pi^2 - t)^2 \, d\sigma_v/dt$ actually passes through zero between the physical region and $t = m_\pi^2$. A straightforward extrapolation to the pole therefore leads to large errors even with the quite accurate data presently available (Devenish and Lyth, 1972).

Quantitative extraction of the pion form factor is possible by means of the fixed-t dispersion relation calculations described in Chapter 4 (Section 4.6). At $Q^2 = 0$ it is known that the forward π^+ cross section is dominated by the pole terms and Δ (1232) contribution, up to the highest measured energies. However, for $Q^2 > 0$ there are resonance wiggles in the second and third resonance region, and the data which show the clearest pion pole contribution are at somewhat higher energies, $W \sim 3$–4 GeV. Here, the scalar cross section σ_s dominates, and a good description of all the forward data (including its rapid t dependence) is given by a simple model (Berends, 1970) in which only the π, N, Δ (1236) contributions are retained in the dispersion relations [Chapter 4, Eqs. (4.85), (4.95)]. Given the relatively well-known nucleon form factors, the only parameter in this model is the pion form factor $F_\pi(Q^2)$, and it has frequently been used to extract values of this form factor from the data (see, e.g., Brown et al., 1973). It has subsequently been verified that higher-resonance contributions (Devenish and Lyth, 1972) and the subtraction constant in $B_3^{(-)}$ [cf. Chapter 4, Eq. (4.95)] (Devenish and Lyth, 1975) have only a small effect.

The most recent determination of F_π by this method is that of Bebek et al. (1976), and references to earlier experiments may be found in this paper. In addition to measurements of π^+ electroproduction on protons over the range $1.2 < Q^2 < 4.0 \, (\text{GeV}/c)^2$, these authors also measured both $\pi^\pm$ electroproduction on deuterium, allowing a small correction for the isoscalar proton contribution to be made. (The pion pole contributes solely to the isovector amplitude.) This correction was also applied to two of the earlier experiments (Brown et al., 1973; Bebek et al., 1974) covering the overlapping range $0.2 \le Q^2 \le 2.0 \, (\text{GeV}/c)^2$. A good fit to the combined data was obtained with the simple pole form

$$F_\pi(Q^2) = \frac{N}{1 + Q^2/m_v^2} \qquad 0 \le Q^2 \le 4 \, (\text{GeV}/c)^2 \tag{3.2}$$

with

$$N = 1.029 \pm 0.035, \qquad M_v^2 = 0.451 + 0.025 \, (\text{GeV}/c)^2 \tag{3.3}$$

N was allowed to differ from unity, because of experimental normalization errors of the order of 7%, and the central prediction is compared to the

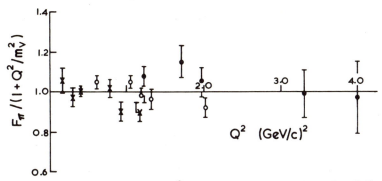

Fig. 7. Results on the pion form factor $F_\pi(Q^2)$ divided by the single pole form, Eqs. (3.2) and (3.3). Solid circles are the results of Bebek *et al.* (1976), and crosses and open circles the earlier results of Brown *et al.* (1973) and Bebek *et al.* (1974), as corrected by Bebek *et al.* (1976) (see text). Statistical errors only are shown.

data in Fig. 7. The implied value of the rms charge radius

$$r_\pi = 0.730 \pm 0.024 \, \text{fm} \tag{3.4}$$

is in excellent agreement with the value

$$r_\pi = 0.78 \pm 0.09 \, \text{fm} \tag{3.5}$$

obtained from direct measurements of πe scattering (Adylov *et al.*, 1974). The simple pole form also fits in well with observations in the timelike region, where it is clear that the ρ-meson gives rise to the dominant, but not the only contribution (see the article of Close and Cottingham in the companion volume).

4. Electroproduction

Pion photoproduction in the resonance region provides important constraints and checks on the theory of the electromagnetic interactions of hadrons by the determination of both the magnitudes and signs of the radiative decay amplitudes of the baryon resonances (see Chapter 2 in this volume, by Moorhouse), and pion electroproduction in the resonance region provides information on the internal structure of the baryon resonances via the Q^2 dependence of their transition form factors. This information is very relevant for models that make specific statements about the internal wave functions of the baryons, in particular, constituent models like the naive quark model.

Although a lot of data for the total inelastic cross section $eN \to eX$ in the resonance region have existed for some time, this has provided little quantitative information on the properties of the baryon resonances, other than the $p_{33}(1230)$, which completely dominates its energy region. At higher energies there is a considerable nonresonant background present, which effectively conceals the details of the resonance behavior. The most fruitful source of information on the structure of the baryon resonances is the exclusive process of single-pion electroproduction $eN \to eN\pi$.

A considerable quantity of coincidence data now exists on the reactions $ep \to ep\pi^0$ and $ep \to en\pi^+$, and the application of fixed-t dispersion relations to the analysis of these data has yielded a general picture of the behavior of the transition form factors of the leading resonances $p_{33}(1230)$, $d_{13}(1510)$, $s_{11}(1535)$, $f_{15}(1690)$, and $f_{37}(1950)$. As yet, no information is available on neutral states nor are the proton data sufficient to allow investigation of the transition form factors of the more weakly coupled resonances in the charged mode, but the data situation is improving rapidly and these results will become available in due course. Thus, the present situation is an interim one, and while the general features that we will discuss are likely to be qualitatively correct, some detailed quantitative changes should be expected.

Before dealing with the resonances in detail we discuss two related topics: the relation between electroproduction in the resonance region and deep inelastic scattering, and electroproduction at threshold.

4.1. Resonance Electroproduction, Duality, and Scaling

An interesting application of the total cross-section measurements in the resonance region is to obtain information from them, via duality arguments, on deep inelastic electroproduction. For this we require only the total cross section for the absorption of virtual photons, not the detailed resonance/background structure. The total cross section can be presented in two ways, one emphasizing the connection to photoproduction with real photons, and the other the correspondence to elastic electron–proton scattering.

We can write the total cross section for inelastic electron scattering ($eN \to eX$) as [Chapter 4, Eq. (5.15)]

$$\frac{d^2\sigma}{d\Omega_e \, dE_e} = \Gamma\sigma_v \tag{4.1}$$

where the quantity Γ can be regarded as the flux of virtual photons per unit element of electron phase space, and σ_v is the total cross section for the absorption of virtual photons. This consists of two parts and conventionally is written as (Hand, 1963)

$$\sigma_v = \sigma_t + \varepsilon\sigma_l \tag{4.2}$$

where σ_t and σ_l are, respectively, the total absorption cross sections for photons with transverse and longitudinal polarization, the quantity ε representing the degree of transverse polarization of the photon. The cross sections σ_t and σ_l depend on the energy loss ν of the scattered electron and the four-momentum transfer Q^2. In the photoproduction limit $Q^2 \to 0$, the longitudinal cross sections $\sigma_l(\nu, Q^2)$ vanishes, and the transverse cross section $\sigma_t(\nu, Q^2) \to \sigma_{\text{tot}}$ ($\nu = E_\gamma$, $Q^2 = 0$).

The correspondence to the Rosenbluth formula for elastic electron proton scattering is best displayed by writing

$$\frac{d^2\sigma}{d\Omega_e \, dE_e} = \sigma_{\text{Mott}}\{W_2(\nu, Q^2) + 2W_1(\nu, Q^2)\tan^2(\theta_e/2)\} \qquad (4.3)$$

with

$$\sigma_{\text{Mott}} = \frac{\alpha^2 \cos^2(\theta_e/2)}{4E_0^2 \sin^4(\theta_e/2)} \qquad (4.4)$$

Here E_0 is the primary electron energy and θ_e the electron scattering angle. The *structure functions* W_1 and W_2 are related to σ_t and σ_l by

$$W_1(\nu, Q^2) = \frac{K}{4\pi^2\alpha}\sigma_t(\nu, Q^2) \qquad (4.5a)$$

$$W_2(\nu, Q^2) = \frac{K}{4\pi^2\alpha}\frac{Q^2}{Q^2 + \nu^2}\{\sigma_t(\nu, Q^2) + \sigma_l(\nu, Q^2)\} \qquad (4.5b)$$

where K is the equivalent photon energy, $(s - M^2)/2M$.

The Bjorken scaling hypothesis (Bjorken, 1969) states that

$$\lim_{\nu, Q^2 \to \infty} W_1(\nu, Q^2) = F_1(\omega) \qquad (4.6a)$$

$$\lim_{\nu, Q^2 \to \infty} \nu W_2(\nu, Q^2) = F_2(\omega) \qquad (4.6b)$$

where $\omega = 2M\nu/Q^2$ is kept fixed. A full discussion of this topic is given in the article by Landshoff and Osborn in the companion volume, and for our present purposes we need only note that the scaling requirements (4.6) seem to be satisfied reasonably well.

Historically, scaling was first demonstrated for νW_2 as a function of ω for $Q^2 > 1$ $(\text{GeV}/c)^2$ and $W > 2$ GeV which are surprisingly low values for a property that is supposed to be asymptotic. However, it was then shown by Bloom and Gilman (1970) that if one considered νW_2 as a function of

$$\omega_{\text{BG}} = \frac{2M\nu + M^2}{Q^2} = 1 + \frac{W^2}{Q^2} \qquad (4.7)$$

then scaling works to quite low values of the hadronic mass W, where there are the prominent resonant peaks in the total electroproduction cross section, in the sense that the scaling curve is an average over the resonance peaks. This was extended further by Rittenberg and Rubinstein (1971) using the variable

$$\omega_w = \frac{2M\nu + a^2}{Q^2 + b^2} \qquad (4.8)$$

with a and b suitably chosen constants. It is then possible for $(\omega/\omega_w)\nu W_2$ to scale as a function of ω_w down to $Q^2 = 0$ again in the sense that the scaling curve averages the resonance peaks.

Although the question of duality and the extent of scaling are really separate and independent aspects of the problem, in practice they are linked by the present level of data, which allows a free choice of variable. A proper understanding of these problems would tell us what variable to choose, but at the moment the only requirement (other than being compatible with the data and the imposed duality/scaling requirements) is that asymptotically $(\nu \to \infty, Q^2 \to \infty)$ it should reduce to $2M\nu/Q^2$.

The adequacy of the simple form (4.8) has been confirmed by Brasse et al. (1972) using an extensive compilation of total cross sections for photoproduction and electroproduction on protons and by Moritz et al. (1972), using their own data (which were not included in the compilation of Brasse et al., 1972). One difficulty with these analyses is that for most ν, Q^2 values the cross sections are known only for one given electron scattering angle, so that the two structure functions νW_2 and W_1 cannot be determined separately.

The standard approach to this problem is to rephrase our ignorance in terms of the ratio R of longitudinal and transverse cross sections

$$R = \sigma_l/\sigma_t \qquad (4.9)$$

which has been determined (or equivalently νW_2 and W_1 separated) for some values of ν, Q^2, and to make simplifying assumptions on this. Asymptotically,

$$F_2(\omega) = [1 + R(\omega)]\frac{2MF_1(\omega)}{\omega} \qquad (4.10)$$

where $R(\omega)$ is the asymptotic form of $R(\nu, Q^2)$.

Experimentally, it appears that $R(\nu, Q^2)$ has a simple behavior (hence its usefulness), although it is not determined particularly well. Two expressions for R are generally considered: Either it is constant or it has the form $R = Q^2/\nu^2$. Moritz et al. (1972) considered only the first of these possibilities, taking $R = 0.2$, while Brasse et al. (1972) tried both, their choice for the constant being $R = 0.18$.

In practice, the results are not too sensitive to the choice of R. It is clear from these analyses that a variable ω_w of the form (4.8) is a satis-

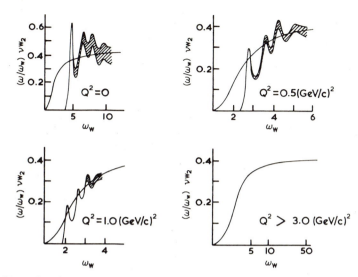

Fig. 8. Comparison between data in the resonance region and the scaling function (ω/ω_w) νW_2. The experimental bands enclose all relevant data points.

factory scaling variable, which satisfies local duality in the resonance region and generates the scaling asymptotic functions correctly, and allows the smooth connection between photoproduction and electroproduction to be established beyond doubt. The experimental uncertainties about R are reflected mainly in the parameter a^2 of Eq. (4.8), which shows variations of more than 30%. [Brasse *et al.* (1972) find it in the range 1.43–1.95 GeV^2, while Moritz *et al.* (1972) prefer it to be $\approx$1.3 GeV^2. Statistical errors are $\approx$0.06 GeV^2.] On the other hand, the parameter b^2 is somewhat better determined, lying between 0.4 and 0.5 GeV^2, the statistical errors in this case being $\approx$0.02 GeV^2. A typical comparison between the data in the resonance region and the scaling function $(\omega/\omega_w)\nu W_2$ is given in Fig. 8. A similar comparison holds for W_1.

As we shall see in Section 4.3, resonances have form factors that fall rapidly with Q^2, so it is at first sight surprising that they can be consistent with a scaling limit curve, which is characterized by having no dependence on Q^2 (the well-known decrease of νW_2 at fixed ω for low Q^2 is a purely dynamical effect, due to the necessary vanishing of νW_2). To see how this works, take $s = M_R^2$, the mass squared of a given resonance, and vary Q^2. If $F(Q^2)$ is the excitation form factor of the resonance, then its contribution to νW_2 in the narrow resonance approximation is

$$\nu W_2 = 2M\nu[F(Q^2)]^2 \, \delta(s - M_R^2)$$
$$= (M_R^2 - M^2 + Q^2)[F(Q^2)]^2 \, \delta(s - M_R^2) \qquad (4.11)$$

Form factors generally appear to fall off as some power for large Q^2,

$$F(Q^2) \sim (1/Q^2)^{n/2} \tag{4.12}$$

so that for large Q^2 and fixed s

$$\nu W_2 \sim (1/Q^2)^{n-1} \tag{4.13}$$

Now as Q^2 increases, a given resonance is pushed down toward $\omega = 1$ [$\omega = 2M\nu/Q^2 = 1 + (M_R^2 - M^2)/Q^2$], where νW_2 can be parameterized by some power behavior

$$\nu W_2 \underset{\omega \to 1}{\to} c(\omega - 1)^p \underset{\substack{Q^2 \\ \text{large}}}{\sim} [1/Q^2]^p \tag{4.14}$$

(Obviously, this resonance movement occurs for the scaling variables ω_{BG} and ω_w, and is clearly visible in Fig. 8.)

If we require that Eqs. (3.13) and (3.14) be consistent and that the resonances build up the scaling-limit curve locally, then we must have

$$n = p + 1 \tag{4.15}$$

i.e., all resonance bands (including the nucleon) that are to follow the scaling limit curve as $Q^2 \to \infty$ must have the same power of falloff in Q^2 for large Q^2, and this power is related to the power with which νW_2 rises at threshold.

The result (4.15) was first derived by Drell and Yan (1970) for the elastic form factor in the parton model. In our present discussion we have obtained it for the elastic and inelastic form factors by requiring that the resonances build up the scaling limit curve locally. We shall see in Section 4.3 that the resonance excitation form factors do indeed appear to have approximately the same behavior at large Q^2 as the nucleon elastic form factor. However, it is worth noting that the apparent dipole behavior of the latter (i.e., $n = 4$) leads to the result $\nu W_2 \sim (\omega - 1)^3$ near $\omega = 1$, while the present data on inelastic electron scattering appear to favor $(\omega - 1)^4$.

Although we have primarily considered resonance contributions, there is clearly a considerable nonresonant background contributing to the total cross sections in the resonance region, suggesting that there is a large contribution of Pomeron exchange in the deep continuum. If one wishes to treat the resonances alone, then it is necessary to consider the difference between the total cross sections for protons and for neutrons, to which diffraction should not contribute.

This study has been made by von Gehlen et al. (1972) and by Kowalski et al. (1973). They performed a detailed analysis of the difference (ω/ω_w) $(\nu W_2^p - \nu W_2^n)$, which is the imaginary part of the forward virtual Compton scattering amplitude with isospin 1 in the t-channel. For real photons, they generated the structure functions from the results of multipole analyses of single-pion photoproduction and an explicit theoretical model for multipion production, while for virtual photons they used the experimental total cross sections for protons and neutrons. These analyses demonstrate clearly that the difference $(\omega/\omega_w)(\nu W_2^p - \nu W_2^n)$ is entirely generated by resonances in the real Compton amplitude and that the average in the resonance region is the same as in the deep inelastic region.

4.2. Threshold Electroproduction and the Nucleon Axial Vector Form Factor

Near the threshold for charged pion production the slope of the total cross section is of particular interest, since it can be related to the axial vector form factor of the nucleon, $G_A(Q^2)$. The relationship was first demonstrated by Nambu and Shrauner (1962) in the framework of chirality conservation. They showed that in the exact soft-pion limit, i.e., for pions of zero four-momentum (implying zero mass), the transverse s-wave multipole at threshold is given by

$$E_{0+} \sim \left[1 + \frac{Q^2}{4M^2}\right]^{1/2} g_{\pi NN} \left[\frac{G_A(Q^2)}{G_A(0)} + \frac{Q^2}{2M^2 + Q^2} G_M^n\right] \qquad (4.16)$$

where G_M^n is the neutron magnetic form factor and $g_{\pi NN}$ is the pion nucleon coupling constant. The same result may be derived equivalently by the application of PCAC and current-algebra equal-time commutators (Riazuddin and Lee, 1966; Adler and Gilman, 1966; Furlan *et al.*, 1966; see also the discussion by Lyth in Chapter 4 of this volume and, in particular, that of Paver in the companion volume).

The total cross section σ_v of Eq. (4.1) has been given in terms of the individual multipoles by Eq. (5.26) of Chapter 4. From this we can write the threshold cross section as

$$\frac{K_L \sigma_v}{q} = |E_{0+}|^2 + \varepsilon \frac{|k^2|}{|\mathbf{k}|^2} |S_{0+}|^2 \equiv A_1 \qquad (4.17)$$

where q is the c.m. pion momentum, K_L the equivalent photon momentum $[(s - M^2)/2M]$ and we have followed convention by absorbing a factor of $4\pi W/M$ in the multipole amplitudes. At threshold the scalar electroproduction multipole, S_{0+}, is proportional to the neutron electric form factor G_E^n in the soft-pion limit.

In practice, the threshold cross section is obtained by extrapolation from a range of energies close to threshold. If we make the reasonable assumption that only s- and p-wave multipoles contribute significantly to the cross section, then there are contributions from E_{0+} and S_{0+}, which tend to constant values as $q \to 0$, and from E_{1+}, M_{1+}, M_{1-}, S_{1+}, S_{1-}, which all show a leading linear dependence on q near threshold. Expanding the differential cross section in powers of q up to q^2 then gives

$$\frac{4\pi}{\Gamma} \frac{K_L}{q} \frac{d^3\sigma}{dE_3 \, d\Omega_e \, d\Omega_\pi} = A_1 + A_2 q \cos \theta_\pi + A_3 q^2 \cos^2\theta_\pi$$
$$+ A_4 q \sin \theta_\pi \cos \phi_\pi$$
$$+ A_5 q^2 \sin^2\theta_\pi \cos 2\phi_\pi + A_6 q^2$$
$$+ A_7 q^2 \sin \theta_\pi \cos \theta_\pi \cos \phi_\pi \qquad (4.18)$$

from which we obtain the total cross-section formula

$$\frac{K_L \sigma_v}{q} = A_1 + (\tfrac{1}{3}A_3 + A_6)q^2 \equiv A_1 + Bq^2 \qquad (4.19)$$

The coefficients A_i can, of course, be expressed explicitly in terms of the individual multipoles and their derivatives (del Guerra *et al.*, 1975), but the only one relevant for our discussion is A_1, which we have defined in Eq. (4.17). Within the present experimental limits the restriction to s- and p-wave multipoles is justified: Although theoretically one expects some small contribution from d waves (or higher), the coefficient of a q^4 term in the total cross section is experimentally consistent with zero (del Guerra *et al.*, 1976).

Unfortunately, the experimental values for the coefficient A_1 cannot be confronted directly by theory, since the theoretical result (4.16) is obtained for unphysical pions and it is necessary to extend this to the physical region. Several approaches have been tried.

The general problem of extrapolating from soft-pion limits to the physical region was considered initially by Fubini and Furlan (1968), who suggested going from the soft-pion point to the threshold using a dispersion relation in q_0 with $\mathbf{q}$ fixed at zero in the Breit frame. General arguments can be used to claim that the corrections that arise are relatively small, but their explicit numerical evaluation is difficult. For electroproduction one writes a subtracted dispersion relation at the Breit physical threshold for the appropriate amplitude, using the soft-pion limit value as the subtraction constant, and attempting to calculate the corrections accurately.

Furlan *et al.* (1969) extended this method to connect the soft-pion point to physical points different from the Breit threshold. Difficulties arise in doing this, since it is necessary to make specific assumptions not only on the equal-time commutator $[Q_5, J_\mu^{\text{e.m.}}(0)]$, but also on the equal-time commutator $[A_0(x), J_\mu^{\text{e.m.}}(0)]$ and on moving from the Breit threshold new terms appear whose evaluation is not straightforward. Indeed, the difficulty in evaluating these terms restricts the applicability of their result to a region close to the Breit threshold, where these contributions can be neglected. The other corrections that appear are tackled in the way proposed by Fubini and Furlan (1968).

Benfatto *et al.* (1972) have argued against this approach, partly because of the problems mentioned above, and partly on the grounds that the extrapolation of the soft-pion limit amplitude (used as the subtraction constant) may not be sufficiently smooth for the corrections to be small. To overcome this latter difficulty, they split the electroproduction amplitude in two and write separate subtracted dispersion relations at the Breit threshold for the individual parts. For that part whose extrapolation is expected to be well behaved, they use the soft-pion limit as the subtraction constant; otherwise, they use essentially the normal Born terms; i.e., their subtraction constant is effectively determined by the requirements of gauge invariance. This approach appears to give better control over the corrections, and one can move away from the Breit threshold. However, there is a price to pay for this, namely, the inclusion of the pion form factor $F_\pi(Q^2)$.

A somewhat different approach has been adopted by Dombey and Read (1973), who note that in the conventional dynamical approach to low-energy photo- and electroproduction via fixed-t dispersion relations (as discussed by Lyth in Chapter 4 of this volume), the requirements of current algebra are not explicitly imposed. Indeed, if, for charged pion production, the threshold multipoles E_{0+} were assumed to be dominated by the pole terms alone, they would be proportional to the proton charge form factor, rather than the nucleon axial vector form factor as suggested by current algebra in the soft-pion limit. This latter behavior is presumably realized when the other contributions to the dispersion relations are explicitly taken into account. Alternatively, and more phenomenologically, one may replace the Born term poles by the full Born approximation, where the latter is calculated using pseudovector rather than pseudoscalar coupling for the πNN vertex. An immediate consequence is that, in addition to the pole terms, the derivative nature of the coupling gives rise to a contact term for $\gamma N \to \pi^\pm N$, which can be identified with the G_A part from the equal-time commutator of the PCAC analysis. This identification provides the extrapolation from the soft-pion limit to the physical region, and threshold electroproduction is described primarily by the nucleon axial vector form factor $G_A(Q^2)$ and the pion form factor $F_\pi(Q^2)$.

It should be noted that it is possible to obtain a perfectly adequate description of threshold electroproduction via the normal dispersion relations with pseudoscalar Born terms (von Gehlen, 1969, 1970; Devenish and Lyth, 1975), although the direct contact with the PCAC result is lost. In practice, the corrections to the Born terms in normal dispersion theory, coming from s- and u-channel continuum contributions, simulate the phenomenological contact term.

The reaction $ep \to en\pi^+$ near threshold has been measured by Amaldi
et al. (1970, 1972), Brauel *et al.* (1973, 1974), and del Guerra *et al.* (1975,
1976). The results of all three sets of experiments are consistent, although
those of Amaldi *et al.* are not strictly comparable with the other two since
their value of the photon polarization parameter ε (0.75–0.84) was dis-
tinctly different from that of Brauel *et al.* (0.98) and del Guerra *et al.*
(0.96).

In Fig. 9 we show the results for the A_1 coefficient (the threshold
cross-section slope), together with the predictions of the fixed-t dispersion
relation model of Devenish and Lyth (1975). In their experiment del
Guerra *et al.* (1975, 1976) were able to determine, in addition, the
coefficients A_4 and A_5 of Eq. (4.18) and the coefficient B of Eq. (4.19).
These are shown in Fig. 10, together with the predictions of Devenish and
Lyth (1975) and Dombey and Read (1973).

The experimental values for A_1 can now be used to determine the
nucleon axial vector form factor by selecting one of the relevant theoretical
models. The results for $G_A(Q^2)/G_A(0)$ using the model of Benfatto *et al.*
(1972, 1973) are shown in Fig. 11: The results for the model of Dombey

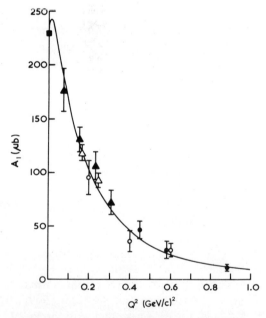

Fig. 9. Experimental results for the coefficient A_1 of Eqs. (4.18) and (4.19), and the predictions of the fixed-t dispersion relation model of Devenish and Lyth. The data shown are as
follows: ($\triangle$) Amaldi *et al.* (1970, 1972); ($\bigcirc$) Brauel *et al.* (1973, 1974); ($\blacktriangle$) del Guerra *et al.*
(1975); ($\bullet$) del Guerra *et al.* (1976); ($\blacksquare$) photoproduction limit.

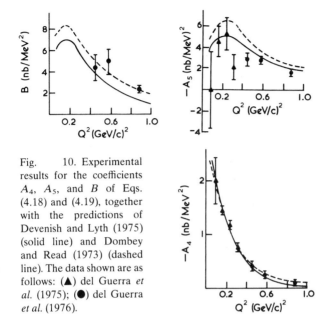

Fig. 10. Experimental results for the coefficients A_4, A_5, and B of Eqs. (4.18) and (4.19), together with the predictions of Devenish and Lyth (1975) (solid line) and Dombey and Read (1973) (dashed line). The data shown are as follows: (▲) del Guerra *et al.* (1975); (●) del Guerra *et al.* (1976).

and Read (1973) are not significantly different, although somewhat larger. Also shown in Fig. 11 are the results of fitting G_A to a dipole form:

$$\frac{G_A(Q^2)}{G_A(0)} = \left(1 + \frac{Q^2}{M_A^2}\right)^{-2} \tag{4.20}$$

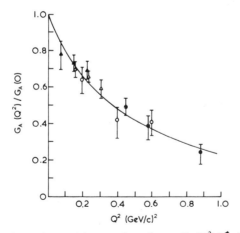

Fig. 11. Results for the nucleon axial vector form factor $G_A(Q^2)/G_A(0)$. The data are as follows: (△) Amaldi *et al.* (1970, 1972); (○) Brauel *et al.* (1973, 1974); (▲) del Guerra *et al.* (1975); (●) del Guerra *et al.* (1976). The curve is a fit to the form $(1 + Q^2/M_A^2)^{-2}$, with $M_A = 0.96 \text{ GeV}/c^2$.

The best fit with a dipole gives $M_A = 0.96 \pm 0.03 \, \text{GeV}/c^2$. Using the model of Dombey and Read (1973) yields a dipole mass $M_A = 1.12 \pm 0.03 \, \text{GeV}/c^2$. These are to be compared with the most recent results obtained from quasielastic neutrino scattering (Perkins, 1976), where dipole fits to the differential cross section give $M_A = 0.84 \pm 0.11 \, \text{GeV}/c^2$ and dipole fits to the total cross section give $M_A = 0.98 \pm 0.13 \, \text{GeV}/c^2$.

In view of the theoretical uncertainties and experimental difficulties, the overall agreement is very satisfactory.

4.3. Pion Electroproduction in the Resonance Region

The earliest electroproduction experiment investigated the reaction $ep \rightarrow e\pi^+ n$ by detecting only the π^+ (Panofsky et al., 1955), which is practically equivalent to the real photoproduction experiment $\gamma p \rightarrow \pi^+ n$, owing to the dominance of forward electron scattering. The first measurements of the total virtual γp cross section were made only when the final electron was detected (Panofsky and Alton, 1958), and of the differential cross sections for $\gamma_v p \rightarrow \pi^0 p$ (Akerloff et al., 1965) and $\gamma_v p \rightarrow \pi^+ n$ (Akerloff et al., 1967) when the final electron and charged hadron were detected in coincidence.

Extensive and accurate measurements have since been made both on the total cross section (Brasse et al., 1968; Bartel et al., 1968a, 1968b, 1971; Albrecht et al., 1968, 1969; Bloom et al., 1969, 1970; Moritz et al., 1972; Alder et al., 1972b; Köbberling et al., 1974; Stein et al., 1975) and on the differential $\gamma_v p \rightarrow \pi^0 p$ cross sections (Albrecht et al., 1971a, 1971b; Siddle et al., 1971; Shuttleworth et al., 1972; Alder et al., 1972a, 1975a; Bätzner et al., 1974). In contrast measurements on the differential $\gamma_v p \rightarrow \pi^+ n$ cross sections (Bätzner et al., 1973, quoted by Clegg, 1973; Evangelides et al., 1974; Alder et al., 1976) have been less numerous and less extensive. At the time of writing, more comprehensive measurements of the $\gamma_v p \rightarrow \pi^+ n$ differential cross sections are under way, as are further extensions of the $\gamma_v p \rightarrow \pi^0 p$ data. Data on electroproduction on neutrons is restricted to deuteron/hydrogen ratios at the three resonance peaks (Bleckwenn, 1971; Köbberling et al., 1974), although measurements on the differential $\gamma_v n \rightarrow \pi^- p$ cross sections are currently in preparation. Important measurements have been made on the reaction $\gamma_v p \rightarrow \eta^0 p$ (Kummer et al., 1973; Beck et al., 1974; Alder et al., 1975b), and there are also interesting results on the reaction $\gamma_v p \rightarrow \pi^- \Delta^{++}$ (Joos et al., 1976).

Although the total cross-section measurements are the most accurate and extend to much larger momentum transfers than the coincidence data [to a Q^2 value of $10 \, (\text{GeV}/c)^2$ as opposed to a maximum of

1.56 $(\text{GeV}/c)^2$], they do not provide very much information on the detailed behavior of the resonances, with the possible exception of the $p_{33}(1230)$, which is isolated. At higher energies there are many overlapping resonances, apparently with different dependences on Q^2, so that the total cross sections can only provide some kind of average behavior.

Nonetheless, a study of the total cross sections does provide a useful general guide: In particular, the range of data is such that it allows a separation of the longitudinal and transverse cross sections. All existing data on $\gamma_v p$ total cross sections in the resonance region have been fitted in momentum transfer and energy by Brasse *et al.* (1976) for three different ranges of the polarization ε: $\varepsilon \geqslant 0.9$, $0.9 > \varepsilon > 0.6$, $\varepsilon \leqslant 0.6$.

The results show a faster decrease of the first resonance with increasing Q^2 compared to the nonresonant background in its vicinity, and compared to the two other resonance peaks around 1.5 and 1.67 GeV. In contrast, the second peak increases slightly in relation to the background at the same energy, whereas for the third one the peak/background ratio remains practically constant. The position of the third peak clearly moves to higher energies with increasing Q^2, a point noted first by Stein *et al.* (1975).

A comparison of the fits for different ranges of ε indicates that longitudinal photon contributions to the cross sections in the resonance region are in general small. Possible exceptions are in the ranges $1.3 < W < 1.5\,\text{GeV}^2$ for $Q^2 > 2.0\,(\text{GeV}/c)^2$ and $W > 1.6\,\text{GeV}$ for $Q^2 > 0.5\,(\text{GeV}/c)^2$, where a contribution of σ_l up to 20% is possible.

As an example of the high quality of the total cross-section data, we show typical results (Gayler, 1971; May, 1971) for σ_v of Eq. (4.1) at $Q^2 = 0.6\,(\text{GeV}/c)^2$ through the resonance region in Fig. 12.

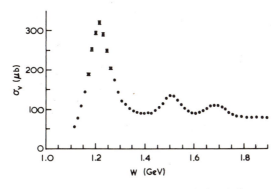

Fig. 12. Total hadronic cross section for the absorption of virtual photons at a momentum transfer of $0.6\,(\text{GeV}/c)^2$. The data are from Gayler (1971) and May (1971).

For the differential cross sections we express the complete differential electroproduction cross section in terms of virtual photoproduction differential cross sections with respect to pion variables in the pion nucleon c.m. Analogously to Eq. (4.1) we have

$$\frac{d^5\sigma}{dE_e \, d\Omega_e \, d\Omega_\pi} = \Gamma \frac{d\sigma_v}{d\Omega_\pi} \tag{4.21}$$

and $d\sigma_v/d\Omega_\pi$ may be expressed in the form

$$\frac{d\sigma_v}{d\Omega_\pi} = A + \varepsilon B + \varepsilon C \sin^2 \theta_\pi \cos 2\phi_\pi + [2\varepsilon(1+\varepsilon)]^{1/2} D \sin \theta_\pi \cos \phi_\pi$$
$$\tag{4.22}$$

Existing experiments on virtual differential cross sections have been performed only for very restricted ranges of the photon polarization parameter ε, so it is not possible to separate A and B in Eq. (4.22). If we use the symbol $\bar{A}$ for $A + \varepsilon B$, and restrict ourselves to s, p, d, and f waves, then

$$\frac{d\sigma_v}{d\Omega_\pi} = \bar{A}_0 + \bar{A}_1 \cos \theta_\pi + \bar{A}_2 \cos^2\theta_\pi + \bar{A}_3 \cos^3\theta_\pi + \bar{A}_4 \cos^4\theta_\pi$$

$$+ \varepsilon (C_0 + C_1 \cos \theta_\pi + C_2 \cos^2\theta_\pi) \sin^2\theta_\pi \cos 2\phi_\pi$$

$$+ [2\varepsilon(1+\varepsilon)]^{1/2} (D_0 + D_1 \cos \theta_\pi + D_2 \cos^2\theta_\pi$$

$$+ D_3 \cos^3\theta_\pi) \sin \theta_\pi \cos \phi_\pi \tag{4.23}$$

The coefficients $\bar{A}_0, \ldots, D_3$ are independent of angle and at a given energy and momentum transfer depend only on the individual multipoles. The expansion (4.23) is sufficient to describe the differential $\gamma_v p \to \pi^0 p$ cross sections over the whole range of energies at which data exist (1.135 GeV $\leq W \leq$ 1.745 GeV), but it cannot be applied directly to the differential $\gamma_v p \to \pi^+ n$ cross sections because of the pion pole contribution.

Typical results for the coefficients of the differential $\gamma_v p \to \pi^0 p$ cross sections at $Q^2 = 0.6$ (GeV/c)^2 (Siddle et al., 1971; Alder et al., 1972a; Shuttleworth et al., 1972) are given in Fig. 13. For $W < 1.6$ GeV the data only require s, p, and d waves, so the coefficients $\bar{A}_4$, C_2, and D_3 are not shown.

In the first resonance region, the data are sensitive only to s and p waves. With this restriction the nonzero angular coefficients have the following multipole decomposition (Siddle et al., 1971; Alder et al.,

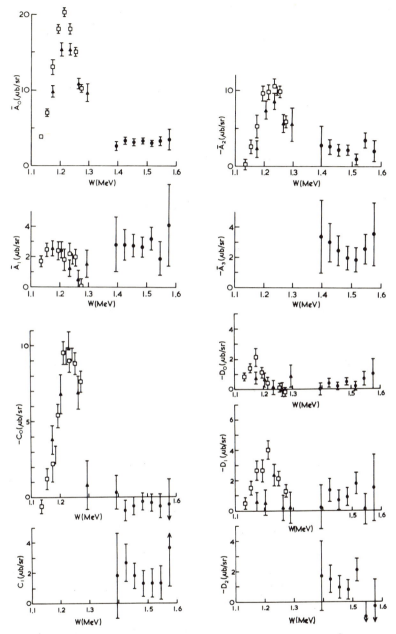

Fig. 13. The coefficients of the angular expansion (4.23) of $d\sigma_v/d\Omega_\pi$ for π^0 electroproduction at a momentum transfer of 0.6 $(\text{GeV}/c)^2$. The data are as follows: (□) Alder *et al.* (1972*a*); (▲) Siddle *et al.* (1971); (●) Shuttleworth *et al.* (1972).

1972a; Shuttleworth et al., 1972):

$$\bar{A}_0 = \{|E_{0+}|^2 + |M_{1-}|^2 + \tfrac{5}{2}|M_{1+}|^2 + \tfrac{9}{2}|E_{1+}|^2 - 3\,\mathrm{Re}(M_{1+}E_{1+}^*)$$
$$+ \mathrm{Re}[(3E_{1+} + M_{1+})M_{1-}^*]$$
$$+ \varepsilon\,(|k^2|/|\mathbf{k}|^2)[|S_{0+}|^2 + |S_{1-}|^2 + 4|S_{1+}|^2 - 4\,\mathrm{Re}(S_{1+}S_{1-}^*)]\}$$

$$\bar{A}_1 = \{2\,\mathrm{Re}[E_{0+}(3E_{1+} + M_{1+})^*] + \varepsilon\,(|k^2|/|\mathbf{k}|^2)2\,\mathrm{Re}[S_{0+}(4S_{1+} + S_{1-})^*]\}$$

$$\bar{A}_2 = \{-\tfrac{3}{2}|M_{1+}|^2 + \tfrac{9}{2}|E_{1+}|^2 + 9\,\mathrm{Re}(M_{1+}E_{1+}^*) - 3\,\mathrm{Re}[(3E_{1+} + M_{1+})M_{1-}^*]$$
$$+ \varepsilon\,(|k^2|/|\mathbf{k}|^2)[12|S_{1+}|^2 + 12\,\mathrm{Re}(S_{1+}S_{1-}^*)]\} \qquad (4.24)$$

$$C_0 = \{-\tfrac{3}{2}|M_{1+}|^2 + \tfrac{9}{2}|E_{1+}|^2 - 3\,\mathrm{Re}(M_{1+}E_{1+}^*) + 3\,\mathrm{Re}[(E_{1+} - M_{1+})M_{1-}^*]\}$$

$$D_0 = -\{|k^2|/|\mathbf{k}|^2\}^{1/2}\,\mathrm{Re}[S_{0+}(3E_{1+} - M_{1+} + M_{1-})^* - E_{0+}(2S_{1+} - S_{1-})^*]$$

$$D_1 = -6\{|k^2|/|\mathbf{k}|^2\}^{1/2}\,\mathrm{Re}[S_{1+}(E_{1+} - M_{1+} + M_{1-})^* + S_{1-}E_{1+}^*]$$

4.3.1. The First Resonance Region

It is immediately apparent from the coefficients $\bar{A}_0$, $\bar{A}_2$, and C_0 of Fig. 13 that the $p_{33}(1230)$ dominates the differential $\gamma_v p \to \pi^0 p$ cross section. For real photons, the resonance excitation is almost purely magnetic dipole (see the articles by Moorhouse and by Donnachie and Shaw in this volume), and this remains true in electroproduction. From Eqs. (4.24) we can see that pure magnetic dipole production (M_{1+}) would require the ratio of the coefficients $\bar{A}_0:\bar{A}_2:C_0$ to be $5:-3:-3$, which holds within the accuracy of the data. Estimates of the other multipoles can be made by neglecting all terms that are second order in any amplitude other than the dominant M_{1+} (Siddle et al., 1971). This shows clearly that the electric quadrupole excitation of the $p_{33}(1230)$ remains negligible over the whole range of Q^2 studied, not being observed within the limit

$$|E_{1+}|/|M_{1+}| < 5\% \qquad (4.25)$$

On the other hand, scalar excitation is observed, but is weak:

$$|S_{1+}|/|M_{1+}| \lesssim 5\%-10\% \qquad (4.26)$$

The scalar excitation of the resonance can be seen in the resonance structure apparent in the coefficient D_1 (Fig. 13), which can only come from S_{1+}, M_{1+} interference. The γNN^* form factor appears to decrease slightly faster than the nucleon form factor.

The nonzero $\bar{A}_1$ and D_0 clearly demand contributions other than purely resonant ones: Within the spirit of our simplifying assumption, $\bar{A}_1$ requires the presence of an E_{0+} contribution and D_0 requires the presence of an S_{0+} contribution.

To go beyond these general qualitative comments requires some theoretical guidance in the form of fixed-t dispersion relations (see Chapter 4, Section 4.6). There have been many such calculations (Zagury, 1966, 1968; Adler, 1968; von Gehlen, 1969, 1970; Crawford, 1971; Devenish and Lyth, 1972), and the most recent of these successfully describes the entire $\gamma_v p \to \pi^0 p$ cross section in the first resonance region over the momentum transfer range $0 \le Q^2 \le 1 \, (\text{GeV}/c)^2$. This analysis, together with the total cross-section analysis of Brasse *et al.* (1976) (which extends out to a much higher-momentum transfer), confirms that the transition form factor of the $p_{33}(1230)$ falls off faster than the nucleon form factor.

The very limited data on the differential $\gamma_v p \to \pi^+ n$ cross section (Bätzner *et al.*, 1973, quoted by Clegg, 1973), mostly lie within the range of the theoretical predictions. There is a large nonresonant background in this channel which contributes significantly to the total $\gamma_v p$ cross section. This is given well by dispersion-relation calculations at $Q^2 = 0$. Although extensive theoretical investigations have not been made for $Q^2 > 0$, the agreement seems to be maintained, at least on the low side of the resonance where the background is dominated by the pole terms (Zagury, 1966, 1968; Adler, 1968; von Gehlen, 1969, 1970; Crawford, 1971).

4.3.2. The Higher-Resonance Region

The first experiments on the differential $\gamma_v p \to \pi^0 p$ and $\gamma_v p \to \pi^+ n$ cross sections in the higher-resonance region were performed only recently, and at present a large amount of data are being accumulated. Even so, there are not likely to be enough data in the foreseeable future to determine the amplitudes in a model-independent way. This is understandable when one remembers that a single value of Q^2 will require more data (owing to the larger number of amplitudes) than the photoproduction ($Q^2 = 0$) case, for which even now after many years of experimentation a model-independent analysis has not been performed above the second resonance region. However, the present data are sufficient for a fixed-t dispersion relation fit to produce meaningful results.

Although the situation is too complex to allow a naive interpretation of the kind we applied in the first resonance region, there are some exceptional features of the data which give strong pointers to what is happening.

One of the most striking features is the data on $\gamma_v p \to \pi^+ n$ in the forward direction (Fig. 14). One sees a large background with wiggles at the second and third resonance positions, which disappear when $Q^2 = 0$. The photoproduction data are quite featureless with no evidence of any resonance structure, while at a momentum transfer of $0.4 \, (\text{GeV}/c)^2$ the resonance structure is already very pronounced. The lack of structure in photoproduction is known to be due to the absence of an helicity-$\frac{1}{2}$ amplitude in the resonance excitation (the helicity-$\frac{3}{2}$ amplitude gives a zero contribution in the forward direction because of angular momentum conservation) and the development of structure with increasing Q^2 implies the presence of a large helicity-$\frac{1}{2}$ amplitude for resonance excitation. It must be predominantly transverse, since we know from the total cross-section results that the longitudinal cross section remains comparatively small.

A second striking feature is provided by η production in the second resonance region. Because the $s_{11}(1520)$ resonance is the only state in the second resonance region with a large branching ratio into ηN, a measurement of $\gamma_v p \to \eta p$ isolates it. The data (Kummer *et al.*, 1973; Beck *et al.*, 1974; Alder *et al.*, 1975*b*) rise initially with increasing Q^2, suggesting a nonnegligible contribution from the longitudinal amplitude, but the dominant behavior is a steady decrease with Q^2 indicating a dominant transverse amplitude. What is surprising about this reaction is that the decrease with Q^2 is so slow: If the interpretation we have indicated is

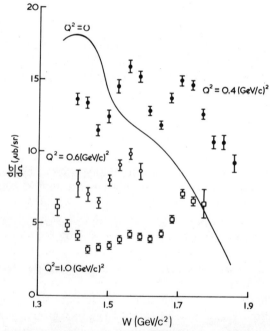

Fig. 14. The differential cross section $d\sigma_v/d\Omega_\pi$ at 0^0 for π^+ electroproduction at momentum transfers of $0.4\ (\mathrm{GeV}/c)^2$ [($\bullet$) Evangelides *et al.* (1974)], $0.6\ (\mathrm{GeV}/c)^2$[(∇) Alder *et al.* (1975)], and $1.0\ (\mathrm{GeV}/c)^2$ [($\triangle$) Alder *et al.* (1975)]. The curve is the average of the photoproduction data.

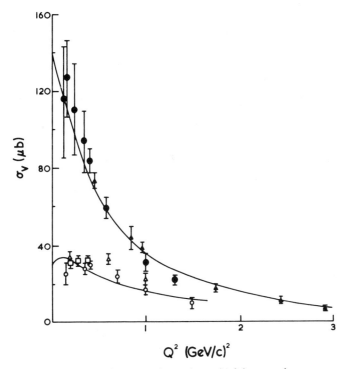

Fig. 15. Comparison of the total cross section at the peak of the second resonance region and the total cross section for η production. The data for the former are as follows: (▲) Breidenbach (1970); (●) Kobberling *et al.* (1974); and for the latter they are the following: (○) Kummer *et al.* (1973); (□) Beck *et al.* (1974); (△) Alder *et al.* (1975a).

correct, one deduces a transverse S_{0+} transition form factor, which falls by only about a factor of 2 between $Q^2 = 0$ and $Q^2 = 1$ $(\text{GeV}/c)^2$. This is markedly different from what we would expect on the basis of our information on the nucleon form factor, the transition form factor of the $p_{33}(1230)$, and the behavior of the total cross section in the region of the second resonance. Figure 15 shows a comparison of the total cross section at the peak of the second resonance region [dominated by the $d_{13}(1520)$ near $Q^2 = 0$] and the total cross section for η production. The difference in the rate of decrease is quite apparent, and shows clearly that the content of the resonance peak at $Q^2 = 2$ $(\text{GeV}/c)^2$ is very different from that at $Q^2 = 0$.

4.3.3. Amplitude Analysis

A fit to the single-pion production data, the single-η production data, and the total cross-section peak heights has been performed within the framework of fixed-t dispersion relations by Devenish and Lyth (1975).

They fitted all energies and all momentum transfers simultaneously, choosing form factors that are specifically free of kinematic singularities (Theis and Hertel, 1970; Scadron and Jones, 1973), and constraining the resonance couplings to reduce to their photoproduction values at $Q^2 = 0$. Reasonably definitive results are obtained for the $p_{33}^+(1230)$, $d_{13}^+(1510)$, $s_{11}^+(1520)$, $f_{15}^+(1690)$, and $f_{37}^+(1920)$. As we have already discussed, no information is available on the neutral states. In terms of the conventional electric $(E_{l\pm})$, magnetic $(M_{l\pm})$ and scalar $(S_{l\pm})$ multipoles, their results can be summarized as follows.

$p_{33}^+(1230)$. The magnetic dipole amplitude remains dominant, but it does not appear to be of dipole form, decreasing somewhat more rapidly. The electric quadrupole excitation remains small, $|E_{1+}|/|M_{1+}| \lesssim 0.02$ for $Q^2 < 1.2 \, (\mathrm{GeV}/c)^2$ and the scalar amplitude is also small, but finite, $|S_{1+}|/|M_{1+}| \simeq 0.06$. These results do not differ much from the qualitative conclusions drawn from direct inspection of the data.

$d_{13}^+(1510)$. The E_{2-} amplitude decreases rapidly, falling by at least a factor of 5 between $Q^2 = 0$ and $Q^2 = 1 \, (\mathrm{GeV}/c)^2$. On the other hand, the M_{2-} amplitude decreases very little in this interval with the consequence that the helicity-$\frac{3}{2}$ amplitude becomes comparable with the helicity-$\frac{1}{2}$ amplitude by $Q^2 \simeq 1.5 \, (\mathrm{GeV}/c)^2$, whereas it is negligible at $Q^2 = 0$. The scalar amplitude is never very significant.

$s_{11}^+(1520)$. The E_{0+} amplitude is well defined and decreases only slowly (by about a factor of 2) in the interval $0 < Q^2 < 1 \, (\mathrm{GeV}/c)^2$. The S_{0+} amplitude is small, but its detailed behavior is not completely clear.

$f_{15}^+(1690)$. The situation here is very similar to that pertaining to the $d_{13}^+(1510)$, with the E_{3-} amplitude decreasing very slowly, if at all, and the S_{3-} amplitude being small and decreasing rapidly.

$f_{37}^+(1920)$. The M_{3+} amplitude probably decreases more slowly than the dipole form. Both the E_{3+} and S_{3+} amplitudes are small, and the latter decreases rapidly.

Devenish et al. (1975) have looked into the problem of trying to unify this apparently random behavior of the transition form factors. They define constraint free $N - N_J^*$ transition form factors and argue that the most likely candidates for possible universal behavior should be sought among them, and not among the physical helicity or multipole form factors, since the latter have a known underlying kinematic Q^2 structure resulting from constraints at thresholds and pseudothresholds. They show that suitably chosen constraint-free form factors have a common Q^2 dependence, with scalar contributions suppressed asymptotically, and in particular a simple parameterization for the three leading resonances p_{33}, d_{13}, f_{15} accounts quite well for the Q^2 dependence of the transverse multipole form factors.

4.3.4. The Reaction $\gamma_v p \to \pi^- \Delta^{++}$

This reaction provides an alternative source of information on the nuclear axial vector form factor, via a threshold theorem analagous to that for $\gamma_v p \to \pi^+ n$. The threshold theorem for $\gamma_v p \to \pi^- \Delta^{++}$ was derived and extensively discussed by Adler and Weisberger (1968). A possible advantage of this latter reaction is that the $G_A(Q^2)$-dependent equal-time commutator term given by current algebra is the dominant term for a good energy range above threshold.

The reaction has been measured by Joos *et al.* (1976) for $1.3 < W < 1.5$ GeV, and for five momentum transfers between 0.35 and $1.0 \,(\text{GeV}/c)^2$. The cross section is found to rise approximately linearly from threshold up to $W \simeq 1.5$ GeV, which is consistent with the expected strong dominance of the equal-time commutator. The dominance of the commutator term is further supported by the observed Δ^{++} production and decay angular distributions.

Interpreting the cross section in this way produces an axial vector form factor not significaantly different from that obtained by the reaction $\gamma_v p \to \pi^+ n$. Explicitly in terms of a dipole fit, the preferred dipole mass is $M_A = 1.16 \pm 0.03 \,\text{GeV}/c^2$.

5. Form Factors and the Quark Model

Since the quark model and Melosh $SU(6)_W$ provide a good description of the $N^* N\gamma$ vertex for real photons, it is natural to ask whether they are equally effective in describing the behavior of these vertices as a function of Q^2 for spacelike photons. There are two main regions of interest: (i) large Q^2, where models are required to explain the observed asymptotic behavior (e.g., the apparent Q^{-4} dependence of the elastic nucleon form factors and the apparent Q^{-2} dependence of the pion form factor); (ii) small Q^2 [say $Q^2 < 2 \,(\text{GeV}/c)^2$], where models are required to explain the detailed structure of the resonance amplitudes.

A model like $SU(6)_W$ can say nothing about the Q^2 dependence of transition form factors. At best, it can be used, as in photoproduction (see the article by Hey in this volume) to correlate the different amplitudes at fixed Q^2. Such an analysis has been carried out by Avilez and Cocho (1974), but because of the restricted information available in pion electroproduction such an analysis does not have the same significance as its photoproduction counterparts.

In principle, the quark model does provide the necessary framework for investigating the Q^2 dependence of form factors, although in practice its success lies in explaining some general qualitative features rather than in providing a detailed quantitative description.

For large Q^2, where it is possible to make asymptotic approximations, it now appears that the quark model, suitably formulated, can indeed give rise to Q^{-4} dependence of the elastic nucleon form factor and the Q^{-2} dependence of the pion form factor. A typical example of such a calculation is that of Alabiso and Schierholz (1975), who obtain their result using the Blankenbecler–Sugar equation at infinite momentum with a two-body interaction corresponding to vector gluon exchange. In this calculation (which is rather general) the experimental asymptotic behavior of the pion and nucleon form factors are obtained naturally by the requirement that the pion and nucleon have an underlying (nonrelativistic) quark–antiquark and three-quark structure, respectively.

There have been numerous attempts to tackle the problem of the transition form factors at small momentum transfers, say, $Q^2 < 2$ $(\text{GeV}/c)^2$ (Thornber, 1968; Fujimura et al., 1970; Ravndal, 1971; Gonzales and Watson, 1972; Abdullah and Close, 1972; Lipes, 1972; Kellett, 1974; Ono, 1976). They have served to highlight the many difficulties inherent in this approach. In addition to the obvious problems of containment and the nature of the wave functions in the hadron rest frame there are serious technical difficulties associated with boosting the wave function of one of the hadrons to nonzero momentum and with the spinor part of the quark wave function, which develops unwanted extra factors. In addition to these practical problems, there are difficulties of principle with timelike states in those models that use four-dimensional oscillators. The various models can be used to provide phenomenological fits to observed transition form factors, but they have little predictive power *ab initio*.

There is, however, one general prediction made by the quark model, which was first noted by Close and Gilman (1972). In both relativistic and nonrelativistic quark models, the helicity amplitudes for photoexcitation of the $d_{13}(1510)$ on protons can be shown to be of the form (see Chapter 2 in this volume, Section 3.1.2)

$$A_{1/2}^p = a(1 - \mathbf{k}^2/k_0^2)F(\mathbf{k}^2) \qquad (5.1a)$$

$$A_{3/2}^p = -2aF(\mathbf{k}^2) \qquad (5.1b)$$

where $\mathbf{k}$ is the photon 3-momentum ($k_0 = |\mathbf{k}|$ at $k^2 = 0$), a is a constant depending on the quark magnetic moment, and $F(\mathbf{k}^2)$ is some form factor. This is in good agreement with experiment, since in photoproduction $A_{3/2}^p$ is the dominant amplitude and $A_{1/2}^p$ is extremely small (Chapter 2, Section 2.5). This result was first obtained by Copley et al. (1969) within the framework of a nonrelativistic quark model with a harmonic oscillator potential. As we go over to spacelike photons, k^2 gets larger, so that $A_{1/2}^p$

increases relative to $A^p_{3/2}$, and in this simple model we would expect the $d_{13}(1510)$ to be electroproduced predominantly in the $A^p_{1/2}$ state at a momentum transfer of less than 1 $(\text{GeV}/c)^2$. An identical argument holds for the $f_{15}(1690)$.

The exact rate at which this change takes place depends rather critically on the wave functions, but the existence of the change is a common feature of quark models. In general, the quark model calculations predict too rapid a change. This is illustrated in Fig. 16, which is a comparison of the helicity asymmetries $(A^p_{1/2} - A^p_{3/2})/(A^p_{1/2} + A^p_{3/2})$ for the $d_{13}(1510)$ and $f_{15}(1690)$ from the analysis of Devenish and Lyth (1975), compared with the quark models of Ravndal (1971) and Ono (1976). Devenish *et al.* (1975) have suggested that since the slow change observed experimentally is largely due to the underlying relativistic kinematic constraint structure, then the essentially nonrelativistic quark model results could be much improved if care were taken to incorporate the correct relativistic constraint structure.

The difficulties faced by the quark model certainly appear to be rather fundamental. For example, the amplitude for photoexcitation of the $s_{11}(1520)$ on protons, analogous to Eq. (5.1), has the form

$$A = 3^{1/2}a\xi^2(1 + \mathbf{k}^2/2k_0^2)F(\mathbf{k}^2) \tag{5.2}$$

where ξ^2 is a parameter mixing the $^2\{8\}_{1/2}$ quark state with the $^4\{8\}_{1/2}$ state. The expressions (5.1) and (5.2) can be used to obtain a quark model

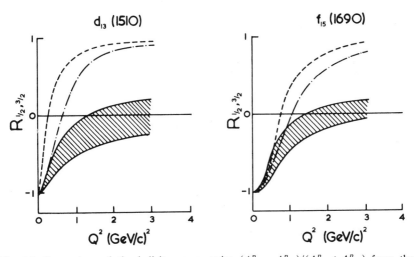

Fig. 16. Comparison of the helicity asymmetries $(A^p_{1/2} - A^p_{3/2})/(A^p_{1/2} + A^p_{3/2})$ from the analysis of Devenish and Lyth (1975) with the quark models of Ravndal (1971) (dashed line) and Ono (1976) (dotted line).

prediction for the s_{11}/d_{13} production ratio, normalizing to photoproduction data. This ratio increases somewhat up to $Q^2 = 0.5$ $(\text{GeV}/c)^2$, as does the data, but then decreases at higher values of Q^2, in strong disagreement with data.

The subject is still very much in a developing stage, and as yet there is no clear theoretical framework within which to fit the quite striking results coming from pion electroproduction. Hopefully, when a large number of transition form factors have been measured, systematic features will emerge, themselves suggesting a theoretical hypothesis.

References

Abdullah, T., and Close, F. E. (1972), *Phys. Rev.* **D5**, 2332.
Adler, S. L. (1968), *Ann. Phys. (N.Y.)* **50**, 189
Adler, S. L., and Gilman, F. J. (1966), *Phys. Rev.* **152**, 1460.
Adler, S. L., and Weisberger, W. I. (1968), *Phys. Rev.* **169**, 1392.
Adylov, G. T., Aliev, F. K., Bardin, D. Yu., Gajewski, W., Kulakov, B. A., Micelmacher, G. V., Niczyporuk, N., Nigmanov, T. S., Tsyganov, E. N., Turala, M., Vudopianov, A. S., Wala, K., Dally, E., Drickey, P., Liberman, A., Shepherd, P., Tompkins, J., Buchanan, D., and Poirier, J. (1974), *Phys. Lett.* **51B**, 402.
Akerlof, C. W., Ash, W. W., Berkelman, K., and Tingcr, M. (1965), *Phys. Rev. Lett.* **14**, 1036.
Akerlof, C. W., Ash, W. W., Berkelman, K., Lichtenstein, C. A., Ramanauskas, A., and Siemaom, R. H. (1967), *Phys. Rev.* **163**, 1482.
Alabiso, C., and Schierholz, G. (1975), *Phys. Rev.* **D11**, 1905.
Albrecht, W., Brasse, F. W., Dorner, H., Flauger, W., Frank, K.-H., Gayler, J., Hultschig, H., May, J., and Ganssauge, E. (1968), *Phys. Lett.* **28B**, 225.
Albrecht, W., Brasse, F. W., Dorner, H., Flauger, W., Frank, K.-H., Gayler, J., Hultschig, H., May, J., and Ganssauge, E. (1969), *Nucl. Phys.* **B13**, 1.
Albrecht, W., Brasse, F. W., Dorner, H., Fehrenbach, W., Flauger, W., Frank, K.-H., Gayler, J., Korbel, V., May, J., Zimmerman, P. D., Courau, A., Diaczek, A., Dumas, J. C., Tristram, G., Valentin, J., Aubret, C., Chazelas, E., and Ganssauge, E. (1971a), *Nucl. Phys.* **B25**, 1.
Albrecht, W., Brasse, F. W., Dorner, H., Fehrenbach, W., Flauger, W., Frank, K.-H., Gayler, J., Korbel, V., May, J., Zimmerman, P. D., Courau, A., Diaczek, A., Dumas, J. C., Tristram, G., Valentin, J., Aubret, C., Chazelas, E., and Ganssauge, E. (1971b), *Nucl. Phys.* **B27**, 615.
Alder, J. C., Brasse, F. W., Chazelas, E., Fehrenbach, W., Flauger, W., Frank, K.-H., Ganssauge, E., Gayler, J., Korbel, V., Krechlok, W., May, J., Merkwitz, M., and Zimmerman, P. D. (1972a), *Nucl. Phys.* **B46**, 415.
Alder, J. C., Brasse, F. W., Chazelas, E., Fehrenbach, W., Flauger, W., Frank, K.-H., Ganssauge, E., Gayler, J., Krechlok, W., Korbel, V., May, J., Merkwitz, M., and Zimmerman, P. D. (1972b), *Nucl. Phys.* **B48**, 487.
Alder, J. C., Brasse, F. W., Fehrenbach, W., Gayler, J., Haidan, R., Gloë, G., Goel, S. P., Korbel, V., Krechlock, W., May, J., Merkwitz, M., Schnitz, R., and Wagner, W. (1975a), *Nucl. Phys.* **B91**, 386.
Alder, J. C., Behrens, H., Brasse, F. W., Fehrenbach, W., Gayler, J., Goel, S. P., Haidan, R., Korbel, V., May, J., and Merkwitz, M. (1975b), *Nucl. Phys.* **B99**, 1.

Alder, J. C., Brasse, F. W., Fehrenbach, W., Gayler, J., Goel, S. P., Haidan, R., Korbel, V., May, J., Merkwitz, M., and Nurimba, A. (1976), *Nucl. Phys.* **B105**, 253.

Amaldi, E., Borgia, B., Pistilli, P., Balla, M., di Giorgio, G. V., Giazotto, A., Serbassi, S., and Stoppini, G. (1970), *Nuovo Cimento* **65A**, 377.

Amaldi, E., Beneventano, M., Borgia, B., de Notaristefani, F., Fronderoli, A., Pistilli, P., Sestili, I., and Severi, M. (1972), *Phys. Lett.* **41B**, 216.

Amati, D., Jengo, R., Rubinstein, H. R., Veneziano, C., and Virasoro, M. A. (1968), *Phys. Lett.* **27B**, 38.

Avilez, C., and Cocho, D. (1974), *Phys. Rev.* **D10**, 3638.

Ball, J. S., and Zachariasen, F. (1968), *Phys. Rev.* **170**, 1541.

Bartel, W., Dudelzak, B., Krehbiel, H., McElroy, J., Meyer-Berkhout, U., Schmidt, W., Walther, V., and Weber, G. (1968*a*), *Phys. Lett.* **27B**, 660.

Bartel, W., Dudelzak, B., Krehbiel, H., McElroy, J., Meyer-Berkhout, U., Schmidt, W., Walther, V., and Weber, G. (1968*b*), *Phys. Lett.* **28B**, 148.

Bartel, W., Büsser, F. W., Dix, W. R., Felst, R., Harms, D., Krehbiel, H., Kuhlmann, P. E., McElroy, J., Meyer, J., and Weber, G. (1971), *Phys. Lett.* **35B**, 181.

Bartel, W., Büsser, F. W., Dix, W. R., Felst, R., Harms, D., Krehbiel, H., Kuhlman, P. E., McElroy, J., Meyer, J., and Weber, G. (1973), *Nucl. Phys.* **B58**, 429.

Bartl, A., and Urban, P. (1966), *Acta Phys. Austriaca* **24**, 139.

Bartoli, B., Felicetti, F., and Silvestrini, V. (1972), *Riv. Nuovo Cimento* **2**, 241.

Bassompierre, G., Binder, G., Dalpiaz, P., Dalpiaz, P. F., Gissinger, G., Jacquey, S., Peroni, C., Schneegans, M. A., and Tecchio, L. (1977), CERN Preprint, April 1977; submitted to *Phys. Lett.* **B**.

Bätzner, K., Beck, U., Becks, K. H., Drees, J., Knop, G., Kolanoski, H., Leenen, M., Moser, K., Nietzel, Ch., Schlösser, E., and Stier, H. E. (1974), *Nucl. Phys.* **B76**, 1.

Bebek, C. J., Brown, C. N., Herzlingler, M., Holmes, S., Lichtenstein, C. A., Pipkin, F. M., Sisterson, L. K., Andrews, D., Berkelman, K., Cassel, D. G., and Hartill, D. L. (1974), *Phys. Rev.* **D9**, 1229.

Bebek, C. J., Brown, C. N., Herzlinger, M., Holmes, S. D., Lichtenstein, C. A., Pipkin, F. M., Raither, S., and Sisterson, L. K. (1976), *Phys. Rev.* **D13**, 25.

Beck, U., Becks, K. H., Burkert, V., Drees, J., Dresbach, B., Gerhardt, B., Knop, G., Kolanoski, H., Leenen, M., Moser, K., Müller, H., Nietzel, C. H., Päsler, J., Rith, K., Rosenberg, M., Sauerwein, R., Schlösser, E., and Stier (1974), *Phys. Lett.* **51B**, 103.

Benaksas, D., Drickey, D., and Frèrejacque, D. (1966), *Phys. Rev.* **148**, 1327.

Benfatto, G., Nicolo, F., and Rossi, G. C. (1972), *Nucl. Phys.* **B50**, 205.

Benfatto, G., Nicolo, F., and Rossi , G. C. (1973), *Nuovo Cimento* **14A**, 425.

Berard, R. W., Buskirk, F. R., Dally, E. B., Dyer, J. N., Maruyama, X. K., Topping, R. L., and Traverso, T. J. (1973), *Phys. Lett.* **47B**, 355.

Berends, F. A. (1970), *Phys. Rev.* **D1**, 2590.

Bincer, A. M. (1960), *Phys. Rev.* **118**, 855.

Bjorken, J. D. (1969), *Phys. Rev.* **179**, 1547.

Bleckwenn, J., Klein, H., Moritz, J., Schmidt, K. H., and Wegener, D. (1971), *Nucl. Phys.* **B33**, 475.

Bloom, E. D., and Gilman, F. J. (1970), *Phys. Rev. Lett.* **25**, 1140.

Bloom, E. D., Cottrell, R. L., Coward, D. H., de Staebler, H., Drees, J., Miller, G., Mo, L. W., Taylor, R. E., Friedmann, J. I., Hartmann, G. C., and Kendall, H. W. (1969), SLAC-PUB 653.

Bloom, E. D., Buschorn, G., Cottrell, R. L., Coward, D. H., de Staebler, H., Drees, J., Jordan, C. L., Miller, G., Mo, L. W., Piel, H., Taylor, R. E., Breidenbach, M., Ditzler, W. R., Friedman, J. I., Hartmann, G. C., Kendall, H. W., and Pucher, J. S. (1970), SLAC-PUB 795.

Bluvstein, R. E., Cheshkov, A. A., and Dubovik, V. M. (1973), *Nucl. Phys.* **B64**, 407.
Borkowski, F., Simon, G. G., Walther, V. H., and Wending, R. D. (1975), *Z. Phys.* **A275**, 29.
Breidenbach, M. (1970), MIT report No. 2098-635.
Brasse, F. W., Engler, J., Ganssauge, E., and Schweitzer (1968), *Nuovo Cimento* **55A**, 679.
Brasse, F. W., Chazelas, E., Fehrenbach, W., Frank, K.-H., Ganssauge, E., Gayler, J.,
    Korbel, V., May, J., Merkwitz, M., Rittenberg, V., and Rubinstein, H. R. (1972), *Nucl.
    Phys.* **B39**, 421.
Brasse, F. W., Flauger, W., Gayler, J., Goel, S. P., Haidan, R., Merkwitz, M., and Wriedt, H.
    (1976), DESY Report No. 76/11.
Brauel, P., Büsser, F. W., Canzler, Th., Cords, D., Dix, W. R., Felst, R., Grindhammer, G.,
    Kollmann, W. D., Krehbiel, H., Meyer, J., and Weber, G. (1973), *Phys. Lett.* **45B**, 389.
Brauel, P., Busser, F. W., Canzler, Th., Cords, D., Dix, W. R., Felst, R., Grindhammer, G.,
    Kollmann, W. D., Krehbiel, H., Meyer, J., and Weber, G. (1974), *Phys. Lett.* **50B**, 507.
Brodsky, S. J., and Farrar, G. R. (1973), *Phys. Rev. Lett.* **31**, 1153.
Brown, C. N., Canizares, C. R., Cooper, W. E., Eisner, A. M., Feldman, G. J., Lichtenstein,
    C. A., Litt, L., Lockeretz, W., Montana, V. B., and Pipkin, F. M. (1973), *Phys. Rev.* **D8**,
    92.
Buchanan, C. D., and Yearian, M. R. (1965), *Phys. Rev. Lett.* **15**, 303.
Cabibbo, N., and Gatto, R. (1961), *Phys. Rev.* **124**, 1577.
Castellano, M., Di Guigno, G., Humphrey, J. W., Sussi-Palmieri, E., Troise, G., Trova, U.,
    and Vitale, S. (1973), *Nuovo Cimento* **14A**, 1.
Chan, L. H., Chen, K. W., Dunning, J. R., Ramsey, N. F., Walker, J. K. and Wilson, R.
    (1966), *Phys. Rev:* **141**, 1298; **147**, 1174.
Ciafaloni, M., and Menotti, D. (1968), *Phys. Rev.* **173**, 1575.
Clegg, A. B. (1974), *Proceedings of the 6th International Symposium on Electron and Photon
    Interactions at High Energies*, Bonn, August 1973 (Amsterdam, North-Holland), p. 49.
Close, F. E., and Gilman, F. J. (1972), *Phys. Lett.* **38B**, 541.
Conversi, M., Massam, T., Muller, Th., and Zichichi, A. (1965), *Nuovo Cimento* **40**, 690.
Conversi, M., Paoluzzi, L., Ceradini, F., d'Angelo, S., Ferrer, M. L., Santonico, R., Grilli, M.,
    Spillanti, P., and Valente, V. (1974), *Phys. Lett.* **52B**, 493.
Copley, L. A., Karl, G., and Obryk, E. (1969), *Nucl. Phys.* **B13**, 303.
Crawford, R. L. (1971), *Nucl. Phys.* **B28**, 573.
Del Guerra, A., Giazotto, A., Giorgi, M. A., Stefanini, A., Botterill, D. R., Braben, D. W.,
    Clarke, D., and Norton, P. R. (1975), *Nucl. Phys.* **B99**, 253.
Del Guerra, A., Giazotto, A., Giorgi, M. A., Stefanini, A., Botterill, D. R., Montgomery, H.
    E., Norton, P. R., and Matone, G. (1976), *Nucl. Phys.* **B107**, 65.
Deo, B. B., and Singh, L. P. (1974), *Phys. Rev.* **D10**, 308.
Devenish, R. C. E., and Lyth, D. H. (1972*a*), *Phys. Rev.* **D5**, 47.
Devenish, R. C. E., and Lyth, D. H. (1972*b*), *Nucl. Phys.* **B43**, 228.
Devenish, R. C. E., and Lyth, D. H. (1975), *Nucl. Phys.* **B93**, 109.
Devenish, R. C. E., Eisenschitz, T. S., and Körner, J. G. (1975), DESY No. 75/48.
Di Vecchia, P., and Drago, F. (1969), *Nuovo Cimento Lett.* **1**, 917.
Dombey, N. (1969), *Rev. Mod. Phys.* **41**, 236.
Dombey, N., and Read, B. J. (1973), *Nucl. Phys.* **B60**, 65.
Drell, S. D. (1967), Proceedings of the International Symposium on Electron and Photon
    Interactions at High Energies, Stanford, 1967.
Drell, S. D., and Pagels, H. R. (1965), *Phys. Rev.* **140**, B397.
Drell, S. D., and Silverman, D. (1968), *Phys. Rev. Lett.* **20**, 1325.
Drell, S. D., and Zachariasen, F. (1961), *Electromagnetic Structure of Nucleons* (Oxford,
    Oxford University Press).
Drell, S. D., and Yan, T. M. (1970), *Phys. Rev. Lett.* **24**, 181.

Drickey, D. J., and Hand, L. N. (1962), *Phys. Rev. Lett.* **9**, 521.
Elias, J. E., Friedman, J. I., Hartmann, G. C., Kendall, H. W., Kirk, P. N., Sogard, M. R., and Van Speybroeck, L. P. (1969), *Phys. Rev.* **177**, 2075.
Evangelides, E., Meaburn, R., Allison, J., Dickinson, B., Ibbotson, M., Lawson, R., Montgomery, H. E., Baxter, D., Foster, F., Hughes, G., Kummer, P. S., Lyth, D. H., Siddle, R., and Devenish, R. C. E. (1974), *Nucl. Phys.* **B71**, 381.
Fermi, E., and Marshall, L. (1947), *Phys. Rev.* **72**, 1139.
Feschbach, H., and Lomon, E. (1967), *Rev. Mod. Phys.* **39**, 611.
Feschbach, H., and Lomon, E. (1968), *Ann. Phys. (N.Y.)* **48**, 94.
Foldy, L. L. (1958), *Rev. Mod. Phys.* **30**, 473.
Frampton, P. (1970), *Phys. Rev.* **D1**, 1341.
Frazer, W. R. (1959), *Phys. Rev.* **115**, 1763.
Frazer, W. R., and Fulco, J. R. (1959), *Phys. Rev. Lett.* **2**, 365.
Frazer, W. R., and Fulco, J. R. (1960), *Phys. Rev.* **117**, 1603; **119**, 1420.
Fubini, S., and Furlan, G. (1968), *Ann. Phys. (N.Y.)* **48**, 322.
Fujimura, K., Kobayashi, T., and Namiki, M. (1970), *Prog. Theor. Phys.* **43**, 73.
Furlan, G., Jengo, R., and Remiddi, E. (1966), *Nuovo Cimento* **44A**, 427.
Furlan, G., Paver, N., and Verzegnassi, C. (1969), *Nuovo Cimento* **62A**, 519.
Galster, S., Klein, H., Moritz, J., Schmidt, K. H., Wegener, D., and Bleckwenn, J. (1971), *Nucl. Phys.* **B32**, 221.
Gayler, J. (1971), thesis, DESY Report No. F21-71/2.
Gonzales, M. A., and Watson, P. J. S. (1972), *Nuovo Cimento* **12A**, 889.
Gounaris, G. J. (1971), *Phys. Rev.* **D4**, 2788.
Gourdin, M. (1966), *Diffusion des electrons de haute énergie* (Paris, Manon et Cie).
Gourdin, M. (1974), *Phys. Rep.* **11C**, 29.
Greenhut, G. K. (1969), *Phys. Rev.* **184**, 1860.
Grosstete, B., Drickey, D., and Lehman, P. (1966), *Phys. Rev.* **141**, 1425.
Hamada, T., and Johnston, J. D. (1962), *Nucl. Phys.* **34**, 382.
Hand, L. N. (1963), *Phys. Rev.* **129**, 1834.
Hanson, K. M., Dunning, J. R., Goitein, M., Kirk, T., Price, L. E., and Wilson, R. (1973), *Phys. Rev.* **D8**, 753.
Hartill, D. L., Barish, B. C., Fong, D. G., Gomez, R., Pine, J., Tollestrup, A. V., Maschke, A. W., and Zipf, T. F. (1969), *Phys. Rev.* **184**, 1415.
Havens, W. W., Rabi, I. I., and Rainwater, L. J. (1947), *Phys. Rev.* **72**, 634.
Hofstadter, R. (1956), *Rev. Mod. Phys.* **28**, 215.
Hofstadter, R. (1963), *Electron Scattering and Nuclear and Nucleon Structure* (W. A. Benjamin, New York).
Hofstadter, R., and McAllister, R. W. (1955), *Phys. Rev.* **98**, 217.
Höhler, G. (1976), in *Lecture Notes in Physics*, Ed. J. Ehlers, K. Hepp, and H. A. Weidenmuller (Berlin, Springer Verlag).
Höhler, G., and Pietarinen, E. (1975*a*), *Phys. Lett.* **53B**, 471.
Höhler, G., and Pietarinen, E. (1975*b*), *Nucl. Phys.* **95B**, 210.
Höhler, G., Sabba-Stefanescu, I., Borkowski, F., Simon, G. G., Walther, V. H., and Wendling, R. D. (1976*a*), "Compilation of Electron Nucleon Scattering Data," Karlsruhe-Mainz report.
Höhler, G., Sabba-Stefanescu, I., Borkowski, F., Simon, G. G., Walther, V. H., and Wendling, R. D. (1976*b*), Karlsruhe-Mainz preprint.
Hulthén, L., and Sugawara, M. (1957), *Handbuch der Physik* (Berlin, Springer Verlag), Vol. I, p. 39.
Jengo, R., and Remiddi, E. (1969), *Nuovo Cimento Lett.* **1**, 922.

Joos, P., Ladage, A., Meyer, H., Söding, P., Stein, P., Wolf, G., Yellin, S., Chen, C. K., Knowles, J., Martin, D., Scarr, J. M., Skillicorn, I. O., Smith, K., Benz, C., Drews, G., Hoffmann, D., Knobloch, J., Kraus, W., Nagel, H., Rabe, E., Sander, C., Schlatter, W. D., Spitzer, H., and Wacker, K. (1976), DESY report No. 76/09.

Kellett, B. H. (1974), *Ann. Phys. (N.Y.)* **87**, 60.

Kirk, P. N., Breidenbach, M., Friedman, J. I., Hartmann, G. C., Kendall, H. W., Buschhorn, G., Coward, D. H., De Staebler, H., Early, R. A., Litt, J., Minten, A., Mo, L. W., Panofsky, W. K. H., Taylor, R. E., Barish, B. C., Loken, S. C., Mar, J., and Pine, J. (1973), *Phys. Rev.* **D8**, 63.

Kirkman, H. C., Railton, R., Rutherglen, J. G., Watson, A. S., Brookes, G. R., Hogg, W. R., Lewis, G. M., Prentice, M. N., Smith, K. M., Combley, F. H., Eaton, G. H., Freeland, J. H., Galbraith, W., and Shaw, J. E. (1970), *Phys. Lett.* **32B**, 519.

Köbberling, M., Moritz, J., Schmidt, K. H., Wegener, D., Zeller, D., Bleckwenn, J., and Heimlich, F. H. (1974), *Nucl. Phys.* **B82**, 201.

Koester, L., Nistler, W., and Waschkowski, W. (1975), University of Munich preprint.

Kowalski, H., Römer, H., and Rubinstein, H. R. (1973), *Nucl. Phys.* **B59**, 589.

Krohn, V. E., and Ringo, C. E. (1973), *Phys. Rev.* **D8**, 1305.

Kummer, P. S., Ashburner, E., Foster, F., Hughes, G., Siddle, R., Allison, J., Dickinson, B., Evangelides, E., Ibbotson, M., Lawson, R. S., Meaburn, R. S., Montgomery, H. E., and Shuttleworth, W. G. (1973), *Phys. Rev. Lett.* **30**, 873.

Landshoff, P. V. (1974), *Proceedings of the XVII International Conference on High Energy Physics*, London, July 1974. Ed. J. R. Smith (Chilton, SRC Rutherford Laboratory), V-57.

Lee, P. S., Shaw, G. L., and Silverman, D. (1974), *Phys. Rev.* **D10**, 2251.

Lipes, R. G. (1972), *Phys. Rev.* **D5**, 2849.

Love, A., and Rankin, W. A. (1970), *Nucl. Phys.* **B21**, 261.

Matveef, V. A., Muradyan, R. M., and Tavkhelidze, A. V. (1973), *Nuovo Cimento Lett.* **7**, 719.

May, J. (1971), thesis, DESY report No. F21-71/3.

McGee, I. (1966), *Phys. Rev.* **151**, 772.

Mo, L. W., and Tsai, Y. S. (1969), *Rev. Mod. Phys.* **41**, 205.

Moritz, J., Schmidt, K. H., Wegener, D., Bleckwenn, J., and Engles, E. (1972), *Nucl. Phys.* **B41**, 336.

Nambu, Y., and Shrauner, E. (1962), *Phys. Rev.* **128**, 862.

Ono, S. (1976), preprint, Techniche Hochschule, Aachen.

Panofsky, W. K. H., and Allton, E. A. (1958), *Phys. Rev.* **110**, 1155.

Panofsky, W. K. H., Newton, C. M., and Yodh, G. B. (1955), *Phys. Rev.* **98**, 751.

Perkins, D. H. (1976), *Proceedings of the 1975 International Symposium on Lepton and Photon Interactions at High Energies*, Stanford University, August 21–27, 1975. Ed. W. T. Kirk (Stanford, Stanford Linear Accelerator Center), p. 571.

Ravndal, F. (1971), *Phys. Rev.* **D4**, 1466.

Riazuddin, and Lee, B. W. (1966), *Phys. Rev.* **146**, 1202.

Rittenberg, V., and Rubinstein, H. R. (1971), *Phys. Lett.* **35B**, 50.

Scadron, M. D., and Jones, H. F. (1973), *Ann. Phys. (N.Y.)* **81**, 1.

Shaw, G. (1972), *Phys. Lett.* **39B**, 255.

Shuttleworth, W. J., Sofair, A., Siddle, R., Dickinson, B., Ibbotson, M., Lawson, R., Montgomery, H. E., Hellings, R. D., Allison, J., Clegg, A. B., Foster, F., Hughes, G., and Kummer, P. S. (1972), *Nucl. Phys.* **B45**, 428.

Siddle, R., Dickinson, B., Ibbotson, M., Lawson, R., Montgomery, H. E., Nuthakki,, V. P. R., Tumer, O. T., Shuttleworth, W. J., Sofair, A., Hellings, R. D., Allison, J., Clegg, A. B., Foster, F., Hughes, G., Kummer, P. S., and Fannon, J. (1971), *Nucl. Phys.* **B35**, 93.

Stein, S., Atwood, W. B., Bloom, E. D., Cottrell, R. L. A., de Staebler, H., Jordan, C. L., Piel, H. G., Prescott, C. Y., Siemann, R., and Taylor, R. E. (1975), SLAC-PUB-1528.

Taylor, R. E. (1976), *Proceedings of the 1975 International Symposium on Lepton and Photon Interactions at High Energies*, Stanford University, August 21–27, 1975. Ed. W. T. Kirk (Stanford Linear Accelerator Center), p. 679.

Theis, W. R., and Hertel, P. (1970), *Nuovo Cimenta* **66**, 152.

Thornber, N. S. (1968), *Phys. Rev.* **169**, 1096.

von Gehlen, G. (1969), *Nucl. Phys.* **B9**, 17.

von Gehlen, G. (1970), *Nucl. Phys.* **B20**, 102.

von Gehlen, G., Rubinstein, H. R., and Wessel, H. (1972), *Phys. Lett.* **42B**, 365.

Wilson, R. (1971), *Proceedings of the Fifth International Symposium on Electron and Photon Interactions at High Energies*, Cornell University, August 1971. Ed. N. B. Mistry (Cornell, Laboratory of Nuclear Studies), p. 97.

Zagury, N. (1966), *Phys. Rev.* **145**, 112; **150**, 1406(E).

Zagury, N. (1968), *Phys. Rev.* **165**, 1934(E).

6

High-Energy Photoproduction: Nondiffractive Processes

J. K. Storrow

1. Introduction

1.1. Photon Beams as Hadron Probes

At high energies hadronic two-body reactions can be classified according to the quantum numbers exchanged into (i) vacuum exchange (elastic scattering), (ii) meson exchange, (iii) baryon exchange, and (iv) exotic exchange reactions. Each class has its own characteristic energy dependence. Photoproduction reactions can be classified in exactly the same way, except that with a photon beam exotic reactions cannot occur. "Elastic" photoproduction reactions, such as $\gamma p \to \rho^0 p$, are discussed by Leith (Chapter 7), and in this chapter we will concentrate on meson exchange and baryon exchange reactions. Except that the cross sections are typically smaller by a factor of around 200, the gross features (s dependence, t dependence, etc.) of photoproduction reactions are similar to those of the corresponding hadronic reactions, and so it is natural to try to understand them in the same theoretical framework as is applied to hadronic processes, i.e., Regge poles and cuts. Despite the two major disadvantages of photon beams, namely, the much smaller cross sections and the mixed isospin of the photon, they have several compensating advantages over hadron beams

J. K. Storrow • Department of Theoretical Physics, University of Manchester, Manchester, M13 9PL, England

and are a very useful tool for investigating high-energy mechanisms, as we shall see.

Phenomenological studies of hadronic reactions have concentrated on the scattering of pseudoscalar mesons off nucleons. Various systematics of such reactions have been tabulated (Fox and Quigg, 1973). However, these reactions are not the whole story. Unnatural parity exchange is not allowed in these reactions and so to study these exchanges one must go to higher-spin processes such as $\pi N \to \rho N$ or photoproduction. At high energies polarized photon beams enable a particularly clean separation of unnatural parity exchange contributions to be made. The cross section for photons polarized perpendicular (parallel) to the reaction plane, $d\sigma_\perp/dt$ ($d\sigma_\parallel/dt$), receives contributions from only natural (unnatural) parity exchange in the limit $s \to \infty$. This was proved by Stichel (1964) for $\gamma N \to \pi N$ and extended to any spin-zero photoproduction process by Ader *et al.* (1968*a*) and by Ravndal (1970). Thus, by measuring the cross section, $d\sigma/dt$, and the polarized beam asymmetry $\Sigma = (d\sigma_\perp/dt - d\sigma_\parallel/dt)/(d\sigma_\perp/dt + d\sigma_\parallel/dt)$ one can separate natural and unnatural parity exchanges. Also, the spin of the photon means that different exchanges appear in different guises in photoproduction. For example, in $KN \to KN$, ω exchange appears to be strongly affected by cuts, whereas ρ exchange is remarkably polelike. This is understood on the basis of the different helicity characteristics of the two exchanges. The situation is expected to be reversed in photoproduction, and this appears to be the case. Another useful feature of high-spin reactions is that they might provide indications as to whether it is simpler to look at high-energy reactions from a t-channel point of view or the currently popular s-channel point of view. A common feature of the s-channel approaches is that the t dependence of amplitudes depends only on the net helicity transfer $n = |\lambda_c - \lambda_d - \lambda_a + \lambda_b|$ and not on the individual helicities $\{\lambda_i\}$. Only in reactions involving three spinning particles would this different helicity dependence show up.

In view of the above, higher-spin reactions are clearly of interest. However, a practical lesson that has been learned from studies of pseudoscalar meson–nucleon scattering is that our theoretical understanding of hadronic reactions is so poor that to make any progress to a phenomenological understanding we need to be able to unravel amplitudes. This can be done by making enough polarization measurements to enable amplitudes to be constructed directly from the data in a model-independent way as has been done in $\pi N \to \pi N$ (Halzen and Michael, 1971). Alternatively, a reasonable amount of polarization data can be supplemented by either (i) assumptions about the nature of the exchanges, or (ii) finite-energy sum rules if the low-energy phase shift analysis is good enough. This latter method was used with success in $\pi N \to \pi N$ (Barger and Phillips, 1969) before the complete set of data became available. It is the good polarization data at high energies and the existence of multipole analyses up to 1.5 GeV that have caused a great deal of interest in pion photoproduction. By comparison, although there are a great deal of data on $\pi N \to \rho N$ as regards the ρ density-matrix elements, the multipole analyses for this reaction are not of the same quality as in photoproduction, and, in addition, there is always the problem of subtracting the background under the wide ρ peak in order to obtain a ρ cross section.

In this article we will review nondiffractive photoproduction, largely in the light of the above discussion. We will be looking to the excellent photoproduction data to shed light on the mechanisms that control high-energy nondiffractive scattering. However, before we do that we must discuss which aspects of photoproduction are expected to be different from a typical hadronic reaction. The constraint of gauge invariance must be obeyed—this is particularly important when we use models, as we shall see. Also, the possible existence of fixed poles in the J plane (because of the linearity of the t-channel unitarity condition) complicates the application of finite-energy sum rules. In electroproduction processes, we have the additional complication of a varying external mass and the longitudinal photon contribution. However, we have the additional bonus of being able to study the effects of a continuously varying external mass. We first turn our attention to the general features of the photoproduction data.

Throughout this article the nucleon mass will be denoted by m and the pion mass by μ.

1.2. General Features of the Data

Having outlined our basic theoretical expectations, we now turn to general features of the data. We first discuss the forward direction. Here hadronic reactions have an energy dependence between s^{-1} and s^{-2}. Photoproduction reactions show an s^{-2} falloff, but in a most striking manner. All reactions scale as k_γ^{-2}, i.e., $k_\gamma^2 \, d\sigma/dt$ is a universal function of t for each reaction. This scaling behavior sets in at a relatively low energy—between 1 and 3 GeV depending on the process. In Regge pole language we would say that the effective trajectory, $\alpha_{\text{eff}}(t)$, defined by

$$\frac{d\sigma}{dt} = F(t)(s - m^2)^{2\alpha_{\text{eff}}(t)-2} \tag{1.1}$$

is approximately equal to zero for all values of t. This scaling means that we can represent all the data for one reaction by a single curve $(s - m^2)^2 \, d\sigma/dt$ and we show this for all pseudoscalar meson photoproduction reactions in Fig. 1. This is called a Diebold plot (Diebold, 1970). The cross sections differ by three orders of magnitude near $t = 0$ and have very different angular distributions. Both of these differences can be ascribed to the fact that the reactions have different exchanges, and so it is surprising that those differences do not show up in the energy dependences, as was first stressed by Harari (1969). For $|t| > 0.7$ $(\text{GeV}/c)^2$ the differences between the various reactions disappear, all showing universal $k_\gamma^{-2} e^{3t}$ behavior. This falloff in t is slower than is typical for hadronic reactions, which go as e^{8t} approximately. We will discuss deviations from scaling when we discuss individual reactions in Section 3. The most notable features of the angular distributions are the pronounced structure for very small t observed in $\gamma p \to \pi^- \Delta^{++}$, $\gamma n \to \pi^- p$ and $\gamma p \to \pi^+ n$, the reactions in which π exchange is allowed, and the dip in $\gamma p \to \pi^0 p$ around $t = -0.5$ $(\text{GeV}/c)^2$. This latter dip has played an important, if confusing, role in the seemingly endless search for dip systematics at high energies.

In the backward direction we again have scaling, but as k_γ^{-3}. The corresponding Diebold plot for small u is shown in Fig. 2, and here the differences between different cross sections is less pronounced.

The scaling of photoproduction cross sections, or lack of shrinkage in Regge language, was originally thought to indicate that photon-induced

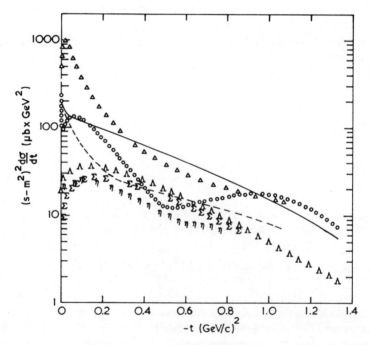

Fig. 1. Diebold plot for forward pseudoscalar photoproduction (Diebold, 1970). $(s - m^2)^2 \, d\sigma/dt$ is plotted against t for various reactions: solid line $\gamma p \to \pi^+ n$; dashed line, $\gamma n \to \pi^- p$; (O) $\gamma p \to \pi^0 p$; ($\triangle$) $\gamma p \to \pi^- \Delta^{++}$; ($\eta$) $\gamma p \to \eta^0 p$; (Λ) $\gamma p \to K^+ \Lambda$; (Σ) $\gamma p \to K^+ \Sigma$.

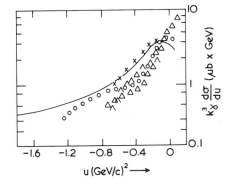

Fig. 2. Diebold plot for backward pseudoscalar photoproduction (after Harari, 1969). $k_\gamma^3 \, d\sigma/dt$ is plotted against u for various reactions: solid line $\gamma p \to n\pi^+$; (○) $\gamma p \to p\pi^0$; (△) $\gamma p \to \Delta^{++}\pi^-$; (∧) $\gamma p \to \Lambda^0 K^+$; (×) $\gamma p \to \Sigma^0 K^+$.

reactions were not amenable to description by Regge poles. However, as stressed by Harari (1971a), only particularly simple hadronic reactions such as $\pi^- p \to \pi^0 n$ show canonical shrinkage. Reactions with many exchanges and/or several helicity amplitudes do not show such simple behavior, and we will find that photon reactions do not seem to present any particular problems, except perhaps at larger momentum transfers.

1.3. Structure and Scope of the Article

In this article we review high-energy nondiffractive photoproduction, taking the low-energy cutoff to be $E_\gamma = 2$ GeV, where the characteristic peaking of $d\sigma/dt$ towards the forward and backward direction sets in. Most of our discussion will be on the phenomenological interpretation of these forward and backward peaks and the light they shed on high-energy exchange mechanisms. Enough formalism is given for this purpose. This means that we give the full formalism for forward $\gamma N \to 0^- \tfrac{1}{2}^+$ reactions, as here we have a great deal of high-energy polarization data and low-energy multipole analysis. For these reactions we also discuss the relation between amplitudes and observables from the point of view of possible direct amplitude determination from the data.

For other forward reactions, such as $\pi\Delta$, and backward reactions, not enough data are available to make such elaboration worthwhile, and we content ourselves with giving a general discussion, including only the minimum amount of formalism necessary for an understanding of the data.

Throughout the article photons are treated as hadron probes in their own right. No discussion of the vector dominance model is given, and no tests of the model against data are included. Certain ideas pertaining to VMD cannot be avoided, for example the fact that the photon is predominantly isovector, but beyond such simple ideas, no discussion is given. This is contained in the chapter by Donnachie and Shaw in the companion volume.

In structure the article falls into two parts—Section 2, in which the theory and formalism of photon-induced reactions is discussed, and Section 3, where the ideas introduced in Section 2 are used to obtain a phenomenological understanding of the data. General conclusions are presented in Section 4.

2. Formalism and Theory

2.1. Amplitudes for $\gamma N \to 0^{-\frac{1}{2}+}$

The scattering amplitude for the reaction $\gamma N \to \pi N$ can be expressed in terms of four invariant amplitudes, the CGLN amplitudes, A_1, A_2, A_3, and A_4 (Chew *et al.*, 1956). These are convenient because of their simple analyticity and crossing properties—they are expected to satisfy Mandelstam analyticity and are either crossing even or odd. To elucidate the crossing properties further, one must make an isospin decomposition of A_i as given in Chapter 2. The result is that $A_{1,2,4}^{(+,0)}$ and $A_3^{(-)}$ are even and $A_3^{(+,0)}$ and $A_{1,2,4}^{(-)}$ are odd. Because of the above properties of the A_i amplitudes one can write down fixed-t dispersion relations for them. We disperse in the crossing-symmetric variable ν, defined by

$$\nu = (s - u)/4m \qquad (2.1)$$

and obtain, for a crossing-even amplitude,

$$\text{Re } A_i(\nu, t) = \frac{B_i \nu_B}{\nu_B^2 - \nu^2} + \frac{2}{\pi} \int_{\nu_0}^{\infty} \nu' \frac{\text{Im } A_i(\nu', t)}{\nu'^2 - \nu^2} d\nu' \qquad (2.2)$$

and, for a crossing-odd amplitude,

$$\text{Re } A_i(\nu, t) = \frac{B_i \nu}{\nu_B^2 - \nu^2} + \frac{2\nu}{\pi} \int_{\nu_0}^{\infty} \frac{\text{Im } A_i(\nu', t)}{\nu'^2 - \nu^2} d\nu' \qquad (2.3)$$

The coefficients of the Born terms, B_i, can be expressed in terms of e^2, g^2 and the anomalous magnetic moments of proton and neutron. Fixed-t dispersion relations (FTDR's) such as these have been used in analyses of high-energy data by dividing the integral at $\nu = \nu_c$ corresponding to around $E_\gamma = 1.5$ GeV, using multipole analyses to calculate the integral from threshold to $\nu = \nu_c$, and taking a Regge-type parameterization for $\nu > \nu_c$. For discussing forward high-energy reactions it is more convenient to use not the CGLN amplitudes A_i, but a closely related set, F_i, given by

$$F_1 = A_1 - 2mA_4, \qquad F_2 = A_1 + tA_2$$
$$F_3 = 2mA_1 - tA_4, \qquad F_4 = A_3 \qquad (2.4)$$

These amplitudes have simple crossing and analyticity properties and also have definite parity in the t-channel—F_1 and F_2 are respectively natural and unnatural parity t-channel amplitudes to all orders in s, F_3, and F_4, respectively, natural and unnatural t-channel amplitudes to leading order in s (or ν).

Although the A_i and F_i amplitudes are convenient for theoretical considerations such as FTDR's and finite-energy sum rules (FESR's), for discussing data in terms of models it is convenient to use helicity amplitudes (Jacob and Wick, 1959). These are convenient because experimental observables can be expressed simply in terms of them and also they are reasonably simple theoretically in the sense that the phase of a Regge-pole contribution to them is given by the usual signature factor. The helicity amplitudes we use here are N, S_1, S_2, D, where N is a no-flip amplitude, S_1 and S_2 are single flip, and D is double flip. They are defined in terms of helicity labels by

$$N = T^1_{+-} = T^{-1}_{-+}, \qquad S_1 = T^1_{--} = T^{-1}_{++}$$
$$S_2 = T^1_{++} = T^{-1}_{--}, \qquad D = T^1_{-+} = -T^{-1}_{+-} \tag{2.5}$$

in the notation $T^{\lambda_\gamma}_{\lambda_i \lambda_f}$, where λ_γ, λ_i, λ_f are the helicities of the photon, initial nucleon, and final nucleon, respectively. The amplitudes are normalized by

$$\frac{d\sigma}{dt} = |N|^2 + |S_1|^2 + |S_2|^2 + |D|^2 \tag{2.6}$$

and their relation to the helicity amplitudes of other authors is given in an appendix by Barker *et al.* (1974).

The relation between the helicity amplitudes and the CGLN amplitudes (or the F_i) is rather complicated and is given in Chapter 2. For a general discussion of high-energy scattering, it is sufficient to consider the asymptotic crossing matrix in the limit $s \to \infty$. This is:

$$\begin{bmatrix} F_1 \\ F_2 \\ F_3 \\ F_4 \end{bmatrix} \simeq -\frac{4\pi^{1/2}}{(-t)^{1/2}} \begin{bmatrix} 2m & (-t)^{1/2} & -(-t)^{1/2} & 2m \\ 0 & (-t)^{1/2} & (-t)^{1/2} & 0 \\ t & 2m(-t)^{1/2} & -2m(-t)^{1/2} & t \\ 1 & 0 & 0 & -1 \end{bmatrix} \begin{bmatrix} S_1 \\ N \\ D \\ S_2 \end{bmatrix} \tag{2.7}$$

We can see immediately that F_1 and F_4 contribute mainly to s-channel helicity flip and F_2 and F_3 to no-flip and double flip.

For reactions such as $\gamma N \to K\Lambda$ with unequal-mass baryons in the initial and final states, the definitions of the F_i's change and so do the crossing matrices. Essentially, where $2m$ occurs in the matrix it is replaced by $(m + m_\Lambda)$ and additional terms proportional to $(m - m_\Lambda)$ occur. Clearly, at high energies our statement concerning the helicity amplitudes to which the various F_i amplitudes contribute is still a reasonable approximation. Helicity amplitudes for this process are defined in exactly the same way. Details of the kinematics are given in Donnachie (1972) and Levy *et al.* (1973*a*).

The final type of amplitude we will find convenient to use is the transversity amplitude (Kotanski, 1966). These have the spin quantization axis perpendicular to the scattering plane, and because of this the relation between amplitudes and observables is even simpler than for helicity amplitudes. These amplitudes are useful in deriving criteria for complete sets of measurements. They are defined as follows:

$$b_1 = \tfrac{1}{2}[(S_1 + S_2) + i(N - D)], \qquad b_3 = \tfrac{1}{2}[(S_1 - S_2) - i(N + D)]$$
$$b_2 = \tfrac{1}{2}[(S_1 + S_2) - i(N - D)] \qquad b_4 = \tfrac{1}{2}[(S_1 - S_2) + i(N + D)]$$

$$(2.8)$$

Theoretically, these amplitudes are not convenient, since they have kinematic singularities on the boundary of the physical region, which can only be removed if a linear combination of amplitudes is used. Also, they have complicated phases even in simple cases, due to the factors of i occurring in Eq. (2.8).

2.2. Relation between Amplitudes and Observables

When a measurement is made we obtain a number, an observable, which is a sum or difference of bilinear products of amplitudes. We can define 16 of them, since there are four amplitudes, but they will not all be independent. We adopt the usual Basel convention with the z axis being the beam direction and the y axis the normal to the reaction plane (Fig. 3). The z' axis is in the direction of the scattered meson. We now define the 16

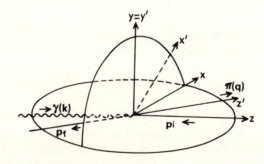

Fig. 3. Definition of axes. If $\mathbf{k}$ is the incoming photon momentum and $\mathbf{q}$ the outgoing meson momentum (both in the c.m. system), then the axes are defined by $\mathbf{z} = \mathbf{k}/|\mathbf{k}|$, $\mathbf{y} = \mathbf{k} \times \mathbf{q}/|\mathbf{k} \times \mathbf{q}|$, $\mathbf{x} = \mathbf{y} \times \mathbf{z}$, $\mathbf{z}' = \mathbf{q}/|\mathbf{q}|$, $\mathbf{y}' = \mathbf{y}$, $\mathbf{x}' = \mathbf{y} \times \mathbf{z}'$.

Table 1. Observables

Usual symbol	Helicity representation	Transversity representation	Experiment required[a]	Type
$d\sigma/dt$	$\|N\|^2 + \|S_1\|^2 + \|S_2\|^2 + \|D\|^2$	$\|b_1\|^2 + \|b_2\|^2 + \|b_3\|^2 + \|b_4\|^2$	$\{-;-;-\}$	
$\Sigma \, d\sigma/dt$	$2\text{Re}(S_1 S_2^* - ND^*)$	$\|b_1\|^2 + \|b_2\|^2 - \|b_3\|^2 - \|b_4\|^2$	$\{L(\tfrac{1}{2}\pi, 0); -; -\}$	
			$\{-; y; y\}$	
$T \, d\sigma/dt$	$2\text{Im}(S_1 N^* - S_2 D^*)$	$\|b_1\|^2 - \|b_2\|^2 - \|b_3\|^2 + \|b_4\|^2$	$\{-; y; -\}$	S
			$\{L(\tfrac{1}{2}\pi, 0); 0; y\}$	
$P \, d\sigma/dt$	$2\text{Im}(S_2 N^* - S_1 D^*)$	$\|b_1\|^2 - \|b_2\|^2 + \|b_3\|^2 - \|b_4\|^2$	$\{-;-; y\}$	
			$\{L(\tfrac{1}{2}\pi, 0); y; -\}$	
$G \, d\sigma/dt$	$-2\text{Im}(S_1 S_2^* + ND^*)$	$2\text{Im}(b_1 b_3^* + b_2 b_4^*)$	$\{L(\pm\tfrac{1}{4}\pi); z; -\}$	
$H \, d\sigma/dt$	$-2\text{Im}(S_1 D^* + S_2 N^*)$	$-2\text{Re}(b_1 b_3^* - b_2 b_4^*)$	$\{L(\pm\tfrac{1}{4}\pi); x; -\}$	BT
$E \, d\sigma/dt$	$\|S_2\|^2 - \|S_1\|^2 - \|D\|^2 + \|N\|^2$	$-2\text{Re}(b_1 b_3^* + b_2 b_4^*)$	$\{c; z; -\}$	
$F \, d\sigma/dt$	$2\text{Re}(S_2 D^* + S_1 N^*)$	$2\text{Im}(b_1 b_3^* - b_2 b_4^*)$	$\{c; x; -\}$	
$O_x \, d\sigma/dt$	$-2\text{Im}(S_2 D^* + S_1 N^*)$	$-2\text{Re}(b_1 b_4^* - b_2 b_3^*)$	$\{L(\pm\tfrac{1}{4}\pi); -; x'\}$	
$O_z \, d\sigma/dt$	$-2\text{Im}(S_2 S_1^* + ND^*)$	$-2\text{Im}(b_1 b_4^* + b_2 b_3^*)$	$\{L(\pm\tfrac{1}{4}\pi); -; z'\}$	BR
$C_x \, d\sigma/dt$	$-2\text{Re}(S_2 N^* + S_1 D^*)$	$2\text{Im}(b_1 b_4^* - b_2 b_3^*)$	$\{c; -; x'\}$	
$C_z \, d\sigma/dt$	$\|S_2\|^2 - \|S_1\|^2 - \|N\|^2 + \|D\|^2$	$-2\text{Re}(b_1 b_4^* + b_2 b_3^*)$	$\{c; -; z'\}$	
$T_x \, d\sigma/dt$	$2\text{Re}(S_1 S_2^* + ND^*)$	$2\text{Re}(b_1 b_2^* - b_3 b_4^*)$	$\{-; x, x'\}$	
$T_z \, d\sigma/dt$	$2\text{Re}(S_1 N^* - S_2 D^*)$	$2\text{Im}(b_1 b_2^* - b_3 b_4^*)$	$\{-; x; z'\}$	TR
$L_x \, d\sigma/dt$	$2\text{Re}(S_2 N^* - S_1 D^*)$	$2\text{Im}(b_1 b_2^* + b_3 b_4^*)$	$\{-; z; x'\}$	
$L_z \, d\sigma/dt$	$\|S_1\|^2 + \|S_2\|^2 - \|N\|^2 - \|D\|^2$	$2\text{Re}(b_1 b_2^* + b_3 b_4^*)$	$\{-; z; z'\}$	

[a]Notation is $\{P_\gamma; P_T; P_R\}$, where P_γ is the polarization of the beam, $L(\theta)$ is the beam linearly polarized at angle θ to the scattering plane, C is the circularly polarized beam; P_T is the direction of target polarization; P_R is the component of recoil polarization measured. In the case of the single-polarization measurements we also give the equivalent double-polarization measurement.

observables in terms of both helicity and transversity amplitudes in Table 1. The precise relation between observables and the experiments we consider is as follows.

Polarized Beam—Polarized Target:

$$\frac{d\sigma}{dt} = \frac{d\sigma}{dt}\bigg|_{\text{unpolarized}} \{1 - P_T \Sigma \cos(2\phi) + P_x[-P_T H \sin(2\phi) + P_\odot F]$$

$$- P_y[-T + P_T P \cos(2\phi)] - P_z[-P_T G \sin(2\phi) + P_\odot E]\} \qquad (2.9)$$

where (P_x, P_y, P_z) is the polarization of the target, P_T is the transverse polarization of the beam at an angle ϕ to the reaction plane, and $P_\odot$ is the degree of right circular polarization of the beam.

Beam-Recoil:

$$\rho_f \frac{d\sigma}{dt} = \frac{d\sigma}{dt}\bigg|_{\text{unpolarized}} [1 + \sigma_y P - P_T \cos(2\phi)(\Sigma + \sigma_y T)$$

$$- P_T \sin(2\phi)(O_x \sigma_x + O_z \sigma_z) - P_\odot(C_x \sigma_x + C_z \sigma_z)] \qquad (2.10)$$

Target-Recoil:

$$\rho_f \frac{d\sigma}{dt} = \frac{d\sigma}{dt}\bigg|_{\text{unpolarized}} \{1 + \sigma_y P + P_x(T_x \sigma_x + T_z \sigma_z)$$

$$+ P_y(T + \Sigma\sigma_y) - P_z(L_x\sigma_x - L_z\sigma_z)\} \qquad (2.11)$$

$$\rho_f = \tfrac{1}{2}(1 + \boldsymbol{\sigma} \cdot \mathbf{P}_f) \qquad (2.12)$$

where ρ_f is the density matrix of the recoil nucleon, and $\mathbf{P}_f$ is its polarization.

These relations have been considered by many authors (Worden, 1972; Berger and Fox 1970*b*; Goldstein *et al.*, 1974; Barker *et al.*, 1974, 1975). Various of them also give observables in terms of parity-conserving *t*-channel amplitudes. Goldstein *et al.* (1974) also consider triple polarization measurements. However, since we can obtain all bilinear products $b_i^* b_j$ from the 16 observables given, we see that complete information can be obtained without measuring a triple polarization. [This is a special case of a general result proved by Simonius (1967).] For this reason we do not consider such experiments.

The advantage of working in the transversity representation can be seen from Table 1. The observables can be divided into four classes of four characterized by the following experimental set-up:

(i) The set *S* consisting of $d\sigma/dt$ and the three single-polarization measurements.

The remaining 12 can be measured in double-polarization experiments and can be divided into three classes, depending on which two particles are polarized:

(ii) Set *BT*, working with a polarized beam and polarized target.

(iii) Set *BR*, working with a polarized beam and measuring the polarization of the recoil baryon.

(iv) Set *TR*, working with a polarized target and measuring the recoil polarization.

The experimentally relevant division is reflected in the structure of the observables in terms of transversity amplitudes: the set *S* giving the moduli of the transversity amplitudes, the set *TR* all being of the form Re(Im) $\{b_1 b_2^* \pm b_3 b_4^*\}$, etc. The reason observables take such a simple form in the transversity basis is, of course, that the direction of spin quantization is taken to be $\mathbf{k} \times \mathbf{q}$, the normal to the scattering plane. Another advantage of the simplicity of the transversity representation is in the consideration of bounds. These are useful in deciding which experiments to do in the light of previous measurements. They have been considered in detail by Goldstein *et al.* (1974) and Barker *et al.* (1975).

In terms of the classes we have defined, the bounds have a certain symmetry. There are the following bounds within the set S:

$$|P \pm T| \le 1 \pm \Sigma, \qquad |T \pm \Sigma| \le 1 \pm P, \qquad |P \pm \Sigma| \le 1 \pm T \qquad (2.13)$$

Also, all double-polarization observables are bounded by the set S as follows:

$$|X_{BT}| \le \min\{(1 - \Sigma^2)^{1/2}, (1 - T^2)^{1/2}\}$$

where

$$X_{BT} = G, H, E, \text{ or } F$$

$$|X_{BR}| \le \min\{(1 - \Sigma^2)^{1/2}, (1 - P^2)^{1/2}\}$$

where

$$X_{BR} = O_x, O_z, C_x, \text{ or } C_z$$

$$|X_{TR}| \le \min\{(1 - P^2)^{1/2}, (1 - T^2)^{1/2}\}$$

where

$$X_{TR} = T_x, T_z, L_x \text{ or } L_z \qquad (2.14)$$

In addition, if one double polarization has already been measured then the following more stringent bounds between two observables of a given set and the set S are useful:

$$\max\{(G^2 + E^2), (H^2 + F^2), (G^2 + H^2), (E^2 + F^2)\} \le \min\{(1 - \Sigma^2), (1 - T^2)\},$$

$$\max\{(O_x^2 + O_z^2), (C_x^2 + C_z^2), (O_x^2 + C_x^2), (O_z^2 + C_z^2)\} \le \min\{(1 - \Sigma^2), (1 - P^2)\}$$

$$\max\{(T_x^2 + T_z^2), (L_x^2 + L_z^2), (T_x^2 + L_x^2), (T_z^2 + L_z^2)\} \le \min\{(1 - P^2), (1 - T^2)\}$$

$$(2.15)$$

$$\max\{|G \pm F|, |E \pm H|\} \le 1 \pm P$$

$$\max\{|T_z \pm L_x|, |T_x \pm L_z|\} \le 1 \pm \Sigma \qquad (2.16)$$

$$\max\{|O_x \pm C_z|, |O_z \mp C_x|\} \le 1 \pm T$$

These bounds (and many more) are easy to prove, particularly using transversity amplitudes. They can be very useful. For example, at high energies many reactions have Σ near 1 due to the dominance of natural parity exchange, and so quantities bounded by $1 - \Sigma$ or $(1 - \Sigma^2)^{1/2}$ are very restricted.

2.3. Amplitude Analysis in $\gamma N \to 0^{-}\frac{1}{2}^{+}$?

Since amplitude analysis has been the source of most of the progress in hadronic physics in the last few years, albeit almost entirely owing to the efforts of the experimentalists, it is worth examining the possibility of amplitude analysis in photon-induced reactions. We consider the reactions $\gamma N \to 0^{-}\frac{1}{2}^{+}$ since for higher-spin reactions the number of measurements required is prohibitive. For $\gamma N \to 0^{-}\frac{1}{2}^{+}$ there are four (complex) amplitudes and so we need to make seven measurements at a given E and t to determine the amplitudes up to an overall phase, which cannot normally be determined experimentally. In special cases, such as $\gamma p \to \pi^{0}(\eta^{0})p$ near the forward direction where we have interference with the purely electromagnetic Primakoff mechanism, one can obtain information on this overall phase, but we will ignore this possibility for the moment.

The experimentally relevant question is as follows: What are the conditions for three double-polarization measurements (excluding those double-polarization measurements that are equivalent to single-polarization measurements) to determine the amplitudes up to an overall phase (and quadrant ambiguities) when taken in conjunction with $d\sigma/dt$, Σ, P, and T? Barker *et al.* (1975) have derived a necessary and sufficient condition, which can be stated economically in terms of the three sets defined earlier, BT, TR, and BR. It is that any three measurements give complete information, provided that they are not all taken from the same set. There are no systematic rules for choosing measurements to resolve the quadrant ambiguities that arise because we are solving quadratic equations. It is normally possible to resolve them by making one further measurement, but it must be chosen carefully. It is possible to make three further measurements and still be left with an ambiguity.

We now turn our attention to the practical applications of these rules to particular reactions in which the set S either is known or could easily be completed:

(*a*) $\gamma p \to \pi^{0}p$. Here it is difficult to measure a recoil polarization since it involves a rescattering. In obtaining P this can be avoided (and has been) by measuring $\{L(\frac{1}{2}\pi, 0); y; 0\}$, and so the set S is known. Also, G and H are being measured at the time of writing. To obtain a complete set one must measure any component of recoil polarization in the xz plane with either a target polarized in the xz plane or a beam either circularly polarized or linearly polarized at an angle $\frac{1}{4}\pi$ to the reaction plane, and any such (i.e., any measurement from the set TR or BR) measurement would suffice. However, the component of recoil polarization in the z' direction is not measurable by rescattering.

(*b*) $\gamma p \to K^{+}\Lambda$. In this case the decay of the Λ gives all three components of recoil polarization. Thus, in this case one would avoid the set BT and do any three measurements from TR and BR, though not all from the same set.

(*c*) $\gamma p \to \pi^{+}n$. Since some measurement of recoil polarization in a double-polarization experiment is required, it would appear that the difficulty of rescattering experiments for neutrons would rule out the possibility of amplitude analysis in this reaction.

(d) $\gamma n \to \pi^- p$. This is same as (a) with the additional difficulties of using a polarized deuterium target.

We see that at least for the reactions $\gamma p \to \pi^0 p$ and $\gamma p \to K^+ \Lambda$ the measurement of a complete set is certainly feasible and—to judge from what has been learned in $\pi N \to \pi N$—worthwhile, especially since many features of theoretical interest, such as the zero structure of amplitudes, are unaffected by quadrant ambiguities. After this exhortation to experimentalists we turn our attention to theoretical methods of analyzing incomplete sets of data.

2.4. Current Ideas in High-Energy Scattering

2.4.1. Forward Reactions

It is generally accepted that Regge-pole exchange is the dominant mechanism in high-energy scattering and that, in order to explain details of amplitudes, Regge cut corrections are needed. However, no theory of Regge cuts gives a satisfactory explanation of all of the features of high-energy scattering. Because of these difficulties, there has been an increasing trend in recent years to study the s-channel properties of amplitudes (presumably including cut corrections) and try to build up systematic rules for them. We will discuss the implications of these points of view for pseudoscalar photoproduction.

First, we discuss Regge poles. We do not consider the Reggeization of the regularized t-channel parity-conserving helicity amplitudes which is given in Cohen-Tannoudji et al. (1968), Ader et al. (1968b) and Donnachie (1972), being content to quote the result. Each Regge pole, with trajectory function $\alpha_j(t)$ gives a contribution to each F_i amplitude to which it is allowed to contribute, equal to

$$F_i(\nu, t) = \beta_{ji}(t)\xi_\alpha(\nu/\nu_0)^{\alpha_j(t)-1} \tag{2.17}$$

where β_{ji} is the residue function, ν_0 a scale parameter [normally taken to be 1 (GeV)^2] and

$$\xi_\alpha = \frac{1 \pm e^{-i\pi\alpha}}{\sin(\pi\alpha)} \tag{2.18}$$

is the appropriate signature factor.

In photoproduction the requirement of gauge invariance must be satisfied, and in general any individual exchange can be written in a manifestly gauge-invariant form. This is not the case when a Born term is involved (by a Born term we mean the exchange of a particle that also appears among the external ones). It turns out that the sum of all Born terms is gauge invariant (Ball, 1961; Horn and Jacob, 1968; Bardeen and Tung, 1968). This will turn out to be important when we discuss pion exchange.

In high-spin processes the phenomena of conspiracy and evasion arise, which impose constraints on the Regge residues $\beta_{ji}(t)$. In this problem they arise because the amplitudes $F_2(v, t)$ and $F_3(v, t)$ satisfy a constraint equation at $t = 0$:

$$F_3(v, 0) = 2mF_2(v, 0) \qquad (2.19)$$

which can be derived immediately from Eq. (2.4) (the CGLN amplitudes A_2 and A_4 do not have poles at $t = 0$). However, as stated in Section 2.1, F_2 (F_3) corresponds to unnatural (natural) parity exchange in the t channel, so that, for example, the pion can contribute to F_2, but not to F_3. So the only way the constraint can be satisfied is that there is a natural parity trajectory degenerate with the pion, a pion conspirator π_c, which has a related residue in order that Eq. (2.19) can be satisfied or the pion residue in F_2 vanishes at $t = 0$. This vanishing of the residue is called evasion and is the currently favored solution in the absence of any plausible conspirator candidate for any Regge trajectory. In fact, the realization that cuts are important has led to the rather artificial device of conspiracy being discarded. In view of this, Fox and Quigg (1973) have suggested that the term "evasion" is somewhat of a misnomer and should be replaced by the simpler statement that the Regge residue satisfies the requirements of factorization by vanishing.

There is a further constraint equation to be satisfied at $t = 4m^2$, i.e.,

$$F_3(v, 4m^2) = 2mF_1(v, 4m^2) \qquad (2.20)$$

which is usually ignored since it is not a relation between natural and unnatural parity exchanges and also is very far from the region of interest, viz., the physical region $t \leq 0$.

A more controversial point about Regge residues is whether they have zeros at wrong signature points, so-called nonsense wrong signature zeros (NWSZ's). In hadronic reactions this question has never been settled, essentially because of uncertainties in Regge-cut contributions, though it is probably fair to say that they do not appear universally.

Concerning Regge cuts it can be said unequivocally that there is no satisfactory model for calculating them and currently there is little interest in such model-building, essentially because of the weakness of the theory. In view of this we will not discuss specific models of Regge cuts, but state general expectations. These are that the tip of the cut should have the same value at $t = 0$ as the pole, but a flatter slope in t. The cut contributions to the cross section are expected to have a flatter t-dependence than poles and are expected to interfere destructively with poles. Also, cuts are expected to be much stronger in no-flip amplitudes than flip amplitudes. These expectations are based on convolution and absorption approaches to cuts, in which the central partial waves of the pole amplitude are absorbed. These general features can be produced using heuristic models such as the so-called poor man's absorption model of Williams (1970) or the square cut model of Worden (1972).

The approach that has dominated studies of hadronic and photonic reactions in the last few years is the essentially s-channel approach of studying amplitudes in the impact parameter plane. That this approach gives insight into high-energy reactions was stressed by the Michigan school (Ross *et al.*, 1970), and the philosophy has survived the demise of the Michigan model, which was essentially a Regge-cut model that produced peripheral amplitudes by strongly absorbing structureless* Regge poles. Harari (1971c) combined these ideas with duality in the dual absorption model (DAM), in which only the imaginary parts of amplitudes are peripheral, i.e., for an amplitude $M_n(s, t)$ of net helicity flip n we have

$$\text{Im}\, M_n(s, t) \propto J_n[R(-t)^{1/2}] \qquad (2.21)$$

where R is a characteristic hadronic radius, normally taken to be 1 fm. This peripheral behavior of imaginary parts is well established for vector exchange, but for tensor exchange the situation is unclear (Fox and Quigg, 1973).

Current developments of these ideas are the dual peripheral model of Schrempp and Schrempp (1973a, 1973b, 1974, 1975), in which the scattering is dominated by s-channel Regge poles and the b-universality hypothesis (Brion and Peschanski, 1974; Ader *et al.*, 1975). This latter hypothesis is that impact parameter amplitudes $a_n(b, s)$ corresponding to net helicity flip n and defined by

$$a_n(b, s) = \frac{1}{64\pi q} 2 \int_0^{-\infty} dt\, M_n(s, t) J_n[b(-t)^{1/2}] \qquad (2.22)$$

have, for a given reaction, universal behavior (except for the necessary kinematic factor b^n) given by

$$a_n(b, s) = \lambda_n(s)(-1)^n b^n f(b, s) \qquad (2.23)$$

Both this approach and the dual peripheral model lead to derivative relations between the helicity amplitudes for a given reaction of the form

$$M_n(s, t) = C(s)[(-t)^{1/2}]^n \left(\frac{1}{(-t)^{1/2}} \frac{\partial}{\partial(-t)^{1/2}}\right)^n M_0(s, t) \qquad (2.24)$$

*Structureless in t, central in impact parameter.

which were first proposed as an empirical guess for πN amplitudes by Hogaasen (1971).

This completes our resumé of the ideas that have been found useful in discussing forward photoproduction and hadronic reactions. The special case of fixed poles, which cannot occur in hadronic reactions, will be discussed in Section 2.5. Before that, we discuss backward photoproduction.

2.4.2. Backward Reactions

To discuss Regge-pole exchange in the backward direction we need to introduce parity-conserving u-channel helicity amplitudes, $f_1^{\pm}(u^{1/2}, s)$ and $f_3^{\pm}(u^{1/2}, s)$, where the superscript $\pm$ is the τP of the exchanged particle.* These amplitudes are given in terms of the CGLN invariant amplitudes by†

$$f_1^{\pm}(u^{1/2}, s) = 2(u - m^2)A_1 + (\pm u^{1/2} + m)(tu^{1/2} - \mu^2 m)A_2$$
$$+ m(t - \mu^2)(A_3 + A_4) + (\pm u^{1/2} + m)(u - m^2)(A_3 - A_4)$$

(2.25)

$$f_3^{\pm}(u^{1/2}, s) = (\pm u^{1/2} + m)A_2 + A_3 + A_4 \qquad (2.26)$$

From the above definitions it can be immediately seen that the $f_i^{\pm}(u^{1/2}, s)$ have no singularities except an irremoveable singularity at $u = 0$, and this singularity leads to the unique problems of backward scattering. Since the A_i are functions of u, not $u^{1/2}$, f_i^+ and f_i^- are related by the MacDowell symmetry relations (MacDowell, 1959)

$$f_1^+(u^{1/2}, s) = f_1^-(-u^{1/2}, s) \qquad (2.27)$$

$$f_3^+(u^{1/2}, s) = f_3^-(-u^{1/2}, s) \qquad (2.28)$$

If we apply standard Reggeization techniques to these amplitudes, we obtain the following asymptotic forms:

$$f_1^{\pm}(u^{1/2}, s) \simeq \sum_i \gamma_i^{\pm}(u^{1/2})R(\alpha_i^{\pm}(u^{1/2}), s) \qquad (2.29)$$

$$f_3^{\pm}(u^{1/2}, s) \simeq \sum_i \beta_i^{\pm}(u^{1/2})R(\alpha_i^{\pm}(u^{1/2}), s)s^{-1} \qquad (2.30)$$

where the $\alpha_i^{\pm}$ are the trajectory functions of the Regge poles and the γ_i and

---

*The subscripts are historical and refer to the fact that the $f_1^{\pm}$ correspond to u-channel helicity $\frac{1}{2}$ and $f_3^{\pm}$ to u-channel helicity $\frac{3}{2}$.

†Our definitions are the same as those of Barger and Weiler (1969, 1970), Bajpai and Donnachie (1970), and Berger and Fox (1971), apart from irrelevant factors.

β_i are residue functions, the sum is over Regge poles and the factor R is the standard Regge factor given by

$$R(\alpha, s) = \frac{1 + \tau e^{-i\pi(\alpha - 1/2)}}{\cos(\pi\alpha)} s^{\alpha - 1/2} \qquad (2.31)$$

In order to satisfy the MacDowell symmetry conditions, clearly Regge poles must occur in pairs of opposite parity, with trajectories and residues related by the conspiracy conditions

$$\alpha_i^+(u) = \alpha_i^-(-u^{1/2}) \qquad (2.32)$$

$$\gamma_i^+(u) = \gamma_i^-(-u^{1/2}) \qquad (2.33)$$

$$\beta_i^+(u) = \beta_i^-(-u^{1/2}) \qquad (2.34)$$

first discovered by Gribov (1963). The same problem arises in backward πN scattering. Known baryon trajectories appear to be linear in u, in which case these parity doublet trajectories should have recurrences degenerate in mass. There are no reasonable candidates for parity doublets (Storrow, 1972), and their nonappearance is one of the great puzzles of hadron physics. Three possible explanations have been put forward:

(a) Trajectories contain $u^{1/2}$ terms (Barger and Cline, 1969), making the doublets nondegenerate in mass, or even preventing the unwanted trajectory from passing through physical j values (Barger and Cline, 1969; Storrow and Winbow, 1973b).

(b) The residue function of the unwanted trajectory vanishes at physical j values, thus decoupling the lower mass recurrences (Barger and Cline, 1969), or all recurrences if required (Storrow, 1972; Minkowski, 1970; Halzen and Minkowski, 1971).

(c) Kinematic Regge cuts banish the unwanted trajectory to another sheet (Carlitz and Kislinger, 1970).

Since even with the great amount of data available on backward πN scattering, this problem is not resolved, it is clearly asking too much of the rather sparse photoproduction data to provide any hints. In fact, photoproduction has been studied phenomenologically from all three points of view, as we will see later. The simplest and most popular approach has been to eliminate the lowest recurrence by requiring the residue to vanish at the relevant u value, i.e.,

$$\gamma^{\pm}(u^{1/2} = m_R) = 0 \qquad (2.35)$$

with the choice of sign depending on which parity is present and which absent.

In addition to the MacDowell symmetry relation, there is an additional constraint at the u-channel threshold, $u = m^2$ (Jackson and Hite, 1968):

$$f_1^{\pm}(m, s) = 2m(t - \mu^2)f_3^{\pm}(m, s) \qquad (2.36)$$

which is normally incorporated into the parametrization of residue functions.

All of the considerations involving Regge cuts and s-channel pictures mentioned in the previous subsection can also be applied to backward reactions. However, in view of the uncertainties in how to apply Regge pole theory, and the fact that only cross-section data exist in backward photoproduction, it is hardly worth applying such sophistication there. The only part of the previous section that does not apply to backward scattering is evasion, which does not occur in this case, essentially because there is always a conspiracy.

2.5. Fixed Poles and Finite-Energy Sum Rules

Since finite energy sum rules (FESR's) enable us to use the low-energy multipole analyses to constrain the high-energy amplitudes it is worth studying them in some detail. In photoproduction, the possible existence of fixed poles in the j plane complicates the application of FESR's, and to study this we will derive them from fixed-t dispersion relations (FTDR's). For a crossing-odd amplitude $F_i(\nu, t)$ we have

$$\text{Re } F_i(\nu, t) = \frac{\bar{B}_i \nu}{\nu_B^2 - \nu^2} + \frac{2\nu}{\pi} \int_{\nu_0}^{\infty} \frac{\text{Im } F_i(\nu', t)}{\nu'^2 - \nu^2} \, d\nu' \qquad (2.37)$$

We assume that above some energy, i.e., for $\nu > \nu_c$, we can parametrize $\text{Im } F_i(\nu, t)$ as a sum of Regge terms, i.e.,

$$\text{Im } F_i(\nu, t) = \sum_j \beta_{ji}(t)\nu^{\alpha_j(t)-1} \qquad (2.38)$$

and for $\nu < \nu_c$ $\text{Im } F_i(\nu, t)$ is known from multipole analysis. We thus obtain

$$\text{Re } F_i(\nu, t) = \frac{\bar{B}_i \nu}{\nu_B^2 - \nu^2} + \frac{2\nu}{\pi} \int_{\nu_0}^{\nu_c} \frac{\text{Im } F_i(\nu', t)}{\nu'^2 - \nu^2} \, d\nu' + \frac{2\nu}{\pi} \sum_j \beta_{ji}(t) P \int_{\nu_c}^{\infty} \frac{\nu'^{\alpha_j-1}}{\nu'^2 - \nu^2} \, d\nu' \qquad (2.39)$$

For large ν we can write this as

$$\text{Re } F_i(\nu, t) = \sum_{n=0}^{\infty} \frac{1}{\nu^{2n+1}} \left[\bar{B}_i \nu_B^{2n} + \frac{2}{\pi} \int_{\nu_0}^{\nu_c} \nu'^{2n} \text{Im } F_i(\nu', t) \, d\nu' \right.$$
$$\left. - \frac{2}{\pi} \sum_j \beta_{ji}(t) \frac{\nu_c^{\alpha_j(t)+2n}}{\alpha_j(t)+2n} \right] - \sum_j \beta_{ji}(t)\nu^{\alpha_j(t)-1} \cot[\tfrac{1}{2}\pi\alpha_j(t)] \qquad (2.40)$$

In the hadronic case, the expression in large square brackets must vanish, since asymptotic behavior $\nu^{-(2n+1)}$ corresponds to a fixed pole in the j plane at $j = -n$, which is forbidden by generalized unitarity. In photoproduction we have no such restriction and so the expression in large square brackets must be equated to the residue of the fixed pole at $j = -n$, $R_n^{(i)}(t)$, giving a modified nth-moment FESR

$$\frac{\pi}{2}\bar{B}_i\nu_B^{2n} + \int_{\nu_0}^{\nu_c} \nu'^{2n} \operatorname{Im} F_i(\nu', t)\, d\nu' = \sum_j \beta_{ji}(t)\frac{\nu_c^{\alpha_j(t)+2n}}{\alpha_j(t)+2n} + \frac{\pi}{2} R_n^{(i)}(t),$$

$$n = 0, 1, 2, \ldots \quad (2.41)$$

For a crossing-even amplitude the nth-moment FESR is

$$\frac{\pi}{2}\bar{B}_i\,\nu_B^{2n+1} + \int_{\nu_0}^{\nu_c} \nu'^{2n+1} \operatorname{Im} F_i(\nu', t)\, d\nu' = \sum_j \frac{\beta_{ji}(t)\nu_c^{\alpha_j(t)+1+2n}}{\alpha_j(t)+1+2n} + \frac{\pi}{2}\bar{R}_n^{(i)}(t)$$

$$(2.42)$$

Note that the highest-lying fixed pole in a crossing-even amplitude is at $j = -1$ and so is not expected to be important in the asymptotic behavior of the amplitude, whereas for a crossing-odd amplitude the highest-lying pole is at $j = 0$. We will find the formalism of this section useful when we discuss phenomenological analyses that use FESR's and/or FTDR's.

2.6. High-Spin Photoproduction Reactions

Up to now we have only considered the theory and formalism of reactions like $\gamma N \to \pi N, K\Lambda$, etc., and we now want to consider the possibility of higher spins in the final state. Since for these reactions there is not much data we will confine ourselves to generalities in this section, and introduce any specific formalism when the need arises.

Most of Section 2.4 generalizes in a fairly straightforward manner. Evasion is a general phenomenon, the general result being that, for t small, a helicity amplitude $T_{\lambda_a\lambda_b;\lambda_c\lambda_d}$ for a reaction $a + b \to c + d$ vanishes as

$$T_{\lambda_a\lambda_b;\lambda_c\lambda_d} \alpha [\sin(\tfrac{1}{2}\theta_s)]^{n+x} \quad (2.43)$$

where n is the net helicity flip

$$n = |\lambda_a - \lambda_b - \lambda_c + \lambda_d| \quad (2.44)$$

and x is defined by

$$n + x = |\lambda_a - \lambda_c| + |\lambda_b - \lambda_d| \quad (2.45)$$

The extra factor $[\sin(\tfrac{1}{2}\theta_s)]^x$ (extra in the sense of being more than is required by the conservation of angular momentum) is the evasive factor, which is a direct consequence of factorization, of course.

All of the ideas of Regge cuts, geometrical concepts, and s-channel pictures generalize easily and were, in fact, stated for a helicity amplitude of net helicity flip n in Section 2.4.

In the backward direction there is always the problem of irremoveable singularities at $u = 0$, which leads to MacDowell symmetry relations and the parity doublet problem. The general prescription for constructing combinations of s-channel helicity amplitudes with definite u-channel parity is given by Ader *et al.* (1969), who also give generalized MacDowell symmetry relations. As in pseudoscalar photoproduction, evasion does not occur in the backward direction.

3. Phenomenology

3.1. Forward Photoproduction of Charged Pions

3.1.1. Data

In this section we will consider the reactions $\gamma p \to \pi^+ n$ and $\gamma n \to \pi^- p$, usually referred to as π^+ and π^- photoproduction, respectively. Typical cross-section data are shown in Fig. 4, plotted against $(-t)^{1/2}$ in order that

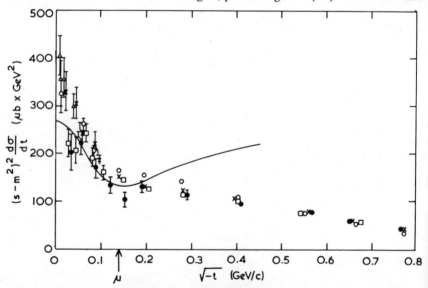

Fig. 4. Plot of $(s - m^2)^2 \, d\sigma/dt$ against $(-t)^{1/2}$ for the reaction $\gamma p \to \pi^+ n$. Data: (○) 5 GeV, (×) 8 GeV, (□) 11 GeV, (●) 16 GeV (all from Boyarski *et al.*, 1968*a*), (△) 4.9 GeV (Heide *et al.*, 1968). Lower-energy data (Bar-Yam *et al.*, 1967; Joseph *et al.*, 1967) have been omitted for clarity. The solid line is the electric Born term approximation.

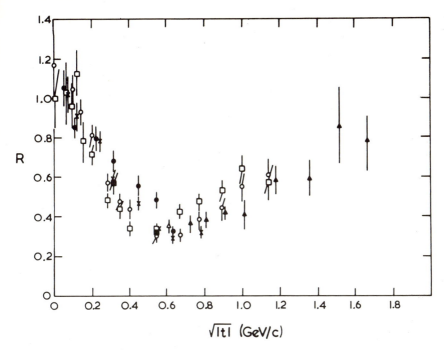

√|t| (GeV/c)

Fig. 5. The π^-/π^+ ratio from deuterium, i.e., the ratio $R = [d\sigma(\gamma d \to \pi^- pp)/dt]/[d\sigma(\gamma d \to \pi^+ nn)/dt]$ plotted against $(-t)^{1/2}$. Data: (▲) 3.4 GeV (Bar-Yam *et al.*, 1967); (●) 3.4 GeV, (×) 5 GeV (both Heide *et al.*, 1968); (○) 8 GeV, (□) 16 GeV (both Boyarski *et al.*, 1968*b*).

the forward spike, in which $d\sigma/dt$ rises by a factor of 2 between $t = -\mu^2$ and $t = 0$, can be seen. The cross section for $\gamma n \to \pi^- p$ is similar near the forward direction, but smaller away from there—it is normally plotted as a ratio of $d\sigma/dt$ ($\gamma p \to \pi^+ n$), the π^-/π^+ ratio R, and is shown in Fig. 5. Both cross sections seem to show a slight change of slope around $t = -0.6$ $(\text{GeV}/c)^2$.

The polarized beam asymmetry Σ, defined by

$$\Sigma = \frac{d\sigma_\perp/dt - d\sigma_\parallel/dt}{d\sigma_\perp/dt + d\sigma_\parallel/dt} \tag{3.1}$$

has been measured for both reactions, and typical results are shown in Figs. 6(a) and (b). For π^+ photoproduction the polarized target asymmetry T has been measured and is shown in Fig. 7.

From the cross-section data for $\gamma p \to \pi^+ n$, one can extract an effective trajectory $\alpha_{\text{eff}}(t)$ (Fig. 8), which shows remarkably little deviation from the k_γ^{-2} scaling behavior discussed in Section 1.2 [i.e., $\alpha_{\text{eff}}(t) \approx 0$]. The π^-/π^+ ratio seems to be independent of energy. Using polarized beam data for

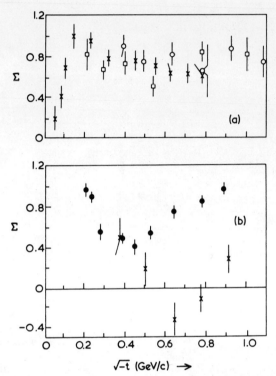

Fig. 6. Polarized beam asymmetry Σ for (a) $\gamma p \to \pi^+ n$ and (b) $\gamma n \to \pi^- p$ (in fact, $\gamma d \to \pi^- p p_s$) plotted against $(-t)^{1/2}$. Data: (a) ($\bigcirc$) 3 GeV, Bar-Yam *et al.* (1970b); ($\times$) 3.4 GeV, Geweniger *et al.* (1969) and Burfeindt *et al.* (1970); ($\square$) 12 GeV, Schwitters *et al.* (1971). The 16-GeV data of Sherden *et al* (1973) have been omitted for clarity. (b) ($\times$) 3 GeV, Bar-Yam *et al.* (1970a); ($\bullet$) 16 GeV, Sherden *et al.* (1973). The 3.4-GeV data of Burfeindt *et al.* (1973) have been omitted for clarity.

$\gamma n \to \pi^- p$ one can extract an α_{eff} for $d\sigma_{\|}(\gamma n \to \pi^- p)/dt$, which, as mentioned earlier, only receives contributions from unnatural parity exchanges (Stichel, 1964). The result obtained (Fig. 9) is remarkably similar to what one would expect from the exchange of a conventional pion Regge trajectory (Sherden *et al.*, 1973).

3.1.2. Extreme Forward Direction $(|t| \le 2\mu^2)$

We first consider the data in the region of the forward spike. It is a general rule in hadron physics that such a rapid change in $d\sigma/dt$ can only be associated with π exchange and must therefore be solely due to the rapid variation of $d\sigma_{\|}/dt$. In the forward direction, kinematics demand that $d\sigma_{\|}/dt$ and $d\sigma_{\perp}/dt$ are equal and presumably $d\sigma_{\perp}/dt$ must be slowly

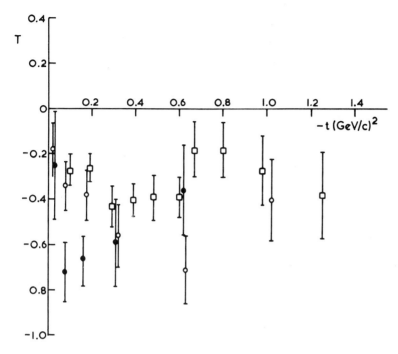

Fig. 7. Polarized target asymmetry, *T*, for $\gamma p \to \pi^+ n$. Data: (○) 5 GeV, (●) 16 GeV (both from Morehouse *et al.*, 1970); (□) 5 GeV (from Genzel *et al.*, 1975). The 3.4-GeV data of Genzel *et al.* (1975) have been omitted for clarity.

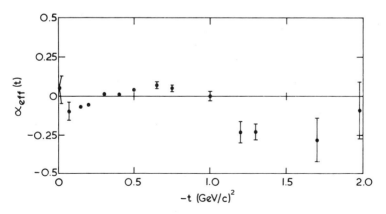

Fig. 8. $\alpha_{\text{eff}}(t)$ for $\gamma p \to \pi^+ n$ (from Fox and Quigg, 1973).

varying over this t range since it does not contain a pion contribution. This rise in $d\sigma/dt$ by a factor of 2 from $t = -\mu^2$ to $t = 0$ [Fig. 10(a)] is due to $d\sigma_\parallel/dt$ varying [Fig. 10(b)], $d\sigma_\perp/dt$ being roughly constant [Fig. 10(c)]. This immediately enables one to predict that Σ will rise from zero at $t = 0$ to almost 1 at $t = -\mu^2$ [Fig. 10(d)], in agreement with experiment. Thus, any reasonable model which explains the forward spike will explain Σ, since the above connection between the two is almost model independent (Harari, 1971a; Jackson and Quigg, 1969, 1970).

We can take these model-independent arguments a step further, using FESR's. The important amplitudes near $t = 0$ are F_2 and F_3, and let us assume that F_2 is dominated by a pole with $\alpha(0) \approx 0$ (presumably the pion). The FESR for F_2 reads

$$\frac{ef\pi}{2\mu} \frac{(t + \mu^2)}{(t - \mu^2)} + \int_{\nu_0}^{\nu_c} \text{Im}\, F_2(\nu, t)\, d\nu = \frac{\beta_2(t)\nu_c^{\alpha(t)}}{\alpha(t)} \tag{3.2}$$

where f is the rationalized pion–nucleon coupling constant: $f = (\mu/2m)g$.

It is an empirical fact (Bietti et al., 1968; Jackson and Quigg 1969, 1970; Harari, 1971a; Barbour et al., 1971) that the left-hand side of the

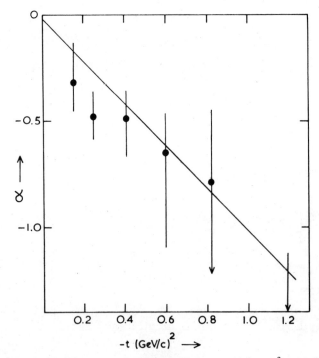

Fig. 9. α_{eff} for $d\sigma_\parallel/dt$ (Sherden et al., 1973). Solid line is $\alpha(t) = t - \mu^2$, i.e., pion trajectory with unit slope.

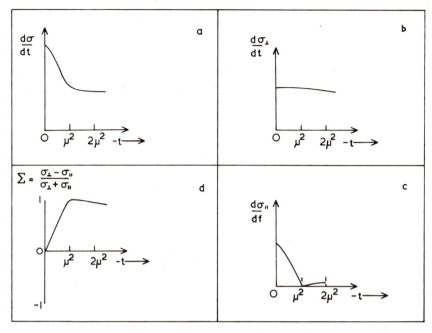

Fig. 10. Schematic representation of (a) $d\sigma/dt$, (b) $d\sigma_\perp/dt$, (c) $d\sigma_\parallel/dt$ and hence (d) Σ at small t values for the reaction $\gamma p \to \pi^+ n$ (from Harari, 1971a).

above equation is dominated by the nucleon contribution, the resonances canceling amongst themselves *near the forward direction*. Therefore, F_2 is approximately given by

$$F_2(t) \simeq \frac{ef(t+\mu^2)}{\mu\nu(t-\mu^2)} \tag{3.3}$$

and since F_3 is slowly varying over this t range and must satisfy the conspiracy relation

$$F_3(0) = 2mF_2(0) \tag{3.4}$$

then F_3 is given by*

$$F_3(t) \simeq -\frac{2mef}{\mu\nu} \tag{3.5}$$

Since

$$\frac{1}{2}\frac{d\sigma_\parallel}{dt} \simeq \frac{|F_2|^2}{32\pi} \tag{3.6}$$

*This can also be obtained from the F_3 FESR (Jackson and Quigg, 1969, 1970; Harari, 1971a).

and

$$\frac{1}{2}\frac{d\sigma_\perp}{dt} \simeq \frac{1}{128\pi m^2}|F_3|^2 \tag{3.7}$$

we obtain, using $d\sigma/dt = \frac{1}{2}(d\sigma_\parallel/dt + d\sigma_\perp/dt)$,

$$\frac{d\sigma}{dt} \simeq \frac{e^2 f^2}{32\pi\mu^2\nu^2}\frac{(1+t^2/\mu^4)}{(1-t/\mu^2)^2} \tag{3.8}$$

which gives a forward spike, in agreement with the data to within 20%, since putting in the numbers for e, f, and μ ($e^2/4\pi = 1/127$, $f^2/4\pi = 0.08$) we find

$$(s-m^2)^2 \, d\sigma(t=0)/dt \simeq 260 \, \mu b \, \text{GeV}^2 \tag{3.9}$$

The vanishing of F_2 at $t = \mu^2$ ensures the correct behavior of $d\sigma_\parallel/dt$ and, of course, Σ.

We now go on to discuss how this behavior can be produced in models. One pion exchange (OPE) needs modification, since the pion pole term, whether Reggeized or elementary, vanishes in the forward direction by factorization (evasion):

$$F_2^\pi(t) \propto \frac{tv^{-1}}{t-\mu^2} \qquad (\text{or } \pi\alpha'\xi_{\alpha_\pi}tv^{\alpha_\pi(t)-1}) \tag{3.10}$$

To produce a spike at small t one must add a smooth background to the pion pole. This background will also contribute to F_3 by Eq. (3.4). The various models introduce the background in different ways.

(*i*) *The Electric Born Term Model.* The basic motivation of this model is the observation that elementary OPE [Fig. 11(a)] is not by itself gauge invariant, and the simplest way to make it gauge invariant is to add the s-channel (or u-channel in the case of $\gamma n \to \pi^- p$) nucleon Born term [Fig.

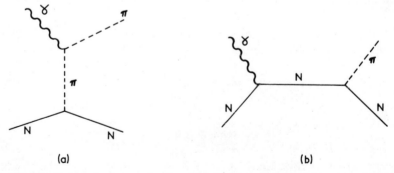

(a)   (b)

Fig. 11. Electric Born term approximation: (a) one-pion exchange and (b) s-channel nucleon Born term.

11(b)] in which the nucleon is treated as a Dirac particle with g factor 2. This gives for $\gamma p \to \pi^+ n$

$$M = \bar{u}(p)\gamma_5\left[\frac{2q_u}{t-\mu^2} + \frac{(\gamma \cdot p + m)}{s-m^2}\gamma_\mu + \frac{\gamma \cdot k}{s-m^2}\gamma_\mu\right]\varepsilon_\mu u(p') \quad (3.11)$$

In fact, the first two terms are sufficient to guarantee gauge invariance (the third term, corresponding to the magnetic moment, is gauge invariant by itself) and this, in fact, was the original proposal of Stichel and Scholz (1964). However, nowadays it is usual to include the magnetic moment term (Boyarski *et al.*, 1968*a*; Diebold, 1970; Harari, 1969, 1971*a*, 1971*b*; Kellett 1970; Barbour *et al.*, 1971; Jackson and Quigg, 1969, 1970). Of course, in a Feynman diagram approach this is automatic. It makes very little difference in the forward direction, which is the only place where either version works. For the cross section for $\gamma p \to \pi^+ n$ we obtain Eq. (3.8), in agreement with the data, and, of course, Σ is predicted correctly since we have essentially merely added a smooth background to F_2 and F_3.

Since we in fact obtained this result in an essentially model-independent manner we can see that there is nothing magical about the electric Born approximation. Its success is simply due to the fact that the amplitude is real and that the resonances do not contribute to FESR's (or FTDR's) in the forward direction. So the magic is in the cancellation, and why this occurs is an unsolved problem, except that it is interesting to note that this agrees with quark model predictions for resonances coupling to the different helicity states in $\gamma N \to \pi N$ (Barbour *et al.*, 1971).

(*ii*) *Poor Man's Absorption Model.* This model—first proposed by Williams (1970) and since advocated by Fox (1969, 1972) and Fox and Quigg (1973) as a universal prescription for *all* π-exchange reactions— specifies that each s-channel helicity amplitude has minimal t-dependence $[(-t)^{1/2}]^n$ and the correct residue at the pion pole. So for a no-flip amplitude N we replace a factorizing (i.e., evasive) elementary π contribution $N \propto t/(t-\mu^2)$ by

$$N \propto \mu^2/(t-\mu^2) \quad (3.12)$$

Since we can rewrite

$$\mu^2/(t-\mu^2)$$

as

$$t/(t-\mu^2)-1$$

we see that we are simply adding a smooth background, which interferes destructively with the pole. Note that the modified amplitude satisfies the constraints of angular momentum conservation near the forward direction, but has no evasive factors. It has the advantage over the previous model in that it can be applied to all π-exchange reactions, not just photoproduction, and that it makes no distinction between Reggeized and elementary pion exchange. It has the disadvantage of being completely obscure theoretically, corresponding to a complicated lack of factorization.

Since D is taken as

$$D \propto t/(t - \mu^2) \tag{3.13}$$

(again the minimum requirements of angular momentum conservation), using the crossing matrix [Eq. (1.7)] we obtain

$$F_2 \propto (t + \mu^2)/(t - \mu^2) \tag{3.14}$$

and since F_3 is given by the conspiracy relation [Eq. (2.19)] we obtain results similar to those of the electric Born term.

(iii) *Regge-Cut Models.* In these models the smooth background is provided by a Regge cut π_c normally produced by absorbing a Regge pole. The cut, being an object of mixed parity, does not vanish in the forward direction, but "self-conspires," i.e., satisfies Eq. (2.19) in accordance with the conservation of angular momentum. With strong cuts one can reproduce the results of PMA—Eq. (3.14) being interpreted as the term proportional to t being modified by a cut $\propto \mu^2$. There is a danger that the cuts have to be so strong that overabsorption occurs, i.e., the low partial waves become negative,* but this can be rectified if A_2 exchange and associated cuts are included (Worden, 1972, 1973).

An interesting property of all these models is that they predict that the π^-/π^+ ratio $\simeq 1$ at $t = 0$, in agreement with the data. This is because the difference between π^+ and π^- is the relative sign of $G = +1$ and $G = -1$ exchanges. Natural and unnatural parity exchanges cannot interfere because they contribute to different amplitudes, ρ and A_2 cannot interfere since they are $\pi/2$ out of phase, and so the only interference is between the ρ pole and the background contribution (π_c) to F_3. However, if the ρ pole factorizes then it evades and so we have $R \simeq 1$ at $t = 0$ (Worden, 1972; Fox and Quigg, 1973). Strong-cut models have a slight problem here, in that they produce a nonzero ρ cut contribution in the forward direction and thus predict that R is slightly greater than 1 in the forward direction and passes through 1 around $t = -2\mu^2$. Both of these effects are hardly seen in the (inconclusive) data. This effect was first pointed out by Kaidalov and Karnakov (1969) and is also discussed by Worden (1972, 1973). However, this is a fine detail, the main aspects of the data being well reproduced by strong-cut models.

---

*If the absorbing function $f(b)$, which multiplies the Regge pole amplitude, is written $f(b) = 1 - C \exp(-b^2/2a)$, where b is impact parameter, $a \simeq 8$ in GeV units, and C defines the strength of the absorption, then overabsorption corresponds to taking $C > 1$ (Worden, 1972, 1973).

Table 2. t-Channel Quantum Numbers in $\pi^{\pm}$ Photoproduction

Parity	Amplitude	$GP(-1)^I$	$G = +1$	$G = -1$
Natural	F_1	1	ρ	A_2
	F_3	1	ρ	A_2
Unnatural	F_2	-1	B	π
	F_4	1	Z_1	A_1

3.1.3. Intermediate-t Region $(2\mu^2 \leq |t| \leq 1 \ (GeV/c)^2)$

Although in the near-forward direction pion exchange is dominant, for moderate t other exchanges become important and we must try and separate their contributions. All exchanges have $I = 1$, since they must have $I \geq 1$ since $|I_3| = 1$ and $I \leq 1$ as they couple to $\bar{N}N$. The G parity exchanged can be ± 1, $G = 1$ coupling to the $I = 0$ component of the photon and $G = -1$ coupling to the isovector component. Both parities can be exchanged. Natural-parity mesons must have $GP(-1)^I = 1$, whereas unnatural-parity mesons can have $GP(-1)^I = \pm 1$, the plus sign corresponding to F_4 and the minus sign to F_2. The possible exchanges are summarized in Table 2, where the symbols ρ, A_2, etc., refer to the quantum numbers exchanged and are not necessarily related to particles. The entry Z_1 refers to the state with $J^{PC} = 2^{--}$, expected in the quark model[*] ($L = 2$, $S = 1$), but not yet observed, though Irving (1976) has claimed evidence for its exchange in $\pi^- p \to \omega^0 n$.

Since the two reactions $\gamma p \to \pi^+ n$ and $\gamma n \to \pi^- p$ are related by line reversal, the exchanges with different G parity enter with different relative signs in the two cases. Also, the polarized beam asymmetry has been measured in both reactions and so we can construct $d\sigma_{\parallel}/dt$ and $d\sigma_{\perp}/dt$ for both reactions. Theoretically, we can express the cross sections in the following way in self-explanatory notation:

$$\frac{d\sigma_{\perp}^{\pi^{\pm}}}{dt} = |\rho(1) \pm A_2(1)|^2 + |\rho(3) \pm A_2(3)|^2 \tag{3.15}$$

$$\frac{d\sigma_{\parallel}^{\pi^{\pm}}}{dt} = |\pi \pm B|^2 + |Z_1 \pm A_1|^2 \tag{3.16}$$

The experimental values of $R_{\parallel}$ and $R_{\perp}$, the π^-/π^+ ratios for photons polarized parallel and perpendicular to the scattering plane, are shown in Fig. 12 (Lübelsmeyer, 1969). It can be seen that $R_{\parallel}$ is consistent with 1 within large errors, whereas $R_{\perp}$ falls well below 1 away from $t = 0$.

---

[*]The 0^{--} recurrence is not allowed in the quark model.

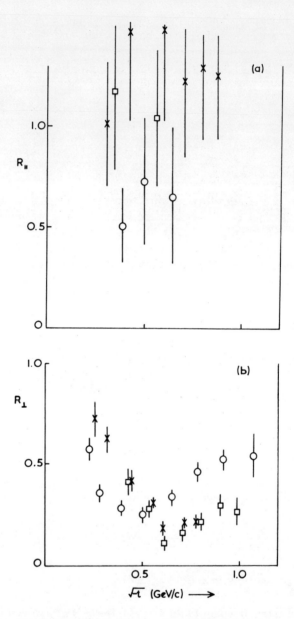

Fig. 12. The π^-/π^+ ratio for (a) photons polarized parallel to the reaction plane, $R_\parallel$ (unnatural-parity exchange), and (b) photons polarized perpendicular to the reaction plane $R_\perp$ (natural-parity exchange). Updated from Lübelsmeyer (1969). Data used: (□) 3 GeV, Bar-Yam *et al.* (1970*a*); (×) 3.4 GeV, Burfeindt *et al.* (1973); (○) 16 GeV, Sherden *et al.* (1973).

Most fits neglect A_1 and Z_1 exchange but include π and B. Since the latter two poles are expected to be exchange degenerate (EXD) and hence $\pi/2$ out of phase, we still obtain no information from $R_\parallel$. However, FESR studies, such as that of Worden (1972) prefer some B exchange, and it is usual to satisfy the requirement of duality, $SU(3)$, and vector dominance and take

$$\text{Im}(\pi) = 3 \, \text{Im}(B) \qquad (3.17)$$

Of course, substantial interference must be present in $d\sigma_\perp/dt$, which cannot be provided by EXD ρ and A_2 poles. However, we know from our studies of the near-forward direction that F_3 has a substantial pion-cut contribution π_c, which is presumably included in "A_2," and in fact the interference between ρ and π_c explains the observed $R_\perp$ well (Worden, 1972; Goldstein and Owens, 1974; Barbour and Moorhouse, 1974). The polarized target asymmetry T for $\gamma p \to \pi^+ n$ is mainly due to the term $\pi_c^*(\rho + A_2)$ and is predicted correctly in this type of approach (Worden, 1972, 1973). A large A_2 contribution, consistent with the EXD prediction

$$\text{Im} \, A_2(1) = 3 \, \text{Im} \, \rho(1) \qquad (3.18)$$

and

$$\text{Im} \, A_2(3) = 3 \, \text{Im} \, \rho(3) \qquad (3.19)$$

helps to produce a correct amplitude at $t = 0$ without the undesirable feature of overabsorption of the pion pole, which tends to occur otherwise (Kane *et al.*, 1970; Worden, 1972, 1973). The good agreement of Eqs. (3.17)–(3.19) with FESR's is also striking (Worden, 1972, 1973).

The fact that the π^-/π^+ ratio is not equal to unity at $t = -0.6$ $(\text{GeV}/c)^2$ has been quoted as evidence that the ρ pole does not have a NWSZ at $\alpha_\rho = 0$ (Kane *et al.*, 1970; Ross *et al.*, 1970). However, these arguments are model dependent in the sense that they depend on the size of the cut contribution and we have seen that in order to explain the π^-/π^+ ratio, cuts are needed.

We have seen that although pole models are completely unable to explain the forward spike and the π^-/π^+ ratio the single addition of a smooth background π_c to F_2 and hence F_3 enables a qualitative understanding of all features of the data to be attained. This background can be introduced by an absorption approach either by convoluting poles or by the simpler prescription of PMA and a quantitative understanding of the data obtained (Worden, 1972, 1973; Goldstein *et al.*, 1974).

The electric Born term model breaks down away from $t = 0$, giving a cross section much too high (see Fig. 4). This can be rectified by introducing a form factor $\approx e^{1.5t}$ into the amplitude (Richter, 1967; Boyarski *et al.*, 1968a; Kellett, 1970), but this is rather arbitrary. It is more illuminating to see why it breaks down by using FTDR's. Away from $t = 0$ the resonances do not cancel with each other (as they did at $t = 0$), but they all cancel against the Born term, reducing the real part of the (real) amplitude (Barbour *et al.*, 1971). Thus, the success of the model at $t = 0$ appears even more of a coincidence.

Efforts to combine fits to the high-energy data with FTDR's in a model-independent way have also been made (Hontebeyrie *et al.*, 1973; Barbour and Moorhouse, 1974), but the results are inconclusive. In both papers the approach is to parameterize the imaginary part of the high-energy amplitude as a Regge pole,

$$\text{Im } A_i(\nu, t) = \beta_i(t)\nu^{\alpha(t)-1} \tag{3.20}$$

and then calculate the real part using a FTDR, using multipole analysis for $\nu < \nu_c$ (a cutoff energy) and Eq. (3.20) for $\nu > \nu_c$. Hontebeyrie *et al.* claim that they can fit all the data with peripheral imaginary parts {i.e., $M_n \propto J_n[b(-t)^{1/2}]$}, whereas Barbour and Moorhouse claim that a fit cannot be achieved with a peripheral imaginary part for the no-flip amplitude.

In fact, in neither of these papers are the fits of the quality one normally expects of model-independent approaches, presumably due to the restrictive parametrization. They only take one term in Eq. (3.20) and this means that the only way to break the phase energy relation is to introduce fixed poles, particularly in F_1, F_2, and F_3 where they are at $j = -1$. This seems bound to lead to poor fits to the data. More model-independent work in this area would seem to be necessary.

As a final point we should note that the flat $\alpha_{\text{eff}} \approx 0$ has in all of the work discussed above been ascribed to the fact that the pion dominates at small t with the ρ and A_2 taking over at moderate $|t|$. This can only be true over a limited energy range, and deviations from the k_γ^{-2} scaling behavior must set in at higher energies. Detailed calculations (Worden, 1973) indicate that these deviations should be significant at $E_\gamma = 70$ GeV.

More experimental work with polarized beams leading to a better determination of $\alpha_{\text{eff}}(t)$ for $d\sigma_\parallel(\gamma n \to \pi^- p)/dt$ [or $d\sigma_\parallel(\gamma p \to \pi^+ n)/dt$] would be interesting. The effective trajectory shown in Fig. 9 is surprisingly similar to that expected from a conventional pion trajectory, especially in view of the fact that large cut corrections were found to be necessary for pion exchange.

3.2. Forward Photoproduction of Neutral Pseudoscalar Mesons

3.2.1. π^0 Photoproduction

Since π exchange is forbidden in π^0 photoproduction by C-parity conservation, it is normally discussed separately from charged pion

Table 3. t-Channel Quantum Numbers in π^0 Photoproduction

Parity	Amplitude	$GP(-1)^I$	$G = +1$ $I = 1$	$G = -1$ $I = 0$
Natural	F_1	1	ρ	ω
	F_3	1	ρ	ω
Unnatural	F_2	-1	B	H
	F_4	1	Z_1	Z_0

photoproduction and considered in conjunction to the closely related η^0 photoproduction reaction. The allowed exchanges (again the symbols refer to quantum numbers, not necessarily poles) are summarized in Table 3. The symbol Z_0 refers to a $I = 0$, $J^{PC} = 2^{--}$ state, the $I = 0$ relation of the Z_1 discussed previously.

Differential cross-section data are plotted in Fig. 13 and show a turnover near the forward direction, reflecting the absence of π exchange and a dip around $t \simeq -0.5$ $(\text{GeV}/c)^2$ followed by a secondary maximum. The data show some deviation from k_γ^{-2} scaling in that shrinkage is observed, and this is reflected in the α_{eff}, shown in Fig. 14, being different from zero.

There are some data on the difficult-to-measure reaction $\gamma n \to \pi^0 n$ and we plot the ratio of the two cross sections

$$R = \frac{d\sigma(\gamma n \to \pi^0 n)/dt}{d\sigma(\gamma p \to \pi^0 p)/dt} \qquad (3.21)$$

in Fig. 15. It shows that the interference between $I = 0$ and $I = 1$ exchanges, which change sign between the two reactions, is small. The polarized beam asymmetry Σ for $\gamma p \to \pi^0 p$ is shown in Fig. 16. It is large and shows that the reaction is dominated by natural parity exchange. All these features are in rough accord with the theoretical expectation of ω dominance, based on the fact that the natural-parity poles should dominate since they are higher lying and that ω exchange should dominate over ρ exchange by a factor $\simeq 3$ in amplitude, since the former couples to the ρ-like photon, whereas the latter couples to the ω-like photon. The turnover near $t = 0$ indicates that the scattering is dominated by the single-flip amplitudes S_1 and S_2, which, since natural parity exchange dominates, are approximately equal (in terms of t-channel amplitudes, F_1 dominates). Both of these conclusions are supported by FESR's (Worden, 1972; Barker *et al.*, 1974). This is also what we would expect if the reaction is dominated by ω exchange since studies of ω exchange in $K(\bar{K})N \to K(\bar{K})N$ indicate that it does not flip nucleon s-channel helicity (Fox and Quigg, 1973; Fox and Hey, 1973; Nagels *et al.*, 1976).

The dip is normally understood as the vanishing of the dominant amplitudes S_1 and S_2. This is what would be expected in a model with NWSZ's and weak cuts since $t = -0.5$ $(\text{GeV}/c)^2$ coincides with the wrong signature point $\alpha = 0$ and cuts should be particularly weak in a flip amplitude. It also would be expected in geometrical, peripheral, or strong-cut approaches, since a peripheral single-flip amplitude is proportional to $J_1[R(-t)^{1/2}]$, and taking $R \simeq 1$ fm this has its first zero at $t \simeq -0.5$ $(\text{GeV}/c)^2$. Thus, the existence of the dip does not distinguish between the two approaches (Harari, 1971b, 1971d).

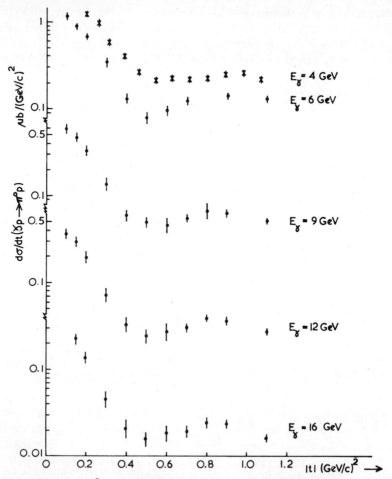

Fig. 13. $d\sigma/dt$ ($\gamma p \to \pi^0 p$) plotted against t for various energies. Data: ($\bullet$) 6, 9, 12, 15 GeV (all R. L. Anderson et al., 1970, 1971a, 1971b); ($\times$) 4 GeV, Braunschweig et al. (1973b). The data of Braunschweig et al. (1970a), Hufton (1973), and Osborne et al. (1972) have been omitted for clarity.

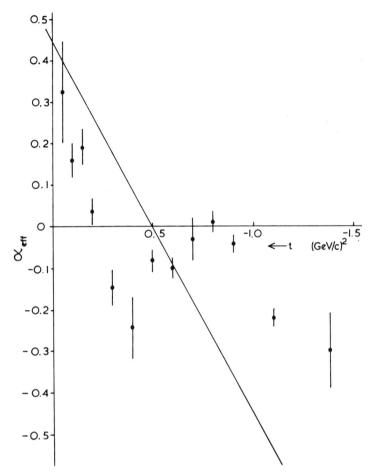

Fig. 14. $\alpha_{\text{eff}}(\gamma p \to \pi^0 p)$ (Barker and Storrow, 1977). Calculated from all of the data listed in the caption to Fig. 13. The solid line is $\alpha(t) = 0.45 + 0.9t$, the traditional ω trajectory.

We can form a qualitative picture of $\gamma p \to \pi^0 p$ as dominated by ω exchange with a small amount of unnatural parity exchange to give $\Sigma \simeq 0.8$. In addition, one needs cuts for several reasons:

(i) To fill in the dip. This cannot be done by unnatural parity exchange as this would imply that the dip deepened with increasing energy, in contradiction to the data. It would also imply that $\Sigma \simeq -1$ in the dip.

(ii) To reproduce the experimental α_{eff}, which is rather higher than an ω-trajectory with conventional slope (Fig. 14) and quite different from the α_{eff} from the kinematically similar πN charge exchange.

(iii) To reproduce the appreciable polarized target asymmetry T [Fig. 17(a)] and recoil proton polarization P [Fig. 17(b)]. In fact $P \simeq T$.

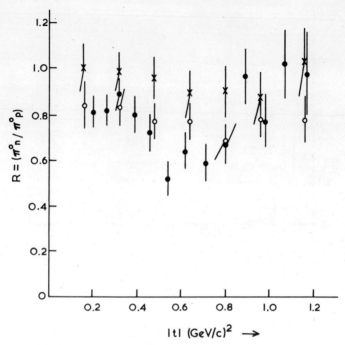

Fig. 15. Neutron–proton ratio for π^0 photoproduction, i.e., the ratio $[d\sigma(\gamma d \to \pi^0 n p_s)/dt/[d\sigma(\gamma d \to \pi^0 p n_s)/dt]$ plotted against t. Data: (●) 4 GeV, Braunschweig *et al.* (1973*a*); (○) 4.7 GeV, (×) 8.2 GeV, Osborne *et al.* (1972).

These ideas have been incorporated into quantitative models with ω, ρ, and B (and H) poles and absorptive cuts in the work of Ross *et al.* (1970), Kane *et al.* (1970), Gault *et al.* (1971), and Worden (1972). Worden's analysis included requiring a fit to FESR's and concludes that models with NWSZ's are preferred. This is rather surprising as the F_1 FESR does not vanish at $t \simeq -0.5$ $(\text{GeV}/c)^2$, in fact vanishing around $t \simeq -0.8$ $(\text{GeV}/c)^2$ (Fig. 18). This is achieved in a model with NWSZ's by having a mainly real cut. This does not shift the minimum in $d\sigma/dt$ from the point $t = -0.5$ $(\text{GeV}/c)^2$, where the pole contribution to F_1 vanishes, since the pole is mainly imaginary here. On the other hand he was completely unable to achieve a fit to the FESR's (particularly the FESR for the dominant F_1 amplitude) with a model without NWSZ's (Worden, 1972). Both of his conclusions as regards the preference for a ω pole with a NWSZ and the reality of the cuts are supported by the analyses of Barker *et al.* (1974) and Barker and Storrow (1977), who fit all the data and the FESR's in a more model-independent way. None of the above model-dependent approaches predicted the polarized target asymmetry correctly, though Collins and Fitton (1974) modified Worden's work by including Regge–Regge cuts and

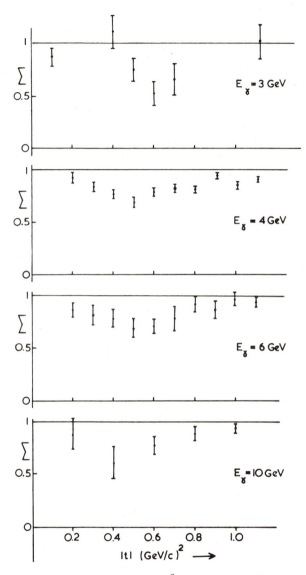

Fig. 16. Polarized photon asymmetry, Σ, for $\gamma p \to \pi^0 p$. Data: 3 GeV (Bellenger *et al.*, 1969), 4, 6, 10 GeV (R. L. Anderson *et al.*, 1971*a*, *b*).

succeeded in fitting all high-energy data, but not FESR's. However, in view of the fact that we now have data for $\gamma p \to \pi^0 p$ on $d\sigma/dt$, Σ, T, and recoil polarization P (i.e., the set S of Section 2.3) and good multipole analyses up to $E_\gamma = 1.5$ GeV, it seems worth trying a model-independent approach. This has been attempted by Barker and collaborators (Barker *et al.*, 1974; Barker

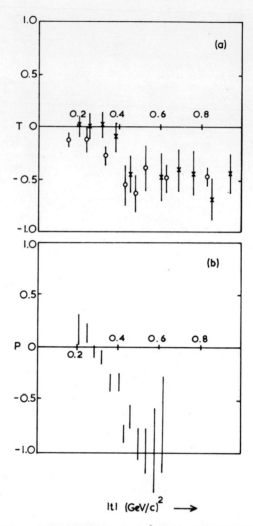

Fig. 17. (a) Polarized target asymmetry for $\gamma p \to \pi^0 p$. Data: (○) 4 GeV, Booth *et al.* (1972), (×) 4 GeV, Bienlein *et al.* (1973). (b) Recoil proton polarization, P, for $\gamma p \to \pi^0 p$. Data from Deutsch *et al.* (1972), $E_\gamma = 2.9 + 5.33 \, |t|$ GeV.

and Storrow, 1977). In view of the sparse neutron data they do not attempt an isospin decomposition and concentrate on the channel $\gamma p \to \pi^0 p$. Their method is to parameterize the imaginary part of the amplitude for $\nu > \nu_c$ (corresponding to $E_\gamma = 1.5$ GeV) as a sum of Regge terms [as in Eq. (2.38)] and calculate the real part by FTDR's. They also fit FESR's. To use both FESR's and FTDR's is rather over-elaborate for high energies, as is clear from Section 2.5—they do it in order to achieve more reliable real parts for

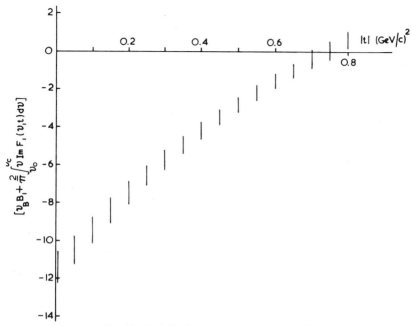

Fig. 18. F_1 FESR from Barker and Storrow (1977).

extrapolation to low energies, a feature which does not concern us here. Their earlier work is a fixed-t approach in the region where most of the polarization data exist, viz., $-0.2 \geqslant t \geqslant -0.7 \, (\text{GeV}/c)^2$, and it has two main conclusions:

(i) The amplitude F_4 (corresponding to the exchange of the hypothetical Z particles) is not negligible below $E_\gamma = 4$ GeV. Indeed, it was found that in order to fit the (large) FESR and the small measured $d\sigma_\parallel/dt$ at 4 GeV it was necessary to give the amplitude a rather steep energy dependence characteristic of an exotic exchange reaction. The approximate equality $P \simeq T$ does not put very severe restrictions on F_4 as $P - T$ measures $\text{Im}(F_2^* F_4)$ and F_2 is also small (Barker *et al.*, 1974). The large FESR is due to well-known resonances and is unlikely to change with new multipole analyses. This feature is confirmed by Barbour and Crawford (1975) for both π^0 and π^+ photoproduction. This amplitude had been set equal to zero in all previous work, largely because of theoretical prejudice—there is no known particle to contribute and also the conventional absorption model using a s-channel helicity-conserving Pomeron does not produce cuts in F_4.

(ii) The zero structure of the amplitudes is difficult to understand from any geometrical point of view. Although $\text{Im} \, S_1$ and $\text{Im} \, S_2$ have a zero as expected at $t = -0.5 \, (\text{GeV}/c)^2$, $\text{Re} \, S_1$ and $\text{Re} \, S_2$ have not, being strongly modified by cuts as mentioned earlier. However, the dual absorption model of Harari (1971c) predicts peripheral behavior only for imaginary parts. Even this model fails for Im N, which has no zero at $t = -0.2$ $(\text{GeV}/c)^2$. Worden (1972) had noticed the absence of a zero in the FESR for the closely related amplitude $-(2A_1 + tA_2)$. This does not rule out a zero in the imaginary part of the amplitude, as we noticed earlier. However, in this case no analysis with a zero in Im N at $t = -0.2$ $(\text{GeV}/c)^2$ gives a satisfactory fit to the FESR's. The unsatisfactory nature of the fits to the zeroth-moment FESR's can be seen explicitly in Fig. 4 of Argyres $et\ al.$ (1973), although they obtain better fits to the higher moments. This reflects the well-known fact (Worden, 1972; Argyres $et\ al.$, 1973) that the higher-mass resonance contributions to Im N have a zero in the appropriate place, whereas the nucleon Born term and the $\Delta(1236)$ do not. Presumably these difficulties for s-channel approaches are an indication of the fact that t-channel effects, such as evasive factors, which are neglected in these models, are important in higher-spin reactions. Of course, these features do not appear in $0^{-\frac{1}{2}+} \rightarrow 0^{-\frac{1}{2}+}$.

While on the subject of the difficulties of s-channel models in π^0 photoproduction we should mention the work of Schrempp and Schrempp (1975), who test derivative relations for several reactions, including $\gamma p \rightarrow \pi^0 p$. They test the amplitudes of Argyres $et\ al.$ (1973) and Worden (1972) in derivative relations and find that the relation linking the $n = 1$ and $n = 2$ amplitudes

$$M_2 = C_1(-t)^{1/2} \frac{\partial}{\partial(-t)^{1/2}} \left[\frac{M_1}{(-t)^{1/2}} \right]$$ (3.22)

works well. However, two comments must be made:

(i) The amplitudes of Argyres $et\ al.$ (1973) not only do not fit the FESR's, but are constrained to have peripheral imaginary parts, which automatically satisfy derivative relations.

(ii) The amplitude M_2 is not well determined. The critical relation, which they do not discuss, is that relating N and S_1, as can be seen from our discussion of the zero structure and from theoretical expectations—it is N that is peculiar and expected to be so if evasive poles are important.

A recent attempt to obtain insight into the mechanisms governing $\gamma p \rightarrow \pi^0 p$ has been made by Barker and Storrow (1977). They have extended their t-independent fit (Barker $et\ al.$, 1974) to all $|t| \leq 1.6$ $(\text{GeV}/c)^2$ and fitted Primakoff-effect data. By requiring Regge poles to be evasive and to have NWSZ's, whereas cuts are taken to be relatively structureless, they obtain an essentially unique pole-cut separation. Their "pole" residues* are in agreement with what is expected from other reactions, but the cuts they find are difficult to understand in an absorptive

---

*Note that as they only consider $\gamma p \rightarrow \pi^0 p$ their "pole" is $\rho + \omega$.

approach, as was found in their earlier work. They have also investigated derivative relations and find that the relation between N and S_1 is not satisfied.

In general, π^0 photoproduction seems to set rather difficult problems for high-energy models, and it would be most interesting to have amplitudes for the reaction, a possibility that is not beyond present techniques, as discussed in Section 2.3.

3.2.2. η^0 Photoproduction

Eta photoproduction is similar to π^0 photoproduction, insofar as π exchange is forbidden. Experimentally, it is different in that $d\sigma/dt$ has no dip at $t = -0.5$ $(\text{GeV}/c)^2$, although it has a turnover near the forward direction (Fig. 19). The cross section approximately follows k_γ^{-2} scaling law, although the rather sparse 8-GeV data seem rather low (Dewire *et al.*, 1971; Wiik, 1971). There is a definite need for better quality $d\sigma/dt$ data, particularly at higher energies, to settle this question. The main theoretical difference between η and π^0 photoproduction is that the roles of the $I = 0$ and $I = 1$ exchanges are reversed, the former coupling to the isoscalar photon and the latter to the isovector photon in $\gamma p \to \eta^0 p$. There is an

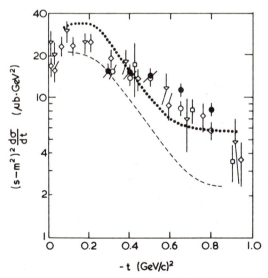

Fig. 19. $(s - m^2)^2 \, d\sigma/dt$ for $\gamma p \to \eta^0 p$ plotted against t. Data: ($\bullet$) 4 GeV, ($\bigcirc$) 8 GeV, both Dewire *et al.* (1971); ($\diamond$) 4 GeV, ($\triangledown$) 6 GeV, both Braunschweig *et al.* (1970*b*); ($\square$) Anderson *et al.* (1970). The data of Bellenger (1968) and R. L. Anderson *et al.* (1968*a*) have been omitted for clarity. The dashed line is the fit of Gault *et al.* (1971) and the dotted line the fit of Worden (1972).

additional complication due to $\eta-\eta'$ mixing. The final answer is that, for a natural-parity amplitude, i.e., F_1 or F_3, if we write

$$A(\gamma p \to \pi^0 p) = A(\omega) + A(\rho) \qquad (3.23)$$

then, using $SU(3)$,

$$A(\gamma p \to \eta^0 p) = \frac{x}{3^{1/2}} \left[\frac{1}{r} A(\omega) + rA(\rho) \right] \qquad (3.24)$$

Here $r = \gamma_\rho^{-1}/\gamma_\omega^{-1} = 3$ in the $SU(3)$ limit and $\simeq 2.8$ empirically, whereas

$$x = \cos \theta - 2^{1/2} \sin \theta \qquad (3.25)$$

where θ is the $\eta-\eta'$ mixing angle. Estimates of the angle put x in the range 1.23–1.55 (Worden, 1972; Gault et al., 1971).

There is a similar formula for the unnatural parity poles. Although in principle these formulas enable us to predict $\gamma p \to \eta p$ from $\gamma p \to \pi^0 p$, in practice it is difficult since the rather poor data for $\gamma n \to \pi^0 n$, in particular the lack of Σ data, make it hard to separate the $I = 0$ and $I = 1$ exchanges. In view of this it is better to analyze π^0 and η^0 photoproduction simultaneously in order to separate exchanges of definite isospin and parity, and this has been done by Irving and Vanryckeghem (1975). They find that the data are consistent at 4 GeV, but the lack of Σ data for $\gamma n \to \pi^0 n$ precludes a definite analysis. In the more traditional approach of predicting $\gamma p \to \eta p$ from $\gamma p(n) \to \pi^0 p(n)$ most model calculations give reasonable results for $d\sigma/dt$ (Worden, 1972; Gault et al., 1971).

Probably the most interesting feature of the η data is the absence of a dip in $d\sigma/dt$ at $t = -0.5$ $(\text{GeV}/c)^2$. This presumably reflects the fact that the $I = 1$ exchanges do not vanish here, a feature we noted earlier in connection with the π^-/π^+ ratio not being unity at this point. It is normally explained in terms of the dominant ρ pole flipping s-channel nucleon helicity (predominantly) and thus contributing mainly to the amplitudes N and D. Then the expected strong cut effects in N wash out the NWSZ in the ρ pole contribution, giving a structureless $d\sigma/dt$. This idea is the basis of most of the successful model fits to $\gamma p \to \eta p$ (Worden, 1972; Gault et al., 1971; Collins and Fitton, 1974). Recently, a small cloud has appeared on the horizon. The strong cuts feed equally into natural parity and unnatural parity exchange amplitudes (conspiracy again) and so successful models for $d\sigma(\gamma p \to \eta p)/dt$ tend to predict rather low values of Σ (< 0.5). However, preliminary data show that Σ is just as high here as in π^0 photoproduction ($\simeq 0.8$). There is obviously much work to be done here, both experimental and theoretical. Barker and Storrow (1977) have analyzed this reaction (and the reaction $\gamma n \to \pi^0 n$) in light of their work on $\gamma p \to \pi^0 p$. By making an isospin decomposition of their pole amplitudes based on knowledge from other reactions, they find that a simple and plausible isospin decomposition of their cut amplitudes can explain the above features of the η data, in contrast to conventional absorption approaches [e.g., Collins and Fitton (1974)].

3.2.3. Primakoff Effect

In Section 3.2.1 we discussed the turnover of the π^0 cross section in the forward direction. However, very close to the forward direction, the cross section rises again (Fig. 20), and this is due to one-photon exchange (Fig. 21), usually called the Primakoff effect (Primakoff, 1951; Morpourgo, 1964). The amplitude for this purely electromagnetic process can be calculated explicitly knowing the π^0 lifetime τ, and measuring π^0 photoproduction off complex nuclei gives the best measurements of the π^0 lifetime. Since the separation of the Primakoff effect from the coherent nuclear and incoherent nuclear production of π^0 is nontrivial, it is hardly surprising that different experiments do not agree (Particle Data Group, 1976). The reader is referred to Lübelsmeyer's talk at the Liverpool Conference for a thorough discussion (Lübelsmeyer, 1969). The same technique has been used to measure the η^0 lifetime. Here the data are in even more striking disagreement (Particle Data Group, 1976) and the reader is referred to the talk of Talman (1974) for a discussion.

The data on the Primakoff effect off protons (Braunschweig *et al.*, 1968, 1970a) (Fig. 20) can be used to obtain phase information on the π^0 photoproduction amplitude. One-photon exchange contributes to the ($n = 1$) amplitude F_1 and the interference between one-photon exchange and

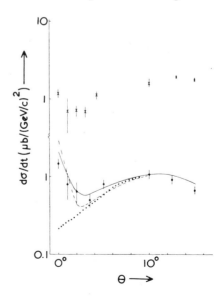

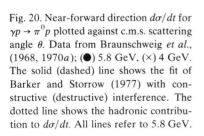

Fig. 20. Near-forward direction $d\sigma/dt$ for $\gamma p \to \pi^0 p$ plotted against c.m.s. scattering angle θ. Data from Braunschweig *et al.*, (1968, 1970a); (●) 5.8 GeV, (×) 4 GeV. The solid (dashed) line shows the fit of Barker and Storrow (1977) with constructive (destructive) interference. The dotted line shows the hadronic contribution to $d\sigma/dt$. All lines refer to 5.8 GeV.

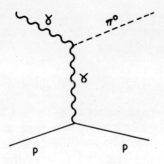

Fig. 21. One-photon exchange (Primakoff effect).

the hadronic amplitude provides us with a constraint. Very close to the forward direction the differential cross section has the following approximate form [ignoring all dependence on $(t - t_0)$ (where t_0 is the minimum momentum transfer) except that in F_1]:

$$\frac{d\sigma}{dt} = \frac{1}{128\pi m^2}\left\{ \left| (t - t_0)^{1/2}\left[F_1(0) + \frac{g}{t}\right]\right|^2 + 4m^2|F_2(0)|^2 + |F_3(0)|^2 \right\} \quad (3.26)$$

where $F_i(0)$ are the forward hadronic amplitudes and g is a coupling constant determined by the π^0 lifetime. In fact

$$|g| = 32m\pi\mu^{-3/2}\alpha^{1/2}(B_{\gamma\gamma}/t)^{1/2} \quad (3.27)$$

where $B_{\gamma\gamma}$ is the branching ratio for $\pi^0 \to \gamma\gamma$.

With data of sufficiently high resolution the different t dependence of the purely hadronic and Primakoff amplitudes should be seen in the cross section. This would enable us to obtain the π^0 lifetime from g, the cut contributions to F_2 and F_3 from $d\sigma/dt|_{t=0}$, and the phase of $F_1(0)$ from the interference with one-photon exchange.

However, in practice, only a vestigial forward spike can be detected in the data, as can be seen from Fig. 20. This is because the resolution is insufficient to separate the forward peak completely and because of the difficulty of measuring near enough the beam direction to see the maximum. However, it can be deduced that the interference is constructive (Braunschweig et al., 1970a)—otherwise the dip is too deep (See Fig. 20.)—and useful information can be obtained on the hadronic amplitudes, particularly as regards pole-cut interference (Barker and Storrow, 1977). The existing data are not very useful for obtaining the π^0 lifetime—the value obtained depends critically on the hadronic model used (Braunschweig et al., 1968, 1970a; Barker and Storrow, 1977). The breakdown of a typical fit into Primakoff and hadronic contributions is shown in Fig. 20.

Table 4. t-Channel Quantum Numbers in K^+ Photoproduction

Parity	Amplitude	Exchange
Natural	F_1	K^*, K^{**}
	F_3	K^*, K^{**}
Unnatural	F_2	$K, Q_B(1^{+-})$
	F_4	$Q_A(1^{++}), Q_Z(2^{--})$

3.3. Forward K^+ Photoproduction

In this section we will discuss the two reactions $\gamma p \to K^+ \Lambda^0$ and $\gamma p \to K^+ \Sigma^0$. They are normally discussed together since they have the same exchanges, shown in Table 4, where the $Q_A(1^{++})$, $Q_B(1^{+-})$, and $Q_Z(2^{--})$ are in the same $SU(3)$ multiplets as the A_1, B, and Z mesons, respectively. In fact, all analyses up to this time have neglected these three mesons, retaining only K^*, K^{**}, and K. The natural-parity mesons are expected to dominate, and in fact do, as can be seen from several features of the data.

(a) The polarized beam asymmetry Σ for the combined reaction $\gamma p \to K^+(\Lambda + \Sigma)$ is large and positive (Fig. 22) (Quinn et al., 1975).

(b) The Σ/Λ ratio is between 0.5 and 1, whereas for pure K exchange the ratio of cross sections would be $1/27$, since, for any exchange

$$\frac{A(\Sigma)}{A(\Lambda)} = 3^{1/2}\frac{(f-d)}{3f+d} \tag{3.28}$$

and for K exchange $f/d \simeq \frac{2}{3}$ (Nagels et al., 1976).

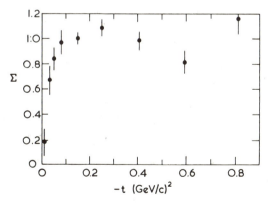

Fig. 22. Polarized photon asymmetry for the combined reaction $\gamma p \to K^+(\Lambda^0 + \Sigma^0)$ plotted against t for $E_\gamma = 16$ GeV. Data from Quinn et al. (1975).

(c)   The effective trajectories α_{eff} for both reactions (Fig. 23) are rather higher than one would obtain from K exchange with canonical slope. The effective trajectories show the nice feature that for Σ photoproduction, in which K exchange should be very small according to Eq. (3.28), α_{eff} is significantly above zero for small $|t|$ (Boyarski et al., 1969b).

Typical cross-section data are shown in Fig. 24. It can be seen that both reactions are structureless, except for a turnover near $t = 0$, which is more pronounced in the Σ reaction. Michael and Odorico (1971) used this feature to determine the f/d ratios for $K^{*}-K^{**}$ exchange for both flip and nonflip amplitudes. They separated the flip (F) and no-flip (NF) contributions by parameterizing the cross sections in the form

$$k_\gamma^2 \frac{d\sigma}{dt} = |NF|^2 - (t - t_0)|F|^2 \qquad (3.29)$$

and find Σ/Λ ratios of 0.69 ± 0.07 for NF and 1.02 ± 0.1 for F. Then a straight substitution into Eq. (3.28) gives $(f/d)_{++} = -2 \pm 0.4$ and $(f/d)_{+-} = 0.27 \pm 0.03$, in excellent agreement with values found by other methods (Nagels et al., 1976). Two comments must be made about this analysis. Firstly, they use effective (i.e., nonevasive) poles, presumably including cut effects, and this is crucial to their flip/no-flip decomposition since pure pole contributions vanish at $t = t_0$. Secondly, they ignore K exchange. This could be important since the K-cut contribution also produces a suppression of the Σ/Λ ratio near the forward direction, being more important

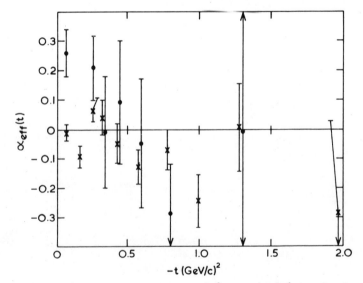

Fig. 23.   Effective trajectories, $\alpha_{eff}(t)$ for $\gamma p \rightarrow K^{+}\Lambda^{0}$ and $\gamma p \rightarrow K^{+}\Sigma^{0}$ plotted against t. From Fox and Quigg (1973). ($\bullet$) $\gamma p \rightarrow K^{+}\Sigma^{0}$, ($\times$) $\gamma p \rightarrow K^{+}\Lambda^{0}$.

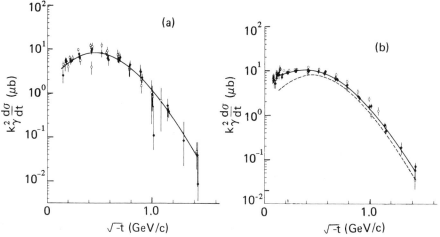

Fig. 24. $k_\gamma^2 \, d\sigma/dt$ in μb plotted against $(-t)^{1/2}$. Data: (○) 5 GeV, (◆) 8 GeV, (●) 11 GeV, (□) 16 GeV, all from Boyarski *et al.*, (1969*b*), for (a) $\gamma p \to K^+ \Sigma^0$ and (b) $\gamma p \to K^+ \Lambda^0$. The solid curves represent the fit of Michael and Odorico (1971) described in the text. The dashed line represents the Σ^0 fit plotted alongside the Λ^0 fit to facilitate visual comparison.

in the Λ reaction. However, Levy *et al.* (1973*a*) and Alonso *et al.* (1972) have analyzed both reactions in models including K^*, K^{**}, and K poles and absorption and are able to explain all features of the data with the usual f/d ratios, even predicting the correct (negative) sign of recoil polarization (Vogel *et al.*, 1972).

The other point of interest in $d\sigma/dt$ is the fact that it is structureless, showing no NWSZ at the point where $\alpha_{K^*} = 0$. This was originally explained by Capella and Tran Thanh Van (1970), who pointed out that one could apply duality arguments to the reactions $K^- p \to \rho^0 \Lambda(\Sigma)$, $K^- p \to \omega \Lambda(\Sigma)$, and $\phi p \to K^+ \Lambda(\Sigma)$ and then vector dominance arguments to obtain exchange degeneracy restrictions on the K^* and K^{**} residues for $\gamma p \to K^+ \Lambda(\Sigma)$. They took the residues as equal (strong exchange degeneracy) and thus the K^* NWSZ is filled in by the K^{**} contribution, and this assumption has also been used in later analyses (Levy *et al.*, 1973*a*; Alonso *et al.*, 1972; Pickering, 1973). This assumption has been criticized by Goldstein *et al.* (1973) and by Goldstein (1974), who point out that if vector dominance is applied correctly, including the ϕ component of the photon, then the exchange degeneracy relation is

$$\text{Im}(K^*) = \tfrac{1}{3}\text{Im}(K^{**}) \qquad (3.30)$$

This observation, though correct, has very little phenomenological significance, since K^* and K^{**} exchanges are assumed to have the same f/d ratios, and any implications on the presence or absence of NWSZ's is complicated by the fact that cuts are known to be important simply from

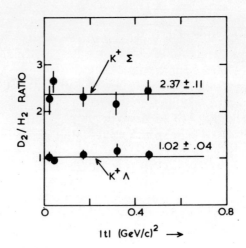

$|t|\ (GeV/c)^2 \longrightarrow$

Fig. 25. The ratio of $K^+\Sigma$ production and $K^+\Lambda$ production off deuterium to production off hydrogen at 5 GeV plotted against t. Data from Boyarski *et al.* (1971).

the nonzero forward cross section. These uncertainties notwithstanding, we appear to have a fairly good qualitative picture of K^+ photoproduction.

As regards interesting further measurements on these reactions, polarized target asymmetry data (T) would be useful for two reasons. Firstly, $T(\Lambda)$ could be compared to $P(\Lambda)$ to see if the neglect of the K_A, K_B, and K_Z trajectories is justified. Also, $T(\Sigma)$ and $T(\Lambda)$ should have opposite signs, as can be seen from Eq. (3.28) with the experimental f/d ratios substituted (Levy *et al.*, 1973*a*). Such a change of sign is observed in the hadronic processes dominated by K^*–K^{**} exchange, $\pi N \to KY$ and $\bar{K}N \to \pi Y$ (Irving *et al.*, 1971).

Cross sections off deuterium have also been measured (Boyarski *et al.*, 1971). These are interesting from the point of view of possible $I = \frac{3}{2}$ exotic exchanges (Harari, 1969). The t-channel isospin decomposition of $\gamma N \to K\Sigma$ is

$$T(\gamma p \to K^+\Sigma^0) = (T_{1/2} + T_{3/2})/2^{1/2}$$
$$T(\gamma n \to K^+\Sigma^-) = T_{1/2} - \tfrac{1}{2}T_{3/2} \tag{3.31}$$

and so in the absence of $I = \frac{3}{2}$ exchanges the $K^+\Sigma$ cross sections off protons and neutrons should obey

$$\frac{d\sigma(\gamma n \to K^+\Sigma^-)/dt}{d\sigma(\gamma p \to K^+\Sigma^0)/dt} = 2 \tag{3.32}$$

and so that $K\Sigma$ photoproduction cross-section ratios of deuterium and hydrogen should be in the ratio $3:1$, ignoring corrections due to deuteron effects. Experimentally, the ratio comes out to be 2.37 ± 0.11 (Fig. 25)

(Boyarski *et al.*, 1971), and as a check on whether deuteron corrections are important the ratio for Λ was measured to be 1.02 ± 0.04 (Fig. 25), whereas isospin conservation gives 1, the Λ having isospin 0. This technique of observing the effects of exotic exchanges by measuring the interference with nonexotic exchanges is obviously much more sensitive than trying to measure them directly in purely exotic exchange reactions. However, it would be nice to have better data on the D_2/H_2 ratio, especially with a tagged beam which would facilitate the Σ/Λ separation.

3.4. Forward $\pi\Delta$ Photoproduction

Data exist for various charge states of the reaction $\gamma N \to \pi\Delta$, and we plot $(s - m^2)^2 \, d\sigma/dt$ against $(-t)^{1/2}$ in Fig. 26 for the charge combination $\gamma p \to \pi^- \Delta^{++}$ for which there are most data (Boyarski *et al.*, 1969a). As can be seen, the momentum transfer dependence is similar to that of $\gamma p \to \pi^+ n$, with a sharp forward peak $(\approx e^{12t})$ in the region $\mu^2 \leqslant |t| \leqslant 0.2 \, (\text{GeV}/c)^2$ and less steep t dependence for larger $|t|$ $[\approx e^{2.5t}$ out to $t = -0.6 \, (\text{GeV}/c)^2$ and $\approx e^{3t}$ for $|t| \geqslant 0.6 \, (\text{GeV}/c)^2]$. An important difference from the $\gamma p \to \pi^+ n$ data is that the sharp rise in $d\sigma/dt$ does not continue to 0^0, but the cross section turns over around $t = -\mu^2$ and then falls sharply between

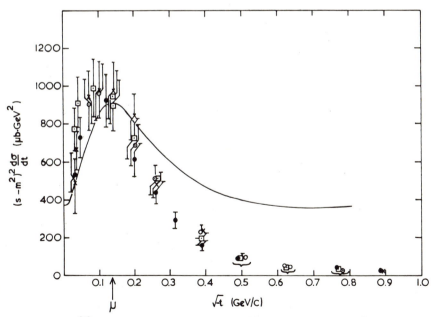

Fig. 26. $(s - m^2)^2 \, d\sigma/dt$ for the reaction $\gamma p \to \pi^- \Delta^{++}$ plotted against $(-t)^{1/2}$. Data: (O) 5 GeV, ($\times$) 8 GeV, ($\square$) 11 GeV, ($\bullet$) 16 GeV, all from Boyarski *et al.* (1969a). The solid line is the electric Born approximation.

$-\mu^2$ and t_0. This structure for small $|t|$ is, of course, interpreted as due to pion exchange. The fact that $d\sigma/dt$ exhibits a dip and not a spike is kinematic, depending on the nonequality of the nucleon and Δ masses. This leads to the pion coupling to helicity flip amplitudes, and in models the dip arises directly as a result of the inequality $(m_\Delta - m) > \mu$ (Broadhurst et al., 1971; Barbour and Malone, 1974). As can be seen from Fig. 26, the energy dependence is given by $(s - m^2)^{-2}$, the α_{eff} being consistent with zero (Fig. 27) (Boyarski et al., 1969a).

As in $\gamma N \to \pi N$, if we interpret the structure for $|t| \lesssim \mu^2$ as due to pion exchange and therefore occurring in $d\sigma_\parallel/dt$, $d\sigma_\perp/dt$ being smooth, we can predict Σ, and, in fact, the prediction (Fig. 28) agrees with the data (Sherden et al., 1974). Another similarity with $\gamma N \to \pi N$ is the fact that gauge-invariant one-pion exchange (Stichel and Scholz, 1964) reproduces the data for small t, although it predicts too high a cross section at larger t (Fig. 26). Again, there is arbitrariness in making one-pion exchange gauge invariant. The complete set of Born diagrams is shown in Fig. 29, where (b) is the contact term. Empirically, it is found that the set of diagrams that produces the best representation of the data is the electric Born approximation [diagrams (a), (b), (c), and (d)], those in which the photon interacts only with the charge of a particle. This set excludes (e) and (f) and all anomalous moment interactions. It is the minimal set of diagrams that includes (a) and is gauge invariant. It has been much studied in the literature (Campbell et al., 1970; Broadhurst et al., 1971; Barbour et al., 1971; Barbour and Malone, 1974). Probably the most illuminating discussion is the dispersion relation analysis of Barbour and Malone (1974). As in the $\gamma N \to \pi N$ case, the amplitude is assumed real at high energies

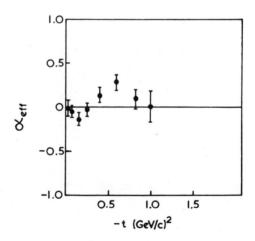

Fig. 27. Effective trajectory $\alpha_{\text{eff}}(t)$ for $\gamma p \to \pi^- \Delta^{++}$ (from Fox, 1972).

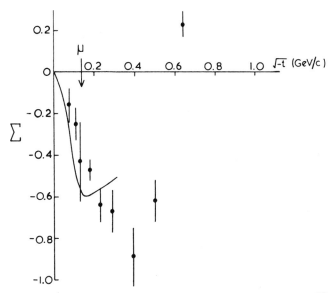

Fig. 28. Polarized photon asymmetry Σ for $\gamma p \to \pi^- \Delta^{++}$ plotted against $(-t)^{1/2}$. Data from Sherden *et al.*, (1974). Curve is prediction arising from assuming that the structure in $d\sigma/dt$ comes from $d\sigma_\parallel/dt$, $d\sigma_\perp/dt$ being smooth (Harari, 1969, 1971*a*).

and so is given by the Born terms plus the resonance contributions to the dispersion integral. The former are given by the Feynman rules applied to Fig. 29 and the latter were calculated from a naive quark model, multipole analysis not being available for $\gamma N \to \pi\Delta$. They find similar results to the $\gamma N \to \pi N$ case.

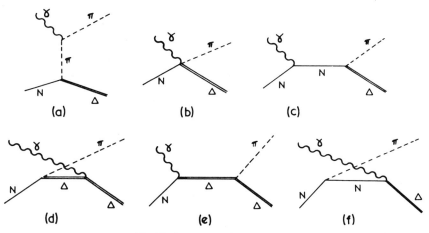

Fig. 29. Born terms for $\gamma p \to \pi\Delta$.

(i) For $|t| \lesssim \mu^2$ the higher resonances cancel amongst themselves and the contributions from diagrams (e) and (f) are small and so the amplitude is given by the electric Born terms to a good approximation. Note that in the $SU(6)$ limit with $m_\Delta = m$ the full Born approximation [i.e., diagrams (a)–(f)] is identically equal to the electric Born approximation.

(ii) Away from the forward direction $(|t| > \mu^2)$ there is, as in the $\gamma N \to \pi N$ case, a strong cancellation between the Born terms, which by themselves are much too big, and the resonance contributions, which all tend to decrease the cross section. Thus, we can understand the successes and failures of the Stichel–Scholz prescription in terms of resonance couplings. However, understanding *how* it works in the forward direction does not enable us to see *why* it works (Moorhouse, 1976).

In a Regge approach one must include cuts in order to produce a non-zero forward cross section. This has been done successfully in an absorptive approach (Goldstein and Owens, 1974) and can be achieved with PMA, which, as in the πN case, gives the same results as the electric Born model (Fox, 1972).

Three other charge combinations, $\gamma p \to \pi^+ \Delta^0$, $\gamma n \to \pi^+ \Delta^-$, and $\gamma n \to \pi^- \Delta^+$, have been measured at 16 GeV and these enable us to do a partial isospin decomposition. In terms of amplitudes of definite t-channel isospin and G parity the six charge states are given by

$$A(\gamma p \to \pi^+ \Delta^0) = -T_1^+ - T_1^- + 3^{1/2} T_2^-$$

$$A(\gamma p \to \pi^0 \Delta^+) = -2^{1/2} T_1^+ + 2(\tfrac{2}{3})^{1/2} T_2^-$$

$$A(\gamma p \to \pi^- \Delta^{++}) = -3^{1/2} T_1^+ + 3^{1/2} T_1^- + T_2^-$$

$$A(\gamma n \to \pi^+ \Delta^-) = 3^{1/2} T_1^+ + 3^{1/2} T_1^- + T_2^-$$

$$A(\gamma n \to \pi^0 \Delta^0) = 2^{1/2} T_1^+ + 2(\tfrac{2}{3})^{1/2} T_2^-$$

$$A(\gamma n \to \pi^- \Delta^+) = T_1^+ - T_1^- + 3^{1/2} T_2^-$$

$$(3.33)$$

where the notation is T_I^G. Near the forward direction the data are compatible with the equalities

$$d\sigma(\gamma p \to \pi^- \Delta^{++})/dt = d\sigma(\gamma n \to \pi^+ \Delta^-)/dt \qquad (3.34)$$

$$d\sigma(\gamma p \to \pi^+ \Delta^0)/dt = d\sigma(\gamma n \to \pi^- \Delta^+)/dt \qquad (3.35)$$

are predicted by one-pion exchange models (T_1^- dominance).

At larger $|t|$ the situation is more complicated and the charge ratios have been studied in Regge models (Goldstein and Owens, 1974) and in the quark model (Barbour *et al.*, 1971). Probably the most interesting point is whether the data demand exotic $I = 2$ exchanges. In the absence of $I = 2$

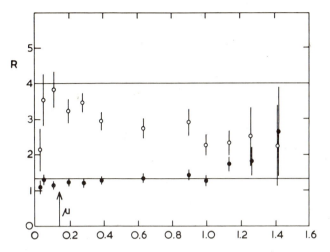

Fig. 30. The deuterium-to-hydrogen ratio for photoproduction of $\pi^{\pm}\Delta$. Data: ($\bigcirc$) π^{+}, ($\bullet$) π^{-} from Boyarski *et al.* (1970).

exchanges then the $\pi^{\pm}$ cross sections off neutrons and protons should satisfy

$$\frac{d\sigma(\gamma n \to \pi^{-}\Delta^{+})/dt}{d\sigma(\gamma p \to \pi^{-}\Delta^{++})/dt} = \frac{1}{3} \qquad (3.36)$$

and

$$\frac{d\sigma(\gamma n \to \pi^{+}\Delta^{-})/dt}{d\sigma(\gamma p \to \pi^{+}\Delta^{0})/dt} = 3 \qquad (3.37)$$

Thus, the cross sections off deuterium and hydrogen should be in·the ratio 4 for π^{+} and $\frac{4}{3}$ for π^{-}, neglecting absorption corrections in deuterium. The data (Boyarski *et al.*, 1970) are shown in Fig. 30, and for π^{+} the ratio is near 3 for $|t| \geqslant 0.15$ $(\text{GeV}/c)^{2}$, indicating an $I = 2$ component at least 15% of the $I = 1$ term.

3.5. Other Forward Reactions

In this section we discuss various forward photoproduction reactions that have been less well studied than those mentioned up to now, giving a brief discussion of the data, followed, if necessary, by comments on possible theoretical implications.

Various pion exchange reactions have been studied, largely with a view to an indirect determination of $\pi\gamma$ decay widths. A good example is $\gamma p \to A_{2}^{+}n$, which has been studied in the SLAC bubble chamber at photon energies E_γ = 4.3, 5.25, and 7.5 GeV (Eisenberg *et al.*, 1969, 1972; Ballam *et al.*, 1969). Assuming that the reaction is dominated by OPE, they deduce a value of the $A_2^+ \to \gamma\pi^+$ width

$$\Gamma(A_2 \to \gamma\pi^+) \simeq 0.5 \text{ MeV} \qquad (3.38)$$

in rough agreement with the value deduced from the measured $A_2 \to \rho\pi$ width using vector dominance (Eisenberg *et al.*, 1969, 1972).

However, the dangers of this assumption of OPE dominance are well illustrated by our next example, $\gamma p \to \rho^-\Delta^{++}$. Early studies of this reaction (Cambridge, 1968; Aachen *et al.*, 1969; Eisenberg *et al.*, 1970, 1972) gave, assuming OPE dominance, values for $\Gamma(\rho^+ \to \pi^+\gamma)$ much larger than that expected from $SU(3)$ using the measured value for $\Gamma(\omega \to \pi\gamma)$ and

$$\Gamma(\rho \to \pi\gamma) \simeq \tfrac{1}{9}\Gamma(\omega \to \pi\gamma) \qquad (3.39)$$

Later studies showed that the energy dependence of $d\sigma/dt$ was not as expected from OPE dominance—fits using the form E_γ^{-a} give $a = 0.6 \pm 0.2$ [compared to the expected $a = 2$ (Ballam *et al.*, 1971)]. Also a recent measurement of the ρ density-matrix elements is not compatible with OPE (Abramson *et al.*, 1976). Current thinking is to believe the $SU(3)$ calculation of the OPE contribution* (i.e., about 16% of the cross section) and use this reaction and the similar reaction $\gamma p \to \rho^+ n$ to study ρ and A_2 exchanges.

This latter reaction and the corresponding line-reversed reaction $\gamma n \to \rho^- p$ provide similar problems as regards estimating $\Gamma(\rho \to \pi\gamma)$ from the cross section assuming OPE dominance. The cross sections are much larger than those expected from $SU(3)$ (Benz *et al.*, 1974). Here we have additional information in that we can obtain an upper bound on $\Gamma(\rho \to \pi\gamma)$ from polarized beam data on $\gamma p \to \rho^0 p$ (Ballam *et al.*, 1973) and the bound is around twice the $SU(3)$ value (Harari, 1971*b*).

Model calculations show that the OPE exchange contribution to $\gamma p \to \rho^+ n$ is so small ($\simeq 11\%$) that the cross section is rather insensitive to $\Gamma(\rho \to \pi\gamma)$ (Clark and Donnachie, 1976). Information on ρ–A_2 exchange degeneracy breaking has been obtained by comparing the line-reversed reactions $\gamma d \to \rho^\pm X$ (Abramson *et al.*, 1976). The ρ and A_2 contributions enter with opposite sign in the two reactions, and so in the limit of exact degeneracy when the ρ and A_2 are $\pi/2$ out of phase the cross sections for the two reactions are equal.

Thus any deviation from this can be ascribed to ρ–A_2 interference and thus a breaking of exchange degeneracy. The data indicate nonequality,

---

*This is more likely to be an underestimate than an overestimate. An estimate of $\Gamma(\rho \to \pi\gamma)$ using the Primakoff effect in $\pi^- A \to \rho^- A$ for a variety of nuclear targets A gives a value about a factor of 3 *smaller* than the $SU(3)$ value (Gobbi *et al.*, 1974).

and with the assumption that OPE can be neglected a value for the trajectory splitting

$$\alpha_{A_2} - \alpha_\rho = -0.11 \pm 0.03 \qquad (3.40)$$

is found, in excellent agreement with that determined at FNAL from measurements of the reactions $\pi^- p \to \pi^0 n$ and $\pi^- p \to \eta^0 n$ (Barnes *et al.*, 1976; Dahl *et al.*, 1976).

Another problem for OPE occurs in the reaction $\gamma n \to \omega^0 \Delta^0$ for which an upper limit of more than twice the calculated OPE contribution was set (Eisenberg *et al.*, 1970). However, a recent measurement of $\gamma p \to \omega \Delta$ is claimed to be compatible with OPE predictions (Abramson *et al.*, 1976). More data are clearly needed here, and, in all of the above reactions, polarized beam data would be helpful.

Clark and Donnachie (1977) have investigated the reactions $\gamma p \to \rho^+ n$, $\gamma p \to \rho^- \Delta^{++}$ and $\gamma N \to \omega \Delta$ using a model with OPE (with PMA) and ρ and A_2 exchanges. They find reasonable agreement with the data, apart from a problem with A_2 exchange.

Other reactions for which only integrated cross sections are available are $\gamma p \to f^0 p$ and $\gamma p \to f' p$ (Aachen *et al.*, 1968), $\gamma p \to A_2^- \Delta^{++}$ (Schacht *et al.*, 1974) and other $\rho \Delta$ charge states, $\rho^+ \Delta^0$ and $\rho^0 \Delta^+$ (Aachen *et al.*, 1969; Eisenberg *et al.*, 1970, 1972).

Turning to strange particle interactions, cross sections for the reactions $\gamma p \to K^+ [\Sigma^0(1385) + \Lambda(1405)]$, $\gamma p \to K^+ \Lambda(1520)$, and $\gamma n \to K^+ \Sigma^-(1385)$ have been measured at 11 GeV (Boyarski *et al.*, 1971). The missing mass technique used does not enable the $\Sigma(1385)$ and $\Lambda(1405)$ signals off protons to be resolved. The most interesting aspect of these data is the gross violations of the $SU(3)$ relations between the $K^+\Sigma(1385)$ and the $\pi\Delta$ cross sections. These relations can be obtained on the assumption that the photon is a U-spin scalar (Levinson *et al.*, 1963) and are

$$d\sigma[\gamma p \to K^+\Sigma^0(1385)]/dt = \tfrac{1}{2} d\sigma(\gamma p \to \pi^+\Delta^0)/dt$$

and

$$d\sigma[\gamma n \to K^+\Sigma^-(1385)]/dt = \tfrac{1}{3} d\sigma(\gamma n \to \pi^+\Delta^-)/dt$$

The scaled cross sections* are compared in Fig. 31 and the $K\Sigma(1385)$ can be seen to be much too small. The violations in the proton reactions could well be much worse since the $K^+\Lambda(1405)$ contribution must increase the apparent $K\Sigma$ cross section. It is difficult to see how this failure of $SU(3)$ can be blamed entirely on the π–K mass difference since it persists to all values of t at which data exist.

*This is necessary as the $\pi\Delta$ data were taken at 16 GeV.

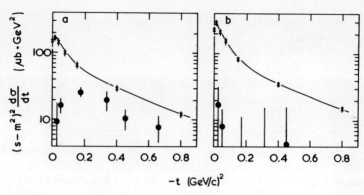

Fig. 31. $SU(3)$ comparison of the scaled cross sections $(s - m^2)^2 \, d\sigma/dt$ in μb $(\text{GeV})^2$ for $\pi\Delta(1236)$ and $K\Sigma(1385)$ photoproduction off (a) protons and (b) neutrons, plotted against t. Off protons the $\Sigma(1385)$ and $\Lambda(1405)$ cannot be separated and so the sum of cross sections is plotted. Data: (a) ($\times$) $\frac{1}{2}(\gamma p \to \pi^+ \Delta^0)$, ($\bullet$) $\gamma p \to K^+ \Sigma^0(1385) + K^+ \Lambda^0(1405)$; (b) ($\times$) $\frac{1}{3}(\gamma n \to \pi^+ \Delta^-)$, ($\bullet$) $\gamma n \to K^+ \Sigma^-(1385)$. Data are from Boyarski *et al.* (1970, 1971). In all cases $\pi\Delta$ data are at 16 GeV and $K\Sigma$ data at 5 GeV. The solid line is to guide the eye through the $\pi\Delta$ data.

3.6. Backward Photoproduction

3.6.1. Backward $\gamma N \to \pi N$

Cross-section data exist for the two reactions $\gamma p \to \pi^+ n$ and $\gamma p \to \pi^0 p$ over a reasonable energy range and are plotted in Fig. 32. The main features of the data are as follows:

(a) The cross sections scale as k_γ^{-3}. In Regge language $\alpha_{\text{eff}} \simeq -\frac{1}{2}$ for all $|u|$ (Fig. 33) with no shrinkage.

(b) The cross sections have no structure in u except for a presumably kinematic dip near the extreme backward direction.

(c) The backward peak can be represented reasonably well by an e^u behavior, a much shallower falloff than normal.

Berger and Fox (1971) made great play of this latter, apparently unique, feature of backward photoproduction. Since their work new data on backward production of ρ^0, ω^0, and f^0 mesons by pions has appeared (Dado *et al.*, 1974; Emms *et al.*, 1974), and this shows a similar slow falloff in u. It is still a complete mystery why some reactions should have such a slow falloff, and in model fits these features must be produced in a purely ad hoc manner (Berger and Fox, 1971).

The lack of shrinkage is interesting, especially in view of the fact that two of the mechanisms for avoiding parity doublets (see Section 2.4.2) give little shrinkage. In the kinematic cut model of Carlitz and Kislinger (1970), in addition to the pole, there is a fixed cut at $j = \alpha(0)$ and its contribution

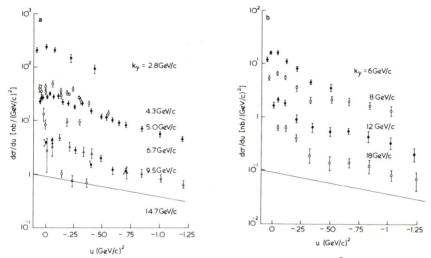

Fig. 32. Backward cross sections $d\sigma/du$ for (a) $\gamma p \to n\pi^+$ and (b) $\gamma p \to p\pi^0$ plotted against u. The solid line in both cases is e^u. Data: (a) R. L. Anderson *et al.* (1968*b*, 1969*a*), (b) Tomkins *et al.* (1969).

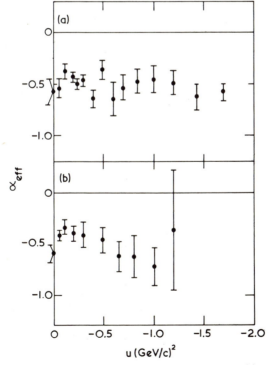

Fig. 33. Effective trajectories, $\alpha_{\text{eff}}(u)$ for the backward reactions (a) $\gamma p \to n\pi^+$ and (b) $\gamma p \to p\pi^0$ plotted against u. Adapted from Berger and Fox (1971).

will not shrink. Such a model has been used to fit the backward photo-production data (Pond and Kumar, 1972; Berger and Fox, 1971). However, as stressed by the latter authors, the model is not adequately tested by the model, although it reproduces a structureless cross section naturally. The main criticism of the model is that it is much less successful in backward $\pi N \to \pi N$ (Berger and Fox, 1970a), and one would hope that one would solve the parity doublet problem by the same mechanism in all reactions.

The same remarks apply to asymmetric trajectories. Since the terms odd in $u^{1/2}$ do not contribute to the energy dependence (Barger and Cline, 1969; Storrow and Winbow, 1973b) they can be made to give little or no shrinkage. However, as an explanation of the parity doublet problem in hadronic reactions they are not very attractive, largely because the natural explanation of the dip at the (NWSZ) point $u = -0.2 \, (\text{GeV}/c)^2$ in $\pi^+ p$ scattering is lost (Storrow and Winbow, 1973a; Storrow, 1975).

Phenomenological studies of backward photoproduction have largely been confined to traditional Regge-pole models with linear trajectories and the parity doublets avoided by the residue vanishing at the appropriate point (Berger and Fox, 1971; Barger and Weiler, 1969, 1970; Bajpai and Donnachie, 1970). In this approach the problems of the structureless cross section and lack of shrinkage must be tackled head-on. The NWSZ dip of the N_α is filled in by a large N_γ contribution. In backward $\pi N \to \pi N$ the N_γ, although present, is fairly small (Berger and Fox, 1970a; Storrow and Winbow, 1973a), but it might be expected to be more important in photoproduction since, as a resonance, it is much more strongly excited in $\gamma N \to \pi N$ than in $\pi N \to \pi N$. Harari (1969) has given an argument based on the application of duality to the exotic reactions $\gamma d \to np$ and $\pi^+ d \to pp$ that the N_α and N_γ should be equally important in backward $\pi N \to \pi N$ and $\gamma N \to \pi N$, but it is not clear how seriously this should be taken. The Δ_δ is not an appropriate choice for filling in the dip for the following reason. The u-channel isospin decomposition for the $\gamma N \to \pi N$ reactions is

$$T(\gamma p \to \pi^0 p) = (N_V - 3^{1/2} N_S + 2\Delta)/3$$

$$T(\gamma p \to \pi^+ n) = (-2^{1/2} N_V - 6^{1/2} N_S + 2^{1/2} \Delta)/3$$

$$T(\gamma n \to \pi^- p) = (2^{1/2} N_V - 6^{1/2} N_S - 2^{1/2} \Delta)/3 \qquad (3.41)$$

where N and Δ denote $I = \tfrac{1}{2}$ and $\tfrac{3}{2}$, respectively, and V and S denote isovector and isoscalar components of the photon, respectively. Thus at any point where the Δ dominates we must have

$$d\sigma(\gamma p \to \pi^0 p)/du \simeq 2 \, d\sigma(\gamma p \to \pi^+ n)/du \qquad (3.42)$$

whereas the experimental cross sections are approximately equal at small

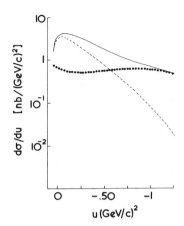

Fig. 34. Isospin breakdown of the backward reaction $\gamma p \to n\pi^+$ at 9 GeV/c. Solid line $d\sigma/du$, dashed line $|N_\alpha + N_\gamma|^2$ and dotted line $|\Delta_\delta|^2$. From Barger and Weiler (1970).

u. In the three analyses mentioned earlier this equality of the cross sections is achieved by having the Δ_δ negligible at small u and taking

$$N_S/N_V = -0.1 \qquad (3.43)$$

The lack of shrinkage is produced by the higher-lying Δ trajectory becoming important at larger $|u|$. A typical isospin breakdown is shown in Fig. 34. Various criticisms can be made of the above interpretation of the data, which is clearly not unique. (The isospin decomposition has been studied in detail by Bajpai, 1970a, 1970b.) The flatter u dependence of Δ exchange is not found in $\gamma p \to \Delta^{++}\pi^-$, which is pure $I = \frac{3}{2}$ exchange and which has $d\sigma/du \propto e^{3u}$, approximately (see Section 3.6.2). The peculiar behavior of the $I = \frac{1}{2} : I = \frac{3}{2}$ ratio is not found in backward $\pi N \to \rho N$, where all three charge states have been measured (Haber $et\ al.$, 1974) and indications are that $|T_{1/2}/T_{3/2}| \simeq 2$ independent of u, similar to $\pi N \to \pi N$. A comparison of backward $\pi N \to \omega N$ and $\pi N \to \rho N$ would lead one to expect $|N_S/N_V| \simeq 0.23$. Also the ratio of the ρ to π photoproduction cross sections would imply a bigger Δ contribution near $u = 0$ (see Section 3.6.4). All in all one can see that although these models fit the available data they are not in accord with what we expect from other reactions. Everything hinges on the lack of shrinkage, and we know from forward photoproduction that cuts can affect this. Clearly more phenomenological work is required with a view to understanding the data with a less ad hoc model by using information culled from other reactions. Better data, particularly measurements of the reaction $\gamma n \to \pi^- p$, and a better measurement of the ratio of the π^+ and π^0 reactions at larger $|u|$ in order to test whether Eq. (3.42) is satisfied, would help to test whether the isospin decomposition is as in Fig. 34, with the Δ dominating here. This is emphasized in a recent study of backward pion photoproduction by Triantafillopoulos (1977).

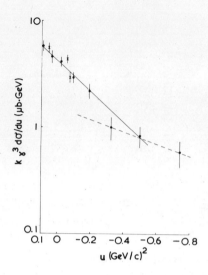

Fig. 35. Scaled backward cross section $k_\gamma^3 d\sigma/du$ for the reaction $\gamma p \to \Delta^{++}\pi^-$ plotted against u. Data: ($\times$) 4.5 GeV, ($\bullet$) 5.3 GeV (R. L. Anderson *et al.*, 1969*a*). The solid line is $e^{3.4u}$, the dotted line $e^{1.3u}$ (Berger and Fox, 1971).

3.6.2. Backward $\pi\Delta$ Photoproduction

Although there is only data on the reaction $\gamma p \to \Delta^{++}\pi^-$ at one energy $(E_\gamma = 5.28 \text{ GeV})$ it is this charge state that is particularly interesting as only Δ exchange is allowed. The cross section shows e^{3u} behavior for $|u| \leq 0.3 \,(\text{GeV}/c)^2$ with some indication of flattening out to an e^u fall-off at larger $|u|$ (Fig. 35). It would be interesting to see whether this is confirmed by better data over a larger range of $|u|$. Also, as advocated by Berger and Fox (1971), data over a reasonable energy range to test whether there is shrinkage or not, would be valuable for deciding between models. The reason is that traditional pole models with linear trajectories would predict canonical shrinkage for $\gamma p \to \Delta^{++}\pi^-$, whereas fixed cut models predict no shrinkage.

3.6.3. Backward $K\Sigma$ and $K\Lambda$ Photoproduction

There is data on the reactions $\gamma p \to K^+\Sigma^0$ and $\gamma p \to K^+\Lambda^0$ at $E_\gamma = 4.3 \text{ GeV}$ and it is shown in Fig. 36. As in the forward direction the most interesting aspect of the data is the $\Sigma:\Lambda$ ratio. The dominant exchanges are expected to be the $(\Lambda_\alpha\Lambda_\gamma)$ and $(\Sigma_\beta\Sigma_\delta)$ exchange degenerate trajectories, since the $(\Sigma_\alpha\Sigma_\gamma)$ trajectory couples weakly to the $\bar{K}N$ vertex (Nagels *et al.*, 1976). The two possibilities of Λ or Σ dominance give very different predictions for the $\Sigma:\Lambda$ ratio, Λ dominance giving

$$R = \frac{d\sigma(\gamma p \to K^+\Sigma^0)/du}{d\sigma(\gamma p \to K^+\Lambda^0)/du} = 3 \tag{3.44}$$

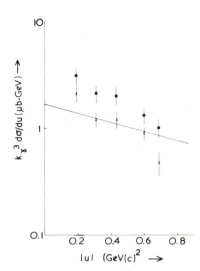

Fig. 36. Scaled cross sections $k_\gamma^3 d\sigma/du$ for the backward reactions $\gamma p \to K^+\Lambda$ and $\gamma p \to K^+\Sigma^0$ in μb GeV at $E_\gamma = 4.3$ GeV plotted against u. The solid line is e^u and the data are from R. L. Anderson *et al.* (1969b): (●) $\gamma p \to \Sigma^0 K^+$, (×) $\gamma p \to \Lambda^0 K^+$.

whereas Σ dominance gives

$$R = \tfrac{1}{3} \tag{3.45}$$

Experimentally it is found that

$$R = 1.7 \pm 0.15 \tag{3.46}$$

(Anderson *et al.*, 1969b), implying that both reactions are dominated by Λ exchange with some Σ exchange. This is exactly what is found in backward KN scattering, where the cross section for $K_L^0 p \to K_S^0 p$, in which only Σ exchange is allowed, is about 20% of that for $K^+ p \to K^+ p$, in which both Λ and Σ exchange are allowed (Brandenberg *et al.*, 1973, 1974).

3.6.4. Other Backward Photoproduction Reactions

The reactions $\gamma p \to \eta^0 p$ and $\gamma p \to (\rho^0 + \omega^0)p$ have been measured at SLAC by missing-mass technique (Tompkins *et al.*, 1969), though the data are rather poor. It might be hoped that the $\eta : \pi$ ratio could give some information on the size of the N_γ contribution for the following reason. For any given $I = \tfrac{1}{2}$ exchange the ratio of cross sections (for a given beam particle a)

$$R_a = \frac{d\sigma(ap \to \eta^0 p)/du}{d\sigma(ap \to \pi^0 p)/du} = \frac{(4\alpha - 1)^2}{3} \tag{3.47}$$

where $\alpha = f/(f + d)$ is related to the f/d ratio. Since $f_\alpha = 0.45$ and $f_\gamma = 0.69$, a measurement of R_γ might give information on the $N_\alpha N_\gamma$

breakdown. Experimentally, $R_\gamma = 0.56 \pm 0.15$ (Tompkins *et al.*, 1969) whereas for pion beams $R_\pi = 0.45 \pm 0.11$ (Boright *et al.*, 1970) so the trend is in the right direction, indicating a larger N_γ contribution in photon-induced reactions (Triantafillopoulos, 1977).

The $(\rho + \omega)$ data, the missing-mass technique not enabling a separation between ρ and ω, which are expected to have around the same size cross sections, have now been superseded by some good data on $\gamma p \to \rho p$ from NINA (Fig. 37) (Clifft *et al.*, 1976). The ρ^0 cross section is about a factor of 2 larger than the π^0 cross section and shows the usual shallow falloff, in fact $\propto \exp\{(1.4 \pm 0.2)u\}$. The fact that the cross section is bigger than the π^0 cross section is interesting from the point of view of the isospin decomposition in backward photoproduction. Data for the reactions $pp \to d\pi^+$ and $pp \to d\rho^+$ (pure $I = \frac{1}{2}$ exchange) show that the cross sections are equal (Allaby *et al.*, 1969) and thus the π and ρ couple equally to $I = \frac{1}{2}$ exchanges, whereas data on $\pi^- p \to \rho^- p$ and $\pi^- p \to \pi^- p$ show that the ρ cross section is bigger by around a factor of 2 (E. W. Anderson *et al.*, 1969) and so the ρ couples more strongly than the π to the Δ. Thus the most natural explanation of the larger ρ cross section in photon-induced reactions is that there is significant Δ exchange here (Triantafillopoulos, 1977).

The f^0 cross section is three times bigger than the π^0 cross section and is consistent with e^u u dependence (Clifft *et al.*, 1976). The fact that the cross section is large can be understood if we apply triple Regge theory to the inclusive reaction $\gamma p \to pX$ with s, M^2 and (s/M^2) all large (M is the

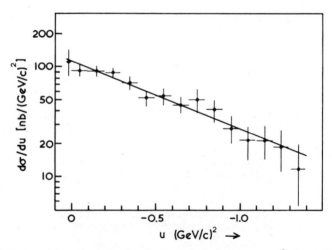

Fig. 37. Differential cross section for the backward reaction $\gamma p \to \rho^0 p$ in $\mathrm{nb}/(\mathrm{GeV}/c)^2$ at $E_\gamma = 3.5$ GeV plotted against u. The solid line is a fitted exponential $111e^{1.4u}$, and the data are from Clifft *et al.* (1976).

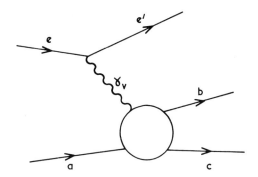

Fig. 38. The one-photon approximation for electroproduction.

mass of the X system). The M^2 dependence is given by

$$\frac{d^2\sigma}{dt\,d(M^2/s)} = \sum_{i,j,k} G_{ijk}(u)\left(\frac{M^2}{s}\right)^{\alpha_k(0)-\alpha_i(u)-\alpha_j(u)} s^{\alpha_k(0)-1} \qquad (3.48)$$

where α_k is a meson trajectory and α_i and α_j are baryon trajectories. Thus taking $\alpha_k(0)=1$, $\alpha_i=\alpha_j=\alpha_{N_\alpha}\simeq-0.4$ we can see that the leading M^2 dependence will be $(M^2)^{1.8}$. The data are taken between $E_\gamma = 2.8$ and 4.8 GeV, which is much too low for triple Regge theory, but one can hopefully appeal to duality and apply the theory on the average.

There is also information on density-matrix elements (Clifft *et al.*, 1976) for both $\gamma p \to \rho^0 p$ and $\gamma p \to f^0 p$, but as yet there has been no theoretical analysis.

Probably the most interesting reaction for which there are no data is $\gamma d \to np$. Since this is pure $I = \frac{1}{2}$ exchange, models that are dominated by Regge poles give shrinkage whereas those dominated by cuts do not. Data would be needed over a reasonable range of u and E to be really useful (Berger and Fox, 1971).

3.7. Electroproduction

3.7.1. Formalism

The procedure of extracting a virtual photon cross section $d\sigma/dt\,d\phi$ for the reaction $\gamma_v + a \to b + c$ from the observed electroproduction cross section using the one-photon approximation (Fig. 38) is described in Lyth's article. Here we regard the electrons as providing a beam of virtual photons of known polarization ε, mass* $-q^2$, and energy $\nu = E - E'$. The

---

*$q^2 = -4EE' \sin^2(\theta_e/2)$ is the k^2 used by Lyth.

center-of-mass energy W and momentum transfer t are defined as usual:

$$W^2 = m^2 + q^2 + 2m\nu \tag{3.49}$$

$$t = (p_\gamma - p_b)^2 \tag{3.50}$$

and the virtual photon cross section can be written in the following way:

$$2\pi \frac{d\sigma}{dt\,d\phi}(s, t, q^2) = \frac{d\sigma_U}{dt} + \varepsilon\,\frac{d\sigma_L}{dt} + \varepsilon\,\cos(2\phi)\,\frac{d\sigma_P}{dt} + [2\varepsilon(\varepsilon + 1)]^{1/2}\cos(\phi)\,\frac{d\sigma_I}{dt} \tag{3.51}$$

where $d\sigma_U/dt$ is the total contribution of transversely polarized photons [analogous to $\frac{1}{2}(d\sigma_\parallel/dt + d\sigma_\perp/dt)$ in photoproduction], $d\sigma_L/dt$ is the total contribution of longitudinal photons, $d\sigma_P/dt$ is the contribution due to polarization effects of transversely polarized photons [analogous to $\frac{1}{2}(d\sigma_\parallel/dt - d\sigma_\perp/dt)$ in photoproduction], and $d\sigma_I/dt$ is the interference term between longitudinal and transverse photons. Note that whereas $d\sigma_U/dt + \varepsilon\,d\sigma_L/dt$, $d\sigma_P/dt$, and $d\sigma_I/dt$ can be isolated experimentally using the ϕ dependence of the cross section, a separation of $d\sigma_U/dt$ and $d\sigma_L/dt$ requires data at very different electron scattering angles. This is experimentally very difficult because of the rapid decrease of Γ with increasing θ_e, and so $d\sigma_U/dt$ and $d\sigma_L/dt$ are usually not separated.

As a theoretical aside, we note that

$$\frac{d\sigma_\parallel}{dt} = \frac{d\sigma_U}{dt} + \frac{d\sigma_P}{dt} \tag{3.52}$$

and $d\sigma_L/dt$ receive contributions from only unnatural parity exchanges at high energies, whereas

$$\frac{d\sigma_\perp}{dt} = \frac{d\sigma_U}{dt} - \frac{d\sigma_P}{dt} \tag{3.53}$$

receives contributions from only natural parity exchanges (Kramer, 1974).

3.7.2. Charged Pion Electroproduction

The reaction for which there is most data above the resonance region (Brown *et al.*, 1971a, 1971b, 1973; Kummer *et al.*, 1971; Sofair *et al.*, 1972; Driver *et al.*, 1971a, 1971b, 1971c; Bebek *et al.*, 1976; Brauel *et al.*, 1976a) is π^+ electroproduction:

$$\gamma_v + p \rightarrow \pi^+ + n \tag{3.54}$$

and the different cross sections can be separated (Fig. 39). We can immediately see that the reaction is dominated by $d\sigma_U/dt + \varepsilon\,d\sigma_L/dt$ and this combination increases rapidly with q^2, eventually doubling in magnitude

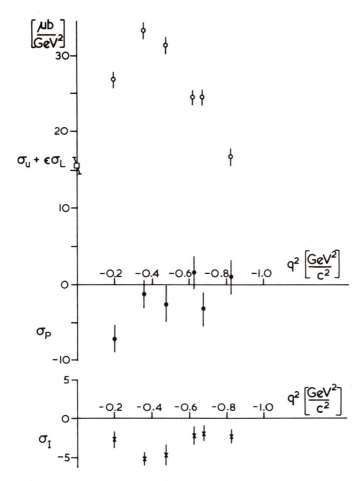

Fig. 39. q^2 dependence of the $\gamma_v p \rightarrow \pi^+ n$ differential cross-section components $d\sigma_U/dt + \epsilon d\sigma_L/dt$, $d\sigma_P/dt$, and $d\sigma_I/dt$. The data shown correspond to $W = 2.2$ GeV and $t - t_0 = -0.01$ $(\text{GeV}/c)^2$ (Driver *et al.*, 1971*a*): (○) $d\sigma_U/dt + \epsilon d\sigma_L/dt$, (●) $d\sigma_P/dt$, (×) $d\sigma_I/dt$, (□) photoproduction.

before decreasing again. In any reasonable model $d\sigma_U/dt$ decreases monotonically with $(-q^2)$ and so the rise is presumably caused by the longitudinal cross section, which vanishes at $q^2 = 0$. This statement can be made more quantitative if we assume that $\sigma_{\parallel}/\sigma_{\perp} \lesssim 0.1$ as in photoproduction. This implies that $|d\sigma_U/dt| \simeq |d\sigma_P/dt|$, and under this assumption $d\sigma_L/dt$ can be separated (Driver, 1971*b*). The result is that $d\sigma_L/dt$ is bigger than the photoproduction cross section near the forward direction—it exceeds it by a factor of 3 at $\theta_{\pi\gamma_v} = 0$ and $q^2 = -0.26$ $(\text{GeV}/c)^2$. It also falls sharply with t in the small-t region—it goes as $\exp[(15 \pm 3)t]$.

This large $d\sigma_L/dt$ and its sharp falloff in t is what we would expect. In charged photoproduction the pion exchange contribution was very important for small t, despite the pole component vanishing at $t = 0$ owing to the evasive factor. The physical reason for the existence of this factor is the fact that angular momentum conservation does not allow a photon with helicity ± 1 to couple to the exchanged pion at $\theta_{\pi\gamma} = 0$. However, pion exchange can couple to a longitudinal photon at $\theta_{\pi\gamma_v} = 0$ and so the pion contribution to $d\sigma_L/dt$ will be correspondingly larger, with no evasive factor. The increased importance of pion exchange in electroproduction, as compared to photoproduction, can be seen from the fact that the cross-section ratio $(\gamma_v d \rightarrow \pi^+ nn_s)/(\gamma_v d \rightarrow \pi^- pp_s)$ is closer to unity for a larger range of t (Fig. 40) than in photoproduction, signifying that the isoscalar–isovector interference is smaller in the electroproduction case (Bebek *et al.*, 1976; Brauel *et al.*, 1976*b*; Wolf, 1976).

Not surprisingly, pion exchange models that were successful in photoproduction can be modified to fit electroproduction. The electric Born approximation can be modified by the inclusion of the nucleon and pion form factors—in fact, electroproduction provides the best measurement of the pion form factor, as discussed by Donnachie, Lyth, and Shaw in Chapter 5 and in the review by Gourdin (1974). The model works reasonably well for small t, but as in photoproduction, dispersion relation calculations show

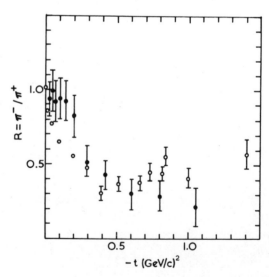

Fig. 40. The π^-/π^+ cross-section ratio $(\gamma_v d \rightarrow \pi^- pp_s)/(\gamma_v d \rightarrow \pi^+ nn_s)$ at $q^2 = 0$ and $-0.7\,(\text{GeV}/c)^2$ plotted against t: ($\bigcirc$) $q^2 = 0$, ($\bullet$) $q^2 = -0.7\,(\text{GeV}/c)^2$ $\varepsilon = 0.86$, $s = 4.8\,\text{GeV}^2$, $150° < \phi < 210°$. From Brauel *et al.* (1976*b*); also Wolf (1976).

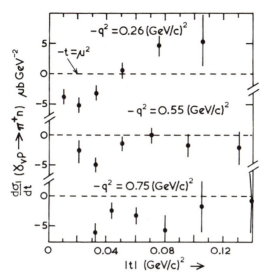

Fig. 41. Interference cross section $d\sigma_I/dt$ for $\gamma_v p \to \pi^+ n$ plotted against t for $-q^2 = 0.26$, 0.55, and 0.75 $(\text{GeV}/c)^2$. From Irving (1975); data from Driver et al. (1971a, b).

(Berends and Gastmans, 1972; Devenish and Lyth, 1972; Manweiler and Schmidt, 1971; Gutbrod and Kramer, 1972) that higher resonance contributions are more important at large $|t|$ and are, in fact, needed to fit the data. Also, as in photoproduction, the PMA cut prescription (Irving, 1975) is identical to the electric Born approximation. In fact, the analysis of Irving (1975) shows that neither fits the fine details of the data. The crucial feature is that $d\sigma_I/dt$, the interference contribution, is proportional to the amplitude for transverse photons, and this has an evasive pion pole and a smooth background, which we will denote as π_c, and which is produced in models exactly as in photoproduction. Around $t = -\mu^2$ these cancel and so $d\sigma_I/dt$ vanishes. Both models predict that the t value at which this occurs is independent of q^2, whereas in the data the position of the zero moves out with increasing $|q^2|$ (Fig. 41). The natural interpretation of this is that the cut is increasing relative to the pole as q^2 becomes more negative (Irving, 1975). Irving also points out this is in accordance with the trend as one goes from $q^2 = m_\rho^2$ in $\pi N \to \rho N$ to $q^2 = 0$ in $\gamma N \to \pi N$. Vanryckeghem (1974) also finds that absorptive effects are larger in electroproduction than in photoproduction.

The above discussion was concerned with data for $|t| < 0.2$ $(\text{GeV}/c)^2$. Recently, data at larger $|t|$ have appeared (Brauel et al., 1976a) and show that $d\sigma/dt\,d\phi$ is roughly independent of q^2 for $-q^2 > 0.7$ $(\text{GeV}/c)^2$ in this t range.

3.7.3. π^0 Electroproduction

As discussed in Section 3.3, the π^0 cross section has a dip at $t =$ -0.5 (GeV/c)^2 and the interest in π^0 electroproduction is in studying what happens to this dip. Different approaches have different explanations of the dip. The Regge school associate it with a NWS point $\alpha = 0$ for the ω trajectory, and this would imply that the dip position is independent of q^2. The geometrical school associate the dip with the first zero of $J_1[R - (t)^{1/2}]$ with R, the radius, taken as around 1 fm. Theoretically, the interaction radius of the photon is expected to decrease with increasing $-q^2$ (Bjorken *et al.*, 1971; Nieh, 1972), and this should manifest itself in the slope parameter in ρ electroproduction decreasing with $-q^2$. Although individual experiments support this (Ballam *et al.*, 1974; Dakin *et al.*, 1973a, 1973b; Eckardt *et al.*, 1973), when taken together they form a very confusing picture (Wolf, 1976), though they are not in contradiction with theoretical expectations. If the interaction radius decreases, then the zero will move to larger $|t|$ as q^2 becomes negative. Thus, π^0 electroproduction was to be the crucial test of dip mechanisms (Harari, 1971b, 1971d). Data have since been taken at $\phi = 90°$ (Brasse *et al.*, 1975), where $d\sigma_I/dt$ vanishes, so

$$\frac{d\sigma}{dt\,d\phi}\bigg|_{(\phi=90°)} = \frac{1}{4\pi}\left[(1+\varepsilon)\frac{d\sigma_\perp}{dt}+(1-\varepsilon)\frac{d\sigma_\parallel}{dt}+2\varepsilon\frac{d\sigma_L}{dt}\right] \quad (3.55)$$

Since unnatural parity exchange is known to be small in π^0 electroproduction it seems likely that $d\sigma_\parallel/dt$ and $d\sigma_L/dt$ are small in electroproduction and so, to a reasonable approximation,

$$\frac{d\sigma_\perp}{dt} \simeq \frac{4\pi}{1+\varepsilon}\frac{d\sigma}{dt\,d\phi}\bigg|_{(\phi=90°)} \quad (3.56)$$

In Fig. 42 we show the quantity on the right-hand side of Eq. (3.56) plotted against t for different q^2. The secondary maximum, and with it the dip, disappear remarkably quickly with q^2, contrary to the simple prediction of either approach. Barker and Storrow (1977) [also see Donnachie (1976)] have suggested that this might be another example of the q^2 dependence of cut effects. Using their pole-cut separation of the photoproduction amplitudes and a similar prescription for the q^2 dependence of pole and cut effects as that given by Irving (1975) in the charged pion case, they are able to reproduce the data at the larger q^2 values. It should be pointed out that the lowest q^2 data corresponds to an equivalent photon energy of around 3 GeV and at this energy the experimental evidence for a secondary maximum in the photoproduction data is not overwhelming (Braunschweig, 1970a). The same phenomenon is expected to happen in

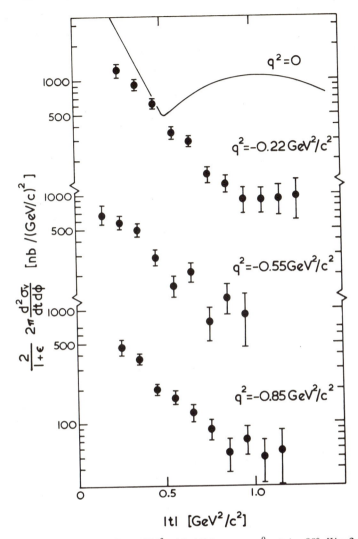

Fig. 42. The cross section $[4\pi/(1 + \varepsilon)](d^2\sigma_v/dt\, d\phi)$ for $\gamma_v p \to \pi^0 p$ at $\phi = 90°$, $W = 2.55$ GeV is plotted against t for various values of q^2. The cross section $d\sigma_\perp/dt$ for real photons is shown as the solid line. From Brasse *et al.* (1975).

the resonance region. In photoproduction the dip is present for $s^{1/2} > 1680$ MeV and first appears associated with the dominance of the helicity-$\frac{3}{2}$ amplitude in the $f_{15}(1688)$ resonance. Partial wave analysis of electroproduction predicts that as we go from $q^2 = 0$ the helicity-$\frac{3}{2}$ amplitude becomes smaller and the helicity $\frac{1}{2}$ increases, and for the f_{15}, helicity $\frac{1}{2}$, no dip is predicted at $t = -0.5$ (GeV/$c)^2$ (Moorhouse, 1976). Presumably, duality relates the two phenomena.

3.7.4. $K^+\Lambda$ and $K^+\Sigma^0$ Electroproduction

The reactions

$$\gamma_v + p \to K^+ + \Lambda^0 \qquad (3.57)$$

$$\gamma_v + p \to K^+ + \Sigma \qquad (3.58)$$

have been measured (Azemoon *et al.*, 1975; Brown *et al.*, 1972*b*), and the data for $W = 2.2$ GeV and t in the range $0.1 \le |t| \le 0.18$ $(\text{GeV}/c)^2$ are plotted for different q^2 (including $q^2 = 0$) in Fig. 43. The two reactions show qualitatively different behavior, the Σ^0 cross section decreasing rapidly $[\sim(1 - q^2/m_\rho^2)^{-2}]$ whereas the Λ data are fairly constant. This is in agreement with naive expectations. In the Λ case, K exchange will be significant, particularly in the longitudinal cross section (compare the discussion on π exchange in Section 3.7.2). In the Σ case, since $g_{KN\Sigma}^2 \ll g_{KN\Lambda}^2$, K exchange will be much less important, just as in photoproduction (Levy

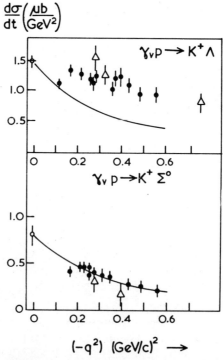

Fig. 43. The differential cross section $d\sigma/dt$ for $\gamma_v p \to K^+\Lambda$ and $\gamma_v p \to K^+\Sigma^0$ plotted as a function of q^2. Data with $W = 2.2$ GeV and $0.1 \le |t| \le 0.18$ $(\text{GeV}/c)^2$ have been selected from Azemoon *et al.* (1975) (●) and Brown *et al.* (1972*b*) (△). The solid line in both cases has q^2 dependence $\propto (1 - q^2/m_\rho^2)^{-2}$ and is normalized to go through the photoproduction data point (○). From Wolf (1976).

et al., 1973*a*, 1973*b*). In fact, the situation may be rather more complicated. Bartl and Majerotto (1975) have studied both photo- and electroproduction of $K^+\Lambda$ and $K^+\Sigma^0$ and claim that, for a quantitative description of the data, K, K^*, K^{**} and either K_B or $K_A - K_Z$ exchange is required.

3.7.5. $\pi\Delta$ Electroproduction

In this reaction the best-studied charge state is

$$\gamma_v + p \to \pi^+ + \Delta^0 \tag{3.59}$$

and the different cross sections have been separated (Driver *et al.*, 1971*d*; Dammann *et al.*, 1973; Brown *et al.*, 1972*a*) and are shown in Fig. 44. As can be seen, $d\sigma_U/dt + \varepsilon\, d\sigma_L/dt$ is the dominant term and its q^2 dependence suggests the presence of a significant longitudinal part, though not as much as in the π^+ channel. Vector dominance model calculations (Bartl and Schildnecht, 1972) indicate that $d\sigma_L/dt$ is around 50% of the forward cross section. This relative smallness of $d\sigma_L/dt$ compared to the π^+n case is a kinematic effect—the large t_0 makes the pion pole remote from the physical region—and should disappear at higher energies. The electric Born model is an adequate representation of the data, though there are theoretical uncertainties in the q^2 dependence.

The cross section for $\gamma_v p \to \pi^-\Delta^{++}$ has been measured (Dammann *et al.*, 1973), but more interesting is the measurement of the four charge combinations

$$\gamma_v + p \to \pi^+ + \Delta^0 \tag{3.60}$$

$$\gamma_v + p \to \pi^- + \Delta^{++} \tag{3.61}$$

$$\gamma_v + n \to \pi^+ + \Delta^- \tag{3.62}$$

$$\gamma_v + n \to \pi^- + \Delta^+ \tag{3.63}$$

at $q^2 = -0.15\ (\text{GeV}/c)^2$ and $W = 2.15$ GeV (Brown *et al.*, 1972*a*). If there is no isoscalar–isovector interference then we should have

$$d\sigma(\gamma_v + p \to \pi^+ + \Delta^0)/d\Omega = d\sigma(\gamma_v + n \to \pi^- + \Delta^+)/d\Omega \tag{3.64}$$

and

$$d\sigma(\gamma_v + p \to \pi^- + \Delta^{++})/d\Omega = d\sigma(\gamma_v + n \to \pi^+ + \Delta^-)/d\Omega \tag{3.65}$$

These equations are not quite satisfied, presumably indicating the presence of a small component of isoscalar photon coupling (Brown *et al.*, 1972*a*).

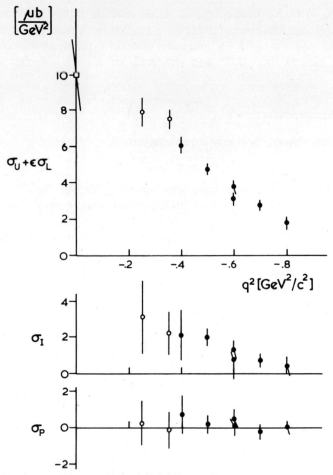

Fig. 44. Differential cross-section components $d\sigma_U/dt + \varepsilon d\sigma_L/dt$, $d\sigma_P/dt$ and $d\sigma_I/dt$ for the reaction $\gamma_v p \to \pi^+ \Delta^0$ plotted against q^2 for $s = 5.5\ \mathrm{GeV}^2$, $t - t_0 = -0.05\ (\mathrm{GeV}/c)^2$. Data: ($\square$) $d\sigma/d(t - t_0)$ (photoproduction), ($\bigcirc$) data scaled from $s = 4.84$ to $5.5\ (\mathrm{GeV})^2$, ($\bullet$) $s = 5.5\ (\mathrm{GeV})^2$. From Driver et al. (1971d).

4. Conclusions

In general we have found that the understanding of high-energy photoproduction reactions is good, probably almost on a par with current understanding of hadronic reactions. In forward reactions, the lack of shrinkage has been understood in terms of the low-lying unnatural-parity exchanges being dominant for small $|t|$, with the higher-lying natural-parity exchanges taking over at larger $|t|$. This interpretation is nicely confirmed by the fact that the two reactions where unnatural-parity exchanges are expected to be suppressed, $\gamma p \to \pi^0 p$ and $\gamma p \to K^+ \Sigma^0$, show shrinkage at small $|t|$ with effective trajectories significantly above zero here (Figs. 14 and 23). It would be nice to have this feature confirmed in $\gamma p \to \eta^0 p$. The critical test will come in the next energy regime, where the unnatural-parity exchanges should die away and significant deviations from k_γ^{-2} scaling appear. The above discussion really only applies for $|t| \lesssim 0.7 \, (\mathrm{GeV}/c)^2$. For $|t|$ greater than this value it appears that effective trajectories in photoproduction reactions are significantly higher than those in hadronic reactions, which may be an indication that new dynamical mechanisms are coming into play here. Fixed poles have been suggested by Fox (1972), but this seems unlikely as fixed poles at $j = 0$ are needed to fit the energy dependence, and for $\gamma p \to \pi^0 p$ these can only occur in the unnatural-parity exchange amplitudes, which are known to be small. A more likely explanation is that cuts play a different role in photoproduction, and this is supported by the large size of the cuts found in $\gamma p \to \pi^0 p$.

The magical success of the electric Born model in $\pi^\pm N$ and $\pi^\pm \Delta$ photoproduction near the forward direction has now been clarified by dispersion relation calculations and the realization that it is equivalent to the poor man's absorption prescription of Williams (1970). The latter is probably equally mysterious theoretically, but it is a universal prescription and so it avoids assigning a special role to either photon reactions or elementary pion exchange. The fact that it is equally valid for Reggeized pion exchange is an advantage in view of the shrinkage observed in $d\sigma_\parallel(\gamma n \to \pi^- p)/dt$ (Fig. 9). The dispersion relation calculations (Barbour *et al.*, 1971; Barbour and Malone, 1974) show quite clearly how the electric Born model works in the forward direction. Why the resonances couple in this peculiar way is as big a mystery as ever (Moorhouse, 1976).

As mentioned earlier, the critical experiments for current models of photoproduction are at higher energies, but there is much to be done at existing energies. Probably the most useful would be a complete set of measurements on either $\gamma p \to \pi^0 p$ or $\gamma p \to K^+ \Lambda$ in order to obtain amplitudes in a model-independent way. As discussed in Section 2.3, both are technically feasible and the lesson we have learned from hadronic reactions is that the great experimental effort involved is worthwhile. Failing this, polarized target measurements of the comparatively little-studied $K \Lambda$ channel would be very useful.

In view of the fact that the indications are that the effects of exotic exchange have been seen in both $\pi \Delta$ and $K \Sigma$ channels (Sections 3.3 and 3.4) follow-up experiments would be useful. To improve the data it would be desirable to use coincidence techniques and/or a tagged beam.

Compared to forward photoproduction, backward reactions are not very well understood. This is not surprising since this is the state of affairs in hadronic reactions. Theoretical progress on the parity doublet problem is clearly needed. As in the forward direction, there are hints that different mechanisms are emerging at moderate momentum transfers, around $u = -1.0 \, (\text{GeV}/c)^2$. The explanation of the lack of shrinkage is much less plausible than in the forward direction. Higher-energy studies here are probably out of the question because of the low cross sections, but even at existing energies it would be useful to see whether the reactions $\gamma d \to np$ and $\gamma p \to \pi^- \Delta^{++}$ exhibit shrinkage, as they isolate $I = \frac{1}{2}$ and $I = \frac{3}{2}$ exchanges, respectively. Also data off neutrons would provide a useful check on the rather implausible isospin decomposition found in most models.

The study of radiative decays, $\gamma\gamma$ modes using one-photon exchange (Primakoff effect) and $\pi\gamma$ using pion exchange, is a useful way of measuring decay rates which are hard to measure directly. However, as discussed in Section 3.5, their usefulness is limited by our lack of knowledge of the other exchanges.

The field of electroproduction is still in its infancy. The general features of all reactions are in reasonable qualitative accord with extrapolations from photoproduction, and the apparent q^2 dependence of cut effects (Irving, 1975; Vanryckeghem, 1974; Barker and Storrow, 1977) opens up very interesting possibilities for the future.

It is clear that by and large the mechanisms governing high-energy photo- and electroproduction are the same as those governing hadron reactions. This means that the special features of photoproduction, e.g., the existence of good multipole analysis and the easy separation of natural and unnatural parity exchanges using polarized beams, make them very useful in the study of high-energy mechanisms. Also, electroproduction offers us a unique opportunity to study reaction mechanisms as a function of a continuously varying external mass.

Acknowledgments

I wish to thank my colleagues in the theory Group at Manchester, in particular Ian Barker, for useful advice and discussions. I am also grateful to Ms. J. Heather Kimber for showing patience and forebearance above and beyond the call of duty in typing the manuscript.

References

Aachen–Berlin–Bonn–Hamburg–Heidelberg–Munich Collaboration (1968), *Phys. Rev.* **175**, 1669.
Aachen–Berlin–Bonn–Hamburg–Heidelberg–Munich Collaboration (1969), *Phys. Rev.* **188**, 2060.

Abramson, J., Andrews, D. E., Busnello, R., Harvey, J., Lobkowicz, F., May, E. N., Nelson, C. A., Singer, M., Thorndike, E. H., and Nordberg, M. E. (1976), *Phys. Rev. Lett.* **36**, 1432.

Ader, J. P., Capdeville, M., Cohen-Tannoudji, G., and Salin, Ph. (1968*a*), *Nuovo Cimento* **56A**, 952.

Ader, J. P., Capdeville, M., and Navelet, H. (1968*b*), *Nuovo Cimento* **56A**, 315.

Ader, J. P., Capdeville, M., and Rimpault, M. (1969), *Nuovo Cimento* **59A**, 40.

Ader, J. P., Peschanski, R., Lacaze, R., Cohen-Tannoudji, G., and Gilain, C. (1975), *Nuovo Cimento* **27A**, 385.

Allaby, J. V., Binon, F., Diddens, A. N., Duteil, P., Klovning, A., Meunier, R., Peigneux, J. P., Sacharidis, E. J., Schlüpmann, K., Spighel, M., Stroot, J. P., Thorndike, A. M., and Wetherell, A. M. (1969), *Phys. Lett.* **29B**, 198.

Alonso, J. L., Schiff, D., and Tran Thanh Van, J. (1972), *Nuovo Cimento Lett.* (*Ser.* 2) **5**, 27.

Anderson, E. W., Bleser, E. J., Blieden, H. R., Collins, G. B., Garelick, D., Menes, J., Turkot, F., Birnbaum, D., Edelstein, R. M., Hien, N. C., McMahon, T. J., Mucci, J., and Russ, J. (1969), *Phys. Rev. Lett.* **22**, 102.

Anderson, R. L., Gustavson, D., Johnson, J., Ritson, D., Jones, W. G., Kreineck, D., Murphy, F., and Weinstein, R. (1968*a*), *Phys. Rev. Lett.* **21**, 384.

Anderson, R. L., Gustavson, D., Johnson, J., Ritson, D., Weinstein, R., Jones, W. G., and Kreineck, D. (1968*b*), *Phys, Rev. Lett.* **21**, 479.

Anderson, R. L., Gustavson, D., Johnson, J., Overman, I., Ritson, D., and Wiik, B. H. (1969*a*), *Phys. Rev. Lett.* **23**, 721.

Anderson, R. L., Gustavson, D., Johnson, J., Overman, I., Ritson, D., and Wiik, B. H. (1969*b*), *Phys. Rev. Lett.* **23**, 890.

Anderson, R. L., Gustavson, D., Johnson, J., Ritson, D., Wiik, B. H., Jones, W. G., Kreineck, D., Murphy, F., and Weinstein, R. (1970), *Phys. Rev.* **D1**, 27.

Anderson, R. L., Gustavson, D., Johnson, J., Overman, I., Ritson, D., Wiik, B. H., Talman, R., and Worcester, D. (1971*a*), *Phys. Rev. Lett.* **26**, 30.

Anderson, R. L., Gustavson, D., Johnson, J., Overman, I., Ritson, D., Wiik, B. H., and Worcester, D. (1971*b*), *Phys. Rev.* **D4**, 1937.

Argyres, E. N., Contogouris, A. P., Holden, J. P., and Svec, M. (1973), *Phys. Rev.* **D8**, 2068.

Azemoon, T., Damman, I., Driver, C., Lüke, D., Specht, G., Heinloth, K., Ackermann, H., Gausauge, E., Janata, F., and Schmidt, D. (1975), *Nucl. Phys.* **B95**, 77.

Bajpai, R. (1970*a*), *Nuovo Cimento Lett. Ser.* 1 **4**, 754.

Bajpai, R. (1970*b*), *Nuovo Cimento Lett. Ser.* 1 **4**, 950.

Bajpai, R., and Donnachie, A. (1970), *Nucl. Phys.* **B17**, 453.

Ball, J. S. (1961), *Phys. Rev.* **124**, 2014.

Ballam, J., Chadwick, G. B., Guiragossian, Z. G. T., Levy, A., Menke, M., Seyboth, P., and Wolf, G. (1969), *Phys. Lett.* **30B**, 421.

Ballam, J., Chadwick, G. B., Levy, A., Menke, M., Seyboth, P., Shapira, A., Eisenberg, Y., Haber, B., Ronat, E. E., Stahl, Y., and Yekutieli, G. (1971), *Phys. Rev. Lett.* **26**, 995.

Ballam, J., Chadwick, G. B., Eisenberg, Y., Kogan, E., Moffeit, K. C., Seyboth, P., Skillicorn, I. O., Spitzer, H., Wolf, G., Bingham, H. H., Fretter, W. B., Podolsky, W. J., Rabin, M. S., Rosenfeld, A. H., and Smadja, G. (1973), *Phys. Rev.* **D7**, 3150.

Ballam, J., Bloom, E. D., Carroll, J. T., Chadwick, G. B., Cottrell, R. L. A., Della Negra, M., De Staebler, H., Gershwin, L. K., Keller, L. P., Mestayer, M. D., Moffeit, K. C., Prescott, C. Y., and Stein, S. (1974), *Phys. Rev.* **D10**, 765.

Barbour, I. M., and Crawford, R. (1975), "An Analysis of Single Pion Photoproduction at High and Low Energies," Glasgow University preprint.

Barbour, I. M., and Malone, W. (1974), *Nucl. Phys.* **B82**, 477.

Barbour, I. M., and Moorhouse, R. G. (1974), *Nucl. Phys.* **B69**, 637.

Barbour, I. M., Malone, W., and Moorhouse, R. G. (1971), *Phys. Rev.* **D4**, 1521.
Bardeen, W. A., and Tung, W.-K. (1968), *Phys. Rev.* **173**, 1423.
Barger, V., and Cline, D. (1969), *Phenomenological Theories of High Energy Scattering* (New York, Benjamin).
Barger, V., and Phillips, R. J. N. (1969), *Phys. Rev.* **187**, 2210.
Barger, V., and Weiler, P. (1969), *Phys. Lett.* **30B**, 105.
Barger, V., and Weiler, P. (1970), *Nucl. Phys.* **B20**, 615.
Barker, I. S., Donnachie, A., and Storrow, J. K. (1974), *Nucl. Phys.* **B79**, 431.
Barker, I. S., Donnachie, A., and Storrow, J. K. (1975), *Nucl. Phys.* **B95**, 347.
Barker, I. S., and Storrow, J. K. (1977), "Regge Poles and Cuts in π^0 Photoproduction and Related Reactions," Manchester University preprint.
Barnes, A. V., Mellema, D. J., Tollestrup, A. V., Walker, R. L., Dahl, O. I., Johnson, R. A., Kenney, R. W., and Pripstein, M. (1976), *Phys. Rev. Lett.* **37**, 76.
Bartl, A., and Majerotto, W. (1975), *Nucl. Phys.* **B90**, 285.
Bartl, A., and Schildnecht, D. (1972), *Nucl. Phys.* **B36**, 28.
Bar-Yam, Z., de Pagter, J., Hoenig, M. M., Kern, W., Luckey, D., and Osborne, L. S. (1967), *Phys. Rev. Lett.* **19**, 40.
Bar-Yam, Z., de Pagter, J., Dowd, J., and Kern, W. (1970*a*), *Phys. Rev. Lett.* **24**, 1078.
Bar-Yam, Z., de Pagter, J., Dowd, J., Kern, W., Luckey, D., and Osborne, L. S. (1970*b*), *Phys. Rev. Lett.* **25**, 1053.
Bebek, C. J., Brown, C. N., Herzlinger, M., Holmes, S. D., Lichtenstein, C. H., and Pipkin, F. M. (1976), *Phys. Rev.* **D13**, 25.
Bellenger, D., Deutsch, S., Luckey, D., Osborne, L. S., and Schwitters, R. (1968), *Phys. Rev. Lett.* **21**, 1205.
Bellenger, D., Bordelon, R., Cohen, K., Deutsch, S. B., Lobar, W., Luckey, D., Osborne, L. S., Pothier, E., and Schwitters, R. (1969), *Phys. Rev. Lett.* **23**, 540.
Benz, P., Braun, O., Butenschön, H., Gall, D., Idshok, U., Kiesling, C., Knies, G., Müller, K., Nellen, B., Schiffer, R., Schlamp, P., Schnackers, H. J., Söding, P., Stiewe, J., and Storim, F. (1974), *Nucl. Phys.* **B79**, 10.
Berends, F. A., and Gastmans, R. (1972), *Phys. Rev.* **D5**, 205.
Berger, E. L., and Fox, G. C. (1970*a*), *Nucl. Phys.* **B26**, 1.
Berger, E. L., and Fox, G. C. (1970*b*), Argonne preprint No. ANL/HEP 7023.
Berger, E. L., and Fox, G. C. (1971), *Nucl. Phys.* **B30**, 1.
Bienlein, H., Braunschweig, W., Dinter, H., Erlewein, W., Frese, H., Knütel, J., Lübelsmeyer, K., Mango, S., Meyer-Wachsmuth, H., Morehouse, C. C., Paul, L., Schmitz, D., and Schultz Von Dratzig, A. (1973), *Phys. Lett.* **46B**, 131.
Bietti, A., Di Vecchia, P., Drago, F., and Paciello, M. L. (1968), *Phys. Lett.* **27B**, 296.
Bjorken, J. D., Kogut, J., and Soper, D. (1971), *Phys. Rev.* **D1**, 1382.
Booth, P. S. L., Court, G. R., Craven, B., Gamet, R., Hayman, P. J., Holt, J. R., Hufton, A. P., Jackson, J. N., Norem, J. H. and Range, W. H. (1972), *Phys. Lett.* **38B**, 339.
Boright, J. P., Bowen, D. R., Groom, D. E., Orear, J., Owen, D. P., Pawlicki, A. J., and White, D. H. (1970), *Phys. Lett.* **33B**, 615.
Boyarski, A. M., Bulos, F., Busza, W., Diebold, R., Ecklund, S. D., Fischer, G. E., Rees, J. R., and Richter, B. (1968*a*), *Phys. Rev. Lett.* **20**, 300.
Boyarski, A. M., Bulos, F., Busza, W., Diebold, R., Ecklund, S. D., Fischer, G. E., Rees, J. R., and Richter, B. (1968*b*), *Phys. Rev. Lett.* **21**, 1767.
Boyarski, A. M., Bulos, F., Busza, W., Diebold, R., Ecklund, S. D., Fischer, G. E., Murata, Y., Richter, B., and Williams, W. S. C. (1969*a*), *Phys. Rev. Lett.* **22**, 148.
Boyarski, A. M., Diebold, R., Ecklund, S. D., Fischer, G. E., Rees, J. R., Murata, Y., Richter, B., and Williams, W. S. C. (1969*b*), *Phys. Rev. Lett.* **22**, 1131.
Boyarski, A. M., Diebold, R., Ecklund, S. D., Fischer, G. E., Murata, Y., Richter, B., and Sands, M. (1970), *Phys. Rev. Lett.* **25**, 695.

Boyarski, A. M., Diebold, R., Ecklund, S. D., Fischer, G. E., Murata, Y., Richter, B., and Sands, M. (1971), *Phys. Lett.* **34B**, 547.

Brandenberg, G. W., Johnson, W. B., Leith, D. W. G. S., Loos, J. S., Luste, G. J., Matthews, J. A. J., Moriyasu, K., Smart, W. M., Winkelman, F. C., and Yamartino, R. J. (1973), *Phys. Rev. Lett.* **30**, 145.

Brandenberg, G. W., Johnson, W. B., Leith, D. W. G. S., Loos, J. S., Luste, G. J., Matthews, J. A. J., Moriyasu, K., Smart, W. M., Winkelman, F. C., and Yamartino, R. J. (1974), *Phys. Rev.* **D9**, 1939.

Brasse, F. W., Fehrenbach, W., Flauger, W., Gayler, J., Goel, S. P., Haidan, R., Köte, U., Korbel, V., Kreineck, D., Ludwig, J., May, J., Merkwitz, M., Mess, K.-H., Schmüser, P., and Wiik, B. H. (1975), *Phys. Lett.* **58B**, 467.

Brauel, P., Canzler, T., Cords, D., Felst, R., Grindhammer, G., Helm, M., Kollmann, W.-D., Krehbiel, H., and Schädlich, M. (1976*a*), *Phys. Lett.* **65B**, 181.

Brauel, P., Canzler, T., Cords, D., Felst, R., Grindhammer, G., Helm, M., Kollmann, W. D., Krehbiel, H., and Schädlich, M. (1976*b*), *Phys. Lett.* **65B**, 184.

Braunschweig, M., Braunschweig, W., Husmann, D., Lübelsmeyer, K., and Schmitz, D. (1968), *Phys. Lett.* **26B**, 405.

Braunschweig, M., Braunschweig, W., Husmann, D., Lübelsmeyer, K., and Schmitz, D. (1970*a*), *Nucl. Phys.* **B20**, 191.

Braunschweig, W., Erlewein, W., Frese, H., Lübelsmeyer, K., Meyer-Wachsmuth, H., Schmitz, D., Schultz von Dratzig, A., and Wessels, G. (1970*b*), *Phys. Lett.* **33B**, 236.

Braunschweig, W., Dinter, H., Erlewein, W., Frese, H., Lübelsmeyer, K., Meyer-Wachsmuth, H., Morehouse, C. C., Schmitz, D., Schultz Von Dratzig, A., and Wessels, G. (1973*a*), *Nucl. Phys.* **B51**, 157.

Braunschweig, W., Erlewein, W., Frese, H., Lübelsmeyer, K., Meyer-Wachsmuth, H., Schmitz, D., and Schultz Von Dratzig, A. (1973*b*), *Nucl. Phys.* **B51**, 167.

Brion, J. P., and Peschanski, R. (1974), *Nucl. Phys.* **B81**, 484.

Broadhurst, D. J., Dombey, N., and Read, B. J. (1971), *Phys. Lett.* **34B**, 95.

Brown, C. N., Canizares, C. R., Cooper, W. E., Eisner, A. M., Feldman, G. J., Lichtenstein, C. A., Litt. L., Lockeretz, W., Montana, V. B., and Pipkin, F. M. (1971*a*), *Phys. Rev. Lett.* **26**, 987.

Brown, C. N., Canizares, C. R., Cooper, W. E., Eisner, A. M., Feldman, G. J., Lichtenstein, C. A., Litt. L., Lockeretz, W., Montana, V. B., and Pipkin, F. M. (1971*b*), *Phys. Rev. Lett.* **26**, 991.

Brown, C. N., Canizares, C. R., Cooper, W. E., Eisner, A. M., Feldman, G. J., Lichtenstein, C. A., Litt. L., Lockeretz, W., Montana, V. B., and Pipkin, F. M. (1972*a*), Contributed paper No. 274, in *Proceedings of the 1971 International Symposium on Electron and Photon Interactions at High Energies*, Cornell University, Ithaca, New york, Ed. N. B. Mistry (Cornell University Laboratory of Nuclear Studies).

Brown, C. N., Canizares, C. R., Cooper, W. E., Eisner, A. M., Feldman, G. J., Lichtenstein, C. A., Litt. L., Lockeretz, W., Montana, V. B., Pipkin, F. M., and Hicks, N. (1972*b*), *Phys. Rev. Lett.* **28**, 1086.

Brown, C. N., Canizares, C. R., Cooper, W. E., Eisner, A. M., Feldman, G. J., Lichtenstein, C. A., Litt, L., Lockeretz, W., Montana, V. B., and Pipkin, F. M. (1973), *Phys. Rev.* **D8**, 92.

Burfeindt, H., Buschhorn, G., Geweniger, C., Heide, P., Kotthaus, R., Wahl, H., and Wegener, K. (1970), *Phys. Lett.* **33B**, 509.

Burfeindt, H., Buschhorn, G., Geweniger, C., Kotthaus, R., Skronn, H. J., Wahl, H., and Wegener, K. (1973), *Nucl. Phys.* **B59**, 87.

Cambridge Bubble Chamber Group (1968), *Phys. Rev.* **169**, 1081.

Campbell, J. A., Clark, R. B., and Horn, D. (1970), *Phys. Rev.* **D2**, 217.

340   J. K. Storrow

Capella, A., and Tran Thanh Van, J. (1970), *Nuovo Cimento Lett. Ser.* 1 **4**, 1199.

Carlitz, R., and Kislinger, M. (1970), *Phys. Rev. Lett.* **24**, 186.

Chew, G. F., Goldberger, M. L., Low, F. E., and Nambu, Y. (1956), *Phys. Rev.* **106**, 1345.

Clark, M., and Donnachie, A. (1976), "A Model for the Reaction $\gamma p \to \rho^+ n$," Manchester University preprint.

Clark, M., and Donnachie, A. (1977), *Nucl. Phys.* **B25**, 493.

Clifft, R. W., Dainton, J. B., Gabathuler, E., Littenberg, L. S., Marshall, R., Rock, S. E., Thompson, J. C., Ward, D. L., and Brookes, G. R. (1976), *Phys. Lett.* **64B**, 213.

Cohen-Tannoudji, G., Morel, A., and Navelet, H. (1968), *Ann. Phys. (N.Y.)* **46**, 239.

Collins, P. B. D., and Fitton, A. (1974), *Nucl. Phys.* **B68**, 125.

Dado, S., Engler, A., Kraemer, R. W., Toaff, S., Weisser, F., Diaz, J., Dibianca, F., Fickinger, W., and Robinson, D. K. (1974), *Phys. Lett.* **50B**, 275.

Dahl, O. I., Johnson, R. A., Kenney, R. W., Pripstein, M., Barnes, A. V., Mellema, D. J., Tollestrup, A. V., and Walker, R. L. (1976), *Phys. Rev. Lett.* **37**, 80.

Dakin, J. T., Feldman, G. J., Lakin, W. L., Martin, F., Perl, M. L., Petraske, E. W., and Toner, W. T. (1973*a*), *Phys. Rev. Lett.* **30**, 142.

Dakin, J. T., Feldman, G. J., Lakin, W. L., Martin, F., Perl, M. L., Petraske, E. W., and Toner, W. T. (1973*b*), *Phys. Rev.* **D8**, 687.

Damman, I., Driver, C., Heinloth, K., Hofmann, G., Janata, F., Karow, P., Lüke, D., Schmidt, D., and Specht, G. (1973), *Nucl. Phys.* **B54**, 355.

Deutsch, M., Golub, L., Kijewski, P., Potter, D., Quinn, D. J., and Rutherfoord, J. (1972), *Phys. Rev. Lett.* **29**, 1752.

Devenish, R. C. E., and Lyth, D. H. (1972), *Phys. Rev.* **D3**, 2752.

Dewire, J., Gittelman, B., Loe, R., Loh, E. C., Ritchie, D. J., and Lewis, R. A. (1971), *Phys. Lett.* **37B**, 326.

Diebold, R. (1970), in *Proceedings of the Boulder Conference on High Energy Physics*, Boulder, Colorada, August 18–22, 1969, Ed. K. T. Mahanthappa, W. D. Walker, and W. E. Brittin (Boulder, Colorado Assoc. University Press).

Donnachie, A. (1972), *High Energy Physics*, Ed. E. H. Burhop (New York, Academic Press), Vol. V.

Donnachie, A. (1976), *Proceedings of the Seventh International Symposium on Lepton and Photon Reactions at High Energies* (1975), Stanford, Ed. W. T. Kirk (Stanford Linear Accelerator, Stanford), p. 473.

Driver, C., Heinloth, K., Höhne, K., Hofmann, G., Karow, P., Rathje, J., Schmidt, D., and Specht, G. (1971*a*), *Phys. Lett.* **35B**, 77.

Driver, C., Heinloth, K., Höhne, K., Hofmann, G., Karow, P., Rathje, J., Schmidt, D., and Specht, G. (1971*b*), *Phys. Lett.* **35B**, 81.

Driver, C., Heinloth, K., Höhne, K., Hofmann, G., Karow, P., Rathje, J., Schmidt, D., and Specht, G. (1971*c*), *Nucl. Phys.* **B30**, 245.

Driver, C., Heinloth, K., Höhne, K., Hofmann, G., Karow, P., Schmidt, D., Specht, G., and Rathje, J. (1971*d*), *Nucl. Phys.* **B32**, 45.

Eckardt, V., Gabauer, H. J., Joos, P., Meyer, H., Naroska, B., Notz, D., Podolsky, W. J., Wolf, G., Yellin, S., Dau, H., Drews, G., Greubel, D., Meincke, W., Nagel, H., and Rabe, E. (1973), *Nucl. Phys.* **B55**, 45.

Eisenberg, Y., Haber, B., Horovitz, B., Peleg, E., Ronat, E. E., Shapira, A., Vishinsky, G., Yekutieli, G., Ballam, J., Chadwick, G. B., Guiragossian, Z. G. T., Levy, A., Menke, M., Seyboth, P., and Wolf, G. (1969), *Phys. Rev. Lett.* **23**, 1322.

Eisenberg, Y., Haber, B., Ronat, E. E., Shapira, A., and Yekutieli, G. (1970), *Phys. Rev. Lett.* **25**, 764.

Eisenberg, Y., Haber, B., Ronat, E. E., Shapira, A., Stahl, Y., Yekutieli, G., Ballam, J.,

Chadwick, G. B., Menke, M., Seyboth, P., Dagan, S., and Levy, A. (1972), *Phys. Rev.* **D5**, 15.

Emms, M. J., Kinson, J. B., Stacey, B. J., Votruba, M. F., Woodworth, P. L., Bell, I. G., Dale, M., Evans, D., Major, J. V., Neat, K., Charlesworth, J. A., and Sekulin, R. L. (1974), *Phys. Lett.* **51B**, 195.

Fox, G. C. (1969), *Proceedings of the Third International Conference on High Energy Collisions*, Stonybrook (New York, Gordon and Breach).

Fox, G. C. (1972), Argonne preprint No. ANL/HEP 7208, Vol. II.

Fox, G. C., and Hey, A. J. G. (1973), *Nucl. Phys.* **B56**, 386.

Fox, G. C., and Quigg, C. (1973), *Ann. Rev. Nucl. Sci.* **23**, 219.

Gault, F. D., Martin, A. D., and Kane, G. L. (1971), *Nucl. Phys.* **B32**, 429.

Genzel, H., Heide, P., Knütel, J., Lierl, H., Mess, K.-H., Schachter, M.-J., Schmüser, P., Sonne, B., and Vogel, G. (1975), *Nucl. Phys.* **B92**, 196.

Geweniger, G., Heide, P., Kötz, H., Lewis, R. A., Schmüser, P., Skronn, H. J., Wahl, H., and Wegener, K. (1969), *Phys. Lett.* **29B**, 41.

Gobbi, B., Rosen, J. L., Scott, H. A., Shapiro, S. L., Stawezynski, L., and Meltzer, C. M. (1974), *Phys. Rev. Lett.* **33**, 1450.

Goldstein, G. R. (1974), *Nucl. Phys.* **B79**, 341.

Goldstein, G. R., and Owens, J. F. (1974), *Nucl. Phys.* **B71**, 461.

Goldstein, G. R., Owens, J. F., and Rutherfoord, J. P. (1973), *Nucl. Phys.* **B53**, 197.

Goldstein, G. R., Owens, J. F., Rutherfoord, J. P., and Moravcsik, M. J. (1974), *Nucl. Phys.* **B80**, 164.

Gourdin, M. (1974), *Phys. Rep.* **11C**, 29.

Gribov, V. N. (1963), *Sov. Phys. JETP* **16**, 1080.

Gutbrod, F., and Kramer, G. (1972), *Nucl. Phys.* **B49**, 461.

Haber, B., Hodous, M. F., Hulsizer, R. I., Kistiakowsky, V., Levy, A., Pless, I. A., Singer, R. A., Wolfson, J., and Yamamoto, R. K. (1974), *Phys. Rev.* **D10**, 1387.

Halzen, F., and Michael, C. (1971), *Phys. Lett.* **B36**, 367.

Halzen, F., and Minkowski, P. (1971), *Nuovo Cimento* **1A**, 59.

Harari, H. (1969), *Proceedings of the Fourth International Symposium on Electron and Photon Interactions at High Energies*, Liverpool, 14–20 September, 1969, Ed. D. W. Braben and R. E. Rand (Danesbury Laboratory).

Harari, H. (1971*a*), *Hadronic Interactions of Electrons and Photons, Proceedings of the Eleventh Scottish Universities Summer School in Physics*, 1970, Ed. J. Cumming and H. Osborne (London, Academic Press).

Harari, H. (1971*b*), *Proceedings of the 1971 International Symposium on Electron and Photon Interactions at High Energies*, Cornell University, Ithaca, New York, Ed. N. B. Mistry (Cornell University, Laboratory of Nuclear Studies).

Harari, H. (1971*c*), *Phys. Rev. Lett.* **26**, 1400.

Harari, H. (1971*d*), *Phys. Rev. Lett.* **27**, 1028.

Heide, P., Kötz, U., Lewis, R. A., Schmuser, P., Skronn, H. J., and Wahl, H. (1968), *Phys. Rev. Lett.* **21**, 248.

Hogaasen, H. (1971), *Phys. Norvegica* **5**, 219.

Hontebeyrie, M., Procureur, J., and Salin, Ph. (1973), *Nucl. Phys.* **B55**, 83.

Horn, D., and Jacob, M. (1968), *Nuovo Cimento* **56A**, 83.

Hufton, A. (1973), thesis, University of Liverpool.

Irving, A. C. (1975), *Nucl. Phys.* **B86**, 125.

Irving, A. C. (1976), *Nucl. Phys.* **B105**, 491.

Irving, A. C., and Vanryckeghem, L. G. F. (1975), *Nucl. Phys.* **B93**, 324.

Irving, A. C., Martin, A. D., and Michael, C. (1971), *Nucl. Phys.* **B32**, 1.

Jackson, J. D., and Hite, G. E. (1968), *Phys. Rev.* **169**, 1248.

Jackson, J. D., and Quigg, C. (1969), *Phys. Lett.* **29B**, 236.

Jackson, J. D., and Quigg, C. (1970), *Nucl. Phys.* **B22**, 301.

Jacob, M., and Wick, G. C. (1959), *Ann. Phys.* (*N.Y.*), **7**, 404.

Joseph, P. M., Hicks, N., Litt, L., Pipkin, F. M., and Russell, J. J. (1967), *Phys. Rev. Lett.* **19**, 1206.

Kaidalov, A. B., and Karnokov, B. M. (1969), *Phys. Lett.* **29B**, 376.

Kane, G. L., Henyey, F., Richards, D. R., Ross, M., and Williamson, G. (1970), *Phys. Rev. Lett.* **25**, 2519.

Kellett, B. H. (1970), *Nucl. Phys.* **B25**, 205.

Kotanski, A. (1966), *Acta Phys. Polonica* **29**, 699.

Kramer, G. (1974), *Acta Phys. Austriaca* **40**, 150.

Kummer, P. S., Clegg, A. B., Foster, F., Hughes, G., Siddle, R., Allison, J., Dickinson, B., Evangelides, E., Ibbotson, M., Lawson, R., Meaburn, R. S., Montgomery, H. E., Shuttleworth, W. J., and Sofair, A. (1971), *Nuovo Cimento Lett. Ser.* 2 **1**, 1026.

Levinson, C. A., Lipkin, H. J., and Meshkov, S. (1963), *Phys. Lett.* **7**, 81.

Levy, N., Majerotto, W., and Read, B. J. (1973*a*), *Nucl. Phys.* **B55**, 493.

Levy, N., Majerotto, W., and Read, B. J. (1973*b*), *Nucl. Phys.* **B55**, 513.

Lübelsmeyer, K. (1969), *Proceedings of the Fourth International Symposium on Electron and Photon Interactions at High Energies*, Liverpool, 14–20 September, 1969, Ed. D. W. Braben and R. E. Rand (Daresbury Laboratory).

MacDowell, S. W. (1959), *Phys. Rev.* **116**, 774.

Manweiler, P. W., and Schmidt, W. (1971), *Phys. Rev.* **D3**, 2752.

Michael, C., and Odorico, R. (1971), *Phys. Lett.* **34B**, 422.

Minkowski, P. (1970), *Nuovo Cimento Lett. Ser.* 1 **3**, 59.

Moorhouse, R. G. (1976), *High Energy Physics: Proceedings of the EPS International Conference*, Palermo (Italy), 23–28 June, 1975, Ed. A. Zichichi (Bologna, Editrice Compositon).

Morehouse, C. C., Borghini, M., Chamberlain, O., Fuzesy, R., Gorn, W., Powell, T., Robrish, P., Rock, S., Shannon, S., Shapiro, G., Weisberg, H., Boyarski, A. M., Ecklund, S. D., Murata, Y., Richter, B., Siemann, R., and Diebold, R. (1970), *Phys. Rev. Lett.* **25**, 835.

Morpourgo, G. (1964), *Nuovo Cimento* **31**, 569.

Nagels, M. M., de Swart, J. J., Nielsen, H., Oades, G. C., Petersen, J. L., Tromberg, B., Gustafson, G., Irving, A. C., Jarlskog, C., Pfeil, W., Pilkuhn, H., Steiner, F., and Tauscher, L. (1976), *Nucl. Phys.* **B109**, 1.

Nieh, H. T. (1972), *Phys. Lett.* **B38**, 100.

Osborne, A. M., Browman, A., Hanson, K., Meyer, W. T., Silverman, A., Taylor, F. E., and Horwitz, A. (1972), *Phys. Rev. Lett.* **29**, 1621.

Particle Data Group (1976), *Rev. Mod. Phys.* **48**, S1.

Pickering, A. R. (1973), *Nucl. Phys.* **B66**, 493.

Pond, P., and Kumar, A. (1972), *Nucl. Phys.* **B36**, 241.

Primakoff, H. (1951), *Phys. Rev.* **81**, 899.

Quinn, D. J., Rutherfoord, J. P., Shupe, M. A., Sherden, D. J., Siemann, R. H., and Sinclair, C. K. (1975), *Phys. Rev. Lett.* **34**, 543.

Ravndal, F. (1970), *Phys. Rev. D* **2**, 1278.

Richter, B. (1967), *Proceedings of the 1967 International Symposium on Electron and Photon Interactions at High Energies*, Stanford Linear Accelerator Center, 5–9 September, 1967 (Stanford, Stanford Linear Accelerator Center).

Ross, M., Henyey, F., and Kane, G. L. (1970), *Nucl. Phys.* **B23**, 269.

Schacht, P., Derado, I., Fries, D. C., Park, J., and Yount, D. (1974), *Nucl. Phys.* **B81**, 205.

Schrempp, B., and Schrempp, F. (1973*a*), *Nucl. Phys.* **B54**, 525.

Schrempp, B., and Schrempp, F. (1973*b*), *Nucl. Phys.* **B60**, 110.

Schrempp, B., and Schrempp, F. (1974), *Nucl. Phys.* **B77**, 453.

Schrempp, B., and Schrempp, F. (1975), *Nucl. Phys.* **B96**, 307.

Schwitters, R. F., Leong, J., Luckey, D., Osborne, L. S., Boyarski, A. M., Ecklund, S. D., Siemann, R., and Richter, B. (1971), *Phys. Rev. Lett.* **27**, 120.

Sherden, D. J., Siemann, R. H., Sinclair, C. K., Quinn, D. J., Rutherfoord, J. P., and Shupe, M. A. (1973), *Phys. Rev. Lett.* **30**, 1230.

Sherden, D. J., Siemann, R. H., Sinclair, C. K., Quinn, D. J., Rutherfoord, J. P., and Shupe, M. A. (1974), Contribution No. 91 to the *Proceedings of the Sixth International Symposium on Electron and Photon Interactions at High Energies*, Bonn, 27–31 August, 1973, Ed. H. Rollnik and W. Pfeil (Amsterdam, North-Holland).

Simonius, M. (1967), *Phys. Rev. Lett.* **19**, 273.

Sofair, A., Allison, J., Dickinson, B., Evangelides, E., Ibbotson, M., Lawson, R., Meaburn, R. S., Montgomery, H. E., Shuttleworth, W. J., Clegg, A. B., Foster, F., Hughes, G., Kummer, P., and Siddle, R. (1972), *Nucl. Phys.* **B42**, 369.

Stichel, P. (1964), *Z. Phys.* **180**, 170.

Stichel, P., and Scholz, M. (1964), *Nuovo Cimento* **34**, 1381.

Storrow, J. K. (1972), *Nucl. Phys.* **B47**, 174.

Storrow, J. K. (1975), *Nucl. Phys.* **B96**, 77.

Storrow, J. K. and Winbow, G. A. (1973*a*). *Nucl. Phys.* **B53**, 62.

Storrow, J. K., and Winbow, G. A. (1973*b*), *Nucl. Phys.* **B54**, 560.

Talman, R. (1974), *Proceedings of the Sixth International Symposium on Electron and Photon Interactions at High Energies*, Bonn, 27–31 August, 1973, Ed. H. Rollnick and W. Pfeil (Amsterdam, North-Holland).

Tompkins, D., Anderson, R., Gittelman, B., Litt, J., Wiik, B. H., Yount, D., and Minten, A. (1969), *Phys. Rev. Lett.* **23**, 725.

Triantafillopoulos, E. (1977), Manchester University, Ph.D. Thesis.

Vanryckeghem, L. G. F. (1974), *Phys. Lett.* **53B**, 272.

Vogel, G., Burfeindt, H., Buschorn, G., Heide, P., Kötz, U., Mess, K.-H., Schmüser, P., Sonne, B., and Wiik, B. H. (1972), *Phys. Lett.* **40B**, 513.

Wiik, B. H. (1971), *Proceedings of the 1971 International Symposium on Electron and Photon Interactions at High Energies*, Cornell University, Ithaca, New York, Ed. N. B. Mistry (Cornell University, Laboratory of Nuclear Studies).

Williams, P. K. (1970), *Phys. Rev.* **D1**, 1312.

Wolf, G. (1976), *Proceedings of the 1975 International Symposium on Lepton and Photon Interactions at High Energies*, Stanford University, 21–27 August, 1975, Ed. W. T. Kirk (Stanford, Stanford Linear Accelerator Center).

Worden, R. P. (1972), *Nucl. Phys.* **B37**, 253.

Worden, R. P. (1973), *Proceedings of the Pion Exchange Meeting*, Ed. G. A. Winbar, Daresbury Study Weekend Series No. 6, Daresbury Laboratory Report No. DNPL/R30.

7

High-Energy Photoproduction: Diffractive Processes

D. W. G. S. Leith

1. Introduction

The interaction of photons with nucleons in the energy range 2.0–30 GeV is observed to have many of the features of purely hadronic processes (i.e., πN, KN, pN collisions). More specifically, the total cross section is almost independent of energy, and the elastic scattering amplitude is mainly imaginary with a sharp t dependence and almost no energy dependence. These features find a natural explanation in the vector dominance model of photoprocesses (Sakurai, 1960), in which the photon is pictured as a superposition of vector mesons which mediate the interaction of the photon with other hadrons. This implies that the photoproduction of vector mesons should be a favored process in γp reactions, and indeed it is observed to account for $\sim 20\%$ of the total cross section. The vector meson production process, the elastic γ–nucleon scattering (Compton scattering), and, through the optical theorem, the total γ–nucleon cross section all display the characteristics of diffractive processes. But before continuing on a detailed review of the properties of these reactions, we first review the general phenomenological features of diffractive scattering.

Diffraction scattering can be discussed in terms of two pictures representing the t-channel or the s-channel points of view. In the t-channel

D. W. G. S. Leith • Stanford Linear Accelerator Center, Stanford University, Stanford, California 94305

(or exchange-channel) picture the scattering is thought to proceed through the exchange of the Pomeron, the basis for this picture being that of Regge exchange models. The s-channel (or direct-channel) picture is seen in geometric or optical terms, where diffraction is generated by the absorption due to the competition among the many open inelastic channels. The target nucleon is seen as an absorbing disk (black or grey), with a specific radius and a specific opacity. From this viewpoint the diffractive reactions are merely the shadow of all the inelastic processes taking place. In both pictures, measurements of the energy and momentum transfer dependence of the diffractive reactions provides information either on the Pomeron amplitude or on the size and opacity of the scatterer.

Unfortunately, beyond these two pictures we have no good theoretical description of the dynamics of diffractive processes and no basic understanding of the Pomeron singularity. Rather, we have a set of phenomenological rules that allows us to identify what we mean by diffraction. Below we list the features that we expect from a diffractive process:

 (i) Cross sections are energy independent, or at most increase no faster than logarithms of the energy.
 (ii) There is a sharp forward peak in the differential cross section, $d\sigma/dt$.
 (iii) The scattering amplitude is mainly imaginary.
 (iv) Particle cross sections are equal to antiparticle cross sections.
 (v) Factorization—i.e., the strength of a given subprocess (e.g., $a \to A$ or $b \to B$ in the reaction $ab \to AB$)—is independent of which other particles are participating in the reaction.
 (vi) The exchange process is characterized by the quantum numbers of the vacuum (in the t channel).
 (vii) Change in parity in the scattering process follows the natural spin-parity series $(-1)^{\Delta J}$ or $P_0 = P_i(-1)^{\Delta J}$, where ΔJ is the spin change and P_0, P_i are the intrinsic parities of the outgoing and incoming particles.
 (viii) The spin structure of the scattering process is dominantly s-channel helicity conserving (SCHC).

Photoproduction is an interesting process for the study of diffraction reactions, as it probes the same s-channel quantum numbers as πN scattering (i.e., $S = 0$, $B = 1$). However, the mixed isoscalar and isovector nature of the photon, and its spin-1 character bring an unusual variety to the study of the scattering process. Furthermore, since polarized beams are relatively easy to prepare, vector meson photoproduction allows a detailed study of the spin characteristics of diffractive reactions. It has also been pointed out (Freund, 1967) that vector meson photoproduction, especially of the ϕ meson, should be a particularly suitable laboratory for study of diffraction, since, in these reactions, the other exchanges (f, A_2, π, ...) are suppressed.

In the following sections we review the experimental results on high-energy diffractive photoproduction. In the preparation of this review, the following excellent articles have been extensively used: Wolf (1972a), Moffeit (1974), and Silverman (1975). Data on the total cross section and on Compton scattering are presented and discussed in Section 2, followed by a detailed look at vector meson photoproduction of ρ, ω, and ϕ in Section 3 and of higher-mass states in Section 4. The vector dominance model is discussed in Section 5, together with a review of the measurements of the photon–vector-meson coupling strengths. Section 6 contains some concluding remarks.

2. Total Photon–Nucleon Cross Section and Compton Scattering

2.1. Total Cross Sections

The total photon–nucleon cross section has been measured from threshold up to ~ 30 GeV on both proton and neutron targets, using a variety of techniques. The most extensive measurements come from counter setups in tagged photon beams at NINA (Armstrong *et al.*, 1972), DESY (Meyer *et al.*, 1970), SLAC (UCSB) (Caldwell *et al.*, 1973), and SERPUKOV (Lebedev) (Belousov *et al.*, 1972). Other measurements have been obtained with monochromatic photon beams at SLAC—both from $e^+ e^-$ annihilation in flight and from the backscattered laser beams—using hydrogen- or deuterium-filled bubble chambers as a detector (Ballam *et al.*, 1972; Bingham *et al.*, 1973). All the above measurements are in good agreement. Early measurements of the total γp cross section were also obtained from an analysis of extensive inelastic electron (Bloom *et al.*, 1969) and muon scattering data (Lakin *et al.*, 1971), where the one-photon exchange approximation was used and the total cross section was extracted by extrapolating the inelastic scattering to $q^2 = 0$. These measurements agree with the directly measured total cross section to within 10–20%. The energy dependence of the total cross section exhibits the well-known resonant structure from N^* formation at low energy, followed by a very slow, smooth falloff as the photon energy increases above 2 GeV. The photon cross sections show behavior very similar to the πN cross sections, but they are much smaller. Figure 1 shows $1/200$ of the average $\pi^+ p$ and $\pi^- p$ cross sections compared to an average of the photon measurements. The similarity is striking.

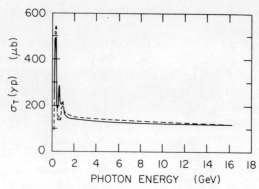

Fig. 1. A comparison of the γp total cross section (solid curve) with (1/200) of the average of the $\pi^+ p$ and $\pi^- p$ total cross sections (broken curve).

At high energies, the forward Compton scattering amplitude and, via the optical theorem, the total cross section, are expected to be dominated by the following $C = +1$ exchanges—$\mathbb{P}$, f, A_2. The Pomeron ($\mathbb{P}$) is expected to give rise to an energy-independent contribution, while the meson exchange contributions (f, A_2) fall off like $E^{-1/2}$. This leads to an expected energy dependence of the form

$$\sigma(\gamma N) = a_0 + a_1 \cdot E^{-1/2} \qquad (2.1)$$

where E is the photon energy. The isoscalar and isovector exchange contributions (T_0 and T_1, respectively) couple with opposite sign to the Compton scattering on protons and neutrons:

$$A_{\gamma p} = T_0 + T_1 \qquad (2.2)$$

$$A_{\gamma n} = T_0 - T_1 \qquad (2.3)$$

Identifying the isospin of the various exchange terms, we may write the cross section on protons and neutrons as

$$\sigma_T(\gamma p) = C_{\mathbb{P}} + (C_f + C_{A_2}) \cdot E^{-1/2} \qquad (2.4)$$

$$\sigma_T(\gamma n) = C_{\mathbb{P}} + (C_f - C_{A_2}) \cdot E^{-1/2} \qquad (2.5)$$

Taking the sum and difference of the total cross sections then allows the isolation of the isoscalar and isovector exchange contributions:

$$\tfrac{1}{2}[\sigma_T(\gamma p) + \sigma_T(\gamma n)] = \operatorname{Im} T_0 = C_{\mathbb{P}} + C_f \cdot E^{-1/2} \qquad (2.6)$$

$$\tfrac{1}{2}[\sigma_T(\gamma p) - \sigma_T(\gamma n)] = \operatorname{Im} T_1 = C_{A_2} \cdot E^{-1/2} \qquad (2.7)$$

In Fig. 2, all the data on direct measurement of the total cross section on protons is shown. A best fit to the form of Eq. (2.1) yields

$$\sigma_T(\gamma p) = (98.7 \pm 3.6) + (65 \pm 10.1) \cdot E^{-1/2} \ \mu\text{b} \qquad (2.8)$$

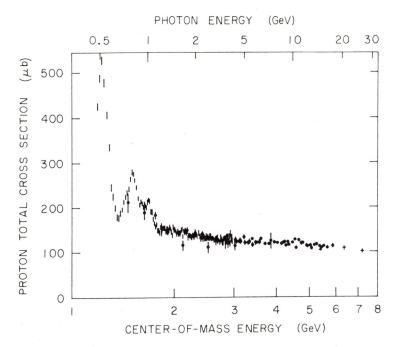

Fig. 2. The total γp cross section from threshold up to 30 GeV.

The measurements of the total cross section on deuterium are given in Fig. 3.

The neutron cross sections are obtained from the deuterium data using the relationship

$$\sigma_T(\gamma n) = K \cdot \sigma_T(\gamma d) - \sigma_T(\gamma p) + \sigma_s + \sigma_w \qquad (2.9)$$

where K is a kinematic factor which takes into account the Fermi motion of the nucleon, σ_s is the screening correction (Franco and Glauber, 1966) to account for the shadowing of one nucleon by the other, and σ_w is a smearing term (West, 1971), which corrects for the fact that the nucleons in the deuterium are slightly off the mass shell and moving. Around 5 GeV the two corrections (σ_s, σ_w), are of comparable size, and together amount to $\sim 3\%$ adjustment to the measured cross sections, with an estimated uncertainty of $\sim 0.5\%$. These corrections are very important since they are of the same magnitude as the difference in the proton and neutron cross sections for energies around 10 GeV. The neutron cross sections, obtained in this manner, are shown in Fig. 4. The best fit to the energy dependence of Eq. (2.1) yields

$$\sigma_T(\gamma n) = (103.4 \pm 6.7) + (33.1 \pm 19.4) \cdot E^{-1/2} \, \mu b \qquad (2.10)$$

The difference between the neutron and proton cross sections is also shown in Fig. 4. A best fit to the SLAC (UCSB) cross sections yields

$$\sigma_T(\gamma p) - \sigma_T(\gamma n) = (18.3 \pm 6.1) \cdot E^{-1/2} \,\mu b \qquad (2.11)$$

However, it is also clear from Fig. 4 that the Serpukov measurements are compatible with no difference between the proton and neutron cross sections. Therefore the result given in Eq. (2.11) should be regarded as an upper limit.

An analysis of the isospin composition of the total cross sections, using Eqs. (2.6) and (2.7), yields the following estimates for the $I = 0$ and 1 exchange contributions (Caldwell *et al.*, 1973; Dominguez *et al.*, 1972):

$$I = 0 \text{ exchange: } \begin{cases} C_P = (101.9 \pm 2.9) & \mu b \\ C_f = (50.9 \pm 8.5) & \mu b \cdot \text{GeV}^{1/2} \end{cases} \qquad (2.12)$$

$$I = 1 \text{ exchange: } C_{A_2} = (9.1 \pm 3) \qquad \mu b \cdot \text{GeV}^{1/2}$$

From these results we may derive the isovector exchange contribution to

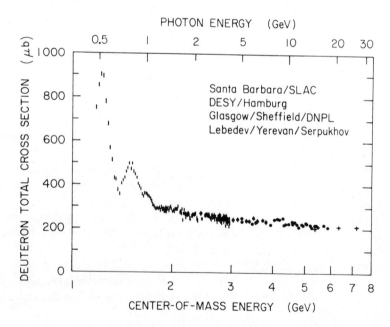

Fig. 3. The total γd cross section from threshold up to 30 GeV.

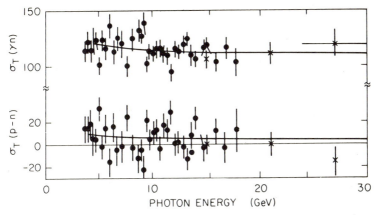

Fig. 4. The γn total cross section derived from the measured γd and γp cross sections. Also shown is the difference between the neutron and proton cross sections.

the forward Compton amplitudes, and find

$$\frac{\text{Im } T_1}{\text{Im } T_0} = \begin{cases} (2 \pm 1)\% & \text{at 6 GeV} \\ (1.6 \pm 0.5)\% & \text{at 25 GeV} \end{cases} \qquad (2.13)$$

In summary, the total photon–nucleon cross section is observed to behave like the πN total cross section, reduced by a factor of ~ 200; the energy dependence exhibits a slow decrease above 2 GeV. The proton cross section is slightly larger than the neutron cross section, implying a small but probably nonzero contribution for isovector exchange to the Compton amplitude ($\lesssim 3\%$ around 6 GeV).

2.2. Compton Scattering

Compton scattering (i.e., elastic photon–nucleon scattering), is not only one of the basic reactions between photons and nucleons, but is also interesting owing to its relation to the total hadronic cross section, σ_T, via the optical theorem, and to the photoproduction of vector mesons through the vector dominance model (Sakurai, 1960). The process is expected to be diffractive and exhibit strong forward peaking in the scattering angular distribution. Furthermore, the comparison of the forward cross section, $(d\sigma/dt)_0$, and σ_T allows a test of forward dispersion relations not possible from πN data, since γp scattering may have extra contributions to the real part of the diffractive amplitude that are not present in πN collisions (Damashek and Gilman, 1970).

The Compton scattering reaction has been well studied over a wide range of energies and momentum transfers, and there is a good agreement

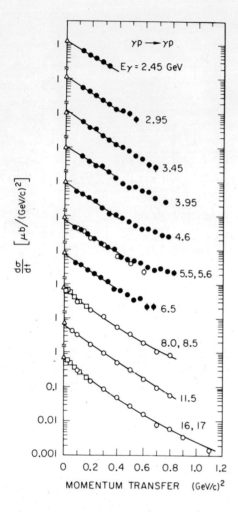

Fig. 5. The differential cross section for Compton scattering for energies in the range 2–18 GeV. The data are from Buschhorn et al. (1970) (●), Anderson et al. (1970a) (○), and Boyarski et al. (1971) (□).

between the different experiments (Buschhorn et al., 1970; Anderson et al., 1970a; Boyarski et al., 1971). The results of these measurements are shown in Fig. 5. The differential cross sections show a sharp forward peak with very little energy dependence of the forward cross section or of the shape of the scattering distribution. The differential cross sections are fitted to the forms

$$\frac{d\sigma}{dt} = \frac{d\sigma}{dt}\bigg|_0 \cdot \exp(At) \qquad (2.13a)$$

$$\frac{d\sigma}{dt} = \frac{d\sigma}{dt}\bigg|_0 \cdot \exp(At + Bt^2) \qquad (2.13b)$$

Table 1. Summary of Compton Scattering Cross Sections

$$\left[\frac{d\sigma}{dt} = \frac{d\sigma}{dt}\bigg|_0 \, exp(At + Bt^2)\right]$$

Experiment	E (GeV)	t Range [(GeV/c)²]	$(d\sigma/dt)_0$ [μb/(GeV/c)²]	A [(GeV/c)⁻²]	B [(GeV/c)⁻⁴]
Buschhorn *et al.* (1970)	2.2–2.7	0.1–0.4	1.26 ± 0.13	5.2 ± 0.5	—
	2.7–3.2	—	1.14 ± 0.11	5.7 ± 0.4	—
	3.2–3.7	—	1.24 ± 0.11	6.2 ± 0.4	—
	3.7–4.2	—	1.02 ± 0.14	5.3 ± 0.5	—
	4.0–5.2	0.06–0.4	0.92 ± 0.09	6.0 ± 0.4	—
	5.0–6.2	—	0.76 ± 0.06	5.5 ± 0.3	—
	6.0–7.0	—	0.76 ± 0.07	5.9 ± 0.4	—
Boyarski *et al.* (1971)	8	0.014–0.17	0.82 ± 0.04	7.7 ± 0.5	—
		0.014–0.8	0.79 ± 0.03	7.6 ± 0.4	2.3 ± 0.5
Anderson *et al.* (1970*a*)	16	0.014–0.17	0.69 ± 0.03	7.9 ± 0.5	—
		0.014–1.1	0.64 ± 0.02	7.3 ± 0.3	1.7 ± 0.3

For momentum transfers less than 0.6 GeV², the simple exponential form [Eq. (2.13a)] describes the data well, but the quadratic form of Eq. (2.13b) is required to fit the data out to $t = 1(\text{GeV}/c)^2$. The results of these fits are summarized in Table 1. The Compton scattering differential cross sections show behavior very similar to that of πN elastic scattering.

The forward amplitude in Compton scattering may be written (Gell-Mann *et al.*, 1954; Damashek and Gilman, 1970),

$$f^0_{\gamma p} = (\boldsymbol{\varepsilon}_f \cdot \boldsymbol{\varepsilon}_i) f_1 + i\boldsymbol{\sigma} \cdot (\boldsymbol{\varepsilon}_f \times \boldsymbol{\varepsilon}_i) f_2 \qquad (2.14)$$

where $\boldsymbol{\varepsilon}_i$, $\boldsymbol{\varepsilon}_f$ are the polarization vectors of the photon before and after the scattering, $\boldsymbol{\sigma}$ is the Pauli spin matrix of the recoil proton, and f_1 and f_2 are the amplitudes for parallel and perpendicular polarization vectors. Now, applying the optical theorem [$\text{Im} f_1 = (k/4\pi)\sigma_T$], the forward Compton cross section, $(d\sigma/dt)|_0(k)$, as a function of photon momentum, k, may be written as

$$\frac{d\sigma}{dt}\bigg|_0 (k) = \frac{\sigma_T^2}{16\pi} + \frac{\pi}{k^2} \cdot |\text{Re} f_1|^2 + \frac{\pi}{k^2} \cdot |f_2|^2 \qquad (2.15)$$

The real part of f_1 can be evaluated using the measured total cross sections and a dispersion relation, assuming Regge behavior for $f_1(k)$:

$$\text{Re} f_1(k) = -\frac{\alpha}{M} + \frac{k^2}{2\pi^2} \int \sigma_T(k') \, dk' \qquad (2.16)$$

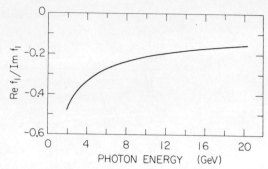

Fig. 6. The calculated ratio of the real to imaginary parts of the forward amplitude for the Compton scattering reaction, $\gamma p \to \gamma p$.

The calculated ratio of the real part to the imaginary part of f_1 is shown in Fig. 6 (Damashek and Gilman, 1970). The calculation implies that the ratio of real to imaginary amplitudes in the forward direction is ~ 0.2 at 5 GeV, and ~ 0.1 at 20 GeV.

A comparison of both sides of Eq. (2.15) is shown in Fig. 7, where the total cross-section data discussed above are compared to the measured forward Compton cross section. Good agreement is obtained assuming $f_2 = 0$, and using f_1 from the dispersion calculation. An estimated upper limit of the f_2 contribution to the forward cross section of 10% is obtained from this comparison.

An estimate of the isovector contribution to Compton scattering may be obtained from a comparison of the cross section on hydrogen and deuterium targets. An example of such a measurement, for photon energies of 8 and 16 GeV, is shown in Fig. 8 (Boyarski et al., 1971). The data show a sharp forward peak from the coherent deuteron scattering with a

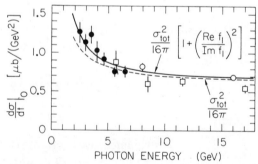

Fig. 7. The energy dependence of the forward differential cross section, $(d\sigma/dt)|_0$, for the Compton scattering process. The curves as calculated from the optical theorem assuming a purely imaginary amplitude (dashed line), and using the calculated real part (solid line). The data are as in Fig. 5.

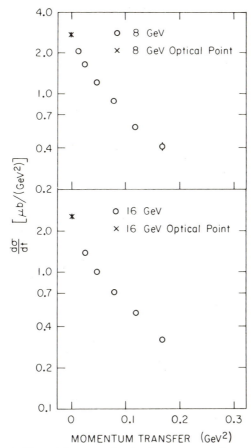

Fig. 8. The measured differential cross section for Compton scattering on a deuterium target, at 8 and 16 GeV.

flatter incoherent contribution at larger t. The deuterium cross section, ignoring spin effects, is given by

$$\frac{d\sigma}{d\Omega}\bigg|_D = 2|T_0|^2\{1 + F(t) + 2G(t)\} + 2|T_1|^2\{1 - F(t)\} \qquad (2.17)$$

where T_0, T_1 are the isoscalar and isovector t-channel exchange amplitudes on nucleons, $F(t)$ is the deuterium form factor, and $G(t)$ is the Glauber scattering term. The cross section on hydrogen is written as follows:

$$\frac{d\sigma}{d\Omega}\bigg|_H = |T_0 + T_1|^2 = |T_0|^2 + |T_1|^2 + 2\,\mathrm{Re}\,T_0 \cdot T_1^* \qquad (2.18)$$

Then,

$$\left[\frac{d\sigma}{dt}\bigg|_D \bigg/ \frac{d\sigma}{dt}\bigg|_H\right] = 2\left\{1 - \frac{2\,\mathrm{Re}\,T_0 \cdot T_1^*}{|T_0 + T_1|^2}\cdot[1 + F(t) + 2G(t)]\right.$$

$$\left. - \frac{2|T_1|^2}{|T_0 + T_1|^2}\cdot[F(t) + G(t)]\right\} \qquad (2.19)$$

The deuterium form factor, $F(t)$, is taken from the electron scattering data, as $F(t) = \exp(56t)$,† and the screening correction term from the calculations of Ogren (1970),

$$G(t) = -0.069\exp(-At/4) + 0.007\exp(-At/2)$$

with $A = 7.8\ \mathrm{GeV}^{-2}$.

The ratio of deuterium and hydrogen cross sections, given in Eq. (2.19), was fitted as a function of t [for $0.014 < t < 0.17\ (\mathrm{GeV}/c)^2$], using the combined 8 and 16 GeV data, and the following ratios were determined:

$$\frac{\mathrm{Re}\,T_0 \cdot T_1^*}{|T_0 + T_1|^2} = -0.049 \pm 0.012$$

$$\frac{|T_1|^2}{|T_0 + T_1|^2} = 0.03 \pm 0.10 \qquad (2.20)$$

A literal interpretation of this result implies that the neutron Compton cross section is larger than the proton cross section, and that the isovector amplitude is purely real. However, a more probable interpretation of these data together with those on the difference between the neutron and proton total cross sections is that the isovector exchange contribution in Compton scattering is small near the forward direction.

3. Photoproduction of Vector Mesons—ρ, ω, ϕ

3.1. Introduction

In this section we discuss the photoproduction of the lowest-lying vector meson states—the ρ, ω, and ϕ mesons. These processes have been extensively studied using a variety of techniques, on both proton and deuteron targets. The experimental situation is summarized in Table 2. The various techniques each have their own peculiar experimental problems. The track chamber experiments allow a clean isolation of the

---

†This parameterization of the form factor is known to fall off too steeply. A more accurate parameterization is given by Franco and Varma (1974), viz.,
$$F(t) = 0.34e^{141.5t} + 0.58e^{26.1t} + 0.08e^{15.5t}$$

Table 2. Summary of Vector Meson Photoproduction Experiments

Experiment	Vector mesons studied	Technique	Beam	Photon energy (GeV)
$\gamma p \to Vp$				
CEA Collab. (1966)	ρ, ω, ϕ	bubble chamber	bremsstrahlung	6
DESY Collab. (1968)	ρ, ω, ϕ	bubble chamber	bremsstrahlung	6
Alexander *et al.* (1972)	ρ, ω, ϕ	bubble chamber	annihilation (quasimonochromatic)	4.3, 5.3, 7.5
Ballam *et al.* (1973)	ρ, ω, ϕ	bubble chamber	backscattered laser (monochromatic)	2.8, 4.7, 9.3
Davier *et al.* (1970)	ρ, ω, ϕ	streamer chamber	bremsstrahlung	16
DESY/AHHM Collab. (1971)	ρ, ω, ϕ	streamer chamber	tagged beam	3–6
Blechschmidt *et al.* (1967)	ρ	counter setup	tagged beam	3–5
McClellan *et al.* (1971*a*; 1971*b*)	ρ, ϕ	counter setup	bremsstrahlung	8.5
Alvensleben *et al.* (1970*b*; 1972)	ρ, ϕ	setup	bremsstrahlung	7
Bulos *et al.* (1970)	ρ	counter setup	annihilation (quasimonochromatic)	9
Giese (1974)	ρ	counter setup	bremsstrahlung	16
Anderson *et al.* (1970*b*)	ρ	counter setup	bremsstrahlung	18
Barish *et al.* (1974)	ρ	counter setup	bremsstrahlung	12
Gladding *et al.* (1973)	ρ, ω	counter setup	tagged beam	3, 4.2
Behrend *et al.* (1971*a*)	ω	counter setup	bremsstrahlung	9
Behrend *et al.* (1975*a*)	ϕ	counter setup	tagged beam	3–7
Berger *et al.* (1972)	ρ, ϕ	counter setup	bremsstrahlung	8.5
$\gamma n \to Vn$				
DESY/ABHHM Collab. (1971)	ρ, ω, ϕ	bubble chamber	bremsstrahlung	6
Eisenberg *et al.* (1972*a, b*; 1976)	ρ, ω	bubble chamber	backscattered laser (monochromatic)	4.3
Alexander *et al.* (1973; 1975)	ρ, ω	bubble chamber	backscattered laser (monochromatic)	7.5
McClellan *et al.* (1971*b*)	ρ, ϕ	counter setup	bremsstrahlung	8
Bulos *et al.* (1970)	ρ	counter setup	annihilation (quasimonochromatic)	9
Giese (1974)	ρ	counter setup	bremsstrahlung	16
Abramson *et al.* (1976)	ω	counter setup	tagged beam	8
Behrend *et al.* (1971*b*)	ω	counter setup	bremsstrahlung	9

exclusive reactions with very little background and they measure the full decay angular distribution. However, they have difficulty measuring the cross section near the forward direction, owing to scanning losses [typically for $t < 0.02$–0.05 $(GeV/c)^2$]. The counter experiments either detect the decay products of the photoproduced meson or the recoil nucleon system and then identify the meson production in the missing mass distribution. The former method detects only part of the decay angular distribution, and the observed angular correlations have to be corrected for the geometrical efficiency of the apparatus. The recoil experiments clearly have no information on the decay correlations. Both counter techniques are also vulnerable to inelastic vector meson production being included as background in the data, since they do not, in general, detect the complete final state. However, the counter experiments are high-statistics, good-resolution measurements with data all the way in to the forward direction.

Since the vector mesons have spin parity 1^- like the photon, one might expect these processes to have large cross sections and be diffractive in character (i.e., to exhibit the properties listed in Section 1 above). Indeed, the production rates are found to be large and decrease slowly with energy above 2 GeV, and with scattering distributions that fall off with an exponential slope. An analysis of the decay distributions of the photoproduced vector mesons, especially from polarized photons, provides a powerful tool for the study of the spin dependence of diffraction scattering.

3.2. Rho Production

3.2.1. Cross Sections

Rho production is the dominant process in the reaction $\gamma N \rightarrow \pi^+ \pi^- N$. The $\pi^+ \pi^-$ mass distribution is shown in Fig. 9 for the 9.3 GeV hydrogen and 4.3 GeV deuterium bubble-chamber experiments (Ballam *et al.*, 1973; Eisenberg *et al.*, 1976). The ρ^0 peak dominates the $\pi^+ \pi^-$ mass spectrum, and what little background exists under the peak is observed to get smaller, the higher the photon energy. The ρ^0 shape is clearly skewed, in that there are too many low-mass events and too few high-mass events when compared to a p-wave Breit–Wigner resonance shape. The skewing, and resultant ρ^0 mass shift, are observed to be dependent on momentum transfer and are most pronounced in the forward direction. Since this effect is not fully understood, and since the ρ^0 is a broad object, defining an absolute cross section is rather difficult. Further, since the backgrounds are energy dependent and the skewing of the ρ^0 shape is t dependent, it is very difficult to compare cross sections from different experimental techniques, measuring different fractions of the decay distribution, over different t ranges, at

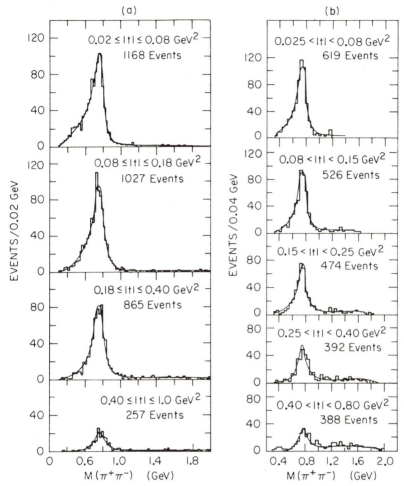

Fig. 9. The dipion mass spectrum for several regions of momentum transfer for two experiments: (a) a 9.3 GeV study of the reaction $\gamma p \to \pi^+\pi^- p$, and (b) a 4.3 GeV experiment measuring $\gamma d \to p n \pi^+ \pi^-$.

different energies. The typical spread in the reported cross sections due to these uncertainties is of order 10–30%.

The above behavior of the $\pi^+\pi^-$ mass spectrum may be described by several models. We discuss the most successful one—the Söding model (Söding, 1966)—which explains the phenomena in terms of an interference between ρ^0 production and the production of pion pairs through the Drell mechanism (Drell, 1961). The model is shown diagramatically in Fig. 10; diagram (a) refers to the ρ^0 production process while (b) and (c) are the Drell terms; diagrams (d) and (e) are rescattering terms introduced to avoid problems with double counting (Bauer and Yennie, 1970; Pumplin, 1970). They ensure that at the ρ^0 mass, the ρ^0 amplitude saturates the unitarity bound. The mass skewing results from the interference of diagram (a) with the other four diagrams. This interference term changes sign from positive to negative in passing through the ρ^0 mass. The various contributions to the cross section are shown in Fig. 11. While the interference term contributes little to the integrated cross section, it does account for the observed ρ^0 mass shift and skewing of the $\pi^+\pi^-$ spectrum. (This model also provides a good description of the t dependence and angular correlation of the photoproduced pion pairs, as we will discuss below.)

Yennie has used this picture in proposing a simple recipe for determining the ρ^0 cross sections free from the problems discussed above, of having to understand the details of the $\pi^+\pi^-$ mass shape for different experiments measuring in different regions of the kinematic variables. He suggests taking the yield of $\pi^+\pi^-$ pairs at the ρ mass (i.e., $M_{\pi\pi} = M_\rho$), where the interference term is zero, and then the cross section is given by

$$\frac{d\sigma}{dt} = \frac{\pi}{2} \cdot \Gamma_\rho \cdot \frac{d^2\sigma}{dM_{\pi\pi} \cdot dt} \qquad (3.1)$$

where Γ_ρ is the ρ width.

There are problems with this recipe; the mass and width of the ρ meson are not precisely known quantities and they influence strongly the determination of the cross section; the presence of $\omega-\rho$ interference effects are ignored and can introduce $\sim 10\%$ differences in the estimated cross section. However, it is a standard method and has served to bring some degree of order into an area of great confusion. We shall use this definition in comparing the various ρ^0 photoproduction experiments discussed below.

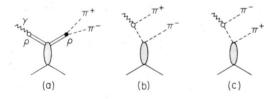

(a)   (b)   (c)

Rescattering Terms

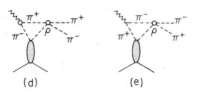

(d)   (e)

Fig. 10. Diagrams describing the Söding model for photoproduction of dipion pairs. For details see text.

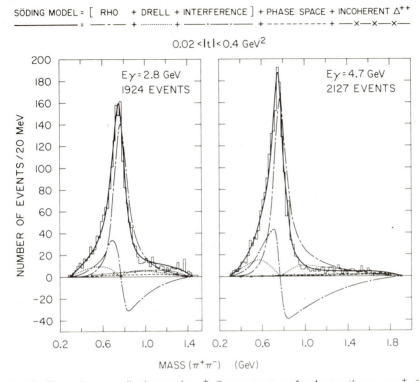

SÖDING MODEL = [RHO + DRELL + INTERFERENCE] + PHASE SPACE + INCOHERENT Δ^{++}

$0.02 < |t| < 0.4 \text{ GeV}^2$

Fig. 11. The various contributions to the $\pi^+\pi^-$ mass spectrum for the reaction $\gamma p \to \pi^+\pi^- p$ at 2.8 and 4.7 GeV, within the framework of the Söding model.

The total ρ^0 production cross section is shown in Fig. 12 from the track chamber experiments. There is good agreement among the different measurements. The energy dependence of the cross section closely resembles that of the average of the elastic $\pi^+ p$ and $\pi^- p$ cross sections. More specifically, from the quark model we expect the $(\rho^0 p)$ system to behave like the average of $\pi^+ p$ and $\pi^- p$ scattering, while from vector dominance we expect the $\gamma \to \rho^0$ transition to be the same as $\rho^0 \to \rho^0$ multiplied by a constant, $(\alpha\pi)^{1/2}/\gamma_\rho$, which measures the strength of the photon–ρ-meson coupling (see Section 5 below). We then may write

$$\sigma(\gamma p \to \rho^0 p) = \frac{\alpha\pi}{\gamma_\rho^2} \sigma^{\text{el}}(\rho^0 p \to \rho^0 p)$$

$$= \frac{\alpha\pi}{\gamma_\rho^2} \frac{1}{2} [\sigma^{\text{el}}(\pi^+ p) + \sigma^{\text{el}}(\pi^- p)] \qquad (3.2)$$

The solid line in Fig. 12 represents the above relation with $(\gamma_\rho^2/4\pi) = 0.65$

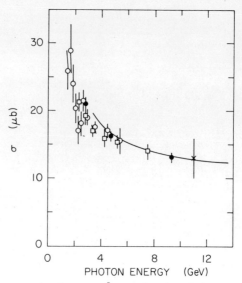

Fig. 12. The total cross section for $\gamma p \to \rho^0 p$ as a function of photon energy. See text for a description of curve. The data are from the ABBHHM collaboration (○), Davies *et al.* (1970) (×), Ballam *et al.* (1973) (●), and Alexander *et al.* (1972) (□).

and using the measured $\pi^\pm p$ cross sections (Giacomelli, 1969). The agreement is good.

Typical differential cross sections for $\gamma p \to \rho^0 p$ are shown in Fig. 13 for 9 and 16 GeV. The agreement between the different experiments is fairly good, with all measurements typically lying within a $\pm 15\%$ band at each

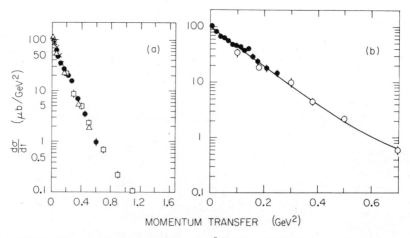

Fig. 13. The differential cross section for $\gamma p \to \rho^0 p$ at 9 and 16 GeV. At 9 GeV the data are from Bulos *et al.* (1970) (×), Ballam *et al.* (1973) (●), Anderson *et al.* (1970a) (□), and McClellan *et al.* (1971a, 1971b), and at 16 GeV the data are from Giese (1974) (●) and Anderson *et al.* (1970b) (○).

energy. The shape of the scattering distribution, for $t < 0.5$ $(\mathrm{GeV}/c)^2$, is well fitted by an exponential form

$$\frac{d\sigma}{dt} = \frac{d\sigma}{dt}\bigg|_0 \exp(At) \qquad (3.3)$$

Experiments measuring over a wide range of momentum transfers [e.g., Anderson et al. (1970b); $0.1 < t < 1.2\,(\mathrm{GeV}/c)^2$] find they require a quadratic term in the exponential. A summary of the forward cross section, $(d\sigma/dt)_0$, and slope parameter, A, for those experiments using the Söding model analysis, is given in Table 3, and the data are plotted in Figs. 14 and 15, respectively.

Table 3. Summary of Forward Cross Section Analyses for $\gamma p \rightarrow \rho^0 p$

Experiment	Photon energy (GeV)	Momentum transfer range $[(\mathrm{GeV}/c)^2]$	Forward cross section, $(d\sigma/dt)_0$ $[\mu\mathrm{b}/(\mathrm{GeV}/c)^2]$	Forward slope A $[(\mathrm{GeV}/c)^{-2}]$
Ballam et al. (1973)	2.8	0.02–0.4	104 ± 6	5.4 ± 0.3
	4.7	0.02–0.4	94 ± 6	5.9 ± 0.3
	9.3	0.02–0.5	86 ± 4	6.6 ± 0.3
Alexander et al. (1972)	2.2	0.06–0.4	134 ± 20	6.4 ± 0.8
	2.7	0.06–0.4	177 ± 26	8.8 ± 1.1
	3.4	0.06–0.4	124 ± 20	7.5 ± 1.2
	4.3	0.06–0.4	101 ± 12	6.5 ± 0.5
	5.2	0.06–0.4	132 ± 17	7.7 ± 0.6
	7.5	0.06–0.4	98 ± 15	7.1 ± 0.6
Gladding et al. (1973)	3.3	0.15–0.7	103 ± 22	7.5 ± 0.7
	4.2	0.15–0.7	102 ± 18	7.4 ± 0.6
McClellan et al. (1971b)	3.9	0	169 ± 14	—
	4.1	0	150 ± 14	—
	4.6	0	140 ± 13	—
	5.6	0	134 ± 6	—
	5.9	0	126 ± 9	—
	6.5	0	109 ± 9	—
	6.9	0	113 ± 10	—
	7.4	0	108 ± 5	—
	8.5	0.00–0.5	103 ± 6	8.1 ± 0.4
Berger et al. (1972)	8.5	0.07–0.52	98 ± 6	7.4 ± 0.5
Alvensleben et al. (1970b)	6.4	0	120 ± 6	—
Bulos et al. (1970)	9	0.00–0.15	113 ± 10	9.3 ± 1.1
Giese (1974)	15	0.00–0.3	104 ± 10	8.7 ± 0.4
	13	0.00–0.3	103 ± 10	8.2 ± 0.3
	11	0.00–0.3	95 ± 9	6.5 ± 0.4
	12	0.00–0.3	99 ± 10	7.5 ± 0.5
	10	0.00–0.3	104 ± 10	7.8 ± 0.5
	8	0.00–0.3	106 ± 10	8.3 ± 0.5

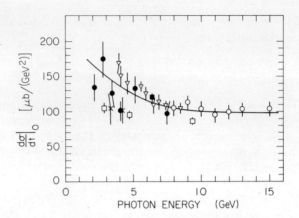

Fig. 14. The energy dependence of the forward differential cross section $(d\sigma/dt)_0$. See text for a description of the curve. The data are from Giese (1974) (○), Ballam *et al.* (1973) (□), Gladding *et al.* (1973) (×), McClellan *et al.* (1971*b*) (▽), Alexander *et al.* (1972) (●), and Alvensleben *et al.* (1970*b*) (■).

The measured forward ρ^0 cross section data in Fig. 14 show good agreement among the different experiments and exhibit a slow falloff with energy similar to that observed for diffractive hadronic reactions. More specifically, we may relate the forward ρ^0 cross section to the measured $\pi^+ p$ and $\pi^- p$ total cross sections using the quark model and vector

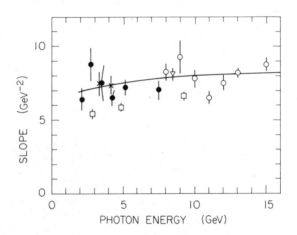

Fig. 15. The slope of the $\gamma p \to \rho^0 p$ differential cross section as a function of energy. The solid curve is the average slope for $\pi^+ p$ and $\pi^- p$ elastic scattering determined over the momentum transfer interval $[0.1 < |t| < 0.4 \ (GeV/c)^2]$. The data are from Giese (1974) (○), Ballam *et al.* (1973) (□), Gladding *et al.* (1973) (×), McClellan *et al.* (1971*b*) (▽), Alexander *et al.* (1972) (●).

Table 4. Slope of the $\gamma p \to \rho^0 p$ *Differential Cross Section for Different Intervals of Momentum Transfer*

Photon energy:	9 GeV		16 GeV	
Momentum transfer interval (exp.)	$(0 < \|t\| < 0.2)$ $(GeV/c)^2$ (Giese, 1974)	$(0.02 < \|t\| < 0.5)$ $(GeV/c)^2$ (Ballam *et al.*, 1973)	$(0 < \|t\| < 0.3)$ $(GeV/c)^2$ (Giese, 1974)	$(0.1 < \|t\| < 1)$ $(GeV/c)^2$ (Anderson *et al.*, 1970*b*)
Slope of cross section, A $[(GeV/c)^{-2}]$	8.3 ± 0.4	6.6 ± 0.3	8.7 ± 0.4	7.5 ± 0.5

dominance [see Eq. (5.5), and the discussion of the vector dominance model in Section 5]. The solid line in Fig. 14 is the result of such a calculation, and provides a good representation of the forward ρ^0 production cross sections, which are measured to be ≈ 125 $\mu b/(GeV/c)^2$ at 4 GeV, and falls to ~ 100 $\mu b/(GeV/c)^2$ at 10 GeV, with an estimated uncertainty of $\sim 10\%$–15%.

The slope of the forward ρ^0 cross section is shown in Fig. 15. It shows very little energy dependence and is in good agreement with the average of the measured $\pi^+ p$ elastic slopes, shown as the solid line. It is important to note the different t ranges measured in the various experiments (see Table 3). Recent studies of hadron elastic scattering distributions at the CERN-ISR (Barbiellini *et al.*, 1972) and at SLAC (Carnegie *et al.*, 1975) have shown that the slope of the scattering distribution for $t \lesssim 0.2$ $(GeV/c)^2$ can be 1–2 units larger than for the region $0.2 \lesssim t \lesssim 0.5$ $(GeV/c)^2$. Examples of this are shown in Fig. 16, where the forward slopes in $\pi^\pm p$, $K^\pm p$, and $p^\pm p$ elastic scattering at 10 GeV, are shown as a function of momentum transfer. For ρ^0 photoproduction, we compare data from the very forward direction $[0 < t < 0.2$ $(GeV/c)^2]$ to larger t measurements $[0.05 < t < 1$ $(GeV/c)^2]$ in Fig. 13, and observe that although the cross sections agree where the data overlap, the measurements of the slope differ by 1–2 units. The situation is summarized in Table 4. This is an indication that forward steepening is present in the ρ^0 photoproduction cross section, just as for other diffractive elastic processes.

3.2.2. Angular Correlation Studies

The systematics of the ρ^0 decay angular distribution have been most beautifully studied in a series of experiments using a linearly polarized photon beam at energies of 2.8, 4.7, and 9.3 GeV (Ballam *et al.*, 1973).

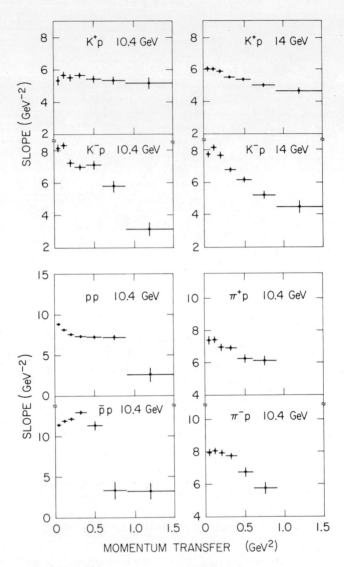

Fig. 16. The logarithmic slope of the elastic differential cross section as a function of momentum transfer for $\pi^+ p$, $K^\pm p$, and $p^\pm p$ scattering.

The definition of the angles used for these studies is shown in Fig. 17, and depending on the coordinates system the z axis is defined as follows:

  (i) γ direction in ρ^0 rest frame, for Gottfried–Jackson;
 (ii) ρ^0 direction in total center-of-mass system, for helicity;
(iii) γ direction in total center-of-mass system, for Adair.

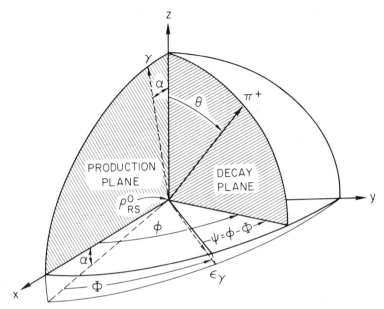

Fig. 17. Definition of the angles used to describe the production and decay of the ρ meson in polarized photoproduction studies.

The full information of the ρ^0 decay angular distribution is contained in nine independent density-matrix elements (Schilling *et al.*, 1971):

$$W(\cos\theta, \phi, \Phi) = (3/4\pi)\{\tfrac{1}{2}(1 - \rho_{00}^0) + \tfrac{1}{2}(3\rho_{00}^0 - 1)\cos^2\theta$$
$$- 2^{1/2}\,\mathrm{Re}\,\rho_{10}^0 \sin 2\theta \cos\phi$$
$$- \rho_{1-1}^0 \sin^2\theta \cos 2\phi - P_\gamma \cos 2\Phi[\rho_{11}^1 \sin^2\theta + \rho_{00}^1 \cos^2\theta$$
$$- 2^{1/2}\,\mathrm{Re}\,\rho_{10}^1 \sin 2\theta \cos\phi - \rho_{1-1}^1 \sin^2\theta \cos 2\phi]$$
$$- P_\gamma \sin 2\Phi[2^{1/2}\,\mathrm{Im}\,\rho_{10}^2 \sin 2\theta \sin\phi$$
$$+ \mathrm{Im}\,\rho_{1-1}^2 \sin^2\theta \sin 2\phi]\}$$

where P_γ is the degree of linear polarization, and the density matrix of the ρ has been split into three parts:

$$\rho_{ik} = \rho_{ik}^0 - P_\gamma \cos 2\phi \cdot \rho_{ik}^1 - P_\gamma \sin 2\phi \cdot \rho_{ik}^2$$

P_γ varies from 95% at 2.8 GeV to 77% at 9.3 GeV.

The photon, owing to its zero rest mass, can only have helicities $\lambda_\gamma = \pm 1$, while the ρ^0 meson may have helicities $\lambda_\rho = \pm 1$ and 0. A study of the decay distribution allows a separation of the various helicity amplitudes, since for $\lambda_\rho = \pm 1$ the decay angular distribution will have the form $\sin^2\theta$, and for $\lambda_\rho = 0$, it will follow $\cos^2\theta$. The decay angular distributions also allow a separation of the natural and unnatural parity t-channel exchange amplitudes to leading order in the photon energy (Schilling *et al.*, 1971). The natural parity exchange process has the pions from ρ^0 decay emerging preferentially in the plane of photon polarization ($\Psi \sim 0^0$), and the unnatural exchange processes have the pions emerging perpendicular to it ($\Psi \sim 90^0$).

Figure 18 shows the distribution of $\cos \theta$ against Ψ in the helicity system, for $\gamma p \to \rho^0 p$ at 4.7 GeV. The data show a beautiful $\sin^2\theta \cos^2\Psi$ correlation, implying that the ρ^0 takes over the photon's polarization with no helicity flip and that the process is dominated by natural parity exchanges.

The relative contribution from natural parity exchange (σ^N) in the t channel is measured by the parity asymmetry

$$P_\sigma \equiv \frac{\sigma^N - \sigma^U}{\sigma^N + \sigma^U} = 2\rho^1_{1-1} - \rho_{00} \tag{3.5}$$

P_σ is shown in Fig. 19 as a function of momentum transfer, for the three

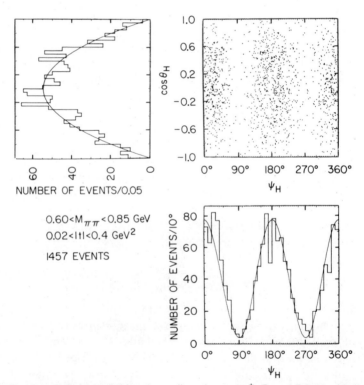

Fig. 18. Decay correlations of the ρ in studies of $\gamma p \to p\pi^+\pi^-$ at 4.7 GeV with linearly polarized photons. The curves are calculated for an s-channel helicity-conserving $\gamma \to \rho^0$ transition and an incident photon polarization of 92%.

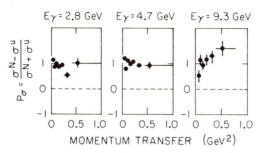

Fig. 19. The parity asymmetry as a function of momentum transfer, for the reaction $\gamma p \to \rho^0 p$ at 2.8, 4.7, and 9.3 GeV.

energies 2.8, 4.7, and 9.3 GeV. Rho production is completely dominated by natural parity exchange. The fraction of unnatural parity exchange is found to be $<5\%$ and consistent with the contribution expected from one-pion-exchange calculations. Similar conclusions were obtained from counter-experiment studies using linearly polarized photons at DESY (Criegee *et al.*, 1970) and Cornell (Diambrini-Palazzi *et al.*, 1970). These experiments were able to show that unnatural exchange contributions were small (i.e., $<10\%$), right down close to threshold for the reaction.

The nine density-matrix elements describing the ρ^0 decay distributions are presented in Fig. 20 as a function of t for the 4.7 GeV study, as evaluated in each of the three coordinate systems—Gottfried–Jackson, Helicity, and Adair. Substantial spin-flip or t-channel helicity-flip contributions are seen in the Gottfried–Jackson and Adair system density-matrix elements while in the helicity system the density-matrix elements are consistent with no flip contributions (i.e., they are consistent with s-channel helicity conservation). The SLAC streamer chamber (Davier *et al.*, 1970) and wire chamber spectrometer (Giese, 1974) confirm this behavior up to the highest energies available (i.e., $E_\gamma = 16$ GeV) and for $t \lesssim 0.5 \ (\text{GeV}/c)^2$. These results imply that in the center-of-mass the ρ behaves like a photon with its spin along its direction of flight.

This tidy picture in which s-channel helicity conservation (SCHC) was a property of the ρ production amplitudes [and perhaps all diffractive amplitudes (Gilman *et al.*, 1970)] was not to be; it certainly is the dominant contribution to the process, but careful and systematic study discovered small helicity-flip amplitudes. The density-matrix elements for the 9.3 GeV experiment are shown in Fig. 21, for the helicity system. The parity asymmetry, P_σ, is close to 1, showing that the ρ meson is predominantly produced by natural parity exchange, and the elements ρ^1_{1-1} and $-\mathrm{Im}\, \rho^2_{1-1}$ are close to 0.5 and all other elements close to zero, as required by SCHC. However, there are small but systematic deviations from zero in the interference terms between helicity-flip and nonflip amplitudes. The relevant

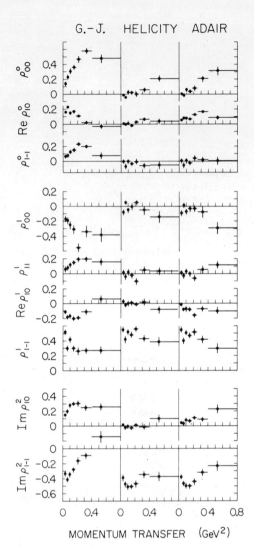

Fig. 20. The ρ spin density-matrix elements as a function of momentum transfer, for the reaction $\gamma\rho \to \rho^0 p$ at 4.7 GeV. The decay correlations are presented for three coordinate systems: Gottfried–Jackson, helicity, and Adair. See text for definitions.

elements are as follows:

intensity of $\Delta\lambda = \pm 1$ amplitude: ρ_{00}^0

intensity of $\Delta\lambda = \pm 2$ amplitude: $\rho_{1-1}^1 + \mathrm{Im}\,\rho_{1-1}^2$

interference between $\Delta\lambda = \pm 1$
and $\Delta\lambda = 0$ amplitudes: $\mathrm{Re}\,\rho_{10}^1 - \mathrm{Im}\,\rho_{10}^2$

$\mathrm{Re}\,\rho_{10}^0$

interference between $\Delta\lambda = \pm 2$
and $\Delta\lambda = 0$ amplitudes: ρ_{1-1}^0

$$ \left. \right\} \qquad (3.6) $$

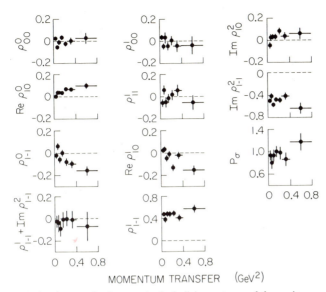

Fig. 21. Rho spin density-matrix elements in the helicity system and the parity asymmetry as a function of momentum transfer, for the reaction $\gamma p \to \rho^0 p$ at 9.3 GeV.

In the forward direction these elements must go to zero from kinematics, but in the range $0.2 < t < 0.8$ $(\text{GeV}/c)^2$ they appear to be systematically nonzero.

Similar studies have been performed for ρ production in deuterium, where they find the same angular correlations. Figure 22 indicates that the

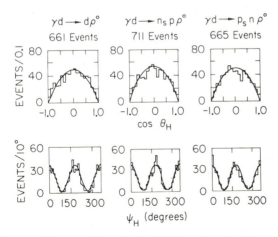

Fig. 22. Rho decay angular correlations in the helicity system, for the reactions $\gamma d \to d\rho^0$, $\gamma d \to n_s p\rho^0$, and $\gamma d \to p_s n\rho^0$ at 4.3 GeV. The curves are calculated for an s-channel helicity-conserving $\gamma \to \rho^0$ transition and an incident photon polarization of 92%.

(a)   (b)

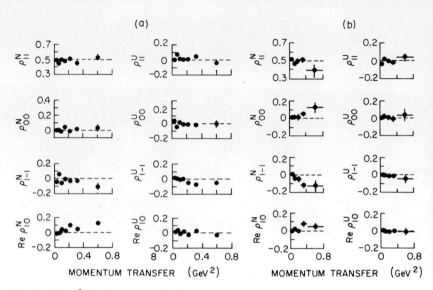

MOMENTUM TRANSFER (GeV²)   MOMENTUM TRANSFER (GeV²)

Fig. 23. The ρ^0 spin density-matrix elements in the helicity system for natural and unnatural parity exchange contributions to the reaction (a) $\gamma p \to \rho^0 p$ at 9.3 GeV and (b) $\gamma d \to pn\rho^0$ at 4.3 GeV.

process is mainly SCHC and natural parity exchange and further detailed analysis of the density matrix elements confirms such a conclusion. This indicates that ρ production on protons and neutrons has the same spin dependence. Since $I = 1$ exchange amplitudes will couple with opposite sign in the proton and neutron reactions [see Eq. (3.11)], and the density-matrix elements for the γp and γn reactions are the same (even for those implying helicity flip), it would appear that the isovector contributions must be small and do not account for the helicity-flip behavior. The $I = 1$ exchange amplitudes are discussed more fully in the next section.

Using the linearly polarized γ beam data, the density-matrix elements, and the cross sections, can be separated (to leading order in energy), into natural and unnatural parity exchange contributions (Schilling et al., 1971):

$$\rho_{ik}^{\overset{N}{U}} = \tfrac{1}{2}[\rho_{ik}^0 \mp (-1)^i \cdot \rho_{ik}^1] \qquad (3.7)$$

Analysis of the hydrogen and deuterium laser beam experiments (Ballam et al., 1973; Eisenberg et al., 1976) for example, shows that the ρ_{ik}^U are close to zero, confirming once more the dominance of natural parity exchange. (See Fig. 23.) The deviations from SCHC in Fig. 21 may now be observed to originate in the natural parity exchange density-matrix elements, ρ_{ik}^N, of Fig. 23.

   In order to check for instrumental bias the density-matrix elements were separately evaluated for data with the photon polarization parallel and normal to the camera axis in the bubble chamber. Since the ρ^0 decays preferentially in the polarization plane, this effectively rotates the asymmetry of the angular distribution by 90° in the chamber. No difference was observed in the two results.

   Given the observation of helicity-flip behavior in the data, it may be identified with either the ρ production or the nonresonant $\pi-\pi$ background. This question was studied in detail using the Söding model (Söding, 1966; Krass, 1967) to describe the mass and t dependence of the dipion angular correlations. Figure 24 shows two of the helicity-flip moments, as examples, with the expected behavior of the nonresonant $\pi-\pi$ background subtracted out. One plainly sees that the s-channel helicity-flip effect is associated with the ρ production.

   Assuming that the ρ production is natural parity exchange, that the helicity-flip amplitudes are small, and that the nonflip amplitude is imaginary, the measured density-matrix elements may be used to estimate the ratio of single and double helicity flip to the dominant nonflip amplitude. The results are shown in Table 5. The different estimates of the strength of the flip term are in fair agreement with each other, and imply that the single-flip amplitude is of same sign as the nonflip amplitude and $\sim 10\%-15\%$ in magnitude, while the double flip is roughly the same size but opposite in sign. Both helicity-flip amplitudes are dominantly natural parity exchange.

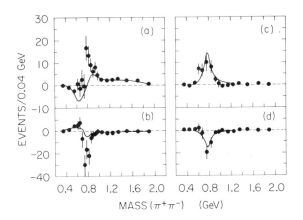

Fig. 24. The reaction $\gamma p \to p\pi^+\pi^-$ at 9.3 GeV. (a), (b) unnormalized moments $\mathrm{Re}\,\rho^0_{10}\,dN/dM_{\pi\pi}$ and $\rho^0_{1-1}\,dN/dM_{\pi\pi}$, respectively, evaluated for $0.2 \leqslant |t| \leqslant 0.8\,(\mathrm{GeV}/c)^2$. The curves were calculated from the Söding model. (c), (d) are the deviations of the measured moments of (a) and (b) respectively, from the Söding model predictions. The curves are p-wave Breit–Wigner ρ^0 shapes, normalized to the area under the experimental points.

D. W. G. S. Leith

Table 5. Summary of Helicity-Flip Amplitudes in Vector Meson Photoproduction (Taken from Kogan, 1975)

Amplitude ratios[a]	Estimator	$\gamma d \to pm\rho^0$ 4.3 GeV	$\gamma p \to p\rho^0$		$\gamma p \to p\omega$[b]	
			4.7 GeV	9.3 GeV	4.7 GeV	9.3 GeV
$|T_{01}|^2/|T_{11}|^2$	ρ_{00}^0	0.05 ± 0.03	0.11 ± 0.03	0.01 ± 0.03		
	ρ_{00}^N	0.07 ± 0.03	0.13 ± 0.04	0.02 ± 0.02	0 ± 0.06	0.11 ± 0.08
Im $T_{01}/|T_{11}|$	$2\mathrm{Re}\,\rho_{10}^0$	0.14 ± 0.04	0.16 ± 0.04	0.14 ± 0.02		
	$2\mathrm{Re}\,\rho_{10}^N$	0.14 ± 0.05	0.14 ± 0.04	0.14 ± 0.02	0.25 ± 0.08	0.02 ± 0.14
	$-\mathrm{Re}\,\rho_{10}^1 - \mathrm{Im}\,\rho_{10}^2$	0.11 ± 0.05	0.08 ± 0.05	0.11 ± 0.03		
$|T_{-11}|^2/|T_{11}|^2$	$\rho_{1-1}^1 + \mathrm{Im}\,\rho_{1-1}^2$	0.09 ± 0.07	0.10 ± 0.06	-0.03 ± 0.07		
	ρ_{1-1}^0	-0.14 ± 0.03	-0.08 ± 0.03	-0.12 ± 0.02		
Im $T_{-11}/|T_{11}|$	ρ_{1-1}^N	-0.12 ± 0.03	-0.01 ± 0.03	-0.05 ± 0.02	-0.05 ± 0.05	-0.04 ± 0.07

[a] The nucleon helicities in the amplitudes listed are $\frac{1}{2}\frac{1}{2}$ (or $-\frac{1}{2}-\frac{1}{2}$).
[b] Natural parity exchange only and $0.15 < |t| < 0.6\,(\mathrm{GeV}/c)^2$.

There are two facts that argue for the isospin of the helicity-flip amplitude being zero:

(a) If the exchange were $I = 1$, the sign of the helicity-flip amplitude on neutrons and protons would be opposite, while it is observed to be the same.

(b) If the exchange were $I = 1$, we would expect a much bigger effect for helicity flip in $\gamma p \to \omega p$, since

$$\frac{\mathrm{Re}\,\rho_{10}^N(\gamma \to \omega)}{\mathrm{Re}\,\rho_{10}^N(\gamma \to \rho)} \sim \frac{\gamma_\omega^2}{\gamma_\rho^2} = 7 \pm 1$$

Therefore, we would expect very large $\mathrm{Re}\,\rho_{00}^1$ and ρ_{1-1}^0 in the natural parity exchange part of omega production. Instead one observes (see Table 5) an amplitude similar in size to that found for ρ production, as one would expect for $I = 0$ exchange.

The magnitude of the helicity-flip amplitude does not change much from 4 to 9 GeV, although an $s^{-1/2}$ behavior cannot be ruled out, owing to the large errors. It is tempting to relate the helicity-flip behavior to the diffractive Pomeron exchange amplitude, especially as similar features have been identified in elastic πN scattering, for the nucleon vertex.

A Saclay group (Cozzika *et al.*, 1972) studying the A and R polarization parameters in πN elastic scattering have shown that the t-channel, $I = 0$ nucleon helicity-flip amplitude F_{+-}^0 is of order $\sim 10\%$ of the nonflip F_{++}^0 amplitude for $0.2 < t < 0.8 \ (\mathrm{GeV}/c)^2$ and is roughly independent of energy. The ratio of flip to nonflip amplitude as a function of momentum transfer, is shown in Fig. 25 for both the $\gamma \to \rho^0$ and $\pi \to \pi$ analyses. The similarity is strong, indicating that helicity flip is a common property of diffractive processes and that it has the same characteristic for the meson or the nucleon vertex.

In summary, ρ^0 photoproduction is dominated by natural parity exchanges in the t-channel and the $\gamma \to \rho^0$ transition mainly conserves the s-channel helicity of the incoming photon. A small helicity-flip amplitude is observed, also dominated by natural parity exchanges, and is $\sim 10\%$ of the nonflip term. The angular correlations are found to be the same for ρ^0 production on neutron or proton targets.

3.2.3. Amplitude Structure in $\gamma N \to \rho^0 N$—Isolation of $I = 1$ Exchanges

The contribution of isovector t-channel exchange to the process $\gamma N \to \rho^0 N$ may be estimated by comparing the production cross sections from deuterium and hydrogen targets. Four separate methods for extracting the $I = 1$ exchange amplitude are summarized below.

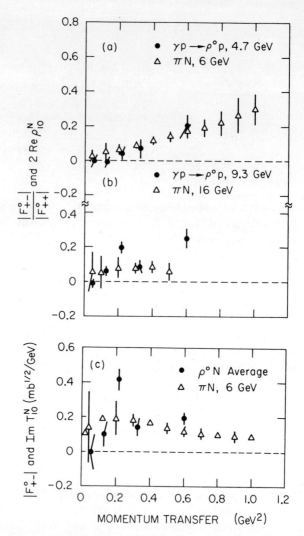

Fig. 25. The ratio of helicity-flip to nonflip amplitudes in ρ photoproduction and πN elastic scattering. (a), (b) 2 Re ρ_{10}^N for $\gamma p \to \rho^0 p$ at 4.7 and 9.3 GeV and $|F_{+-}^0|/|F_{++}^0|$ for $\pi N \to \pi N$ at 6 and 16 GeV/c. (c) $|F_{+-}^0|$ at 6 GeV/c and an average of Im T_{10}^N from the photoproduction experiments normalized by the VDM $\gamma \rho$ coupling.

(*i*) *Comparison of Forward Cross Sections.* The forward cross section, $(d\sigma/dt)_0$, for ρ production on deuterium or hydrogen may be written

$$\left.\frac{d\sigma}{dt}(\gamma d - \rho^0 d)\right|_0 = 4|T_0|^2(1 - G) \qquad (3.8)$$

where G is the shadowing correction (Franco and Glauber, 1966), and

$$\frac{d\sigma}{dt}(\gamma p \to \rho^0 p)\bigg|_0 = |T_0 + T_1|^2 \qquad (3.9)$$

For no $I = 1$ exchange, the ratio of expressions (3.8) and (3.9) should be 3.72. Otherwise, for an $I = 1$ exchange contribution one may obtain from R the following:

$$\frac{2T_1 \cdot T_0^*}{|T_0|^2} + \left|\frac{T_1}{T_0}\right|^2 = 2\left|\frac{T_1}{T_0}\right| \cdot \cos \Delta\phi + \left|\frac{T_1}{T_0}\right|^2 \qquad (3.10)$$

where $\Delta\phi$ is the relative phase between the isoscalar and isovector amplitudes at $t = 0$.

Given that ρ^0 production is dominated by natural parity exchange in the t channel we can identify the isoscalar amplitude with $(P + f)$ exchange, and the isovector amplitude with A_2 exchange. The absolute phase of T_0 has been measured experimentally by DESY-MIT (Alvensleben et al., 1970a) and NINA (Biggs et al., 1971) groups by observing the interference between Bethe–Heitler production and leptonic decay of the ρ meson. They find the phase of T_0 to be $102°$–$106°$. If T_1 were taken to have the phase of the A_2 Regge trajectory it would be $\sim 135°$, and hence $\cos \Delta\phi = 0.84$. If, however, we assume that absorption or other effects may alter this phase, we might still expect T_1 to lie in the second quadrant and hence $\cos \Delta\phi > 0.2$. Applying these assumptions on $\Delta\phi$ to Eq. (3.10) we can estimate $|T_1/T_0|$.

Table 6 shows the results of such a calculation for the three bubble chamber and two counter experiments studying ρ production on deuterium. For the counter experiments only the highest-energy data (i.e., those near the end point of the bremsstrahlung spectrum) were used in order to remove possible confusion due to inelastic background contributions. The ratio of the deuterium to hydrogen cross sections near the forward cross section is shown in Fig. 26 from the highest-energy SLAC experiment (Giese, 1974). The data from all five experiments are in fair agreement, and imply that the isovector exchange amplitude, T_1, is small.

(*ii*) *Comparison of Closure and Coherent Cross Sections.* We may write the helicity-nonflip amplitude for ρ production on protons and neutrons as

$$\begin{aligned}
f(\gamma p \to \rho^0 p) &= T_0 \exp(At/2) + T_1(t) \\
f(\gamma n \to \rho^0 n) &= T_0 \exp(At/2) - T_1(t)
\end{aligned} \qquad (3.11)$$

where T_0, T_1 are the isoscalar and isovector exchange amplitudes, as before. The Glauber multiple-scattering theory then gives the differential

Table 6. *Isospin Contributions to the Reaction* $\gamma N \to \rho^0 N$. *See Text for Explanation of the Methods*[a]

Method	Measurement	Experiment					
		Eisenberg et al. (1976) (4.3 GeV)	Benz et al. (1974) (1.8–5.3 GeV)	Alexander et al. (1975) (7.5 GeV)	McClellan et al. (1971b) (6–8 GeV)	Giese (1974) (9 GeV)	Giese (1974) (14–16 GeV)
1. Comparison of $\gamma d \to \rho^0 d$ and $\gamma p \to \rho^0 p$	$\left\|\dfrac{T_1}{T_0}\right\|$	<0.05	<0.12	<0.12	<0.07	<0.14	<0.03
2. Comparison of $\gamma d \to \rho^0 d + \rho^0 pn$ and $\gamma d \to \rho^0 d$	$\left\|\dfrac{T_1}{T_0}\right\|^2$	$0.07^{+0.10}_{-0.07}$	—	—	—	—	0 ± 0.04
3. Comparison of $\gamma d \to p_s n \rho^0$ and $\gamma d \to n_s p \rho^0$	$2\left\|\dfrac{T_1}{T_0}\right\| \cos \Delta\phi$	0.02 ± 0.04	—	0 ± 0.05	—	—	—
4. Comparison of $\gamma n \to p \rho^-$ and $\gamma p \to p \rho^0$	$\left\|\dfrac{T_1}{T_0}\right\|^2$	<0.03	<0.05	—	—	—	—

[a] The table is based on 1 standard deviation limit. Method 1 assumes the Regge phase for $\cos \Delta\phi$ (i.e., $\cos \Delta\phi = 0.84$). I am indebted to E. Kogan for his help in preparing this table.

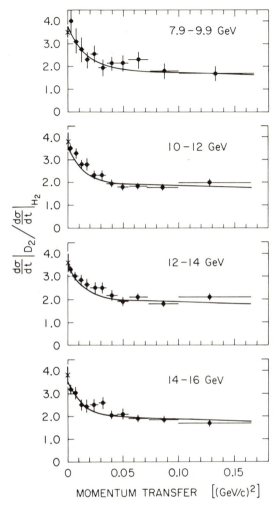

Fig. 26. The measured ratio of ρ photoproduction from deuterium and hydrogen targets as a function of momentum transfer, for several photon energies. The solid line is a fit to the data to determine the amount of $I = 1$ t-channel exchange amplitude (see text).

cross section for coherent production from deuterium, following Eisenberg *et al.* (1976):

$$d\sigma(\gamma d \to \rho^0 d)/dt = 4|T_0|^2[S(t/4)\exp(At) - S(t/4) \cdot G \cdot \exp(3At/4)]$$

(3.12)

where G is the shadowing correction (Franco and Glauber, 1966) and $S(t)$ is the deuteron form factor.

The total closure cross section for ρ production, including both coherent and breakup contributions, is written

$$d\sigma(\gamma d \to \rho^0 pn)/dt = 4|T_0|^2 \exp(At) \cdot \tfrac{1}{2} \cdot [1 + S(t)] + 4|T_1|^2 \cdot \tfrac{1}{2} \cdot [1 - S(t)]$$
$$- 4|T_0|^2 \cdot G \cdot \exp(3At/4) \qquad (3.13)$$

Knowing $|T_0|^2$ from the forward coherent deuteron cross section, the total ρ production data may be fitted to Eq. (3.13) and an estimate of $|T_1/T_0|^2$ obtained. The closure and coherent cross sections from the 4.3 GeV γd bubble-chamber experiment (Eisenberg *et al.*, 1976) are shown in Fig. 27 together with their fit to Eq. (3.13). The results of these calculations are given in Table 6 for experiments at 4.3 and 16 GeV, and also indicate that the isovector contribution is small.

(*iii*) *Neutron–Proton Cross Section Difference.* Another method of estimating the isovector contributions is to compare the production cross section on neutrons and on protons, as we did for Compton scattering and

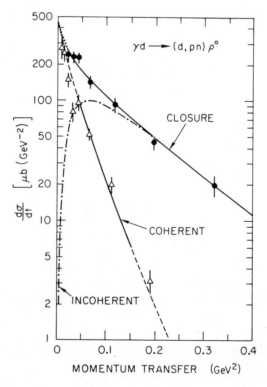

Fig. 27. The differential cross section for the photoproduction of ρ mesons from deuterium, showing both the total production and the coherent cross section.

the total photon cross section in Section 2 above. Here, we have

$$\frac{\sigma(\gamma p \to \rho^0 p) - \sigma(\gamma n \to \rho^0 n)}{\sigma(\gamma p \to \rho^0 p) + \sigma(\gamma n \to \rho^0 n)} \approx \frac{2 \operatorname{Re} T_1 \cdot T_0^*}{|T_0|^2} \tag{3.14}$$

The comparison may be performed from the data of the deuterium bubble-chamber experiments (Eisenberg *et al.*, 1976), but it must be made at a value of momentum transfer where the recoil nucleon can be clearly kinematically separated from the spectator nucleon. The analysis is done for recoil momenta greater than 280 MeV/c. The results are shown in Table 6, and they again show that T_1 is small.

(*iv*) *Exchange Contribution to* $\gamma n \to \rho^- p$. Yet another independent limit on the isovector contribution may be obtained from the cross section for the charge-exchange process, $\gamma n \to \rho^- p$.

The principal t-channel exchanges contributing to this reaction are π^+, ρ^+, A_2^+. If we assume that the natural parity exchanges do not interfere destructively, we can obtain an upper limit for the A_2^+ contribution, and from $SU(2)$, a limit on the A_2^0 contribution to ρ^0 photoproduction. The cross-section data are summarized in Fig. 28 and the resulting limits recorded in Table 6.

In summary, all the methods in Table 6 imply that the isovector contribution to $\gamma N \to \rho^0 N$ is small and less than 5%–10% of the isoscalar amplitude. This is in good agreement with the estimates obtained in Section 2 for the isovector contribution to the related reaction of Compton scattering and the total hadronic cross section, of $\approx 3\%$ at 6 GeV and $\approx 1.5\%$ at 20 GeV.

3.2.4. Amplitude Structure in $\gamma N \to \rho^0 N$—Isolation of f and $\mathbb{P}$ Exchanges

The ρ production process may be considered dominated by f^0 and Pomeron exchanges in the t channel, since both unnatural parity exchanges and $I = 1$ natural parity exchanges have been shown to be small (see Sections 3.2.2 and 3.2.3). Further, since there is substantial energy dependence of both the total ρ cross section and the forward differential cross section in the 2–5 GeV range, it may be expected that the meson exchange amplitude is appreciable.

A separation of the f and $\mathbb{P}$ amplitudes has been performed by Chadwick *et al.* (1973) utilizing the dual absorption model (Harari, 1971) as a guide to the structure of the exchange amplitudes. A more detailed description of this work may be found in the thesis of Kogan (1975). This analysis followed a similar separation of the f and $\mathbb{P}$ exchange amplitudes in πN scattering by Davier (1972).

The ρ production is parameterized in terms of two components—a central Pomeron term and a peripheral f exchange term:

$$\mathbb{P}(s, t) = iC_p \exp[A_p(s)t]$$

$$\mathrm{Im}\, f(s, t) = (C_f/s^{1/2}) \exp[A_f(s)t] \cdot J_0[R(-t)^{1/2}] \qquad (3.15)$$

and the cross section at a given energy is written

$$d\sigma(\gamma p \to \rho^0 p)/dt = |\mathbb{P}(t) + f(t)|^2$$
$$= |\mathbb{P}(t)|^2 + 2\mathbb{P}(t) \cdot \mathrm{Im}\, f(t) \qquad (3.16)$$

neglecting the $|f(t)|^2$ term, which decreases like $1/s$.

The ρ production and Compton scattering cross sections may be well represented with this model, as shown in Fig. 29. The logarithmic slope of the Pomeron contribution to the ρ^0 differential cross section as a function of energy, as determined from this analysis, is given in Fig. 30 together with

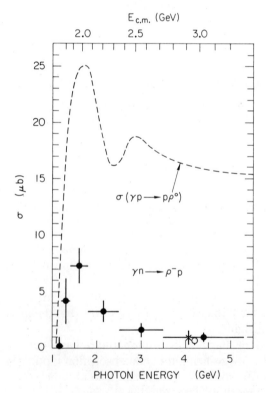

Fig. 28. The cross section as a function of energy for the charge exchange reaction $\gamma n \to \rho^- p$. The data are from Benz *et al.* (1974) (●), Eisenberg *et al.* (1972a) (×), Eisenberg *et al.* (1975) (○).

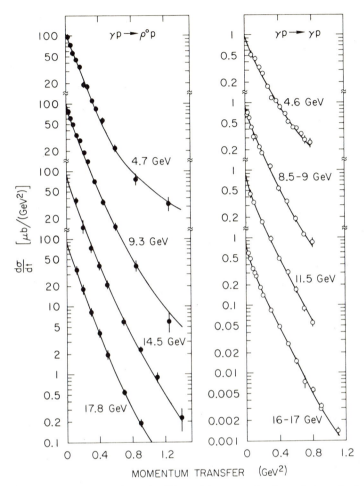

Fig. 29. Dual absorption model fits to the differential cross section for ρ photoproduction and Compton scattering over the energy range 4–18 GeV.

the Pomeron slope from Davier's πN analysis. For comparison the slope of the differential cross section for the reaction $\gamma p \rightarrow \phi p$ is also shown.

It is interesting to note that the lack of energy dependence in the shape of the ρ^0 differential cross section (Fig. 15) is the result of a conspiracy between the falling energy dependence of the steep peripheral f-exchange amplitude and the rapid shrinking of the Pomeron amplitude. This is the same result as found for πN scattering.

It is also interesting to note that the ratio of f exchange to Pomeron exchange required for these fits (Chadwick *et al.*, 1973) is the same for Compton scattering, ρ photoproduction, and for the πN elastic scattering.

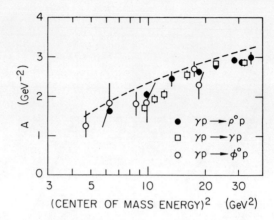

Fig. 30. The energy dependence of the slope of the diffractive contribution to the differential cross section for $\gamma p \to \rho^0 p$, $\gamma p \to \phi p$, and $\gamma p \to \gamma p$, as determined from the dual absorption model fits.

3.2.5. Summary

In summary, the ρ photoproduction reaction is observed to be diffractive, dominantly isoscalar natural parity exchange and with a spin structure that is mainly s-channel helicity conserving. The small 10%–15%, helicity-flip contribution stems from $I = 0$, natural parity exchanges and is probably associated with Pomeron exchange.

3.3. Omega Production

Omega production is experimentally much more difficult to study than the ρ, since in addition to the problem of lack of precise knowledge of the incoming photon beam energy, one has to deal with the three-body $(\pi^+ \pi^- \pi^0)$ decay mode of the vector meson. The various experiments studying ω production on proton and neutron targets are listed in Table 2. The ω mass distribution from two experiments—the 9.3-GeV bubble-chamber experiment (Ballam *et al.*, 1973), and a counter experiment (Gladding *et al.*, 1973), are shown in Fig. 31, where a rather clean signal is observed in each case. Typical backgrounds are estimated to be $\sim 10\%$–15%.

The cross section, as a function of energy, for ω production in the reaction

$$\gamma p \to \omega p$$

is shown in Fig. 32. There is good agreement among the different measurements. Unlike the ρ cross section, the ω data show a very rapid

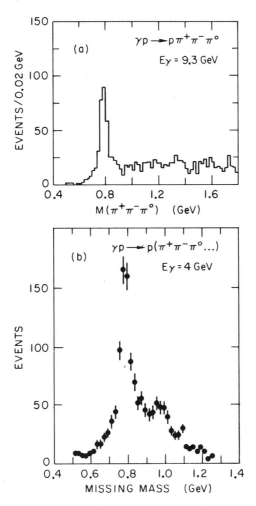

Fig. 31. Omega photoproduction as measured in two experiments: (a) a bubble-chamber study at 9.3 GeV identifying the reaction $\gamma p \to p\pi^+\pi^-\pi^0$ and displaying the $(\pi^+\pi^-\pi^0)$ effective mass, and (b) the missing mass recoiling from the proton in a counter experiment using a tagged photon beam of 4 GeV.

decrease of the cross section in the energy range from 2 to 6 GeV and then become almost constant above 5 GeV. The linearly polarized photon experiment (Ballam *et al.*, 1973) allows a separation into the contributions from natural (σ^N), and unnatural (σ^U), parity exchanges in the *t*-channel

$$\sigma^{\substack{N \\ U}} = \tfrac{1}{2}(1 \pm P_\sigma) \cdot \sigma \qquad (3.17)$$

where P_σ is the parity asymmetry defined in Eq. (3.5) above. The rapid fall

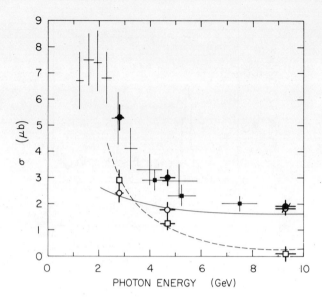

Fig. 32. The energy dependence of the cross section for the reaction $\gamma p \to p\omega$. The cross sections σ^N and σ^U are represented by the solid and broken curves, respectively.

is almost entirely accounted for by the σ^U component of the cross section, which has essentially disappeared by 10 GeV. The natural parity cross section, by contrast, is almost constant, decreasing by only 20% from 2 to 10 GeV.

The behavior of the ω cross section can be understood as a superposition of a strong pion exchange part and a diffractive part. The pion exchange contribution is about half the total ω cross section for energies ~ 4 GeV, but owing to the steep energy dependence this term is negligible by 10 GeV.

The differential cross sections are sharply forward peaked, with a slight steepening near the forward direction (see Fig. 33). The cross sections from different experiments agree within $\sim 10\%$–15% except for the new Cornell data (Abramson *et al.*, 1976), which appears to be systematically $\sim 30\%$ lower than the other measurements. When the differential cross section is separated into the natural $(d\sigma^N/dt)$, and unnatural $(d\sigma^U/dt)$ components, using a t-dependent form of (3.16), the natural parity exchange component is found to be well represented by a simple exponential form

$$\frac{d\sigma^N}{dt} = \frac{d\sigma}{dt}\bigg|_0 \exp At$$

The slope of the ω cross section at small t agrees well with that found for ρ^0

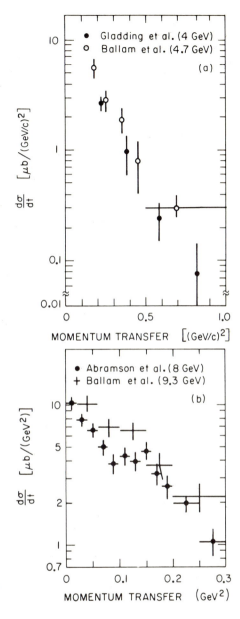

Fig. 33. The differential cross section for the reaction $\gamma p \to p\omega$ at 4 and 9 GeV.

photoproduction in Section 3.2, namely, $\sim 7-8$ $(\text{GeV}/c)^{-2}$. The forward cross-section measurements are summarized in Table 7, where data from both hydrogen and coherent deuterium experiments are included. We quote, where possible forward cross sections evaluated using the same assumption for the slope of $d\sigma^N(\gamma p \to \omega p)/dt$ when fitting the measured

Table 7. Forward Cross Section, $(d\sigma/dt)|_o$ for the Reaction $\gamma N \to \omega N$

Measurement	2.8 GeV (Ballam et al., 1973)	4.5 GeV (Ballam et al., 1973) (Gladding et al., 1973) (Eisenberg et al., 1976)	5.7 GeV (Braccini et al., 1970)	6.8 GeV (Behrend et al., 1971a, b) (Abramson et al., 1976)	7.5 GeV (Alexander et al., 1975)	9.3 GeV (Ballam et al., 1973)
$|T_0 + T_1|^2$	14.5 ± 5.1	13 ± 3	—	~7.5	—	13.5 ± 2.1
(from hydrogen)		12.6 ± 2.4				
$|T_0|^2$ (from deuterium)	—	18.5 ± 4.5	—	~6.2	11.1 ± 2.3	—
$|T_0|^2$ (from complex nuclei)	—	—	16 ± 3	9.6 ± 1.2	—	—
$|T_0|^2$ [from Eq. (5.4) taking $\sigma(\omega N)$ from the quark model and $(\gamma_\omega^2/4\pi)$ from $e^+ e^-$ experiments]				15.3 μb(GeV/c)^2		

differential cross section. As discussed above in Section 3.2 for the ρ^0, the hydrogen data measure $|T_0 + T_1|^2$, while the coherent deuterium cross section determines $|T_0|^2$, where T_0, T_1 are the t-channel isoscalar and isovector exchange amplitudes respectively. Also included in Table 7 are estimates of $|T_0|^2$ from optical model analyses of ω production on complex nuclei (Behrend *et al.*, 1970, 1971*a*; Braccini *et al.*, 1970).

An independent estimate of $|T_0|^2$ may be obtained from the quark model and vector dominance model. Anticipating Section 5 below, in which these relations are fully discussed, we take Eq. (5.4) and insert $\sigma_T(\omega N)$ and η_ω as 27 mb and -0.2, respectively (using the quark model and our knowledge of $\rho^0 N$ scattering), and the measured photon–omega coupling strength, $\gamma_\omega^2/4\pi = 4.6$ from the storage ring experiments. Such a calculation results in $|T_0|^2 = 15.3 \ \mu b/(GeV/c)^2$, and is entered on the bottom line of Table 7. Except for the Cornell–Rochester measurement, all other data are in tolerable agreement with each other and with this VDM estimate of the ω forward cross section, $|T_0|^2$.

The ω decay angular distributions at low energy (< 5 GeV) show none of the beautiful polarization correlations observed for the ρ, due to the large pion exchange contribution to the production process. However, by 9.3 GeV the familiar $\sin^2\theta \cos^2\psi$ pattern for the ω decay distribution is observed (see Fig. 34), indicating that at this energy the ω is produced mainly by natural parity exchange and that the process approximately conserves the s-channel helicity of the photon. The ω density-matrix elements, ρ_{ik}^N, are consistent with those obtained for ρ^0 production, although more poorly determined due to the smaller production cross section. They indicate that the same production and decay properties pertain for the ω reactions as for the ρ^0, even to the presence of a small helicity-flip contribution to the $\gamma \to \omega$ transition.

The cross sections, differential cross sections, and angular distributions for the reaction

$$\gamma N \to \omega N$$

are well represented by a model which describes the natural parity exchange in terms of the $\rho^0 N$ data and the unnatural parity exchange by a one-pion-exchange calculation (Wolf, 1969; Benecke-Dürr, 1968) in which the ω radiative width was taken to be $\Gamma_{\omega\pi\gamma} = 0.9$ MeV (Barash-Schmidt *et al.*, 1974). The solid lines in Figs. 32 and 34 are the result of this fit.

The isospin decomposition for the reaction $\gamma N \to \omega N$ is especially interesting, as was first pointed out by Harari (1969), since any $I = 1$ contribution to the forward Compton amplitude gets amplified by a large factor in the ω production process. More precisely, Δ, the isovector

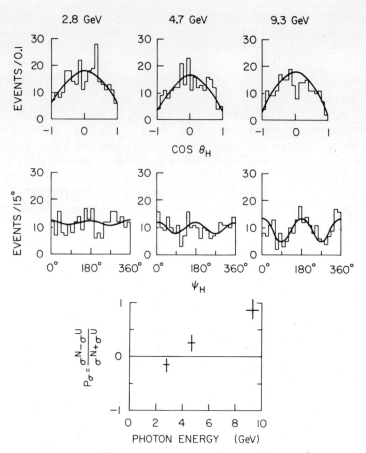

Fig. 34. The ω decay angular distributions in the helicity system, and the parity asymmetry for $\gamma p \to \omega p$ at 2.8, 4.7, and 9.3 GeV, evaluated for the momentum transfer interval $0.02 \le |t| \le 0.3 \ (\text{GeV}/c)^2$.

contribution to $\gamma N \to \omega N$, may be written

$$\Delta = 2\left[\frac{\sigma(\gamma p \to \omega p) - \sigma(\gamma n \to \omega n)}{\sigma(\gamma p \to \omega p) + \sigma(\gamma n \to \omega n)}\right] = 2\left[\frac{\operatorname{Im} T^1_{\gamma p \to \omega p}}{\operatorname{Im} T^0_{\gamma p \to \rho p}}\right]$$

$$= \frac{\gamma_\omega^2}{\gamma_\rho^2} \frac{\operatorname{Im} T^1_{\gamma p \to \gamma p}}{\operatorname{Im} T^0_{\gamma p \to \gamma p}}$$

$$= \frac{\gamma_\omega^2}{\gamma_\rho^2}\left[\frac{\sigma_T(\gamma p) - \sigma_T(\gamma n)}{\sigma_T(\gamma p) + \sigma_T(\gamma n)}\right] \tag{3.18}$$

and

$$\gamma_\omega^2/\gamma_\rho^2 \approx 7$$

In Section 2.1 we estimated, with some reservation, that the isovector amplitude contribution is of order $(3 \pm 1)\%$ at energies around 6 GeV, implying a large isoscalar–isovector interference in ω photoproduction, with a value of $\Delta \sim 0.21$. However, as we mentioned in Section 2.1, this result is very sensitive to the subtraction procedure and the deuterium model used.

We may estimate Δ from the ω photoproduction data through the relations

$$\frac{d\sigma^N}{dt}(\gamma p \to \omega p)\bigg|_0 = |T_0^N + T_1^N|^2 \tag{3.19}$$

(which excludes the pion exchange contribution), and

$$\frac{d\sigma}{dt}(\gamma d \to \omega d)\bigg|_0 = 4(1 - G)|T_0^N|^2 \tag{3.20}$$

where G is the shadowing correction (Franco and Glauber, 1966) and is equal to 0.068. We may then find limits on Δ from

$$\Delta \leqslant \frac{|T_0^N + T_1^N|^2 - |T_0^N|^2}{|T_0^N|^2} \tag{3.21}$$

Using the data summarized in Table 7, the Weizmann group (Eisenberg *et al.*, 1976) find

$$\Delta \leqslant -0.3 \pm 0.3$$

while the Rochester–Cornell group (Abramson *et al.*, 1976) find

$$\Delta = 0.20 \pm 0.12$$

The difference between these estimates stems from the different measured forward cross sections, and as shown in Table 7 the Rochester–Cornell experiment find a much lower cross section than all other experiments. However, neither experiment is sufficiently precise to allow a good determination of the $I = 1$ exchange amplitude, independent of the relative normalization question, and an answer to this interesting question awaits a new experiment.

In summary, the ω photoproduction experiments indicate that at low energies (< 5 GeV) one-pion exchange is an important part of the production process, but that by energies of ~ 10 GeV the ω reaction behaves very much like ρ^0 production, with a cross section ~ 7 times smaller. At these energies the ω is produced via natural parity exchange and is mainly s-channel helicity conserving. The intriguing possibility of using ω production to measure the $I = 1$ exchange contribution to forward Compton scattering, since it is ~ 7 times enhanced in $\gamma N \to \omega N$, has not been

achieved owing to the smallness of the amplitude and the lack of precision
in the current experimental measurements.

3.4. Phi Production

The ϕ photoproduction reaction has been studied in the counter and
track chamber experiments listed in Table 2. The ϕ signal is very clean, as
shown in Fig. 35, where the K^+K^- mass distribution from the 9.3-GeV
bubble-chamber experiment (Ballam *et al.*, 1973) is given. The cross section for ϕ photoproduction as a function of photon energy is shown in Fig.
36 and is about 0.5 μb and rather constant, perhaps rising a little as the
energy increases.

The differential cross section is sharply forward peaked and well
described by the usual exponential form

$$\frac{d\sigma}{dt} = \frac{d\sigma}{dt}\bigg|_0 \exp(At)$$

A compilation of the published differential cross sections prior to Spring
1975 is shown in Fig. 37a, where fair agreement among the different
experiments is observed. There is little indication of any *s* dependence of
the shape or magnitude of the ϕ differential cross section. In Fig. 37b, the
above data are combined with the new small-momentum-transfer, high-
statistics DESY experiment (Behrend *et al.*, 1975) [the previous data are
only plotted for $|t| > 0.4$ $(GeV/c)^2$]. Again agreement between the various
experiments is apparent where the data overlap. However, the small

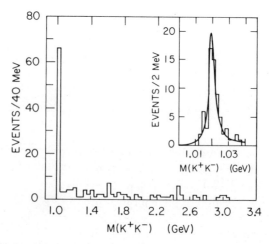

Fig. 35. The effective mass distribution of K^+K^- pairs in the reaction $\gamma p \rightarrow K^+K^-p$ at
9.3 GeV.

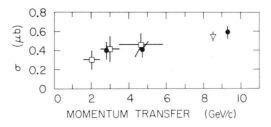

Fig. 36. The energy dependence of the cross section for $\gamma p \to \phi p$.

momentum-transfer measurements exhibit a much steeper falloff of the differential cross section than the larger t measurements, indicating that the cross section is not well described by a simple exponential form.

The energy dependence of the slope of the ϕ differential cross section (i.e., the shrinkage) was studied, prior to the Behrend $et~al.$, experiment by fitting the combined data in Fig. 37a (Moffeit, 1973), to the usual Regge form for the slope, A,

$$A = A_0 + 2\alpha' \ln s \qquad (3.22)$$

(where s is the total center-of-mass energy squared), and yielded the following value for the shrinkage parameter, α':

$$\alpha' = 0.14 \pm 0.09 ~ (\text{GeV}/c)^{-2}$$

This is in good agreement with the results obtained from a SLAC–Wisconsin experiment (Anderson $et~al.$, 1973) studying ϕ production at a fixed t value of 0.6 $(\text{GeV}/c)^2$. The measured cross sections from this experiment are shown in Fig. 38, and an analysis of the s dependence indicates

$$\alpha' = -0.03 \pm 0.13 ~ (\text{GeV}/c)^{-2}$$

(i.e., these measurements confirm the lack of energy dependence of the slope of the ϕ production cross section).

A new fit to the energy dependence of all the available measurements on the ϕ differential cross section (Silverman, 1976) finds, for $|t| < 0.4$ $(\text{GeV}/c)^2$

$$\alpha' = 0.22 \pm 0.27 ~ (\text{GeV}/c)^{-2}$$

The slope of the forward cross section as a function of energy, and the result of this new fit to the energy dependence, are shown in Fig. 39. We will return to the question of the energy dependence of the shape of the ϕ differential cross section later in this section.

The decay angular distribution of the ϕ from the linearly polarized laser beam experiment, although limited in statistics, shows the now familiar $\sin^2\theta \cos^2\Psi$ correlation implying natural parity exchange dominance

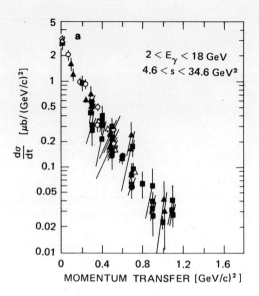

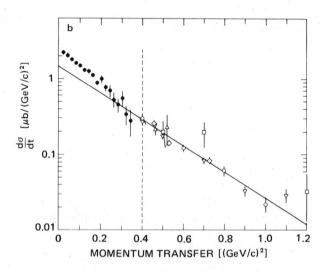

Fig. 37. The differential cross section for the reaction $\gamma p \to \phi p$: (a) a compilation of data prior to April, 1975, i.e., McClellan *et al.* (1971a) (○), Alvensleben *et al.* (1972) (●), Davies *et al.* (1970) (■), Ballam *et al.* (1973) (▲); $2 < E_\gamma < 18$ GeV, $4.6 < s < 34.6$ GeV^2. (b) The new high statistics results of Behrend *et al.* (1975) (●) plotted for $|t| < 0.4$ (GeV/c)^2, with all other measurements shown for $|t| > 0.4$ (GeV/c)^2: Bonn (1974) (◇), SLAC–LRL (1973) (□), SLAC–Wisconsin (1973) (○), Cornell (1972) (△), SLAC–Caltech (1970) (▽).

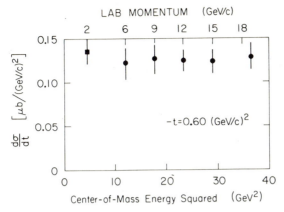

Fig. 38. The differential cross section for $\gamma p \to \phi p$ measured at momentum transfer $t = -0.6$ $(\text{GeV}/c)^2$, as a function of photon energy. The data are from Anderson *et al.* (1973) (●).

and s-channel helicity conservation in the production process (see Fig. 40). Indeed, the density-matrix elements are compatible with those measured in ρ photoproduction. A similar conclusion is obtained from measurements of the asymmetry in the yield of $K^\pm$'s (from ϕ decay), counted in the plane of photon polarization and normal to that plane, by the Wisconsin–SLAC group using polarized photons of energy ~ 8 GeV (Halpern *et al.*, 1972). They find the asymmetry parameter, Σ, is close to unity,

$$\Sigma = 0.985 \pm 0.12$$

implying natural parity exchange and no spin-flip in the ϕ production.

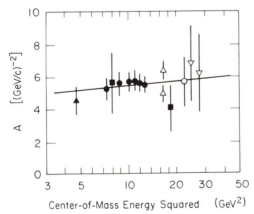

Fig. 39. Slope of the differential cross section for the reaction $\gamma p \to \phi p$, as a function of photon energy. See text for description of curve. The data are from Behrend *et al.* (1975) (●), Anderson *et al.* (1973) (○), Ballam *et al.* (1973) (■), Davies *et al.* (1970) (▽), Barger *et al.* (1972) (△).

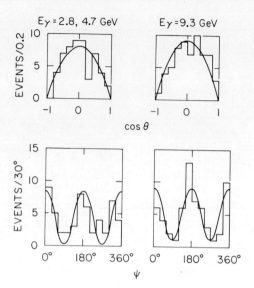

Fig. 40. The reaction $\gamma p \to \phi p$ at 2.8, 4.7, and 9.3 GeV. The decay angular distribution of the K^+K^- pairs in the helicity system, for the momentum transfer interval $0.02 \leqslant |t| \leqslant 0.8\,(\text{GeV}/c)^2$. The curves are calculated for an s-channel helicity-conserving ϕ production amplitude and for photon polarization of 92% and 77% at the two energies.

Photoproduction of ϕ mesons from complex nuclei has been studied by DESY–MIT and Cornell groups (Alvensleben *et al.*, 1972; McClellan *et al.*, 1971*a*). The ϕ is observed to be coherently produced and an analysis of the (K^+K^-) decay distribution is consistent with the SCHC production hypothesis. The photoproduction of ϕ mesons from deuterium and hydrogen targets has been measured for photon energies ~ 8 GeV (McClellan *et al.*, 1971*a*), and the ratio, R, of the cross sections at $t = 0$ is found to be

$$R = 3.6 \pm 0.6$$

compatible with the expected ratio of 3.89 for no $I = 1$ exchange.

Thus the ϕ photoproduction experiments indicate that the ϕ is diffractively produced, with natural parity, $I = 0$ t-channel exchanges and with mainly SCHC.

These observations are in agreement with our expectations for ϕ photoproduction. It has been pointed out on very general grounds that ϕp elastic scattering should proceed only by Pomeron exchange (Freund, 1967). This follows directly from the quark model, in which the ϕ is described in terms of two strange quarks $(\lambda\bar\lambda)$, and is supported by experimental evidence showing the ϕ to be decoupled from nonstrange hadrons. Then a measurement of ϕp elastic scattering should determine the parameters of the Pomeron trajectory rather clearly in comparison to other elastic-scattering processes, which usually involve additional exchange contributions. Since the ϕ-meson photoproduction cross section is related to the elastic scattering of transversely polarized ϕ mesons on protons through VDM via the relation

$$d\sigma(\gamma p \to \phi p)/dt = (\alpha \pi / \gamma_\phi^2) \cdot d\sigma(\phi p \to \phi p)/dt \qquad (3.23)$$

then measurements of the energy dependence of the photoproduction cross section, $d\sigma(\gamma \to \phi)/dt$, should also allow a good understanding of the Pomeron amplitude.

There is one caveat on this picture. The DESY/MIT group (Alvensleben *et al.*, 1971*a*) have measured the real part of the forward amplitude for the $\gamma p \to \phi p$ process by observing the interference between the resonant ϕ production and the Bethe–Heitler process in $\gamma C \to \phi C$, with $\phi \to e^+ e^-$, at 7 GeV. They find that the ϕ amplitude differs from being purely imaginary by $25° \pm 15°$, or in terms of the amplitudes

$$\frac{\mathrm{Re}\, T_\phi}{\mathrm{Im}\, T_\phi} = -0.48^{+0.33}_{-0.45}$$

This may be an indication that the $\gamma \to \phi$ process is not purely due to Pomeron exchange. Unfortunately the above phase measurement is a difficult experiment, and the errors do not allow a firm conclusion.

How do we accommodate our expectations for the behavior of a purely diffractive process with the experimental results on the s and t dependence of the ϕ cross section? Let us consider the two regions of momentum transfer with $|t| < 0.4$ (GeV/$c)^2$, and $|t| > 0.4$ (GeV/$c)^2$, separately.

First, the larger $|t|$ region with $|t| > 0.4$ (GeV/$c)^2$: ϕ photoproduction shows no s dependence over the energy range $s \sim 4$–40 GeV^2. The value of α' obtained from an analysis of the cross sections in this region is

$$\alpha'_\phi = -0.03 \pm 0.13$$

The s dependence for the elastic cross section at $|t| \sim 0.6$ (GeV/$c)^2$ for two hadronic processes (pp and $K^+ p$ elastic scattering), are shown in Fig. 41 for comparison. These processes have exotic quantum numbers in the s channel, and are therefore expected to be dominated by Pomeron exchange (i.e., to be mainly diffractive in character). For energies corresponding to $s > (10$–15) GeV^2 they also exhibit little energy dependence. In fact, the Wisconsin–SLAC value of α'_ϕ agrees well with the shrinkage parameter of other hadronic processes evaluated for $s > 10$ GeV^2, and in the same t range (Leith, 1975a),

$$\alpha'_{\pi^- p} = -0.04 \pm 0.03 \; (\text{GeV}/c)^{-2}$$

$$\alpha'_{K^- p} = 0.00 \pm 0.04 \; (\text{GeV}/c)^{-2}$$

$$\alpha'_{K^+ p} \sim 0.1 \; (\text{GeV}/c)^{-2}$$

$$\alpha'_{pp} = 0.10 \pm 0.06 \; (\text{GeV}/c)^{-2}$$

Therefore the main difference between these purely hadronic processes and the $\gamma \to \phi$ reaction is that at low energies (i.e., $s < 10$ GeV^2) the hadronic reactions begin to show strong shrinkage (see again Fig. 41), while the photoprocess maintains the high-energy behavior. In this same energy region, the total hadronic cross sections exhibit a rapidly falling energy dependence (see Fig. 42), rather than the typical constant or slowly rising cross-section characteristic of our picture of a Pomeron-dominated process. The ϕ cross section, however, is quite constant—see Fig. 36. The

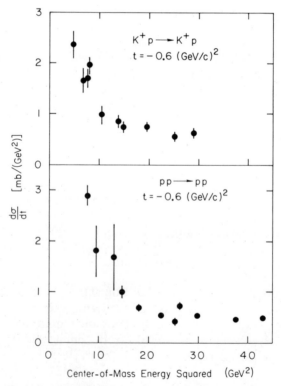

Fig. 41. The differential cross section for elastic $K^+ p$ and pp scattering for fixed momentum transfer $t = 0.6 \; (\text{GeV}/c)^2$, as a function of energy.

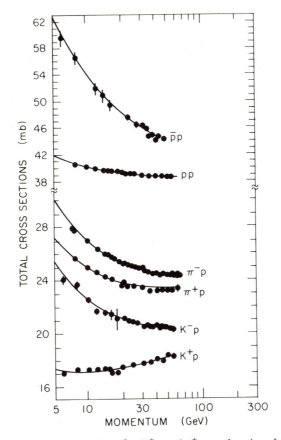

Fig. 42. Total cross section for $\pi^{\pm}p$, $K^{\pm}p$, and $p^{\pm}p$ as a function of energy.

low-energy hadron total cross-section behavior is usually explained in terms of absorptive corrections or cuts (Barger and Phillips, 1971), and perhaps these effects also modify the elastic t dependence. It is entirely possible that the photoproduction of ϕ meson does provide a good picture of the Pomeron at low energies, and that the photoproduction and hadron–hadron scattering data agree at high energies when nondiffractive contributions to the hadronic processes have become small.

The small-$|t|$ region [i.e., $|t| < 0.4$ $(GeV/c)^2$], is more complicated. In this region the ϕ differential cross section displays a steepening in the forward direction and is not well described by a simple exponential form. However, recent high-statistics studies of elastic K^+p and pp scattering around 10 GeV have shown similar behavior in these classically "Pomeron-dominated" reactions (Carnegie *et al.*, 1975). High-energy pp elastic-scattering experiments at the ISR, also observe this feature

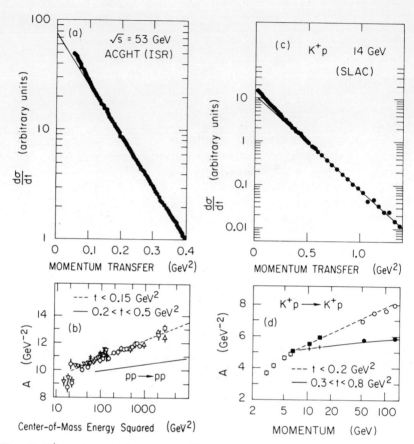

Fig. 43. K^+p and pp elastic scattering, (a), (b) examples of the differential cross section measurements showing an upward curvature of the cross section at small t. (c), (d) the energy dependence of the slope of the elastic differential cross section for K^+p and pp evaluated for two regions of momentum transfer.

(Barbiellini *et al.*, 1972; Leith, 1975*b*). In fact, for both K^+p and pp the slope of the forward-scattering cross section shows qualitatively the same behavior as the ϕ photoproduction; the larger-$|t|$ region shrinks slowly, while the forward direction has a steeper slope and shrinks more rapidly. These data are summarized in Fig. 43. The small-t differential cross sections are characterized by a shrinkage factor α':

$$\alpha'_{pp} = 0.28 \pm 0.03$$

$$\alpha'_{K^+p} \sim 0.5$$

and the analysis of the small-t ϕ photoproduction cross section finds

$$\alpha_\phi \sim 0.22 \pm 0.27$$

To summarize, the ϕ photoproduction experiments find the following:

(i) The production cross section is independent of energy.

(ii) The differential cross section is sharply peaked, with indication of steepening near the forward direction; the energy dependence for the change of the shape of the production angular distribution is poorly determined but quite consistent with that observed for other diffractive dominated processes like K^+p and pp elastic scattering.

(iii) The production process involves natural parity exchanges and conserves s-channel helicity; there is very little $I = 1$ exchange.

(iv) There may be a substantial real part to the forward ϕN scattering amplitudes; measurement of the interference of $\phi \to e^+e^-$ decays with the Bethe–Heitler process yielded $\eta_\phi = -0.48^{+0.33}_{-0.45}$, while the ϕ photoproduction on complex nuclei measurements become consistent with the storage ring data on $(\gamma_\phi^2/4\pi)$ if η_ϕ is around 0.5.

(v) Assuming the $\gamma \to \phi$ reaction is dominated by Pomeron exchange, the slope is around 4 $(\text{GeV}/c)^{-2}$, which implies that the ϕ-nucleon interaction radius is much smaller than either the nucleon–nucleon or pion–nucleon radii. This is just what we expect from the quark model and find experimentally in a study of ϕ production from complex nuclei [i.e., $\sigma(\phi N) \sim 8$–13 mb]. (These experiments will be discussed in Section 5 below.) Indeed, the measured ϕN slope agrees well with that expected from the ratio of the total cross sections and the measured πN, K^+N, pp forward slopes.

(vi) Finally, ϕ photoproduction may well be the way to a better understanding of the diffractive mechanism, but it will require much better experiments over a larger range of momentum transfers, and especially over a wider range of energies.

4. Higher-Mass Vector Mesons

4.1. Introduction

The existence of higher-mass vector meson states (than the ρ^0, ω, and ϕ), has long been predicted by the Veneziano model (Veneziano, 1968) and from analysis of the nucleon form factor (Schumacher and Engle, 1971). Examples of such states are the ρ' at 1300 MeV and ρ'' at 1700 MeV (Shapiro, 1969). There is now experimental support for two isovector states with mass around 1250 and 1600 MeV, and their existence implies that one must also find the associated isoscalar ω', ϕ' states. Below we discuss the evidence for the existence of these higher-mass vector states, and review the characteristics of the photoproduction reactions.

4.2. ρ' (1250)

The first experimental report for such states comes from a study of the missing mass distribution in the reaction

$$\gamma p \to \pi + (\text{missing mass})$$

for photon energies up to 17 GeV (Anderson *et al.*, 1970*b*). An enhancement was observed at a missing mass of 1230 MeV. The SLAC–Berkeley bubble-chamber group (Ballam *et al.*, 1974; Bingham, 1972) the SLAC streamer-chamber group (Davier *et al.*, 1973), and the DESY streamer-chamber collaboration (Rabe, 1971) found an enhancement in ($\pi^+ p^-$ + missing mass) at around the same mass value, when studying the process

$$\gamma p \to p \pi^+ \pi^- + \text{MM}$$

for missing mass, MM, greater than or equal to two pion masses (i.e., not the final state $p\pi^+\pi^-\pi^0$).

The ($\pi^+\pi^-$MM) spectrum is shown in Fig. 44, for the three energies of the bubble-chamber experiment (viz., 2.8, 4.7, and 9.3 GeV). The bump around 1240 MeV is strongly enhanced when $\pi^+\pi^-$ masses in the region 330–660 MeV are selected, as indicated by the shaded region in Fig. 44. This would be characteristic of $\pi^+\pi^-$ from ω decays. From this observation, and from a study of other final states, the bubble-chamber group conclude that the 1240-MeV bump represents an $\omega\pi^0$ decay mode of a resonant state.

The cross section for the production of this object is shown in Fig. 45, where it is observed to be about 1 μb and quite independent of the energy. The differential cross section (see Fig. 46), is found to be sharply peaked, and well represented by an exponential form.

These observations suggests that the $\omega\pi^0$ system is produced diffractively. The absence of any $\omega\pi^-$ enhancement at the same mass for the charge exchange reaction

$$\gamma n \to p \omega \pi^-$$

is a confirmation that nondiffractive contributions are small (Benz *et al.*, 1973).

The isospin, charge conjugation, and G parity of any $\omega\pi^0$ system must be $I^{CG} = 1^{-+}$. The spin and parity have been studied through analysis of the decay angular correlations in the SLAC–Berkeley experiment (Ballam *et al.*, 1973) and both $J^P = 1^-$ and 1^+ assignments are found to be compatible with the data.

Preliminary studies of $e^+e^- \to \omega\pi^0$ from Frascati show indications of an enhancement in the cross section at a center-of-mass energy of ~ 1250 MeV (Conversi *et al.*, 1974). This would be further confirmation of the

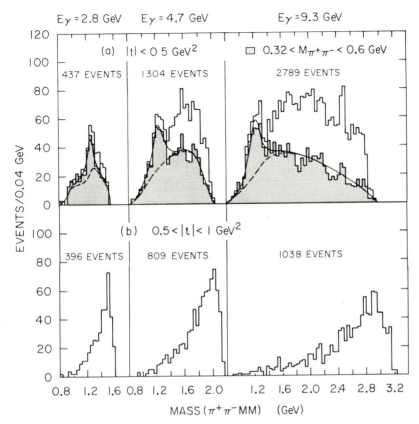

Fig. 44. The reaction $\gamma p \to p\pi^+\pi^-$ + neutrals at 2.8, 4.7, and 9.3 GeV. The effective mass of the ($\pi^+\pi^-$ + neutrals) system for two regions of momentum transfer: (a) $|t| < 0.5$ $(\text{GeV}/c)^2$ and (b) $0.5 \leqslant |t| \leqslant 1.0$ $(\text{GeV}/c)^2$. The shaded region corresponds to events in which the $\pi^+\pi^-$ mass lies in the range $(0.32 \leqslant M_{\pi\pi} \leqslant 0.6 \text{ GeV})$.

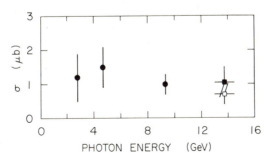

Fig. 45. The energy dependence of the cross section for the reaction $\gamma p \to \rho'(1250)p$. The data are from Ballam *et al.* (1973) (●), Anderson *et al.* (1970*b*) [(□) for $\Gamma = 100$ MeV, (■) for $\Gamma = 150$ MeV].

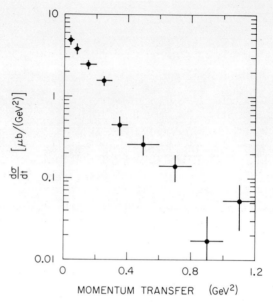

Fig. 46. The differential cross section for the reaction $\gamma p \to \rho'(1250)p$ at 9.3 GeV. The reaction is $\gamma p \to p\pi^+\pi^-$ + neutrals with cuts $1.15 < M_{\pi^+\pi^-MM} < 1.35$ GeV, $0.32 < M_{\pi^+\pi^-} < 0.60$ GeV.

existence of a $J^P = 1^-$ object of mass 1250 MeV decaying into $\omega\pi^0$, since e^+e^- reactions proceed through one-photon exchange.

We call this object the ρ' (1250).

4.3. ρ'' (1600)

The region above the ρ meson was initially scanned for effects in $(\pi^+\pi^-)$ by several counter and track-chamber experiments (Ballam *et al.*, 1972, 1973; Bingham, 1973; Park *et al.*, 1972; Eisenberg *et al.*, 1972a; Hicks *et al.*, 1969; McClellan *et al.*, 1969; Bulos *et al.*, 1971; Alvensleben *et al.*, 1971a) and no clear sign of vector meson production observed. A broad structure, about 200 MeV wide, around 1600 MeV was seen in two of the experiments on complex nuclei (Alvensleben *et al.*, 1971a; Bulos *et al.*, 1971), but the effect was difficult to interpret since the high-mass tail of the ρ^0 mesons is not well understood (see Fig. 47). The 4π system was subsequently studied by several track-chamber experiments (Bingham *et al.*, 1972; Davier *et al.*, 1973) and found to exhibit a clear enhancement around 1600 MeV, as shown in Fig. 48. The cross section for production of this peak was estimated to be between 1 and 1.6 μb and is independent of energy between 9 and 18 GeV. The differential cross section is shown in

Fig. 49 for the streamer-chamber experiment, and is well fit by

$$\frac{d\sigma}{dt} = \frac{d\sigma}{dt}\bigg|_0 \exp At$$

with a slope, A, of (5.7 ± 0.3) $(\text{GeV}/c)^{-2}$.

The 4π bump is identified to be mainly a $\rho^0 \pi^+ \pi^-$ final state. From a study of other final states the isospin of the non-rho–pion pair may be determined. The fact that $\rho^0 \rho^0$ is not observed excludes $I_{2\pi} = 1$, while the ratio $(\rho^0 \pi^0 \pi^0 : \rho^0 \pi^+ \pi^-) \approx 0.5$, favors $I_{2\pi} = 0$ rather than $I_{2\pi} = 2$. This gives an assignment $I = 1$ for the total bump, and since the G parity of a 4π system is $+1$, the change conjugation, C must be -1 [since $G = C(-1)^I$]. We then have $I^{CG} = 1^{-+}$ for the 4π state.

A study of the decay angular distribution from the linearly polarized photon experiment yields the familiar $\sin^2\theta \cos^2\Psi$ correlations we have seen in ρ^0, ω, and ϕ production, where the vector sum of the two π^+ momenta is used as the analyzer for the 4π system decay (Bingham *et al.*, 1972). Examples of the decay correlations are shown in Fig. 50. They imply mainly natural parity exchanges in the t channel and mainly SCHC for the production of the 4π state. In addition, they indicate that the spin parity of the 1600-MeV bump is $J^P = 1^-$. The $I = 0$ exchange character of the production process is further emphasized by the observation of coherent production of ρ'' (1600) on deuterium (Eisenberg *et al.*, 1976).

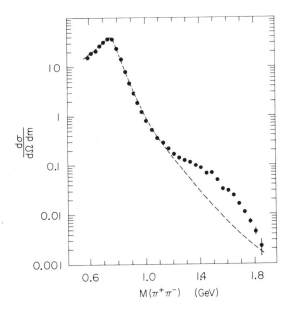

Fig. 47. The dipion mass distribution for the reaction $\gamma C \to \pi^+ \pi^- C$ at 7 GeV photon energy.

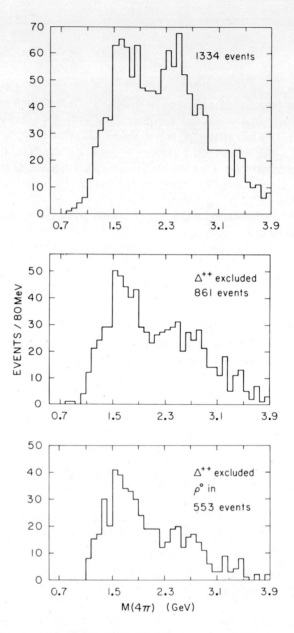

Fig. 48. The four-pion mass distribution for the reaction $\gamma p \to \pi^+ \pi^+ \pi^- \pi^- p$ in the energy range 6–18 GeV.

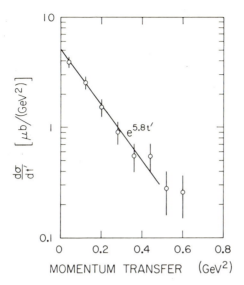

Fig. 49. The differential cross section for the reaction $\gamma p \to \rho''(1600)p$ in the energy range 6.18 GeV, with Δ^{++} excluded, ρ^0 selected and $M(4\pi) < 2.0$ GeV.

Thus, the experimental information from photoproduction reactions indicates that a $(\rho^0 \pi^+ \pi^-)$ enhancement with quantum numbers, $I^{CG} = 1^{-+}$, $J^P = 1^-$, is diffractively produced at energies around 10 GeV. Such a state should be seen in $e^+ e^-$ annihilations through the one-photon-exchange process. The cross section for $e^+ e^- \to \pi^+ \pi^- \pi^+ \pi^-$ is shown in Fig. 51, where it may be observed to rise rapidly from threshold, peak around 1600 MeV, and then fall rapidly as the energy further increases (Bartoli *et al.*, 1970; Barbarino *et al.*, 1972; Ceradini *et al.*, 1973). This bump may be interpreted as confirmation of the production of a vector state in $e^+ e^-$ collisions. The ratio of the $(\pi^+ \pi^- \pi^0 \pi^0 : \pi^+ \pi^+ \pi^- \pi^-)$ cross sections is consistent with that expected for the decay of an $I = 1$ $\rho^0 \pi^+ \pi^-$ system. It

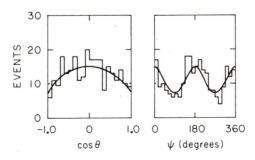

Fig. 50. $\rho''(1600)$ decay angular distributions in the helicity system, for the reaction $\gamma p \to \rho''(1600)p$ at 9.3 GeV.

appears that the e^+e^- and γp experiments are producing the same heavy vector meson, which we call the ρ'' (1600).

Indeed, one finds good agreement between the measured peak cross section in e^+e^- annihilations and that calculated from the photoproduction data using the vector dominance model. We may write

$$\frac{\sigma(\gamma p \to \rho^0 p)}{\sigma(\gamma p \to \rho'' p)_{\downarrow 4\pi}} = \left(\frac{\gamma_{\rho''}^2}{\gamma_\rho^2}\right)\frac{\Gamma_{\rho''}}{\Gamma_{\rho'' \to 4\pi}} \tag{4.1}$$

assuming that the elastic $\rho^0 p$ and $\rho'' p$ cross sections are comparable and that the proces $\rho^0 p \to \rho'' p$ is weak. Similarly we may relate the peak cross sections for $e^+e^- \to \rho^0$ and ρ'' through

$$\frac{\sigma(e^+e^- \to \rho^0)_{\text{peak}}}{\sigma(e^+e^- \to \rho'' \to 4\pi)_{\text{peak}}} = \left(\frac{\gamma_{\rho''}^2}{\gamma_\rho^2}\right)\frac{\Gamma_{\rho''}^2}{\Gamma_\rho \cdot \Gamma_{\rho'' \to 4\pi}}\frac{M_{\rho''}}{M_\rho} \tag{4.2}$$

From Eqs. (4.1) and (4.2), we find

$$\sigma(e^+e^- \to \rho'' \to 4\pi)_{\text{peak}} = \sigma(e^+e^- \to \rho)_{\text{peak}} \cdot \frac{M_\rho}{M_{\rho''}} \cdot \frac{\Gamma_\rho}{\Gamma_{\rho''}}\frac{\sigma(\gamma p \to \rho^0 p)}{\sigma(\gamma p \to \rho'' p)_{\downarrow 4\pi}} \tag{4.3}$$

Substituting the measured quantities on the right-hand side of Eq. (4.3), we find

$$\sigma(e^+e^- \to \rho'' \to 4\pi) = (12\text{--}25) \text{ nb}$$

consistent with the measured cross section of (16 ± 5) nb from the storage ring experiments (Ceradini *et al.*, 1973).

No other strong decay modes of the ρ'' have been observed. The 2π mode is not observed in e^+e^- reactions, while limits of $<20\%$ and $<14\%$

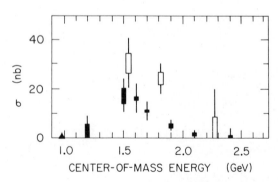

Fig. 51. The cross section for $e^+e^- \to \pi^+\pi^-\pi^+\pi^-$ as a function of energy. The data are from Bartoli *et al.* (1972) ($\square$) and Grilli *et al.* (1973) ($\blacksquare$).

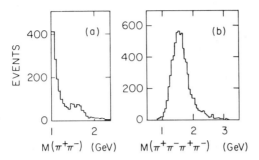

Fig. 52. The dipion and four-pion mass spectra from the reactions $\gamma Be \to \pi^+ \pi^- X$ and $\gamma Be \to \pi^+ \pi^+ \pi^- \pi^- X$ at photon energy in the range 80–100 GeV.

for the ratio of 2π to 4π decay have been set by the photoproduction studies in hydrogen and complex nuclear targets, respectively. However, preliminary results from a high-energy experiment at FNAL (Lee, 1975), have shown a signal in both 2π and 4π in the reactions

$$\gamma Be \to \pi^+ \pi^- X$$
$$\to \pi^+ \pi^+ \pi^- \pi^- X$$

at energies of around 80 GeV. The mass distributions, uncorrected for the apparatus detection efficiency, are shown in Fig. 52. A very rough estimate of the $(2\pi/4\pi)$ branching ratio from these measurements would indicate $R > 0.05$.

A state with the same quantum numbers has been identified in a π–π scattering phase shift analysis (Hyams *et al.*, 1975; Estabrooks and Martin, 1975). The data come from a high-statistics measurement (Grayer *et al.*, 1974) of the reaction

$$\pi^- p \to \pi^+ \pi^- n$$

at 17 GeV. The analysis of the $(\pi$–$\pi)$ angular distribution finds evidence of a p-wave $(\pi$–$\pi)$ state lying below the spin-3 state called the g meson. The resonance has a mass $M = (1590 \pm 20)$ MeV, and width, $\Gamma = (180 \pm 50)$ MeV. The properties of this state, in terms of mass and spin-parity quantum numbers, agree well with those found in photoproduction and $e^+ e^-$ experiments, but it is much narrower and couples more strongly to 2π. However, these differences are probably the result of difficulties in interpretation of the experiments; the photoproduction (4π) bump may not be all resonant (Slattery and Ferbel, 1974), while the ρ'' (1600), found in the π–π scattering studies, is a rather small contribution to the total π–π cross section (see Fig. 53).

A summary of the properties of the ρ'' (1600) is given in Table 8.

Table 8. Summary of $\rho''(1600)$ Properties

Experiment		M (MeV)	Γ (MeV)	I^{CG}	J^P	$\pi^+\pi^-/4\pi^{\pm}$	$K^+K^-/4\pi^{\pm}$
$\gamma A \rightarrow \rho'' A$	(Alvensleben et al., 1971a) (Bulos et al., 1971)	—	—	—	—	<0.14	—
$\gamma p \rightarrow \rho'' p$	(Bingham et al., 1972) (Davier et al., 1973)	1500	500	1^{-+}	1^-	<0.20	<0.04
$\gamma d \rightarrow \rho'' d$	(Alexander et al., 1975)	1570 ± 60	350 ± 90	—	—	—	—
$\gamma Be \rightarrow 2\pi X$ $4\pi X$	(Lee, 1975)	~1600	~500	—	—	>0.05	—
$e^+e^- \rightarrow \pi^+\pi^-\pi^+\pi^-,$ $\pi^+\pi^-\pi^0\pi^0$ $\pi^+\pi^-$	(Ceradini et al., 1973)	~1600	~300	1^{-+}	1^-	not seen $\sim 10^{-2}$	—
$\pi^- p \rightarrow \rho'' n$	(Hyams et al., 1975)	~1590	~200	1^{-+}	1^-	0.25 ± 0.05	—

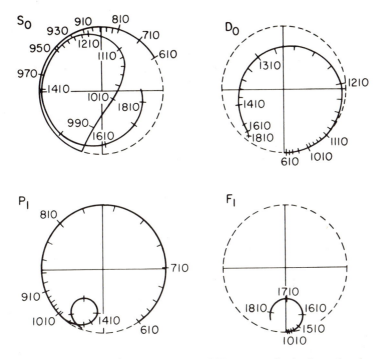

Fig. 53. Argand diagrams of the S-, P-, D- and F-wave amplitudes from the phase-shift analysis of $\pi-\pi$ scattering data.

4.4. "New Particle" Production

Near the end of 1974 there was great excitement at the discovery of a very narrow peak in the e^+e^- system at a mass of 3100 MeV from experiments at Stanford (Augustin *et al.*, 1974) and at Brookhaven (Aubert *et al.*, 1974). The cross section as measured in e^+e^- collisions at SPEAR is shown in Fig. 54. The width of the new particle was found to be (69 ± 15) keV. Shortly afterwards another peak was found at 3700 MeV, with a width of (225 ± 56) keV (Abrams *et al.*, 1975). Such narrow widths at such high mass imply that some new selection rule is at work in the decay of these particles—the Ψ (3100) and Ψ' (3700). The most popular, and currently the most successful, model explaining the existence of these new particles invokes a new quantum number, called charm, and an additional quark—the charmed quark (c) (De Rujula and Glashow, 1975). The Ψ states are then described as being ($c\bar{c}$) states, very much as the ϕ meson is thought of as a system of strange quarks ($\lambda\bar{\lambda}$). Systematic studies of the decay modes of the Ψ and Ψ' have shown that they are heavy vector mesons formed via the one-photon exchange in the e^+e^- annihilation, and

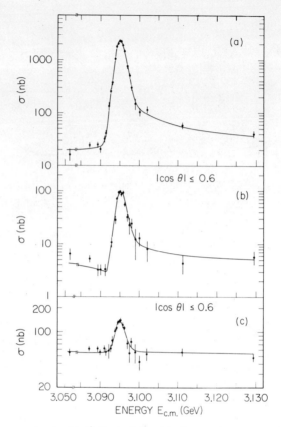

Fig. 54. The cross section for (a) $e^+e^- \to$ hadrons, (b) $e^+e^- \to \mu^+\mu^-$, and (c) $e^+e^- \to e^+e^-$ as a function of energy, in the neighborhood of the $\psi(3100)$ resonance.

that they both have $I^{CG} J^P = 0^{--}1^-$. One would expect these states to be produced diffractively in the photoproduction reaction.

Three experimental groups at Cornell (Gittelman *et al.*, 1975), SLAC (Camerini *et al.*, 1975) and FNAL (Knapp *et al.*, 1975), have observed the photoproduction of these heavy vector mesons, over the energy range (10–100) GeV. In Fig. 55 the effective mass of the detected ($\mu^+\mu^-$) pair is shown for the SLAC and FNAL experiments, and the Ψ (3100) signal is clearly observed.

The forward cross section for the reaction

$$\gamma N \to \Psi N \qquad\qquad (4.4)$$

exhibits a very interesting energy dependence, as shown in Fig. 56. The cross section rises very steeply for energies above ~ 11 GeV, increasing by

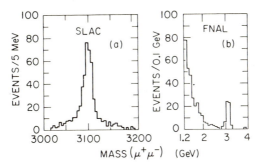

Fig. 55. The $(\mu^+\mu^-)$ mass distribution from the reaction (a) $\gamma p \to \mu^+\mu^- p$ for photon energy of 19 GeV, and (b) $\gamma Be \to \mu^+\mu^- X$ for photon energy in the range 80–100 GeV.

almost two orders of magnitude by ~ 18 GeV. Beyond 20 GeV the growth is quite gradual, increasing by at most a factor of 2 in the interval up to 100 GeV.

The differential cross section is observed to change dramatically through this region of rapid rise in the forward cross section. The measured distributions at 11 GeV (Gittelman *et al.*, 1975) and 19 GeV (Camerini *et al.*, 1975) are shown in Fig. 57. The differential cross sections are well fitted by an exponential form

$$\frac{d\sigma}{dt} = \frac{d\sigma}{dt}\bigg|_0 \exp(At)$$

with the slope, A, equal to 1.25 $(\text{GeV}/c)^{-2}$ at 11 GeV, and $\sim 3 \,(\text{GeV}/c)^{-2}$ at 19 GeV. The highest-energy data are compatible with a slope ~ 4 $(\text{GeV}/c)^{-2}$ (Knapp *et al.*, 1975). This indicates that the production angular distribution is almost isotropic at energies near to the kinematic threshold, and that in the region of the rapid rise of the cross section the distribution becomes quite sharply peaked, but does not change much as the energy is further increased.

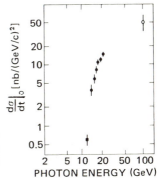

Fig. 56. The energy dependence of the forward cross section for the reaction $\gamma p \to \psi(3100)p$. The data are from Knapp *et al.* (1975) (○), Camerini *et al.* (1975) (●), and Gittelman *et al.* (1975) (■).

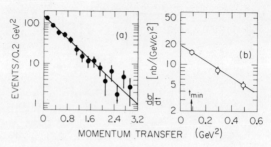

Fig. 57. The differential cross section for the reaction $\gamma p \to \psi(3100)p$ at (a) 11 GeV and (b) 19 GeV.

The SLAC experiment found that inelastic photoproduction of Ψ mesons,

$$\gamma N \to \Psi N'$$

was small, being only $\sim 20\%$–30% of the elastic production.

The photoproduction of the other high-mass particle, the Ψ' (3700), is observed in both the SLAC and FNAL experiments, and the ratio of Ψ to Ψ' production at 19-GeV photon energy is measured to be

$$R = \frac{\sigma(\gamma \to \Psi)}{\sigma(\gamma \to \Psi')} = 6.8 \pm 2.4$$

Several interesting observations may be made in applying the vector dominance model (Section 5) to the above data. Taking the measured forward cross section, $(d\sigma/dt)_0$, and assuming that the ΨN scattering amplitude is purely imaginary we may use Eq. (5.4) to determine $\sigma_T(\Psi N)$, the total ΨN cross section, in the same way we find $\sigma_T(\rho N)$ in Section 5 below. We take the photon–vector-meson coupling strength from the e^+e^- experiments ($\gamma_\Psi^2/4\pi = 2.6$) and find $\sigma_T(\Psi N) \sim 1$ mb. This implies that the ΨN interaction is typically hadronic, but with a much smaller strength than the πN and ρN interaction. Recent measurements comparing the relative A dependence of Ψ photoproduction on complex nuclei, and applying an analysis similar to that discussed in Section 5.2 below, yield an estimate of $\sigma_T(\Psi N) = (3 \pm 1)$ mb, independent of the vector dominance model assumptions (Prepost, 1976; Ritson, 1976).

We may now use this value of $\sigma_T(\Psi N)$, and the measured slope of the production distribution to learn of the ratio σ_{el}/σ_T. From the optical theorem, we have a relationship between the forward elastic cross section and the total cross section

$$\left.\frac{d\sigma}{dt}\right|_0 = \frac{1}{16\pi}\sigma_T^2 \qquad (4.5)$$

If $(d\sigma/dt) = (d\sigma/dt)_0 \exp(At)$, then we may integrate to find the elastic cross section, σ_{el}, and rewrite Eq. (4.5) as

$$\sigma_{el} = \frac{1}{16\pi A} \cdot \sigma_T^2 \qquad (4.6)$$

or

$$\frac{\sigma_{el}}{\sigma_T} = \frac{1}{16\pi A} \cdot \sigma_T \qquad (4.7)$$

This relation (4.6) applies both to ΨN scattering, and, through VDM, to Ψ photoproduction, and for either process we have

$$\frac{\sigma_{el}}{\sigma_T} \sim 10^{-2}$$

This means that most of the ΨN scattering cross section is inelastic. Furthermore, for a 5-nb cross section for reaction (4.4), the total cross section for the "Ψ-like" part of the photon must be of order 0.5 μb, or about half a percent of the total γp cross section. However, we know from the measurements discussed above that inelastic processes involving the Ψ-meson itself are only $\sim 20\%$–30% of the elastic cross section and so negligible in this context. What then, are these inelastic "Ψ-like" photo-processes which must account for $\frac{1}{2}\%$ of all the γp interactions? Within the context of the charmed quark model, this inelastic cross section represents the production of charmed mesons, and one may look at the steep rise of the Ψ-production cross section, seen in Fig. 56, as an indication of the threshold of such a process.

These are intriguing thoughts, and it will be very interesting to watch the results of the photoproduction experiments over the next few years for further insight into the nature of the Ψ and Ψ' mesons and the associated charmed particles.

5. Vector Dominance and the Photon Couplings

5.1. Vector Dominance Model

The origin of the vector dominance model dates back to the early 1960's, when the isovector nucleon form factor was described in terms of a strong $\pi\pi$ resonance later identified as the ρ meson (Frazer and Fulco, 1960; Nambu, 1957) and when Sakurai (1960) suggested that just as the electromagnetic current had a photon associated, so the isospin current and baryon current and hypercharge current had associated vector mesons, and that there would be strong coupling between the "current-associated" particles.

Basically, the electromagnetic interaction of hadrons is described by the coupling of the electromagnetic field to the hadronic electromagnetic current

$$j_\mu^{\text{em}}(x) = j_\mu^I(x) + \tfrac{1}{2} j_\mu^Y(x) \tag{5.1}$$

where $j_\mu^I(x)$ and $j_\mu^Y(x)$ are the zero component of the isospin and the hypercharge currents, respectively. The smallness of the coupling constant, $\alpha = e^2/4\pi$, allows the photoproduction process to be treated in lowest order of the eltromagnetic interaction.

The vector dominance model then connects the hadronic electromagnetic current with the fields of the vector mesons ρ^0, ω, ϕ that have the same quantum numbers as the electromagnetic current, namely, $J = 1$, $P = -1$, $C = -1$, $Y = 0$. This connection can be made via the current field identity (Joos, 1967)

$$j_\mu^{\text{em}}(x) \equiv -\sum_V \frac{m_V^2}{2\gamma_V} V_\mu(x) = -\left[\frac{m_\rho^2}{2\gamma_\rho} \rho_\mu^0(x) + \frac{m_\omega^2}{2\gamma_\omega} \omega_\mu(x) + \frac{m_\phi^2}{2\gamma_\phi} \phi_\mu(x) + \cdots\right] \tag{5.2}$$

where γ_ρ, γ_ω, γ_ϕ are the coupling constants of the electromagnetic current to the vector meson fields $\rho_\mu(x)$, $\omega_\mu(x)$, $\phi_\mu(x)$, respectively, and m_ρ, m_ω, m_ϕ are the masses of the vector mesons.

Initially, the model implied that the three vector mesons ρ, ω, ϕ completely saturated the above identity. The model has since been generalized to include contributions from other higher-mass vector mesons in the summation of Eq. (5.2) and also to reflect the coupling of the photon to the continuum "background" seen in e^+e^- annihilations (Sakurai and Schildknecht, 1972).

The above relationship between the electromagnetic current and the vector meson fields implies that any amplitude involving real or virtual photons may be expressed as a linear combination of vector meson amplitudes each multiplied by a vector meson propagator. The assumption is usually made that the invariant vector meson amplitudes are slowly varying functions of the vector mass, m_V, and that the energy dependence comes from the propagator, not the coupling constants.

More specifically, these arguments allow one to relate vector meson photoproduction to the elastic scattering of transversely polarized vector mesons on nucleons. The relationship is written

$$\frac{d\sigma}{dt}(\gamma N \to VN) = \frac{\alpha}{4} \cdot \frac{4\pi}{\gamma_V^2} \frac{d\sigma}{dt}(VN \to VN) \tag{5.3}$$

where $\gamma_V^2/4\pi$ represents the strength of the photon–vector-meson, V, coupling. This is represented diagrammatically in Fig. 58a. Such a description assumes that off-diagonal terms like $V' \rightarrow V$, where V, V' are different vector mesons, do not exist (i.e., one may neglect processes like those in Fig. 58b).

We may extend Eq. (5.3) by using the optical theorem to relate the forward elastic vector-meson–nucleon scattering to the total cross section, $\sigma_T(VN)$, and write

$$\frac{d\sigma}{dt}(\gamma N \rightarrow VN)\bigg|_0 = \frac{\alpha}{64\pi} \cdot \frac{4\pi}{\gamma_V^2} \cdot (1 + \eta_V^2) \cdot \sigma_T^2(VN) \qquad (5.4)$$

where η_V is the ratio of the real to imaginary forward-scattering amplitude, and $\sigma_T(VN)$, the total cross section, for the vector-meson–nucleon interaction.

The photon–vector-meson coupling can be measured directly in e^+e^- annihilations, where the vector meson is formed via one-photon-exchange (see Fig. 58c). From a measurement of the excitation spectrum in the storage ring experiments one may determine γ_V^2 through the relation

$$\frac{\gamma_V^2}{4\pi} = \frac{\alpha^2 m_V}{12 \cdot \Gamma_{V \rightarrow e^+e^-}} \qquad (5.5)$$

where m_V is the mass of the vector meson, V, and $\Gamma_{e^+e^-}$ is the partial width for its decay into lepton pairs.

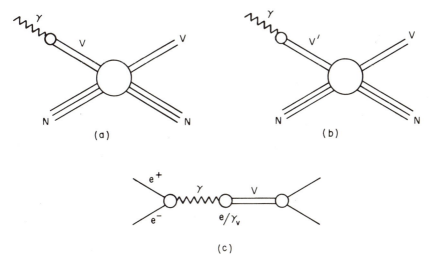

Fig. 58. Schematic diagrams for the vector dominance model.

418   D. W. G. S. Leith

Clearly the photoproduction experiments determine γ_V^2 with the photon on the mass shell $(q^2 = 0)$, while the e^+e^- experiments measure the coupling strength with the vector meson on the mass shell $(q^2 = m_V^2)$. It is the assumption of the VDM that these couplings strengths should be the same. As we shall see below, in Section 5.3, the experiments indicate reasonable agreement with this hypothesis for the ρ^0, ω, and ϕ, but are unable to exclude the possibility of some q^2 dependence. We will discuss this further later.

We may directly test the vector dominance model through a comparison of the Compton scattering process, $\gamma N \to \gamma N$ and vector meson photoproduction. The relationship may be written as

$$\frac{d\sigma}{dt}(\gamma N \to \gamma N) = \frac{\alpha}{4}\left|\sum_V\left[\frac{d\sigma}{dt}(\gamma N \to VN)\cdot\left(\frac{4\pi}{\gamma_V^2}\right)e^{i\delta_V}\right]^{1/2}\right|^2 \tag{5.6}$$

where δ_V relates the phases of the various vector meson amplitudes. Again, using the optical theorem, we can rewrite Eq. (5.6) as

$$\sigma_T(\gamma N) = \sum_V\left[\frac{16\pi^2\alpha}{\gamma_V^2}\cdot\frac{1}{(1+\eta_V^2)}\cdot\frac{d\sigma}{dt}(\gamma N \to VN)\bigg|_0\right]^{1/2} \tag{5.7}$$

The equalities indicated in Eqs. (5.6) and (5.7) should be attained when all of the vector couplings of the photon are identified and included in the summations. Such tests are described in Section 5.4 below.

Finally, we have various predictions for the ratios of the coupling strengths at the $\gamma \to V$ vertex. Application of $SU(6)$ predicts that

$$\frac{1}{\gamma_\rho}:\frac{1}{\gamma_\omega}:\frac{1}{\gamma_\phi} = 3:1:-2^{1/2}$$

which implies that the coupling strengths for $\gamma \to \rho$, ω, ϕ are in the ratio

$$9:1:2 \tag{5.8}$$

There are several calculations of symmetry-breaking schemes (Oakes and Sakurai, 1967; Das et al., 1967) which alter these predicted ratios to

$$9:0.65:1.33$$

and

$$9:1.2:1$$

Further, if one includes the possibility of additional quarks as in the charm scheme, then it has been shown that for charmed quarks with charge $\frac{2}{3}$, the predicted ratio of the couplings, including the $(c\bar{c})$ state, Ψ (Gaillard et al., 1974), would be

$$\rho:\omega:\phi:\Psi = 9:1:2:8 \tag{5.9}$$

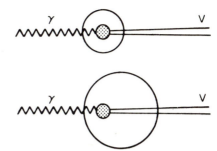

Fig. 59. Schematic diagram for coherent vector meson photoproduction in complex nuclear targets.

In the following sections we will consider these various relationships in the light of the available experimental data, extract the photon–vector-meson coupling strength and see how well the vector dominance model holds up.

5.2. *Experiments on Complex Nuclei*

We have discussed the $\gamma N \to VN$ reaction in Sections 3 and 4 above and have shown that it has the characteristics of a diffractive process. In this section we review experiments studying the coherent photoproduction of vector mesons on complex nuclear targets. These experiments are of interest since they provide information on the vector-meson–nucleon interaction from studies of the absorption in nuclear matter and they also allow an independent measurement of the photon–vector-meson coupling strength. We first discuss the analysis of the experiments and then review the data.

Since for energies above a few GeV almost all the photoproduced vector mesons live long enough to traverse the nucleus and decay in vacuum, it was suggested that by observing the relative yield of vector mesons transmitted through varying path lengths of nuclear matter, the total interaction cross section for the vector mesons on nucleons could be determined (Drell and Trefil, 1966; Ross and Stodolsky, 1966). Such a scheme is shown schematically in Fig. 59, where the variation in nuclear path length is achieved by studying the A dependence of the forward cross section for the reaction

$$\gamma A \to VA$$

More quantitatively, we may write, after applying Glauber multiple-scattering theory (Franco and Glauber, 1966),

$$f_0(\gamma A \to VA) = f_0(\gamma N \to VN) \int d^2 b \int dz\, \rho(\bar{b}, z)$$

$$\times \exp(iq_{\parallel} z + iq_{\perp} \bar{b}) \exp[-\tfrac{1}{2}\sigma_T(VN)(1 - i\eta_V)T] \qquad (5.10)$$

where $\rho(\bar{b}, z)$ is the nuclear density distribution, $q_\parallel$ is the longitudinal momentum transfer necessary to put the vector meson, V, on its mass shell, $\bar{q}_\perp$ is the transverse momentum transfer, with $t = -(q_\perp^2 + q_\parallel^2)$, and

$$T = \int_z^\infty dz' \cdot \rho(\bar{b}, z')$$

This describes the process in which a photon converts into a vector meson at $(\bar{b}, z)$, where $\bar{b}$ is the impact parameter and z is the distance along the incident photon direction. The probability of the conversion is given in terms of the average of the forward photoproduction amplitude for protons and neutrons, $[d\sigma(\gamma N \to VN)/dt]|_0 = |f_0|^2$. The vector meson amplitude is then subject to absorption and refraction in passing through the remaining nuclear matter.

Assuming $\rho(\bar{b}, z)$ is known, the relative A dependence allows a determination of the total vector-meson–nucleon cross section, σ_{VN}, and therefore the forward-scattering amplitude through the relation

$$f_0(VN \to VN) = \frac{ik}{4\pi} \sigma_T(VN)(1 - i\eta_V) \qquad (5.11)$$

where η_V is the ratio of real to imaginary forward-scattering amplitudes. This determination of $f_0(VN \to VN)$ is obtained from the measured A dependence only, and is independent of the vector dominance model. The absolute normalization of the cross section for the reaction determines $f_0(\gamma N \to VN)$ through vector dominance, and allows a determination of the photon–vector-meson coupling strength, $(\gamma_V^2/4\pi)$.

It is interesting to note that the amplitude in Eq. (5.10) differs from an elastic scattering amplitude by the phase factor involving the interference between the longitudinal momentum transfer and the real part of the scattering amplitude. This is an important factor in determining the forward cross section, especially at low photon energies (Swartz and Talman, 1969).

The nuclear density distributions most commonly used in the analysis of these experiments are as follows:

(a) *The Harmonic Oscillator.* For low A nuclei

$$\rho(r) = \rho_0\left(1 + \alpha \frac{r^2}{a_0^2}\right) \exp\left(-\frac{r^2}{a_0^2}\right) \qquad (5.12)$$

where $\alpha = \frac{4}{3}$ for carbon, and $\frac{5}{6}$ for beryllium and a_0 is a parameter of the fit.

(b) *The Wood–Saxon Distribution.* For heavy nuclei

$$\rho(r) = \rho_0\{1 + \exp[(r - R)/\varepsilon]\}^{-1} \qquad (5.13)$$

where R is the nuclear radius and $\varepsilon = 0.545$ F.

Modifications to these distributions to make allowance for two-body correlations have generally been applied in the data analysis (von Bockman, 1969; Bauer, 1971).

There remains the question of what nuclear radius to use in a particular density distribution. The DESY–MIT group (Alvensleben *et al.*, 1970*a*) determined the nuclear radius as a function of A from their measurements of the t distribution in the reaction $\gamma A \to \rho^0 A$. They found $R(A) = (1.12 \pm 0.02) A^{1/3}$ F. Other choices of radius are derived from electron scattering data or from fits to neutron–nucleus, and proton–nucleus total cross section data. The results of the analyses do not strongly depend on which nuclear radii are used.

A summary of the experiments studying the coherent vector meson production on complex nuclei is given in Table 9. The most extensive measurements are those of the DESY–MIT and Cornell groups for ρ^0 photoproduction, and Cornell on ϕ production, while the Rochester–Cornell and DESY groups provide the only heavy nuclei data on ω production. An example of the DESY–MIT measurements on

$$\gamma A \to \pi^+ \pi^- A$$

at 6 GeV is given in Fig. 60, where the $\pi^+ \pi^-$ mass spectrum is shown as a function of momentum transfer for 13 different targets. The Cornell measurement of the differential cross section for ϕ photoproduction in the reaction

$$\gamma A \to K^+ K^- A$$

at 6 GeV, is shown in Fig. 61. The sharp forward peak characteristic of a coherent process is clearly observed. Finally, Fig. 62 shows the differential

Table 9. Experiments on Coherent Vector Meson Production on Complex Nuclei

Group	Photon beam Energy (GeV)	Type	Number of targets used	Mesons studied	References
CEA	2–5	bremsstrahlung	4	ρ	Lanzerotti *et al.*(1968)
DESY–MIT	3–7	bremsstrahlung	14	ρ, ϕ	Alvensleben *et al.* (1970*b*)
Cornell	4–9	bremsstrahlung	10	ρ, ϕ	McClellan *et al.* (1971*b*)
SLAC	5, 7, 9	monochromatic	8	ρ	Bulos *et al.* (1969) and Williams (1973)
DESY	5.7	bremsstrahlung	5	ω	Braccini *et al.* (1970)
Rochester	9	bremsstrahlung	7	ρ, ω	Behrend *et al.* (1970)
	8	tagged	4	ω	Abramson *et al.* (1976)

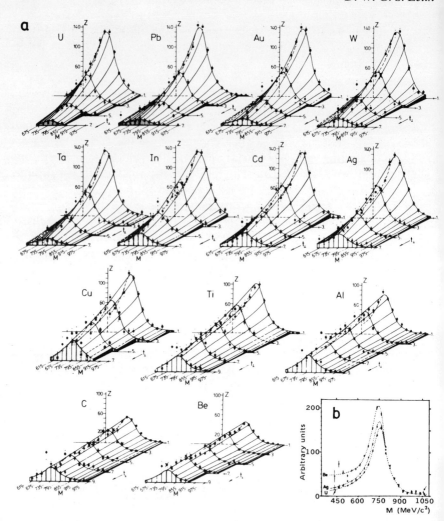

Fig. 60. The dipion mass distribution as a function of momentum transfer for 13 different targets in the reaction $\gamma A \to \pi^+\pi^- A$.

cross section for ω photoproduction from beryllium and copper targets from the Rochester–Cornell experiment.

The analysis of these experiments has proven to be nonunique, in that owing to the very strong correlation between the size of the total vector-meson–nucleon cross section and the ratio of the real to imaginary parts of the forward-scattering amplitude, they have been unable to separately determine all three unknowns $\sigma(VN)$, η_V, $(\gamma_V^2/4\pi)$. In general, if one is known, then the other two quantities are well determined.

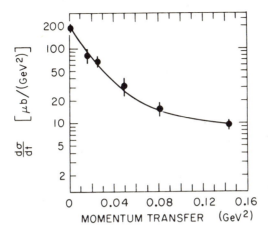

Fig. 61. The differential cross section for the coherent production of ϕ mesons on carbon at 8 GeV.

The analyses of the heavy nuclei experiments are usually expressed as the correlated determination of η_V and $\sigma(VN)$, and subsequently a choice of $\sigma(VN)$ fixes the value of the coupling strength, $(\gamma_V^2/4\pi)$. An example of such an analysis for the ρ^0 data is shown in Fig. 63. The χ^2 contours indicate the strong correlation between η_ρ and $\sigma(\rho N)$ by the narrow valley running across the plot. Roughly speaking it indicates that $\sigma(\rho N) = (32 + \eta_\rho/0.05)$ mb. Also indicated on Fig. 63 are the limits on η_ρ from the

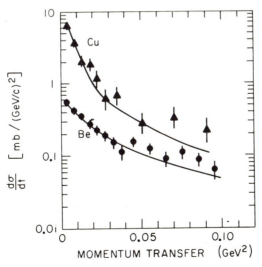

Fig. 62. The differential cross section for the coherent production of ω mesons on beryllium and copper targets for photon energy about 8 GeV.

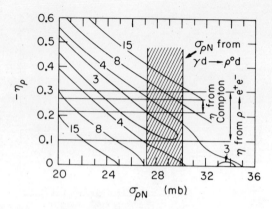

Fig. 63. A χ^2 map of the correlation between the ratio of the real to imaginary forward ρN scattering amplitude, η_ρ, and the total ρN cross section, $\sigma(\rho N)$, as determined from an analysis of coherent ρ photoproduction on 13 different nuclear targets. The data are from Alvensleben *et al.* (1970*b*) and there are 8 degrees of freedom.

interference experiments $(\eta_\rho = -0.2 \pm 0.1)$ (Alvensleben *et al.*, 1970*a*), and from the Compton scattering analysis $(\eta_\rho = -0.24 \pm 0.03)$ (see Section 2.2).

The ρ^0 data have been analyzed by the DESY–MIT (Alvensleben *et al.*, 1970*b*) and Cornell (McClellan *et al.*, 1971*b*) groups, and recently both experiments and analyses were very beautifully reviewed and independently evaluated (Spital and Yennie, 1975). Good agreement is obtained among the different analysis and if η_ρ is taken to be -0.2 for energies around 7 GeV, then

$$\sigma_T(\rho N) = 28 \pm 1.5 \text{ mb}$$

and

$$\gamma_\rho^2/4\pi = 0.61 \pm 0.03 \qquad (5.14)$$

Fortunately, in the case of the ρ^0 data another experiment allows the determination of $\sigma_T(\rho N)$ independent of the question on the size of the real part, η_ρ. These measurements come from a study of coherent ρ^0 production from deuterium at 6, 12, and 18 GeV, and for large momentum transfers $[|t| > 0.6 \ (\text{GeV}/c)^2]$ (Anderson *et al.*, 1971). The requirement that the deuteron remain bound causes the reaction to be dominated by a two-step process in which the ρ^0 is produced on one nucleon and scatters on the other, giving approximately equal recoils to both nucleons. Therefore, at large t values this double scattering amplitude is roughly given by the product of the ρ production and scattering amplitudes $(f_{\gamma \to \rho} \cdot f_{\rho \to \rho})$, and is proportional to $\sigma_T^2(\rho N)/\gamma_\rho$ and independent of η_ρ, while at small t it is proportional to $\sigma_T(\rho N)/\gamma_\rho$. They find that their data are well described at all t values, and at all three energies with $\sigma_T(\rho N) = (28.6 \pm 0.5)$ mb, and $\gamma_\rho^2/4\pi = (0.69 \pm 0.04)$.

The Rochester–Cornell experiments (Behrend *et al.*, 1970; Abramson *et al.*, 1976) provide the best data on ω photoproduction in heavy nuclei. The analysis leans heavily on the knowledge of ρN scattering, and on the one-pion-exchange model description of the nondiffractive contribution to the ω production process. They obtain

$$\sigma(\omega N) = 25.4 \pm 2.7 \text{ mb}$$

and

$$\frac{\gamma_\omega^2}{4\pi} = 7.5 \pm 1.3 \tag{5.15}$$

The analysis of the ϕ data exhibits the same strong correlation between $\sigma(\phi N)$, $(\gamma_\phi^2/4\pi)$, and η_ϕ, discussed above for the ρ^0, only in this case we have no good independent information on either η_ϕ or $\sigma(\phi N)$. A measurement of the ratio of real to imaginary forward ϕ–N amplitudes, η_ϕ, was obtained by observing the interference between $\phi \to e^+e^-$ decays and the Bethe–Heitler pair production process (Alvensleben *et al.*, 1971a). Unfortunately this is a very difficult experiment, and the result—$\eta_\phi = 0.48^{+0.33}_{-0.45}$—does not allow a strong constraint on these analyses. The results of the Cornell analysis (McClellan *et al.*, 1971) are shown in Fig. 64, where both $(\gamma_\phi^2/4\pi)$ and $\sigma(\phi N)$ are plotted against possible values of η_ϕ. If one assumes that the total cross section is equal to the quark model value of 13 mb, then $\eta_\phi \sim -0.22$ and $\gamma_\phi^2/4\pi \sim 6.5$, twice the value found in e^+e^- annihilation experiments. If, on the other hand, $\gamma_\phi^2/4\pi$ is taken consistent with the storage ring value then $\eta_\phi \sim -0.5$ and $\sigma(\phi N) \sim 8$ mb.

The results from these experiments on coherent photoproduction of ρ^0, ω, and ϕ mesons are discussed with the other determinations of the photon coupling strength in Section 5.3 below.

5.3. Photon–Vector-Meson Coupling Strength $(\gamma_V^2/4\pi)$

5.3.1. General

In this section we summarize the various measurements of the photon–vector-meson couplings and evaluate how well the vector dominance model works.

5.3.2. Storage Ring Experiments

The vector mesons are strongly produced in e^+e^- annihilations via the one-photon-exchange process. In Fig. 65 the ratio of the total hadronic

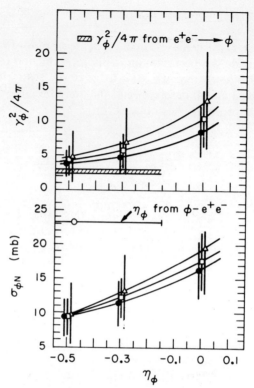

Fig. 64. A two-dimensional plot showing the correlation between the ratio of the real to imaginary forward ϕN scattering amplitude, η_ϕ, and the total ϕN cross section, $\sigma(\phi N)$, and the photon–phi coupling ($\gamma_\phi^2/4\pi$). The data come from an analysis of coherent ϕ photo-production on complex nuclear targets at 6.4 GeV ($\triangle$), 8.3 GeV ($\bullet$), combined ($\square$).

cross section to the point cross section for the production of muon pairs is shown for data from Orsay (Benaksas *et al.*, 1972), Frascati (Salvini, 1974) and SLAC (Augustin *et al.*, 1975). The ρ, ω, ϕ mesons are clearly seen, as are the new narrow states, the Ψ and Ψ' mesons. In Table 10 the leptonic widths obtained from these data are given together with the coupling strength, as calculated via Eq. (5.5).

The couplings of the first four entries in Table 10 are expected to be in the ratio $9:1:2:8$, as discussed in relation (5.9) above. The experimental results indicate

$$9:(1.25 \pm 0.1):(2.04 \pm 0.2):(2.22 \pm 1.1) \tag{5.16}$$

The "old" mesons seem to work fairly well, but the Ψ meson misses by about a factor of 4. This may be an indication of some problems with VDM, or that the charmed quark does not have charge $\frac{2}{3}$, or perhaps, there is after all, some q^2 dependence of the coupling constant, and the values at

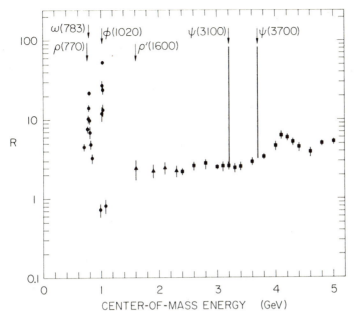

Fig. 65. The ratio, R, of the total cross section for $e^+e^- \to$ hadrons to the point cross section for $e^+e^- \to \mu^+\mu^-$, as a function of energy.

$q^2 = 0$, and for $q^2 = m_\Psi^2$ are indeed different. For the case of the Ψ meson, q^2 is large and of order 10 GeV^2 and so would be sensitive to such a variation. Another possibility is that there are other contributions to the $\gamma N \to \Psi N$ process, such as the off-diagonal terms ($V' \to V$, $V' \neq V$), which have been neglected in the vector dominance model calculations. See, for example Fig. 66. (This possibility will be further discussed below in connection with ϕ photoproduction in Section 5.3.3.)

An interesting observation on the relative vector meson coupling (Yennie, 1975) indicates that the predicted $SU(6)$ relationship for the

Table 10. Vector Meson Couplings, as Obtained from the e^+e^- Storage Rings

Meson	Coupling constant $(\gamma_V^2/4\pi)$	Leptonic width (keV)
ρ	0.53 ± 0.07	6.48 ± 0.9
ω	4.60 ± 0.5	0.76 ± 0.08
ϕ	3.4 ± 0.3	1.34 ± 0.14
Ψ	2.9 ± 0.4	4.8 ± 0.6
Ψ'	7.4 ± 1.0	2.2 ± 0.3

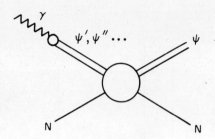

Fig. 66. A schematic diagram for an off-diagonal
vector dominance process.

$\rho : \omega : \phi : \Psi$ couplings [Eq. (5.9)] is rather well obeyed by the leptonic widths if not by the coupling strengths themselves. He showed that

$$\frac{\Gamma_{\rho \to e^+e^-}}{9} : \frac{\Gamma_{\omega \to e^+e^-}}{1} : \frac{\Gamma_{\phi \to e^+e^-}}{2} : \frac{\Gamma_{\Psi \to e^+e^-}}{8}$$

$$= (0.72 \pm 0.1) : (0.76 \pm 0.08) : (0.67 \pm 0.07) : (0.6 \pm 0.09) \quad (5.17)$$

and since $\Gamma_{e^+e^-} \propto m_V/\gamma_V^2$, this may be taken as an indication of some q^2 dependence in the coupling strength, and would indicate corrections to our treatment of the ρ, ω, and ϕ mesons.

The values of $(\gamma_V^2/4\pi)$ obtained from the storage ring measurements, corrected for the finite width of the vector meson line shapes (Gounaris and Sakurai, 1968; Renard, 1970) are listed on the first line of Table 11.

5.3.3. Photoproduction Experiments

The experiments on coherent production, discussed above in section 5.2, provide a measurement of the total cross section, $\sigma(VN)$, from the observed A dependence (given additional information on η_V) while the absolute normalization of the cross sections allows the extraction of the photon coupling strengths $(\gamma_V^2/4\pi)$.

In Table 12, a summary of the various attempts to extract $\sigma(VN)$ are presented. The quark model relates the vector meson cross section to the measured $\pi^\pm p$ data through

$$\sigma(\rho^0 p) = \sigma(\omega p) = \tfrac{1}{2}[\sigma(\pi^+ p) + \sigma(\pi^- p)]$$

$$= 27 \text{ mb}$$

$$\sigma(\phi p) = \sigma(K^+ p) + \sigma(K^- n) - \sigma(\pi^+ p) \quad (5.18)$$

$$= 13 \text{ mb}$$

The double-scattering experiment on deuterium (Anderson *et al.*, 1971), discussed above, provides a good measurement of $\sigma(\rho^0 p)$, independent of assumptions on η_ρ. The analysis of the A dependence of the ρ^0 and ω data,

Table 11. Summary of the Determination of Photon Coupling Strengths

Method	$(\gamma_\rho^2/4\pi)$	$(\gamma_\omega^2/4\pi)$	$(\gamma_\phi^2/4\pi)$	$(\gamma_{\rho''}^2/4\pi)$
1. Storage rings	0.64 ± 0.1	4.60 ± 0.5	2.83 ± 0.2	2.8 ± 0.5
2. Coherent	0.61 ± 0.03	7.50 ± 1.3	10.7 ± 4.1 $(\alpha_\phi = 0)$	
photoproduction	$(\eta_\rho = -0.2)$	$(\eta_\omega = -0.2)$	5.5 ± 2.4 $(\alpha_\phi = -0.25)$	—
on complex			4.3 ± 2.1 $(\alpha_\phi = -0.5)$	
nuclei				
3. Coherent	0.69 ± 0.04	—	—	—
production				
on deutcrium at				
large t				
4. VDM with				
measured				
$[d\sigma(\gamma N$				
$\to VN)/dt]\|_0$				
and $\sigma(VN)$ from				
quark model:			5.8 ± 0.7 $(\alpha_\phi = -0.25)$	
hydrogen data	0.67 ± 0.06	5.3 ± 0.9 (~ 10.2)	6.9 ± 0.8 $(\alpha_\phi = -0.5)$	—
			5.94 ± 1.0 $(\alpha_\phi = -0.25)$	
deuterium data	0.70 ± 0.07	5.0 ± 1.1 (~ 11.3)	7.14 ± 1.1 $(\alpha_\phi = -0.25)$	—
5. As above, but				
taking				
$\sigma_T(\phi N) = 9$ mb			2.8 ± 0.4 $(\alpha_\phi = -0.25)$	
on hydrogen	—	—	3.4 ± 0.5 $(\alpha_\phi = -0.5)$	—
			2.85 ± 0.5 $(\alpha_\phi = -0.25)$	
on deuterium	—	—	3.42 ± 0.5 $(\alpha_\phi = -0.5)$	—
6. Ratio of ω/ρ	—	4.3 ± 0.8	—	3.8 ± 0.8
production on				
hydrogen and				
deuterium and				
knowing $(\gamma_\rho^2/4\pi)$				

Table 12. Vector-Meson–Nucleon Total Cross Section

Method	$\sigma(\rho N)$ (mb)	$\sigma(\omega N)$ (mb)	$\sigma(\phi N)$ (mb)
Quark model	27	27	13
Coherent production on deuterium	28.6 ± 1.4	—	—
A dependence of coherent production	28 ± 1.5	25.4 ± 2.7	12.1 ± 3.0 $(\alpha = -0.25)$ 9.2 ± 2.8 $(\alpha = -0.5)$
VDM $+ d\sigma(\gamma N \to VN)/dt$	28.3 ± 3.5	27.8 ± 5	9.0 ± 1.2 $(\alpha = -0.25)$ 8.3 ± 1.2 $(\alpha = -0.5)$

together with an assumed $\eta_\rho = \eta_\omega \sim -0.2$ yields values of $\sigma(VN)$ in good agreement with the quark model cross sections. The ϕ cross section is given for two values of the ratio of real to imaginary amplitudes, $\eta_\phi = -0.25$ and -0.5. Also shown are estimates of the total cross section obtained using the VDM relation (5.4) and the storage ring value of $(\gamma_V^2/4\pi)$ together with the measured forward cross section on protons, $[d\sigma(\gamma p \to Vp)/dt]|_0$. The data for ρ^0 and ω are in good agreement with each other and with the quark model, while for the ϕ the cross section is 8–9 mb and substantially smaller than the quark model value.

Having determined $\sigma(VN)$, the experiments on coherent production then provide measurement of the couplings $(\gamma_V^2/4\pi)$. These are listed on lines two and three in Table 11. The other entries in Table 11 are obtained from photoproduction experiments on hydrogen and deuterium targets. In line 4 the measured forward cross section for vector meson production is used together with the VDM relation (5.4), the quark model values of the vector-meson–nucleon total cross sections, and an estimate of $\eta_\rho = \eta_\omega = -0.2$ to yield a value of $(\gamma_V^2/4\pi)$. In line 5 these calculations are repeated for the ϕ data, but using $\sigma_T(\phi N) = 9$ mb, rather than the quark model value of 13 mb. Finally, since $\sigma_T(VN)$ and η_V are the same for the ρ and ω mesons, the ratio of the forward ρ^0 and ω production cross section directly measures $(\gamma_\omega^2/4\pi)$ given that $(\gamma_\rho^2/4\pi)$ is known. In line 6 the values for $(\gamma_\omega^2/4\pi)$ obtained from studies of ρ^0 and ω production in hydrogen and deuterium are listed.

Table 11 also includes a column for the coupling of the ρ'' (1600). The photoproduction value is obtained from the ratio of the coherent ρ^0 and ρ'' production in the 7.5-GeV bubble-chamber experiment (Alexander et al., 1975), in which they find

$$R = \frac{d\sigma(\gamma D \to \rho D)/dt}{d\sigma(\gamma D \to \rho'' D)/dt} = 6.0 \pm 1.2 \qquad (5.19)$$

If one assumes that the $\rho''N$ scattering has the same phase and total cross section as ρN scattering, and $\gamma_\rho^2/4\pi = 0.64$ then one obtains $\gamma_{\rho''}^2/4\pi = 3.8 \pm 0.8$. This has to be compared with the value from the storage rings of (2.8 ± 0.5) (Grilli et al., 1973).

The agreement among the different methods on the value of $(\gamma_\rho^2/4\pi)$ (~ 0.65) is very good—perhaps surprisingly good when one considers the number of steps and approximations that go into the evaluation. For $\gamma_\omega^2/4\pi$ (~ 5.0), there is also good agreement (at the 15% level) between the various estimates if one excludes those derived from the Rochester–Cornell experiment. The values of $\gamma_\omega^2/4\pi$ on line 4 based on that experiment were not averaged with all of the other ω data since their measured cross sections were so much lower than the other experiments. (Recall Section

3.3.3 and Table 7.) The values of $\gamma_\omega^2/4\pi$ from this experiment are consequently much higher than the others; they are shown in parentheses in line 4 of Table 11. The values of the coupling constants derived from the forward cross section in hydrogen and deuterium using all other experiments agree well with the storage ring values.

For the ϕ meson the uncertainty in the value of η_ϕ makes it difficult to draw a conclusion. Two clear possibilities emerge:

(a) From the A-dependence studies summarized in Fig. 64, we see that the coupling constant would be consistent with the storage ring determination if η_ϕ were ~ -0.5. This implies that $\sigma(\phi N) \sim 9$ mb (line 2, Table 11 and line 5, Table 12). The data on hydrogen and deuterium, using Eq. (5.4), are also consistent with the storage ring $(\gamma_\phi^2/4\pi)$, if $\sigma(\phi N) \sim 9$ mb (line 5, Table 11).

(b) If one assumes that the quark model cross section $\sigma(\phi N) = 13$ mb is correct, then the heavy nuclei experiments imply that η_ϕ must be ~ -0.2 and $\gamma_\phi^2/4\pi \sim 6$, i.e., about twice the value from the e^+e^- experiments (line 2, Table 11). In this case the forward cross sections from hydrogen and deuterium also yield $\gamma_\phi^2/4\pi \sim 6$ (line 4, Table 11).

Thus a consistent picture is possible, and good agreement with the vector dominance model, if $\sigma(\phi N) \sim 9$ mb. This solution implies a large real part for forward (ϕN) scattering, consistent with the measurements on η_ϕ discussed above. Otherwise, a more conventional picture of nearly imaginary (ϕN) scattering and the canonical quark model cross section, implies that $(\gamma_\phi^2/4\pi)$ is about twice the expected VDM value of 2.83 ± 0.2.

There are other possible explanations of the ϕ problems. It has been suggested (Bauer and Yennie, 1975) that off-diagonal terms have been neglected in the vector dominance relation discussed in Eq. (5.3) and Eq. (5.4) above, and that processes like that shown in Fig. 58b with $V' = \phi$, and $V = \omega$, may contribute. Such processes would mean that the ϕ meson is not a pure strange quark state, but involves a small admixture of the nonstrange quark state. The physical ϕ, ω states were allowed to mix through the relations

$$|\phi\rangle = \cos\beta\,|s\rangle + \sin\beta\,|ns\rangle$$
$$-|\omega\rangle = -\sin\beta\,|s\rangle + \cos\beta\,|ns\rangle \tag{5.20}$$

where $|s\rangle$ denotes a state of strange quarks and $|ns\rangle$ a state of nonstrange quarks. If θ is the mixing angle from the pure $SU(3)$ octet and singlet states $(|\phi^0\rangle, |\omega^0\rangle)$, and θ_q is the magic mixing angle $(\tan\theta_q = 1/2^{1/2}$ or $\theta_q = 35.26^0)$ whereby the physical $|\phi\rangle$, $|\omega\rangle$ states consist of only strange and nonstrange quarks, respectively, then $\beta = (\theta - \theta_q)$. Silverman, extending Bauer and Yennie's original calculation, claims that good agreement can

be obtained for a value of $\beta \sim 8^0$ (Silverman, 1975). However, it is important to remember that β is not a free parameter in such a model, but that
there are strong constraints on the value of $\theta(=\beta+\theta_q)$ from conventional
meson spectroscopy (Barash-Schmidt *et al.*, 1974; Samios *et al.*, 1974).
From these considerations, it would seem unlikely that θ could be more
than a few degrees different from θ_q, and therefore one would expect
$\beta \lesssim 5^0$.

Yet another possibility is that the basic vector dominance assumption
of q^2 independence of the couplings may not be valid, i.e.,

$$\gamma_V^2(q^2=0) \neq \gamma_V^2(q^2=m_V^2)$$

The discussion on the Ψ meson couplings in Section 5.3.2 above may be an
indication that such a q^2 dependence of the vector-meson–photon couplings has to be taken into account. For the purpose of this review, we
merely leave this as an interesting remark.

5.4. Compton Sum Rule Test

In this section we consider yet another check of the vector dominance
model introduced above in Section 5.1, namely, testing the Compton sum
rule through Eqs. (5.6) and (5.7).

First we consider the ratio, R, developed from Eq. (5.7):

$$R = \left[\frac{d\sigma}{dt}(\gamma N \to \gamma N)\bigg|_0\right]^{1/2} \bigg/ \left[\frac{\alpha}{4} \cdot \sum_V \left(\frac{4\pi}{\gamma_V^2}\right) \cdot \frac{d\sigma}{dt}(\gamma N \to VN)\bigg|_0\right]^{1/2} \qquad (5.21)$$

where N refers to production from a proton, or coherent production from
a deuteron. The forward cross-section measurements used to calculate R
are listed in Table 13. For the 9.3-GeV hydrogen data the ρ^0 accounts for
71% of the summation in the denominator, the ω for 9%, the ϕ for 6%,
and the $\rho''(1600)$ for 14%. The value of the ratio R for the three energies
was found to be

$$\text{at } 4.3 \text{ GeV, on deuterium,} \quad R = 1.20 \pm 0.09$$

$$\text{at } 7.5 \text{ GeV, on deuterium,} \quad R = 1.26 \pm 0.9 \qquad (5.22)$$

$$\text{at } 9.3 \text{ GeV, on hydrogen,} \quad R = 1.19 \pm 0.07$$

These values of R imply that $\sim 20\%$ of the vector meson contributions to
the Compton amplitude are still missing.

A similar story emerges in considering the test as a function of
momentum transfer, as indicated in Eq. (5.6). In Fig. 67, the ratio of the

Table 13. Input Data for Compton Sum Rule

Vector meson	$(\gamma_V^2/4\pi)$	$\dfrac{d\sigma}{dt}(\gamma N \to VN)\Big|_0 \ [\mu b/(GeV/c)^2]$		
		9.3 GeV (H)	4.3 GeV (D)	7.5 GeV (D)
ρ	0.64 ± 0.05	100 ± 10^a	437 ± 27^b	327 ± 11^c
ω	4.8 ± 0.5	13.5 ± 2^a	69 ± 17^c	42 ± 9^d
ϕ	2.8 ± 0.2	2.49 ± 0.15^a	11.2 ± 1.3^d	11.2 ± 1.3^d
ρ''	2.8 ± 0.5	15 ± 5^e	50 ± 10^c	50 ± 10^c
Forward Compton cross section $[\mu b/(GeV/c)^2]$		0.79 ± 0.03^a	$3,48 \pm 0.2^f$	3.0 ± 0.14^g

[a] Average of the data presented in Sections 2 and 3 of this review.
[b] Eisenberg *et al.* (1976).
[c] Alexander *et al.* (1975).
[d] McClellan *et al.* (1971).
[e] Wolf (1972).
[f] Derived from the total cross-section measurements of Caldwell *et al.* (1973) using the optical theorem.
[g] Boyarski *et al.* (1971).

predicted Compton cross section from the right-hand side of Eq. (5.6) is divided by the actual measured cross section as a function of t for energies of 3.5 and 16 GeV. The assumption is made that the ratios of real to imaginary parts of the forward vector meson scattering amplitudes are the same, which should be a good assumption for the ρ^0 and ω, and the error introduced in the case of the ϕ is negligible. The storage ring values for $(\gamma_V^2/4\pi)$ are used. The t dependence of both the Compton scattering and the vector meson production are similar, but there must be missing contributions from other vector mesons if the sum rule is to be satisfied. [Note this t-dependent comparison did not include the contributions from the ρ'' (1600).]

Both tests imply that $\sim 20\%$ additional contribution is required beyond ρ^0, ω, ϕ, and ρ'' to satisfy the sum rule, and verify the VDM relation. (The contributions of Ψ and Ψ' are negligible.)

There are many claims on this missing 20%. Yennie (1975b) estimates that there should be a contribution from nonresonant 2π final states included in the summation on the denominator of Eq. (5.21), which would add another 10%. There are also the contributions from the higher-mass vector continuum, which have been estimated at $\sim 20\%$ (Sakurai and Schildknecht, 1972). So the naive vector dominance picture in which the ρ^0, ω, ϕ mesons completely saturate the photon's vector field couplings is clearly wrong, but the generalized picture, which includes the higher-mass states, may work reasonably well.

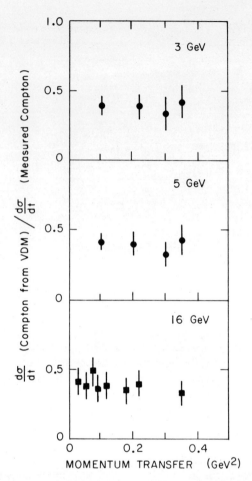

Fig. 67. The ratio of the predicted Compton scattering cross section (from the measured vector meson cross sections and using the vector dominance model), to the measured Compton ($\gamma p \rightarrow \gamma p$) cross section as a function of momentum transfer for three regions of photon energy—3, 5, and 16 GeV.

5.5. Summary

Vector dominance has worked well as a guide to the qualitative features of vector meson photoprocesses and has been especially useful in its naive form of ρ^0 dominance. More quantitatively, we have seen that it works well for the ρ^0 meson, and perhaps the ω meson, but that signs of trouble emerge with the ϕ and the Ψ mesons. It appears that to obtain a good quantitative description of photoreactions the vector dominance model will have to take into account the possibility of the mixing of the vector meson states and of off-diagonal contributions to the scattering process ($V' \to V$, where $V' \neq V$). It is also likely that the VDM will have to incorporate some q^2 dependence of the couplings.

6. Conclusion

We have reviewed the exclusive diffractive reactions of the photon and found that they behave very much like other hadronic processes:

The total cross section shows the same behavior with energy as other meson–nucleon interactions, namely, a rapid falloff with energy as meson exchange processes die out, a flattening out as the diffractive amplitude dominates, and finally we would expect to see the total photon–nucleon cross section rising around energies of 200–300 GeV.

The elastic scattering, and the quasielastic reaction (vector meson photoproduction), behave very much like other meson–nucleon scattering reactions:

  (i) little energy dependence in the cross section,
 (ii) sharply peaked scattering distribution, with an indication of steepening in the very forward direction,
(iii) mainly imaginary forward-scattering amplitude,
(iv) mainly natural parity exchange,
 (v) mainly $I = 0$ exchange,
(vi) mainly s-channel helicity conservation.

The vector dominance model gives a good guide to the systematics of the photon–nucleon interaction, and for the lower mass states allows calculations to 10%–20% precision. However, it appears that additional effects such as mixing of states and/or a q^2 dependence of the coupling will have to be included to allow a good description of the high mass vector meson processes.

Acknowledgment

This work was supported by the Energy Research and Development Administration.

References

Abrams, G. S., Briggs, D., Chinowsky, W., Friedberg, C. E., Goldhaber, G., Hollebeek, R. J., Kadyk, J. A., Litke, A., Lulu, B., Pierre, F., Sadoulet, B., Trilling, G. H., Whitaker, J. S., Wiss, J., Zipse, J. E., Augustin, J.-E., Boyarski, A. M., Breidenbach, M., Bulos, F.,

Feldman, G. J., Fischer, G. E., Fryberger, D., Hanson, G., Jean-Marie, B., Larsen, R. R.,
   Lüth, V., Lynch, H. L., Lyon, D., Morehouse, C. C., Paterson, J. M., Perl, M. L., Richter,
   B., Rapidis, P., Schwitters, R. F., Tanenbaum, W., and Vanucci, F. (1975), *Phys. Rev.
   Lett.* **33**, 1453.
Abramson, J., Andrews, D. E., Harvey, J., Lobkowicz, F., May, E. N., Nelson, C. A., Singer,
   M., and Thorndike, E. M. (1976), Rochester University report No. UR-566 *Phys. Rev.
   Lett.* **36**, 1428.
Alexander, G., Bar-Nir, I., Brandstetter, A., Benary, O., Gandsman, J., Levy, A., Oren, Y.,
   Ballam, J., Chadwick, G. B., Menke, M. M., Eisenberg, Y., Haber, B., Kogan, E., Ronat,
   E. E., Shapira, A., and Yekutieli, G. (1972), *Phys. Rev.* **D5**, 15; **D8**, 1965; **D9**, 644.
Alexander, G., Gandsman, J., Jacobs, L. D., Levy, A., Lissauer, D., and Rosenstein, L. M.
   (1973), *Nucl. Phys.* **B61**, 32; **B68**, 1; **B69**, 445.
Alexander, G., Benary, O., Gandsman, J., Levy, A., Lissauer, D., and Oren, Y. (1975), Tel
   Aviv University preprint No. TAUP-481-75.
Alvensleben, H., Becker, U., Chen, M., Cohen, K. J., Edwards, R. T., Knasel, T. M.,
   Marshall, R., Quinn, D. J., Rohde, M., Sanders, G. H., Schubel, H., and Ting, Samuel C.
   C. (1970*a*), *Phys. Rev. Lett.* **25**, 1377. See also (1971), *Nucl. Phys.* **B25**, 342.
Alvensleben, H., Becker, U., Bertram, William K., Chen, M., Cohen, K. J., Knasel, T. M.,
   Marshall, R., Quinn, D. J., Rohde, M., Sanders, G. H., Schubel, H., and Ting, Samuel C.
   C. (1970*b*), *Nucl. Phys.* **B18**, 333.
Alvensleben, H., Becker, U., Busza, W., Chen, M., Cohen, K. J., Edwards, R. T., Mantsch, P.
   M., Marshall, R., Nash, T., Rohde, M., Sadrozinski, H. F. W., Sanders, G. H., Schubel,
   H., Ting, Samuel C. C., and Wu, Sau-Lan (1971*a*), *Phys. Rev. Lett.* **27**, 444.
Alvensleben, H., Becker, U. J., Bertram, William K., Chen, M., Cohen, K. J., Edwards, R. T.,
   Knasel, T. M., Marshall, R., Quinn, D. J., Rohde, M., Sanders, G. H., Schubel, H., and
   Ting, Samuel C. C. (1971*b*), *Phys. Rev. Lett.* **26**, 273.
Alvensleben, H., Becker, U., Biggs, P., Binkley, M., Busza, W., Chen, M., Cohen, K. J.,
   Coleman, E., Edwards, R. T., Mantsch, P. M., Marshall, R., Nash, T., Quinn, D. J.,
   Rohde, M., Sadrozinski, H. F. W., Sanders, G. H., Schubel, H., Ting, Samuel C. C., and
   Wu, Sau Lan (1972), *Phys. Rev. Lett.* **28**, 66.
Anderson, R. L., Gustavson, D., Johnson, J., Overman, I., Ritson, D., Wiik, B. H., Talman,
   R., Walker, J. K., and Worcester, D. (1970*a*), *Phys. Rev. Lett.* **25**, 1218.
Anderson, R., Gustavson, D., Johnson, J., Ritson, D., Wiik, B. H., Jones, W. G., Kreinick,
   D., Murphy, F., and Weinstein, R. (1970*b*), *Phys. Rev.* **D1**, 27.
Anderson, R. L., Gustavson, D., Johnson, J., Overman, I., Ritson, D. M., Wiik, B. H.,
   Talman, R., Worcester, D. (1971), *Phys. Rev.* **D4**, 3245.
Anderson, R. L., Gottschalk, B., Gustavson, D. B., Ritson, D. M., Weitsch, G. A., Wiik, B.
   H., Halpern, H. J., Prepost, R., Tompkins, D. H. (1973), *Phys. Rev. Lett.* **30**, 149.
Armstrong, T. A., Hogg, W. R., Lewis, G. M., Robertson, A. W., Brookes, G. R., Clough, A.
   S., Freeland, J. H., Galbraith, W., King, A. F., Rawlinson, W. R., Tait, N. R. S.,
   Thompson, J. C., and Tolfree, D. W. L. (1972), *Phys. Rev.* **D5**, 1640; *Nucl. Phys.* **B41**,
   445.
Aubert, J. J., Becker, U., Biggs, P. J., Burger, J., Chen, M., Everhart, G., Goldhagen, P.,
   Leong, J., McCorriston, T., Rhoades, T. G., Rohde, M., Ting, Samuel C. C., Wu,
   Sau-Lan, and Lee, Y. Y. (1974), *Phys. Rev. Lett.* **33**, 1404.
Augustin, J.-E., Boyarski, A. M., Breidenbach, M., Bulos, F., Dakin, J. T., Feldman, G. J.,
   Fischer, G. E., Fryberger, D., Hanson, G., Jean-Marie, B., Larsen, R. R., Lüth, V.,
   Lynch, H. L., Lyon, D., Morehouse, C. C., Paterson, J. M., Perl, M. L., Richter, B.,
   Rapidis, P., Schwitters, R. F., Tanenbaum, W. M., Vannucci, F., Abrams, G. S., Briggs,
   D., Chinowsky, Friedberg, C. E., Goldhaber, G., Hollebeek, R. J., Kadyk, J. A., Lulu,
   B., Pierre, F., Trilling, G. H., Whitaker, J. S., Wiss, J., and Zipse, J. E. (1974), *Phys. Rev.
   Lett.* **33**, 1406.

Augustin, J.-E., Boyarski, A. M., Breidenbach, M., Bulos, F., Dakin, J. T., Feldman, G. J., Fischer, G. E., Fryberger, D., Hanson, G., Jean-Marie, B., Larsen, R. R., Lüth, V., Lynch, H. L., Lyon, D., Morehouse, C. C., Paterson, J. M., Perl, M. L., Richter, B., Schwitters, R. F., Vannucci, F., Abrams, G. S., Briggs, D., Chinowsky, W., Friedberg, C. E., Goldhaber, G., Hollebeek, R. J., Kadyk, J. A., Trilling, G. H., Whitaker, J. S., Zipse, J. E. (1975), Stanford Linear Accelerator Center preprint No. SLAC-PUB-1520.

Ballam, J., Chadwick, G. B., Gearhart, R., Guiragossián, Z. G. T., Murray, J. J., Seyboth, P., Sinclair, C. K., Skillicorn, I. O., Spitzer, H., Wolf, G., Bingham, H. H., Fretter, W. B., Moffeit, K. C. Podolsky, W. J., Rabin, M. S., Rosenfeld, A. H., Windmolders, R., and Milburn, R. H. (1972), *Phys. Rev.* **D5**, 545.

Ballam, J., Chadwick, G. B., Eisenberg, Y., Kogan, E., Moffeit, K. C., Seyboth, P., Skillicorn, I. O., Spitzer, H., Wolf, G., Bingham, H. H., Fretter, W. B., Podolsky, W. J., Rabin, M. S., Rosenfeld, A. H., and Smadja, G. (1973), *Phys. Rev.* **D7**, 3150.

Ballam, J., Chadwick, G. B., Eisenberg, Y., Kogan, E., Moffeit, K. C., Skillicorn, I. O., Spitzer, H., Wolf, G., Bingham, H. H., Fretter, W. B., Podolsky, W. J., Rabin, M. S., Rosenfeld, A. H., Smadja, G., and Seyboth, P. (1974), *Nucl. Phys.* **B76**, 375.

Barash-Schmidt, N., Barbaro-Galtieri, A., Bricman, C., Chaloupka, V., Chew, D. M., Kelly, R. L., Lasinski, T. A., Rittenberg, A., Roos, M., Rosenfeld, A. H., Söding, P., Trippe, T. G., and Uchiyama, F. (1974), "Particle property tables," Lawrence Berkeley Laboratory report.

Barbarino, G., Grilli, M., Iarocci, E., Spillantini, P., Valente, V., Visentin, R., Ceradini, F., Conversi, M., Paoluzi, L., Santonico, R., Nigro, M., Trasatti, L., and Zorn, G. T. (1972), *Nuovo Cimento Lett.* **3**, 689.

Barbiellini, G., Bozzo, M., Darriulat, P., Diambrini-Palazzi, G., De Zorzi, G., Fainberg, A., Ferrero, M. I., Holder, M., McFarland, A., Maderni, G., Orito, S., Pilcher, J., Rubbia, C., Santroni, A., Sette, G., Staude, A., Strolin, P., and Tittel, K. (1972), *Phys. Lett.* **B39**, 663.

Barger, V., Philips, R. (1971), *Nucl. Phys.* **B32**, 93.

Barish, B., Gomez, R., Kreinick, D., Peck, C., Pine, J., Sciulli, F., Sherwood, B., Tollestrup, A., and Young, K. (1972), *Phys. Rev.* **D9**, 566.

Bartoli, B., Coluzzi, B., Felicetti, Silvestrini, V., Goggi, G., Scannicchio, D., Marini, G., Massa, F., Vanoli, F. (1970), *Nuovo Cimento* **70**, 615. See also (1972), *Phys. Rev.* **D6**, 2374.

Bauer, T. (1971), *Phys. Rev.* **D3**, 2671.

Bauer, T., and Yennie, D. R. (1970), *Phys. Rev. Lett.* **25**, 485.

Bauer, T., and Yennie, D. (1975), Cornell preprint.

Behrend, H.-J., Lobkowicz, F., Thorndike, E. H., Wehmann, A. A., and Nordberg, Jr., M. E. (1970), *Phys. Rev. Lett.* **24**, 1246.

Behrend, H.-J., Lee, C. K., Lobkowicz, F., Thorndike, E. H., Wehmann, A. A., Nordberg, Jr., M. E. (1971*a*), *Phys. Rev. Lett.* **26**, 151.

Behrend, H.-J., Lee, C. K., Lobkowicz, F., Thorndike, E. H., Nordberg, Jr., M. E., and Wehmann, A. A. (1971*b*), *Phys. Rev. Lett.* **27**, 65.

Behrend, J.-J., Bodenkamp, J., Hesse, W. P., Fries, D. C., Heine, P., Hirschmann, H., McNeely, Jr., W. A., Markou, A., Seitz, E. (1975), *Phys. Lett.* **B56**, 408.

Belousov, A. S., Budanov, N. P., Govorkov, B. B., Lebedev, A. I., Malinovsky, E. I., Minarik, E. V., Michaclov, I. V., Plaksin, V. P., Rusakov, S. V., Sergienkov, V. I., Tamm, E. I., Chevenkov, P. A., Shaveiko, P. N., Alikanian, A. I., Baiatian, G. L., Markarvian, A. T., Vartanian, G. S., Frolov, A. M., and Samoylov, A. V. (1972), Lebedev preprint.

Benaksas, D., Cosme, G., Jean-Marie, B., Jullian, S., Laplanche, F., LeFrançois, J., Liberman, A. D., Parrour, G., Repellin, J. P., Sauvage, G. (1972), *Phys. Lett.* **B39**, 289; **B42**, 507; **B48**, 155; **B48**, 155; **B40**, 685.

Benecke, J., and Durr, H. P. (1968), *Nuovo Cimento* **56**, 269.

Benz, P., Braun, O., Butenschön, H., Finger, H., Gall, D., Idschok, U., Kiesling, C., Knies, G., Kowalski, H., Müller, K., Nellen, B., Schiffer, R., Schlamp, P., Schnackers, H. J., Schulz, V., Söding, P., Spitzer, H., Stiewe, J., Storim, F., and Weigl, J. (1973), *Nucl. Phys.* **B65**, 158.

Benz, P., Braun, O., Butenschön, H., Gall, D., Idschok, U., Kiesling, C., Knies, G., Müller, K., Nellen, B., Schiffer, R., Schlamp, P., Schnackers, H. J., Söding, P., Stiewe, J., and Storim, F. (1974), *Nucl. Phys.* **B79**, 10.

Berger, C., Mistry, N., Roberts, L., Talman, R., and Walstrom, P. (1972), *Phys. Lett.* **B39**, 659.

Biggs, P. J., Braben, D. W., Clifft, R. W., Gabathuler, E., and Rand, R. E. (1971), *Phys. Rev. Lett.* **27**, 1157.

Bingham, H. H., Fretter, W. B., Podolsky, W. J., Rabin, M. S., Rosenfeld, A. H., Smadja, G., Yost, G. P., Ballam, J., Chadwick, G. B., Eisenberg, Y., Kogan, E., Moffeit, K. C., Seyboth, P., Skillicorn, I. O., Spitzer, H., and Wolf, G. (1972), *Phys. Lett.* **B41**, 635.

Bingham, H. H., Fretter, W. B., Podolsky, W. J., Rabin, M. S., Rosenfeld, A. H., Smadja, G., Ballam, J., Chadwick, G. B., Eisenberg, Y., Gearhart, R., Kogan, E., Moffeit, K. C., Murrary, J. J., Seyboth, P., Sinclair, C. K., Skillicorn, I. O., Spitzer, H., and Wolf, G. (1973), *Phys. Rev.* **D8**, 1277.

Blechschmidt, H., Dowd, J. P., Elsner, B., Heinloth, K., Höhne, K. H., Raither, S., Rathje, J., Schmidt, D., Smith, J. H., and Weber, J. H. (1967), *Nuovo Cimento* **A52**, 1348.

Bloom, E. D., Cottrell, R. L., Coward, D. H., DeStaebler, Jr., H., Drees, J., Miller, G., Mo, L. W., Taylor, R. E., Friedman, J. I., Hartmann, G. C., and Kendall, H. W. (1969), Stanford Linear Accelerator Center preprint No. SLAC-PUB-653.

Boyarski, A. M., Coward, D. H., Ecklund, S., Richter, B., Sherden, D., Siemann, R., and Sinclair, C. (1971), *Phys. Rev. Lett.* **26**, 1600; and (1973), *ibid.* **30**, 1098.

Braccini, P. L., Bradaschia, C., Castaldi, R., Foa, L., Lübelsmeyer, K., and Schmitz, D. (1970), *Nucl. Phys.* **B24**, 173.

Bulos, F., Busza, W., Giese, R., Larsen, R. R., Leith, D. W. G. S., Richter, B., Perez-Mendez, V., Stetz, A., Williams, S. H., Beniston, M., and Rettberg, J., (1969), *Phys. Rev. Lett.* **22**, 490.

Bulos, F., Busza, W., Giese, R., Larsen, R. R., Leith, D. W. G. S., Richter, B., and Williams, S. (1970), reported by D. W. G. S. Leith in *Hadronic Interactions of Electrons and Photons*: Proceedings of the 11th Session of the Scottish Universities Summer School in Physics, 1970, Eds. J. Cumming and H. Osborn (London, Academic Press, 1971), p. 195.

Bulos, F., Busza, W., Giese, R., Kluge, E. E., Larsen, R. R., Leith, D. W. G. S., Richter, B., Williams, S. H., Kehoe, B., Beniston, M., and Stetz, A. (1971), *Phys. Rev. Lett.* **26**, 149.

Buschhorn, G., Griegee, L., Dubal, L., Franke, G., Geweniger, C., Heide, P., Kotthaus, R., Poelz, G., Timm, U., Wegener, K., Werner, H., Wong, M., and Zimmerman, W. (1970), *Phys. Lett.* **B33**, 241.

Caldwell, D. O., Elings, V. B., Hesse, W. P., Morrison, R. J., Murphy, F. V., and Yount, D. E. (1973), *Phys. Rev.* **D7**, 1362.

Camerini, U., Learned, J. G., Prepost, R., Spencer, C. M., Wiser, D. E., Ash, W. W., Anderson, R. L., Ritson, D. M., Sherden, D. J., Sinclair, C. K. (1975). *Phys. Rev. Lett.* **35**, 483.

Carnegie, R. K., Cashmore, R. J., Davier, M., Leith, D. W. G. S., Walden, P., Williams, S. H. (1975), *Phys. Lett.* **B59**, 313.

CEA Bubble Chamber Collaboration (1966), *Phys. Rev.* **146**, 994; **155**, 1468; **156**, 1426; **169**, 1081.

Ceradini, F., Conversi, M., D'Angelo, S., Paoluzi, L., Santonico, R., Elkstrand, K., Grilli, M., Iarocci, E., Spillantini, P., Valente, V., Visentin, R., and Nigro, M. (1973), *Phys. Lett.* **B43**, 341.

Chadwick, G., Eisenberg, Y., and Kogan, E. (1973), *Phys. Rev. D* **8**, 1607.

Conversi, M., Paoluzi, L., Ceradini, F., d'Angelo, S., Ferrer, M. L., Santonico, R., Grilli, M., Spillantini, P., and Valente, V. (1974), paper 137, *IVth International Conference on Experimental Meson Spectroscopy*, Boston, 26–27 April, 1974 (New York, American Institute of Physics).

Cozzika, G., Ducros, Y., Gaidot, A., De Lesquen, A., Merlo, J. P., and Van Rossum, L. (1972), *Phys. Lett.* **B40**, 281.

Criegee, L., Franke, G., Löffler, G., Schüler, K. P., Timm, U., Zimmermann, W., Werner, H., Dougan, P. W. (1970), *Phys. Rev. Lett.* **25**, 1306.

Damashek, M., and Gilman, F. J. (1970), *Phys. Rev.* **D1**, 1319.

Das, T., Mathur, V. S., and Okubo, S. (1967), *Phys. Rev. Lett.* **19**, 470.

Davier, M. (1972), *Phys. Lett.* **B40**, 369.

Davier, M., Derado, I., Drickey, D., Fries, D., Mozley, R., Odian, A., Villa, F., and Yount, D. (1970), *Phys. Rev.* **D1**, 790; and (1972), *Nucl. Phys.* **B36**, 404.

Davier, M., Derado, I., Fries, D., Liu, F., Mozley, R. F., Odian, A., Park, J., Swanson, W. P., Villa, F., and Yount, D. (1973), Stanford Linear Accelerator Center preprint No. SLAC-PUB-1205.

De Lesquen, A., Amblard, B., Beurtey, R., Cozzika, G., Bystricky, J., Deregel, J., Ducros, Y., Fontaine, J. M., Gaidot, A., Hansroul, M., Lehar, F., Merlo, J. P., Miyashita, S., Movchet, J., and Von Rossum, L., (1972), *Phys. Lett.* **B40**, 277.

De Rujula, A., and Glashow, S. L. (1975), *Phys. Rev. Lett.* **34**, 46.

DESY/ABBHHM Collaboration (1968), *Phys. Rev.* **175**, 1669; **188**, 2060.

DESY/AHHM Collaboration (1971); see Wolf (1971) and Moffeit (1973).

Diambrini-Palazzi, G., McClellan, G., Mistry, N., Mostek, P., Ogren, H., Swartz, J., and Talman, R. (1970), *Phys. Rev. Lett.* **25**, 478.

Dominguez, C. A., Gunion, J. F., and Suaya, R. (1972), *Phys. Rev.* **D6**, 1404.

Drell, S. D. (1961), *Rev. Mod. Phys.* **33**, 458.

Drell, S. D., and Trefil, J. S. (1966), *Phys. Rev. Lett.* **16**, 552; **16**, 832.

Eisenberg, Y., Haber, B., Ronat, E. E., Shapira, A., Stahl, Y., Yekutieli, G., Ballam, J., Chadwick, G. B., Menke, M. M., Seyboth, P., Dagan, S., and Levy, A. (1972*a*), *Phys. Rev.* **D5**, 15.

Eisenberg, Y., Haber, B., Kogan, E., Ronat, E. E., Shapira, A., and Yekutieli, G. (1972*b*), *Nucl. Phys.* **B25**, 499; **B42**, 349.

Eisenberg, Y., Haber, B., Kogan, E., Karshon, U., Ronat, E. E., Shapira, A., and Yekutieli, G. (1976), *Nucl. Phys.* **B104**, 61.

Estabrooks, P., and Martin, A. (1975), *Nucl. Phys.* **B95**, 322.

Franco, V., and Glauber, R. J. (1966), *Phys. Rev.* **142**, 1195.

Franco, V., and Varma, G. K. (1974), *Phys. Rev. Lett.* **33**, 44.

Frazer, W. R., and Fulco, J. R. (1960), *Phys. Rev.* **117**, 1603.

Freund, P. (1967), *Nuovo Cimento* **48**, 541.

Gaillard, M. K., Lee, B. W., and Rosner, J. L. (1974), Fermi-Lab report No. 74/86.

Gell-Mann, M., Goldberger, M. L., and Thirring, W. E. (1954), *Phys. Rev.* **95**, 1612.

Giacomelli, G. (1969), CERN report No. CERN-HERA-69-30.

Giese, R. (1974), thesis, Stanford University.

Gilman, F., Pumplin, J., Schwimmer, A., Stodolsky, L. (1970), *Phys. Lett.* **B31**, 387.

Gittelman, B., Hanson, K. M., Larson, D., Loh, E., Silverman, A., Theodisiou, G. (1975), *Phys. Rev. Lett.* **35**, 1616.

Gladding, Gary E., Russell, John J., Tannenbaum, Michael J., Weiss, Jeffrey M., Thomson, Gordon B. (1973), *Phys. Rev.* **D8**, 3721.

Gounaris, G. J., and Sakurai, J. J. (1968), *Phys. Rev. Lett.* **21**, 244.

Grayer, G., Hyams, B., Jones, C., Schlein, P., Weilhammer, P., Blum, W., Dietl, H., Koch, W., Lorenz, E., Lütjens, G., Männer, W., Meissburger, J., Ochs, W., and Stierlin, U. (1974), *Nucl. Phys.* **B75**, 189.

Grilli, M., Iarocci, E., Spillantini, P., Valente, V., Visentin, R., Borgia, B., Ceradini, F., Conversi, M., Paoluzi, L., Santonico, R., Nigro, M., Trasatti, L., and Zorn, G. T. (1973), *Nuovo Cimento* **A13**, 593.

Halpern, H. J., Prepost, R., Tompkins, D. H., Anderson, R. L., Gottschalk, B., Gustayson, D. B., Ritson, D. M., Weitsch, G. A., and Wiik, B. H. (1972), *Phys. Rev. Lett.* **29**, 1425.

Harari, H. (1969), *Proceedings of the 4th International Symposium on Electron and Photon Interactions at High Energies*, Liverpool, 14–20 September, 1969 (Daresbury Laboratory), p. 107.

Harari, H. (1971), *Ann. Phys. (N.Y.)* **63**, 432.

Hicks, N., Eisner, A., Feldman, G., Litt, L., Lockeretz, W., Pipkin, F. M., Randolph, J. K., and Stanfield, K. C. (1969), *Phys. Lett.* **B29**, 602.

Hyams, B., Jones, C., Weilhammer, P., Blum, W., Dietl, H., Grayer, G., Koch, W., Lorenz, E., Lütjens, G., Männer, W., Meissburger, J., Ochs, W., and Stierlin, U. (1975), *Nucl. Phys.* **B100**, 205.

Joos, H. (1967), *Acta Phys. Austriaca Suppl. IV.*

Knapp, B., Lee, W., Leung, P., Smith, S. D., Wijangco, A., Knauer, J., Yount, D., Nease, D., Bronstein, J., Coleman, R., Cornell, L., Gladding, G., Gormley, M., Messner, R., O'Halloran, T., Sarracino, J., Wattenberg, A., Wheeler, D., Binkley, M., Orr, R., Peoples, J., and Read, L. (1975), *Phys. Rev. Lett.* **34**, 1040.

Kogan, E. (1975), thesis, Weizman Institute.

Krass, A. (1967), *Phys. Rev.* **159**, 1496.

Lakin, W. L., Braunstein, T. J., Cox, J., Dieterle, B. D., Perl, M. L., Toner, W. T., Zipf, T. F., and Bryant, H. (1971), *Phys. Rev. Lett.* **26**, 34.

Lanzerotti, L. J., Blumenthal, R. B., Ehn, D. C., Faissler, W. L., Joseph, P. M., Pipkin, F. M., Randolph, J. K., Russell, J. J., Stairs, D. G., and Tenenbaum, J. (1968), *Phys. Rev.* **166**, 1365.

Lee, W. Y. (1976), *Proceedings of the 1975 International Symposium on Electron and Photon Interactions at High Energies*, Stanford University, August 21–27, 1975, Ed. W. T. Kirk (Stanford, Stanford Linear Accelerator Center), p. 213.

Leith, D. W. G. S. (1975*a*), Proceedings of 1974 SLAC Summer Institute, Stanford Linear Accelerator Center report No. SLAC-179.

Leith, D. W. G. S. (1975*b*), Lectures presented at the Canadian IPP International School, McGill University, Montreal and Stanford Linear Accelerator Center preprint No. SLAC-PUB-1646.

McClellan, G., Mistry, N., Mostek, P., Ogren, H., Osborne, A., Silverman, A., Swartz, J., Talman, R., and Diambrini-Palazzi, G. (1969), *Phys. Rev. Lett.* **23**, 718.

McClellan, G., Mistry, N., Mostek, P., Ogren, H., Osborne, A., Swartz, J., Talman, R., and Diambrini-Palazzi, G. (1971*a*), *Phys. Rev. Lett.* **26**, 1593.

McClellan, G., Mistry, N., Mostek, P., Ogren, H., Silverman, A., Swartz, J., and Talman, R. (1971*b*), *Phys. Rev. D* **4**, 2683; see also (1969), *Phys. Rev. Lett.* **22**, 374.

Meyer, H., Naroska, B., Weber, J. H., Wong, M., Heynen, V., Mandelkow, E., and Notz, D. (1970), *Phys. Lett.* **B33**, 189.

Moffeit, K. (1974), *Proceedings of the 6th International Symposium on Electron and Photon Interactions at High Energies*. Bonn, August 27–31, 1973 (Amsterdam, North-Holland), p. 313.

Nambu, Y. (1957), *Phys. Rev.* **106**, 1366.

Oakes, R. J., and Sakurai, J. J. (1967), *Phys. Rev. Lett.* **19**, 1266.

Ogren, H. O. (1970), thesis, Cornell University.

Park, J., Davier, M., Derado, I., Fries, D. C., Liu, F. F., Mozley, R. F., Odian, A. C., Swanson, W. P., Villa, F., Yount, D. (1972), *Nucl. Phys.* **B36**, 404.

Prepost, R. (1976), private communication.

Pumplin, J. (1970), *Phys. Rev.* **D2**, 1859.
Rabe, E. (1971), Diplomarbeit, University of Hamburg.
Renard, F. M. (1970), *Nucl. Phys.* **B15**, 267.
Ritson, D. (1976), Talk presented at 2nd International Conference on New Results in High Energy Physics, Vanderbilt, Stanford Linear Accelerator Center preprint No. SLAC-PUB-1728.
Ross, M., and Stodolsky, L. (1966), *Phys. Rev.* **149**, 1172.
Sakurai, J. J. (1960), *Ann. Phys. (N.Y.)* **11**, 1.
Sakurai, J. J., and Schildknecht, D. (1972), *Phys. Lett.* **B40**, 121.
Salvini, D. (1974), Talk presented to the Italian Physical Society summarizing best Frascati results.
Samios, N. P., Goldberg, M., and Meadows, B. T. (1974), *Rev. Mod. Phys.* **46**, 49.
Schilling, K., Seyboth, P., Wolf, G. (1971), *Nucl. Phys.* **B15**, 397.
Schumacher, G., and Engle, I. (1971), Argonne National Laboratory report No. ANL/HEP-7032.
Shapiro, J. A. (1969), *Phys. Rev.* **179**, 1345.
Silverman, A. (1976), *Proceedings of the 1975 International Symposium on Electron and Photon Interactions at High Energies*, Stanford University, August 21–27, 1975, Ed. W. T. Kirk (Stanford, Stanford Linear Accelerator Center), p. 355.
Slattery, P., and Ferbel, T. (1974), *Phys. Rev.* **D9**, 824.
Söding, P. (1966), *Phys. Lett.* **19**, 702.
Spital, Robin, and Yennie, Donald R. (1975), *Phys. Rev.* **D9**, 138.
Swartz, J., and Talman, R. (1969), *Phys. Rev. Lett.* **23**, 1078.
Veneziano, G. (1968), *Nuovc Cimento* **A57**, 190.
Von Bockman, G., Margolis, B., and Tang, C. L. (1969), *Phys. Lett.* **B30**, 254.
West, G. B. (1971), *Phys. Lett.* **B37**, 509.
Williams, S. (1973), thesis, University of California, Berkeley.
Wolf, G. (1969), *Phys. Rev.* **182**, 1538.
Wolf, G. (1972*a*), *Proceedings of the 1971 International Symposium on Electron and Photon Interactions at High Energies*, Cornell Univ., Aug. 23–37, 1971, Ed. N. B. Mistry (Cornell University, Laboratory of Nuclear Studies), p. 189.
Wolf, G. (1972*b*), DESY report No. 72/61.
Yennie, D. (1975*a*), *Phys. Rev. Lett.* **34**, 239.
Yennie, D. (1975*b*), *Rev. Mod. Phys.* **47**, 311.

Index

Political Parties and the Internet

Can the Internet help to re-engage the public in politics? How are political parties using the Internet as a communication tool? Has politics changed in the information age?

The emergence of new information and communications technologies (ICTs) – in particular the Internet – and their potential impact on politics has sparked widespread interest in how they already have impacted upon party activity and how they might influence the future of political parties. *Political Parties and the Internet* discusses three of the principal areas within this debate:

- party competition and campaigning online;
- internal party democracy;
- the role of parties within democracies.

This book provides an assessment of how political parties are adapting to the rise of new ICTs, and what the consequences of that adaptation will be. It includes case studies from the USA, the UK, Australia, Korea, Mexico, France, Romania and the Mediterranean region.

Rachel Gibson is Deputy Director of the ACSPRI Centre for Social Research and Fellow in the Research School of Social Sciences at the Australian National University. **Paul Nixon** is Senior Lecturer in European Politics at the Institute of Higher European Studies, Haagse Hogeschool, Den Haag, the Netherlands. **Stephen Ward** is Senior Lecturer in Politics at the European Studies Research Institute, University of Salford.